MATHEMATICAL METHODS OF PHYSICS

Second Edition

MATHEMATICAL METHODS OF PHYSICS

Second Edition

JON MATHEWS

R. L. WALKER

California Institute of Technology

THE BENJAMIN/CUMMINGS
PUBLISHING COMPANY
Menlo Park, California • Reading, Massachusetts
London • Amsterdam • Don Mills, Ontario • Sydney

MATHEMATICAL METHODS OF PHYSICS
Second Edition

Standard Book Number 8053–7002–1

Library of Congress Catalog Card Number 71-80659
Manufactured in the United States of America

ISBN 0-8053-7002-1
HIJ-HA-798

Preface

For the last nineteen years a course in mathematical methods in physics has been taught, by the authors and by others, at the California Institute of Technology. It is intended primarily for first-year physics graduate students; in recent years most senior physics undergraduates, as well as graduate students from other departments, have been taking it. This book has evolved from the notes that have been used in this course.

We make little effort to teach physics in this text; the emphasis is on presenting mathematical techniques that have proved to be useful in analyzing problems in physics. We assume that the student has been exposed to the standard undergraduate physics curriculum: mechanics, electricity and magnetism, introductory quantum mechanics, etc., and that we are free to select examples from such areas.

In short, this is a book *about mathematics, for physicists*. Both motivation and standards are drawn from physics; that is, the choice of subjects is dictated by their usefulness in physics, and the level of rigor is intended to reflect current practice in theoretical physics.

It is assumed that the student has become acquainted with the following mathematical subjects:

1. Simultaneous linear equations and determinants;
2. Vector analysis, including differential operations in curvilinear coordinates;
3. Elementary differential equations;
4. Complex variables, through Cauchy's theorem.

However, these should not be considered as strict prerequisites. On the one hand, it will often profit the student to have had more background, and on the other hand, it should not be too difficult for a student lacking familiarity with one of the above subjects to remedy the defect by extra work and/or outside reading. In fact, the subject of differential equations is discussed in the first chapter, partly to begin on familiar ground and partly in order to treat some topics not normally covered in an elementary course in the subject.

A considerable variation generally exists in the amount of preparation different students have in the theory of functions of a complex variable. For this reason, we usually give a rapid review of the subject before studying contour integration (Chapter 3). Material for such a review is presented in the Appendix. Also, there are some excellent and reasonably brief mathematical books on the subject for the student unfamiliar with this material.

A considerable number of problems are included at the end of each chapter in this book. These form an important part of the course for which this book is designed. The emphasis throughout is on understanding by means of examples.

A few remarks may be made about some more or less unconventional aspects of the book. In the first place, the material which is presented does not necessarily flow in a smooth, logical pattern. Occasionally a new subject is introduced without the student's having been carefully prepared for the blow. This occurs, not necessarily through the irrationality or irascibility of the authors, but because that is the way physics is. Students in theoretical physics often need considerable persuasion before they will plunge into the middle of an unfamiliar subject; this course is intended to give practice and confidence in dealing with problems for which the student's preparation is incomplete.

A related point is that there is considerable deliberate nonuniformity in the depth of presentation. Some subjects are skimmed, while very detailed applications are worked out in other areas. If the course is to give practice in doing physics, the student must be given a chance to gain confidence in his ability to do detailed calculations. On the other hand, this is a text, not a reference work, and the material is intended to be covered fully in a year. It is therefore not possible to go into everything as deeply as one might like.

The plan of the book remains unaltered in this Second Edition; we have rewritten a number of sections, though only one (Section 14–6) is new. We have tried to make clearer certain material which has proved particularly difficult for students, such as the discussion of the Riemann P-symbol in Section 7–3 and the Wiener–Hopf integral equation in Section 11–5.

A number of new problems have been added at the ends of chapters, and several new detailed examples are included, such as the triangular drumhead of Section 8–3 and the recursive calculation of Bessel functions in Section 13–3.

Several acknowledgements are in order. The course from which this text evolved was originally based on lectures by Professor R. P. Feynman at Cornell University. Much of Chapter 16 grew out of fruitful conversations with Dr. Sidney Coleman. The authors are grateful to Mrs. Julie Curcio for rapid, accurate, and remarkably neat typing through several revisions.

JON MATHEWS
R. L. WALKER

Pasadena, California
May 1969

Contents

PREFACE   v

**CHAPTER 1 ORDINARY DIFFERENTIAL
EQUATIONS**   1

1-1 Solution in Closed Form   1
1-2 Power-Series Solutions   13
1-3 Miscellaneous Approximate Methods   22
1-4 The WKB Method   27
References   37
Problems   38

CHAPTER 2 INFINITE SERIES   44

2-1 Convergence   44
2-2 Familiar Series   48
2-3 Transformation of Series   50
References   55
Problems   56

CHAPTER 3 EVALUATION OF INTEGRALS   58

3-1 Elementary Methods   58
3-2 Use of Symmetry Arguments   62
3-3 Contour Integration   65
3-4 Tabulated Integrals   75
3-5 Approximate Expansions   80
3-6 Saddle-Point Methods   82
References   90
Problems   91

CHAPTER 4   INTEGRAL TRANSFORMS   **96**

4-1 Fourier Series   96
4-2 Fourier Transforms   101
4-3 Laplace Transforms   107
4-4 Other Transform Pairs   109
4-5 Applications of Integral Transforms   110
References   119
Problems   120

CHAPTER 5   FURTHER APPLICATIONS OF COMPLEX
**   VARIABLES**   **123**

5-1 Conformal Transformations   123
5-2 Dispersion Relations   129
References   135
Problems   135

CHAPTER 6   VECTORS AND MATRICES   **139**

6-1 Linear Vector Spaces   139
6-2 Linear Operators   141
6-3 Matrices   143
6-4 Coordinate Transformations   146
6-5 Eigenvalue Problems   150
6-6 Diagonalization of Matrices   158
6-7 Spaces of Infinite Dimensionality   160
References   163
Problems   163

CHAPTER 7   SPECIAL FUNCTIONS   **167**

7-1 Legendre Functions   167
7-2 Bessel Functions   178
7-3 Hypergeometric Functions   187
7-4 Confluent Hypergeometric Functions   194
7-5 Mathieu Functions   198
7-6 Elliptic Functions   204
References   210
Problems   210

CHAPTER 8   PARTIAL DIFFERENTIAL EQUATIONS   **217**

8-1   Examples   217
8-2   General Discussion   219
8-3   Separation of Variables   226
8-4   Integral Transform Methods   239
8-5   Wiener–Hopf Method   245
References   252
Problems   253

CHAPTER 9   EIGENFUNCTIONS, EIGENVALUES, AND GREEN'S FUNCTIONS   **260**

9-1   Simple Examples of Eigenvalue Problems   260
9-2   General Discussion   263
9-3   Solutions of Boundary-Value Problems as Eigenfunction Expansions   266
9-4   Inhomogeneous Problems. Green's Functions   267
9-5   Green's Functions in Electrodynamics   277
References   282
Problems   282

CHAPTER 10   PERTURBATION THEORY   **286**

10-1   Conventional Nondegenerate Theory   286
10-2   A Rearranged Series   292
10-3   Degenerate Perturbation Theory   293
References   296
Problems   296

CHAPTER 11   INTEGRAL EQUATIONS   **299**

11-1   Clasification   299
11-2   Degenerate Kernels   300
11-3   Neumann and Fredholm Series   302
11-4   Schmidt–Hilbert Theory   305
11-5   Miscellaneous Devices   311
11-6   Integral Equations in Dispersion Theory   316
References   318
Problems   318

CHAPTER 12   CALCULUS OF VARIATIONS   **322**

    12-1   Euler–Lagrange Equation   322
    12-2   Generalizations of the Basic Problem   327
    12-3   Connections between Eigenvalue Problems and
            the Calculus of Variations   333
    References   341
    Problems   341

CHAPTER 13   NUMERICAL METHODS   **345**

    13-1   Interpolation   345
    13-2   Numerical Integration   349
    13-3   Numerical Solution of Differential Equations   353
    13-4   Roots of Equations   358
    13-5   Summing Series   363
    References   368
    Problems   368

CHAPTER 14   PROBABILITY AND STATISTICS   **372**

    14-1   Introduction   372
    14-2   Fundamental Probability Laws   373
    14-3   Combinations and Permutations   375
    14-4   The Binomial, Poisson, and Gaussian
            Distributions   377
    14-5   General Properties of Distributions   380
    14-6   Multivariate Gaussian Distributions   384
    14-7   Fitting of Experimental Data   387
    References   395
    Problems   396

**CHAPTER 15   TENSOR ANALYSIS AND DIFFERENTIAL
                GEOMETRY**   **402**

    15-1   Cartesian Tensors in Three-Space   403
    15-2   Curves in Three-Space; Frenet Formulas   408
    15-3   General Tensor Analysis   416
    References   421
    Problems   421

CHAPTER 16   INTRODUCTION TO GROUPS AND GROUP REPRESENTATIONS   **424**

16-1   Introduction; Definitions   424
16-2   Subgroups and Classes   426
16-3   Group Representations   430
16-4   Characters   433
16-5   Physical Applications   440
16-6   Infinite Groups   449
16-7   Irreducible Representations of SU(2), SU(3),
     and O(3)   457
References   467
Problems   468

APPENDIX   SOME PROPERTIES OF FUNCTIONS OF A COMPLEX VARIABLE   **471**

A-1   Functions of a Complex Variable.   Mapping   471
A-2   Analytic Functions   477
References   483
Problems   483

BIBLIOGRAPHY   **485**

INDEX   **493**

ONE

ORDINARY DIFFERENTIAL EQUATIONS

We begin this chapter with a brief review of some of the methods for obtaining solutions of an ordinary differential equation in closed form. Solutions in the form of power series are discussed in Section 1–2, and some methods for obtaining approximate solutions are treated in Sections 1–3 and 1–4.

The use of integral transforms in solving differential equations is discussed later, in Chapter 4. Applications of Green's function and eigenfunction methods are treated in Chapter 9, and numerical methods are described in Chapter 13.

1–1 SOLUTION IN CLOSED FORM

The *order* and *degree* of a differential equation refer to the derivative of highest order after the equation has been rationalized. Thus, the equation

$$\frac{d^3 y}{dx^3} + x\sqrt{\frac{dy}{dx}} + x^2 y = 0$$

is of third order and second degree, since when it is rationalized it contains the term $(d^3 y/dx^3)^2$.

We first recall some methods which apply particularly to first-order equations.   If the equation can be written in the form

$$A(x)\, dx + B(y)\, dy = 0 \tag{1-1}$$

we say the equation is *separable*; the solution is found immediately by integrating.

EXAMPLE

$$\frac{dy}{dx} + \sqrt{\frac{1 - y^2}{1 - x^2}} = 0$$

$$\frac{dy}{\sqrt{1 - y^2}} + \frac{dx}{\sqrt{1 - x^2}} = 0 \tag{1-2}$$

$$\sin^{-1} y + \sin^{-1} x = C$$

or, taking the sine of both sides,

$$x\sqrt{1 - y^2} + y\sqrt{1 - x^2} = \sin C = C'$$

More generally, it may be possible to integrate immediately an equation of the form

$$A(x, y)\, dx + B(x, y)\, dy = 0 \tag{1-3}$$

If the left side of (1-3) is the differential du of some function $u(x, y)$, then we can integrate and obtain the solution

$$u(x, y) = C$$

Such an equation is said to be *exact*.   A necessary and sufficient condition that Eq. (1-3) be exact is

$$\frac{\partial A}{\partial y} = \frac{\partial B}{\partial x} \tag{1-4}$$

EXAMPLE

$$(x + y)\, dx + x\, dy = 0$$

$$A = x + y \qquad B = x \tag{1-5}$$

$$\frac{\partial A}{\partial y} = \frac{\partial B}{\partial x} = 1$$

The solution is

$$xy + \tfrac{1}{2}x^2 = C$$

Sometimes we can find a function $\lambda(x, y)$, such that

$$\lambda(A\ dx + B\ dy)$$

is an exact differential, although $A\ dx + B\ dy$ may not have been.   We call such a function λ an *integrating factor*.   One can show that such factors always exist (for a first-order equation), but there is no general way of finding them.

Consider the general *linear* first-order equation

$$\frac{dy}{dx} + f(x)y = g(x) \tag{1-6}$$

Let us try to find an integrating factor $\lambda(x)$.   That is,

$$\lambda(x)[dy + f(x)y\ dx] = \lambda(x)g(x)\ dx$$

is to be exact.   The right side is immediately integrable, and our criterion (1-4) that the left side be exact is

$$\frac{d\lambda(x)}{dx} = \lambda(x)f(x)$$

This equation is separable, and its solution

$$\lambda(x) = \exp\left[\int f(x)\ dx\right] \tag{1-7}$$

is the integrating factor we were looking for.

EXAMPLE

$$xy' + (1 + x)y = e^x$$

$$y' + \left(\frac{1 + x}{x}\right)y = \frac{e^x}{x} \tag{1-8}$$

The integrating factor is $\exp\left\{\int[(1 + x)/x]dx\right\} = xe^x$

$$xe^x\left[y' + \left(\frac{1 + x}{x}\right)y\right] = e^{2x}$$

Now our equation is exact; integrating both sides gives

$$xe^x y = \int e^{2x}\ dx = \tfrac{1}{2}e^{2x} + C$$

$$y = \frac{e^x}{2x} + \frac{C}{x}e^{-x}$$

One can often simplify a differential equation by making a judicious change of variable. For example, the equation

$$y' = f(ax + by + c) \tag{1-9}$$

becomes separable if one introduces the new dependent variable,

$$v = ax + by + c$$

As another example, the so-called *Bernoulli equation*

$$y' + f(x)y = g(x)y^n \tag{1-10}$$

becomes linear if one sets $v = y^{1-n}$. (This substitution becomes "obvious" if the equation is first divided by y^n.)

A function $f(x, y, \ldots)$ in any number of variables is said to be *homogeneous* of degree r in these variables if

$$f(ax, ay, \ldots) = a^r f(x, y, \ldots)$$

A first-order differential equation

$$A(x, y)\, dx + B(x, y)\, dy = 0 \tag{1-11}$$

is said to be homogeneous if A and B are homogeneous functions of the same degree. The substitution $y = vx$ makes the homogeneous equation (1-11) separable.

EXAMPLE

$$y\, dx + \left(2\sqrt{xy} - x\right) dy = 0 \tag{1-12}$$
$$y = vx \qquad dy = v\, dx + x\, dv$$
$$vx\, dx + \left(2x\sqrt{v} - x\right)(v\, dx + x\, dv) = 0$$
$$2v^{3/2}\, dx + \left(2\sqrt{v} - 1\right)x\, dv = 0$$

This equation is clearly separable and its solution is trivial.

Note that this approach is related to dimensional arguments familiar from physics. A homogeneous function is simply a dimensionally consistent function, if x, y, $\ldots$ are all assigned the same dimension (for example, length). The variable $v = y/x$ is then a "dimensionless" variable.

This suggests a generalization of the idea of homogeneity. Suppose that the equation

$$A\, dx + B\, dy = 0$$

is dimensionally consistent when the dimensionality of y is some power m of the dimensionality of x. That is, suppose

$$A(ax, a^m y) = a^r A(x, y)$$
$$B(ax, a^m y) = a^{r-m+1} B(x, y) \tag{1-13}$$

Such equations are said to be *isobaric*. The substitution $y = vx^m$ reduces the equation to a separable one.

EXAMPLE

$$xy^2(3y\,dx + x\,dy) - (2y\,dx - x\,dy) = 0 \qquad (1\text{-}14)$$

Let us test to see if this is isobaric. Give x a "weight" 1 and y a weight m. The first term has weight $3m + 2$, and the second has weight $m + 1$. Therefore, the equation is isobaric with weight $m = -\frac{1}{2}$.

This suggests introducing the "dimensionless" variable $v = y\sqrt{x}$. To avoid fractional powers, we instead let

$$v = y^2 x \qquad x = \frac{v}{y^2} \qquad dx = \frac{dv}{y^2} - \frac{2v\,dy}{y^3}$$

Equation (1-14) reduces to

$$(3v - 2)y\,dv + 5v(1 - v)\,dy = 0$$

which is separable.

An equation of the form

$$(ax + by + c)\,dx + (ex + fy + g)\,dy = 0 \qquad (1\text{-}15)$$

where $a, \ldots, g$ are constants, may be made homogeneous by a substitution

$$x = X + \alpha \qquad y = Y + \beta$$

where α and β are suitably chosen constants [provided $af \neq be$; if $af = be$, Eq. (1-15) is even more trivial].

An equation of the form

$$y - xy' = f(y') \qquad (1\text{-}16)$$

is known as a *Clairaut equation*. To solve it, differentiate both sides with respect to x. The result is

$$y''[f'(y') + x] = 0$$

We thus have two possibilities. If we set $y'' = 0$, $y = ax + b$, and substitution back into the original equation (1-16) gives $b = f(a)$. Thus $y = ax + f(a)$ is the general solution. However, we also have the possibility

$$f'(y') + x = 0$$

Eliminating y' between this equation and the original differential equation (1-16), we obtain a solution with no arbitrary constants. Such a solution is known as a *singular solution*.

EXAMPLE

$$y = xy' + (y')^2 \qquad (1\text{-}17)$$

This is a Clairaut equation with general solution

$$y = cx + c^2$$

However, we must also consider the possibility $2y' + x = 0$, which gives

$$x^2 + 4y = 0$$

This singular solution is an *envelope* of the family of curves given by the general solution, as shown in Figure 1–1.   The dotted parabola is the singular solution, and the straight lines tangent to the parabola are the general solution.

There are various other types of singular solutions, but we shall not discuss them here.   See Cohen (C4) or Ince (I2) for more complete discussions and further references.

Next we review some methods which are useful for higher-order differential equations.   An important type is the *linear equation with constant coefficients*:

$$a_n y^{(n)} + a_{n-1} y^{(n-1)} + \cdots + a_1 y' + a_0 y = f(x) \qquad (1\text{-}18)$$

If $f(x) = 0$, the equation is said to be *homogeneous*; otherwise it is *inhomogeneous*.   Note that, if a linear equation is homogeneous, the sum of two solutions is also a solution, whereas this is not true if the equation is inhomogeneous.

The general solution of an inhomogeneous equation is the sum of the general solution of the corresponding homogeneous equation (the so-called *complementary function*) and any solution of the inhomogeneous equation

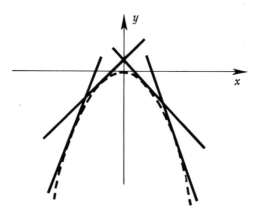

Figure 1–1   Solutions of the differential equation (1-17) and their envelope

(the so-called *particular integral*).   This is in fact true for *any* linear differential equation, whether or not the coefficients are constants.

Solutions of the homogeneous equation [(1-18) with $f(x) = 0$] generally have the form

$$y = e^{mx}$$

Substitution into the homogeneous equation gives

$$a_n m^n + a_{n-1} m^{n-1} + \cdots + a_0 = 0$$

If the n roots are $m_1, m_2, \ldots, m_n$ the complementary function is

$$c_1 e^{m_1 x} + \cdots + c_n e^{m_n x} \qquad (c_i \text{ are arbitrary constants})$$

Suppose two roots are the same, $m_1 = m_2$.   Then we have only $n - 1$ solutions, and we need another.   Imagine a limiting procedure in which m_2 approaches m_1.   Then

$$\frac{e^{m_2 x} - e^{m_1 x}}{m_2 - m_1}$$

is a solution, and, as m_2 approaches m_1, this solution becomes

$$\frac{d}{dm} e^{mx} \bigg|_{m = m_1} = x e^{m_1 x}$$

This is our additional solution.   If three roots are equal, $m_1 = m_2 = m_3$, then the three solutions are

$$e^{m_1 x} \qquad x e^{m_1 x} \qquad x^2 e^{m_1 x}$$

and so on.

Arguments involving similar limiting procedures are frequently useful; see, for example, the discussion on pp. 17 and 18.

A particular integral is generally harder to find.   If $f(x)$ has only a finite number of linearly independent derivatives, that is, is a linear combination of terms of the form x^n, $e^{\alpha x}$, $\sin kx$, $\cos kx$, or, more generally,

$$x^n e^{mx} \cos \alpha x \qquad x^n e^{mx} \sin \alpha x$$

then the method of *undetermined coefficients* is quite straightforward.   Take for $y(x)$ a linear combination of $f(x)$ and its independent derivatives and determine the coefficients by requiring that $y(x)$ obey the differential equation.

EXAMPLE

$$y'' + 3y' + 2y = e^x \tag{1-19}$$

Complementary function:

$$m^2 + 3m + 2 = 0$$
$$m = -1, -2$$
$$y = c_1 e^{-x} + c_2 e^{-2x}$$

Particular integral: Try $y = Ae^x$. Substitution into the differential equation (1-19) gives

$$6A = 1 \qquad A = \tfrac{1}{6}$$

Thus, the general solution is

$$y = \tfrac{1}{6}e^x + c_1 e^{-x} + c_2 e^{-2x}$$

If $f(x)$, or a term in $f(x)$, is also part of the complementary function, the particular integral may contain this term and its derivatives multiplied by some power of x. To see how this works, solve the above example (1-19) with the right-hand side, e^x, replaced by e^{-x}.

There are several formal devices for obtaining particular integrals. If D means d/dx, then we can write our equation (1-18) as

$$(D - m_1)(D - m_2)\cdots(D - m_n)y = f(x) \qquad (1\text{-}20)$$

A formal solution of (1-20) is

$$y = \frac{f(x)}{(D - m_1)\cdots(D - m_n)}$$

or, expanding by partial fraction techniques,

$$y = A_1 \frac{f(x)}{D - m_1} + \cdots + A_n \frac{f(x)}{D - m_n} \qquad (1\text{-}21)$$

What does $f(x)/(D - m)$ mean? It is the solution of $(D - m)y = f(x)$, which is a first-order linear equation whose solution is trivial [see (1-6)].

Alternatively, we can just peel off the factors in (1-20) one at a time. That is,

$$(D - m_2)(D - m_3)\cdots(D - m_n)y = \frac{f(x)}{D - m_1} \qquad (1\text{-}22)$$

We evaluate the right side, divide by $D - m_2$, evaluate again, and so on.

Next, we consider the very important method known as *variation of parameters* for obtaining a particular integral. This method has the useful feature of applying equally well to linear equations with nonconstant coefficients. Before giving a general discussion of the method and applying it to an example, we shall digress briefly on the subject of *osculating parameters*.

Suppose we are given two linearly independent functions $y_1(x)$ and $y_2(x)$. By means of these we can define the two-parameter family of functions

$$c_1 y_1(x) + c_2 y_2(x) \tag{1-23}$$

Now consider some arbitrary function $y(x)$. Can we represent it by an appropriate choice of c_1 and c_2 in (1-23)? Clearly, the answer in general is no. Let us try the more modest approach of *approximating* $y(x)$ in the neighborhood of some fixed point $x = x_0$ by a curve of the family (1-23). Since there are two parameters at our disposal, a natural choice is to fit the *value* $y(x_0)$ and *slope* $y'(x_0)$ exactly. That is, c_1 and c_2 are determined from the two simultaneous equations

$$\begin{aligned} y(x_0) &= c_1 y_1(x_0) + c_2 y_2(x_0) \\ y'(x_0) &= c_1 y'_1(x_0) + c_2 y'_2(x_0) \end{aligned} \tag{1-24}$$

The c_1 and c_2 obtained in this way vary from point to point (that is, as x_0 varies) along the curve $y(x)$. They are called *osculating parameters* because the curve they determine fits the curve $y(x)$ as closely as possible at the point in question.[1]

One can, of course, generalize to an arbitrary number N of functions y_i and parameters c_i. One chooses the c_i to reproduce the function $y(x)$ and its first $N - 1$ derivatives at the point x_0.

We now return to the problem of solving linear differential equations. For simplicity, we shall restrict ourselves to second-order equations. Consider the inhomogeneous equation

$$p(x)y'' + q(x)y' + r(x)y = s(x) \tag{1-25}$$

and suppose we know the complementary function to be

$$c_1 y_1(x) + c_2 y_2(x)$$

Let us seek a solution of (1-25) of the form

$$y = u_1(x)y_1(x) + u_2(x)y_2(x) \tag{1-26}$$

where the $u_i(x)$ are functions to be determined. In order to substitute (1-26) into (1-25), we must evaluate y' and y''. From (1-26),

$$y' = u_1 y'_1 + u_2 y'_2 + u'_1 y_1 + u'_2 y_2 \tag{1-27}$$

Before going on to calculate y'', we observe that it would be convenient to impose the condition that the sum of the last two terms in (1-27) vanish, that is,

$$u'_1 y_1 + u'_2 y_2 = 0 \tag{1-28}$$

---

[1] Latin *osculari* = to kiss.

This prevents second derivatives of the u_i from appearing, for now

$$y' = u_1 y_1' + u_2 y_2' \tag{1-29}$$

and differentiating gives

$$y'' = u_1 y_1'' + u_2 y_2'' + u_1' y_1' + u_2' y_2' \tag{1-30}$$

Note that the condition (1-28) not only simplifies the subsequent algebra but also ensures that u_1 and u_2 are just the osculating parameters discussed above; compare (1-26) and (1-29) with (1-24).

The rest of the procedure is straightforward. If (1-26), (1-29), and (1-30) are substituted into the original differential equation (1-25), and we use the fact that y_1 and y_2 are solutions of the homogeneous equation, we obtain

$$p(x)(u_1' y_1' + u_2' y_2') = s \tag{1-31}$$

(1-28) and (1-31) are now simultaneous linear equations for the u_i', whose solution is straightforward. The student is urged to complete the solution; see Problem 1-26.

EXAMPLE

Consider the differential equation

$$x^2 y'' - 2y = x \tag{1-32}$$

The complementary function, that is, the general solution of

$$x^2 y'' - 2y = 0 \tag{1-33}$$

is most easily found by noting that $y = x^m$ is a natural trial solution.[2] Substitution into (1-33) shows that $m = 2$ or -1, so that the complementary function is

$$c_1 x^2 + \frac{c_2}{x}$$

Therefore, we are led to try to find a solution of (1-32) of the form

$$y = u_1 x^2 + \frac{u_2}{x} \tag{1-34}$$

---

[2] In fact, $y = x^m$ is an obvious trial solution for any linear differential equation of the form

$$c_n x^n y^{(n)} + c_{n-1} x^{n-1} y^{(n-1)} + \cdots + c_1 x y' + c_0 y = 0$$

We differentiate, obtaining

$$y' = 2xu_1 - \frac{1}{x^2} u_2 + x^2 u_1' + \frac{1}{x} u_2'$$

and impose the condition

$$x^2 u_1' + \frac{1}{x} u_2' = 0 \qquad (1\text{-}35)$$

Then

$$y' = 2xu_1 - \frac{1}{x^2} u_2 \qquad (1\text{-}36)$$

and

$$y'' = 2u_1 + \frac{2}{x^3} u_2 + 2xu_1' - \frac{1}{x^2} u_2' \qquad (1\text{-}37)$$

Substituting (1-34) and (1-37) back into the differential equation (1-32), we obtain

$$2x^3 u_1' - u_2' = x$$

Solving this together with (1-35) gives

$$u_1' = \frac{1}{3x^2} \qquad u_2' = -\frac{x}{3}$$

$$u_1 = -\frac{1}{3x} + c_1 \qquad u_2 = -\frac{x^2}{6} + c_2$$

Then the general solution of (1-32) is

$$y = -\frac{x}{2} + c_1 x^2 + \frac{c_2}{x}$$

Changes of variable sometimes help. We shall discuss one general transformation which is particularly useful. Consider the second-order linear equation

$$y'' + f(x)y' + g(x)y = 0 \qquad (1\text{-}38)$$

The substitution

$$y = v(x)p(x) \qquad (1\text{-}39)$$

will give another linear equation in $v(x)$ which may be easier to solve for certain choices of $p(x)$. The resulting equation is

$$v'' + \left(2\frac{p'}{p} + f\right)v' + \left(\frac{p'' + fp' + gp}{p}\right)v = 0 \qquad (1\text{-}40)$$

If we know one solution of the original equation (1-38), we can take p to be that solution and thereby eliminate the term in v from (1-40). This is very useful because we can then find the general solution by two straight-forward integrations.

Alternatively, we may choose

$$p = \exp\left[-\frac{1}{2}\int f(x)\,dx\right] \qquad (1\text{-}41)$$

and eliminate the first derivative term of (1-40). This procedure is helpful as an aid to recognizing equations, and it is especially useful in connection with approximate methods of solution.

EXAMPLE

Bessel's equation is

$$x^2 y'' + xy' + (x^2 - n^2)y = 0 \qquad (1\text{-}42)$$

The substitution $y = u(x)p(x)$ with $p(x) = x^{-1/2}$ gives

$$u'' + \left(1 - \frac{n^2 - \frac{1}{4}}{x^2}\right)u = 0 \qquad (1\text{-}43)$$

This equation, with no first derivative term, is very convenient for finding the approximate behavior of Bessel functions at large x, for example, by the WKB method (Section 1–4).   The details are left as an exercise (Problem 1-41).

In conclusion, we shall briefly mention some other devices for solving differential equations.   For details, refer to a text on differential equations.

If the dependent variable y is absent, let $y' = p$ be the new dependent variable.   This lowers the order of the equation by one.

If the independent variable x is absent, let y be the new independent variable and $y' = p$ be the new dependent variable.   This also lowers the order of the equation by one.

If the equation is homogeneous in y, let $v = \log y$ be a new dependent variable.   The resulting equation will not contain v, and the substitution $v' = p$ will then reduce the order of the equation by one.

If the equation is isobaric, when x is given the weight 1 and y the weight m, the change in dependent variable $y = vx^m$, followed by the change in independent variable $u = \log x$, gives an equation in which the new independent variable u is absent.

Always watch for the possibility that an equation is exact.   Also consider the possibility of finding an integrating factor.   For example, the commonly occurring equation $y'' = f(y)$ can be integrated immediately if both sides are multiplied by y'.

1-2   POWER-SERIES SOLUTIONS

Before discussing series solutions in general, consider a simple (although nonlinear) example:

$$y'' = x - y^2 \tag{1-44}$$

Try

$$y = c_0 + c_1 x + c_2 x^2 + \cdots$$

Then (1-44) becomes

$$2c_2 + 6c_3 x + 12c_4 x^2 + \cdots = x - c_0^2 - 2c_0 c_1 x - (c_1^2 + 2c_0 c_2)x^2 - \cdots$$

Equating coefficients of equal powers of x gives

$$c_2 = -\tfrac{1}{2}c_0^2$$
$$c_3 = \tfrac{1}{6} - \tfrac{1}{3}c_0 c_1$$
$$c_4 = -\tfrac{1}{12}c_1^2 + \tfrac{1}{12}c_0^3, \quad \text{etc.}$$

Suppose, for example, we want the solution with $y = 0$, $y' = 1$ at $x = 0$. Then

$$c_0 = 0, c_1 = 1, c_2 = 0, c_3 = \tfrac{1}{6}, c_4 = -\tfrac{1}{12}, \ldots \tag{1-45}$$

This method of solution is a very useful one, but we have been careless about justifying the method, establishing convergence of the series, etc.   We now briefly outline the general theory of series solutions of *linear* differential equations.[3]

Consider the equation

$$\frac{d^n y}{dx^n} + f_{n-1}(x)\frac{d^{n-1}y}{dx^{n-1}} + \cdots + f_0(x)y = 0 \tag{1-46}$$

---

[3] This theory is discussed more fully, for example, in Copson (C8) Chapter X, or Jeffreys and Jeffreys (J4) Chapter 16.

If $f_0(x), \ldots, f_{n-1}(x)$ are regular (in the sense of complex variable theory; see the Appendix, Section A-2) at a point $x = x_0$, x_0 is said to be an *ordinary point* of the differential equation. Near an ordinary point, the general solution of a differential equation can be written as a Taylor series whose radius of convergence is the distance to the nearest singularity of the differential equation; by a singularity, of course, we mean a point which is not ordinary.

The Taylor series is an ordinary power series

$$y = \sum_{m=0}^{\infty} c_m(x - x_0)^m \tag{1-47}$$

whose coefficients c_m may be conveniently found by substitution in the differential equation, as in the example above.

If x_0 is not an ordinary point, but $(x - x_0)f_{n-1}(x)$, $(x - x_0)^2 f_{n-2}(x)$, ..., $(x - x_0)^n f_0(x)$ are regular at x_0, x_0 is said to be a *regular singular point* of the differential equation. Near a regular singular point, we can always find *at least one* solution of the form

$$y = (x - x_0)^s \sum_{m=0}^{\infty} c_m(x - x_0)^m \qquad \text{with } c_0 \neq 0 \tag{1-48}$$

where the exponent s of the leading term will not necessarily be an integer. The series again converges in any circle which includes no singularities except for x_0.

The algebraic work involved in substituting the series (1-47) or (1-48) in a differential equation is simplified if $x_0 = 0$. Thus it is usually convenient to first make a translation to x_0 as origin, that is, to rewrite the equation in terms of a new independent variable, $z = x - x_0$.

If a point is neither an ordinary point nor a regular singularity, it is an *irregular singular point*.

EXAMPLE

We first consider an example of expansion about an ordinary point: *Legendre's differential equation* is

$$(1 - x^2)y'' - 2xy' + n(n + 1)y = 0 \tag{1-49}$$

The points $x = \pm 1$ are regular singular points. We shall expand about the ordinary point $x = 0$. Try

$$y = \sum_{m=0}^{\infty} c_m x^m = c_0 + c_1 x + c_2 x^2 + \cdots$$

Substitution into the differential equation (1-49) gives

$$c_2 = -\frac{n(n+1)}{2} c_0$$

$$c_3 = \frac{2 - n(n+1)}{6} c_1$$

$$c_4 = -\frac{n(n+1)[6 - n(n+1)]}{24} c_0$$

$$c_5 = \frac{[2 - n(n+1)][12 - n(n+1)]}{120} c_1$$

$$\cdots$$

The general recursion relation is

$$\frac{c_{i+2}}{c_i} = \frac{i(i+1) - n(n+1)}{(i+1)(i+2)} = \frac{(i+n+1)(i-n)}{(i+1)(i+2)} \tag{1-50}$$

Note that this is a two-term recursion relation, that is, it relates two co-efficients only. Their indices differ by two. These facts could have been noticed at the beginning by inspection of the differential equation.

The general solution of our differential equation (1-49) is therefore

$$y = c_0 \left[1 - n(n+1)\frac{x^2}{2!} + n(n+1)(n-2)(n+3)\frac{x^4}{4!} - + \cdots \right]$$

$$+ c_1 \left[x - (n-1)(n+2)\frac{x^3}{3!} \right.$$

$$\left. + (n-1)(n+2)(n-3)(n+4)\frac{x^5}{5!} - + \cdots \right] \tag{1-51}$$

For physical applications, it is necessary to consider the behavior of the infinite series in (1-51) near the singular points $x = \pm 1$. If we let $i \to \infty$ in the recursion relation (1-50), we see that

$$\frac{c_{i+2}}{c_i} \to 1$$

Thus, if we go farther and farther out in either series, we find it more and more resembling a geometric series with x^2 being the ratio of successive terms. The sum of such a series clearly approaches infinity as x^2 approaches 1.[4]

---

[4] This discussion is incomplete. Infinite series are discussed in more detail in Chapter 2, and the series of (1-51) are studied as a special case on p. 47. The series in fact "almost" converge as x^2 approaches 1; their sums go to infinity logarithmically $[y \sim \ln(1 - x^2)]$.

In cases of physical interest, however, we often require solutions $y(x)$ which are finite for $-1 \leqslant x \leqslant +1$. There are two ways we can arrange this:

(1) Let $c_1 = 0$, and choose n to be one of the integers $\ldots$, -5, -3, -1, $0, 2, 4, \ldots$. Then the first series in (1-51) terminates, and the second is absent.

(2) Let $c_0 = 0$, and choose n to be one of the integers $\ldots$, -6, -4, -2, $1, 3, 5, \ldots$. Then the second series in (1-51) terminates, and the first is absent.

We see that if we wish a solution of the Legendre differential equation (1-49) which is finite on the interval $-1 \leqslant x \leqslant +1$, n must be an integer. The resulting solution $y(x)$ is a polynomial which, when normalized by the condition $y(x) = 1$, is known as a *Legendre polynomial* $P_n(x)$.[5]

As an example of the problems that can arise at a regular singular point, consider *Bessel's equation*

$$x^2 y'' + xy' + (x^2 - m^2)y = 0 \tag{1-52}$$

This has a regular singular point at $x = 0$. Thus, a solution of the form

$$y = x^s \sum_{n=0}^{\infty} c_n x^n \qquad (c_0 \neq 0)$$

is guaranteed to exist, and we can see from the differential equation that a two-term recursion relation will be obtained. If we substitute into the differential equation, the coefficient of x^s is

$$c_0(s^2 - m^2) = 0$$
$$s^2 - m^2 = 0 \tag{1-53}$$

This is called the *indicial equation*. Its roots are $s = \pm m$.

Next consider the coefficient of x^{s+1}. It is

$$c_1[(s + 1)^2 - m^2] = 0$$

Thus $c_1 = 0$ except in the single case $m = 1/2$, $s = -m = -1/2$, and in that case we can set $c_1 = 0$, since the terms thereby omitted are equivalent to those which make up the other solution, with $s = +m = +1/2$.

We therefore confine ourselves to even values of n in the sum, writing

$$y = x^{\pm m}(c_0 + c_2 x^2 + c_4 x^4 + \cdots)$$

The recursion relation is easily found to be

$$\frac{c_{n+2}}{c_n} = \frac{-1}{(s + n + 2)^2 - m^2} = \frac{-1}{(n + 2)(2s + n + 2)} \tag{1-54}$$

---

[5] Note that $P_n(x) = P_{-1-n}(x)$.

Thus our solution is

$$y = c_0 x^s \left[1 - \frac{x^2}{(4s+1)} + \frac{x^4}{4 \times 8(s+1)(s+2)} - + \cdots \right] \qquad (1\text{-}55)$$

Suitably normalized, this series is called a *Bessel function*; we shall discuss Bessel functions in more detail in Chapter 7.

If m is not an integer, we have two independent solutions to our equation, namely, (1-55) with $s = \pm m$. If m is an integer (which we may assume positive or zero), we can only choose $s = +m$; for $s = -m$, all denominators in (1-55) beyond a certain term vanish. If we multiply through by $(s+m)$ before setting $s = -m$, to cancel the offending factors, we get a multiple of the solution with $s = +m$, as may be readily verified.

This is a situation reminiscent of our difficulty on p. 7 in obtaining independent solutions of a differential equation with constant coefficients. The resolution of the difficulty is quite analogous. We begin by letting $y(x, s)$ denote the series

$$y(x, s) = x^s \sum_{n=0}^{\infty} c_n x^n$$

with $c_0 = 1$, $c_1 = 0$, and all other coefficients given by (1-54), but without making use of the indicial equation (1-53). That is, we set

$$y(x, s) = x^s \left[1 - \frac{x^2}{(s+2+m)(s+2-m)} \right.$$
$$\left. + \frac{x^4}{(s+2+m)(s+2-m)(s+4+m)(s+4-m)} - + \cdots \right] \qquad (1\text{-}56)$$

If we abbreviate by L the Bessel differential operator,

$$L = x^2 \frac{d^2}{dx^2} + x \frac{d}{dx} + (x^2 - m^2)$$

then the series $Ly(x, s)$ will contain only a term in x^s, because the recursion relation makes the coefficients of all higher powers vanish. We obtain [compare (1-53)]

$$Ly(x, s) = (s - m)(s + m)x^s$$

We see (again) that s must equal $\pm m$ in order that $y(x, s)$ be a solution of Bessel's equation $Ly = 0$. However, if m is a positive integer, we have remarked that

$$[(s + m)y(x, s)]_{s = -m}$$

is a constant multiple of $y(x, m)$, and we must find a second solution. To do so, consider the result obtained by substituting $(s + m)y(x, s)$ into Bessel's equation. We get

$$L[(s + m)y(x, s)] = (s + m)^2(s - m)x^s$$

The derivative of the right side with respect to s vanishes at $s = -m$. Therefore,

$$\left\{\frac{\partial}{\partial s}[(s + m)y(x, s)]\right\}_{s=-m} \tag{1-57}$$

is a solution of Bessel's equation, and is in fact the second solution we were looking for.

EXAMPLE

Let us find a second solution of Bessel's equation for $m = 2$.

$$y(x, s) = x^s\left[1 - \frac{x^2}{s(s + 4)} + \frac{x^4}{s(s + 2)(s + 4)(s + 6)} - + \cdots\right]$$

$$(s + 2)y(x, s) = x^s\left[(s + 2) - \frac{s + 2}{s(s + 4)}x^2 + \frac{x^4}{s(s + 4)(s + 6)}\right. \tag{1-58}$$

$$\left. - \frac{x^6}{s(s + 4)(s + 4)(s + 6)(s + 8)} + - \cdots\right]$$

Remembering that $(d/ds)x^s = x^s \ln x$ and

$$\frac{d}{dx}\left(\frac{uv \cdots}{w \cdots}\right) = \frac{uv \cdots}{w \cdots}\left(\frac{u'}{u} + \frac{v'}{v} + \cdots - \frac{w'}{w} - \cdots\right)$$

we obtain

$$\frac{\partial}{\partial s}[(s + 2)y(x, s)] = (s + 2)y(x, s) \ln x$$

$$+ x^s\left[1 - \frac{s + 2}{s(s + 4)}\left(\frac{1}{s + 2} - \frac{1}{s} - \frac{1}{s + 4}\right)x^2\right.$$

$$\left. + \frac{1}{s(s + 4)(s + 6)}\left(-\frac{1}{s} - \frac{1}{s + 4} - \frac{1}{s + 6}\right)x^4 - + \cdots\right]$$

Setting $s = -2$, and observing that $[(s + 2)y(x, s)]_{s=-2} = -\frac{1}{16}y(x, 2)$, we obtain

$$\left\{\frac{\partial}{\partial s}[(s + 2)y(x, s)]\right\}_{s=-2}$$

$$= -\frac{1}{16}y(x, 2)\ln x + x^{-2}\left(1 + \frac{x^2}{4} + \frac{x^4}{64} + \cdots\right) \quad (1\text{-}59)$$

This solution and $y(x, 2)$ are two independent solutions of Bessel's equation with $m = 2$.

Let us now consider the differential equation

$$\frac{d^2\psi}{dx^2} + (E - x^2)\psi = 0 \quad (1\text{-}60)$$

This is the Schrödinger equation for a one-dimensional quantum-mechanical harmonic oscillator, in appropriate units. If one tries a direct power series solution of (1-60) about $x = 0$, one obtains a three-term recursion relation. These are a little inconvenient,[6] so that it is advantageous to look for a transformation of variables which will lead to a simpler equation.

A trick that often helps in such a situation is to "factor out" the behavior near some singularity or singularities. Where are the singularities of this equation? There are none in the finite z plane, but we must digress for a moment on *singularities at infinity*.

Consider the differential equation

$$y'' + P(x)y' + Q(x)y = 0 \quad (1\text{-}61)$$

Recall that $\begin{cases} x = 0 & \text{is an ordinary point if } P \text{ and } Q \text{ are regular there} \\ x = 0 & \text{is a regular singular point if } xP \text{ and } x^2Q \text{ are regular} \\ & \text{there} \end{cases}$

Let $z = 1/x$. The equation becomes

$$\frac{d^2y}{dz^2} + \left[\frac{2}{z} - \frac{1}{z^2}p(z)\right]\frac{dy}{dz} + \frac{1}{z^4}q(z)y = 0 \quad (1\text{-}62)$$

where

$$p(z) = P(x) \qquad q(z) = Q(x)$$

Then $x = \infty$ is an ordinary point or a singularity of Eq. (1-61) depending on whether $z = 0$ is an ordinary point or a singularity of Eq. (1-62). That

---

[6] Three-term recursion relations can often be treated by continued fraction methods (see Section 7-5).

is, $x = \infty$ is an ordinary point if $2x - x^2 P(x)$ and $x^4 Q(x)$ are regular there, and $x = \infty$ is a regular singular point if $xP(x)$ and $x^2 Q(x)$ are regular there. This concludes our digression.

Using these criteria, we see that our differential equation (1-60) is highly singular at $x = \infty$. For large x, the equation is approximately

$$\frac{d^2\psi}{dx^2} - x^2\psi = 0 \tag{1-63}$$

and the solutions are roughly

$$\psi \sim e^{\pm x^2/2} \tag{1-64}$$

in that, if we substitute either function into the differential equation (1-63), the terms which are dominant at infinity cancel.

In quantum mechanics, physically acceptable solutions must not become infinite at $|x| \to \infty$; therefore, try

$$\psi = ye^{-x^2/2} \tag{1-65}$$

This change of variable does not ensure the desired behavior at infinity, of course, and we shall still have to select solutions $y(x)$ which give this behavior. In fact, we may expect in general $y(x) \to e^{x^2}$ so as to give the divergent solution, $\psi \to e^{+x^2/2}$.

The differential equation (1-60) becomes

$$y'' - 2xy' + (E - 1)y = 0 \tag{1-66}$$

[If we write $E - 1 = 2n$, (1-66) is known as the *Hermite differential equation*.] We can obtain the general solution of this equation in the form of a power series which will converge everywhere, and the recursion relation for the coefficients will contain only two terms. The recursion relation is

$$\frac{c_{m+2}}{c_m} = \frac{(2m + 1) - E}{(m + 1)(m + 2)}$$

and the solution is

$$y = c_0\left[1 + (1 - E)\frac{x^2}{2!} + (1 - E)(5 - E)\frac{x^4}{4!} + \cdots\right]$$

$$+ c_1\left[x + (3 - E)\frac{x^3}{3!} + (3 - E)(7 - E)\frac{x^5}{5!} + \cdots\right] \tag{1-67}$$

If $E = 2n + 1$ where n is an integer, one series terminates after the term in x^n (the even or odd series, depending on n). The resulting polynomial, suitably normalized, is called a *Hermite polynomial*[7] of order n, $H_n(x)$. The other series can be eliminated by setting its coefficient, c_0 or c_1, equal to zero, and the resulting solution (1-65) will go to zero at infinity.

If either series of (1-67) does not terminate, its behavior at large x is determined by the terms far out where the recursion relation is approximately

$$\frac{c_{m+2}}{c_m} \approx \frac{2}{m}$$

Thus either series behaves for large x like e^{x^2}, and $\psi \to e^{x^2/2}$ as expected. A solution ψ, which remains bounded as $x \to \pm\infty$, is therefore only possible if $E = 2n + 1$ with integral n; that is,

$$\psi = \psi_n(x) = H_n(x)e^{-x^2/2}$$

for

$$E = E_n = 2n + 1 \tag{1-68}$$

This is another example of how boundary conditions may impose restrictions on the acceptable values for a constant appearing in a differential equation. The acceptable solutions ψ_n are called *eigenfunctions* of the differential operator $-(d^2/dx^2) + x^2$ belonging to the *eigenvalues* E_n (see Chapter 9).

As another example of series solutions, we consider briefly the *associated Legendre equation*

$$(1 - x^2)y'' - 2xy' + \left[n(n + 1) - \frac{m^2}{1 - x^2}\right]y = 0 \tag{1-69}$$

Note that when $m = 0$, this reduces to Legendre's equation. The origin is an ordinary point, and $x = \pm 1$ are regular singularities.

A straightforward attempt at a power series solution about $x = 0$ again yields a three-term recursion relation. Let us try to factor out the behavior at $x = \pm 1$. Let $x = \pm 1 + z$ and write the approximate form of the differential equation valid for $|z| \ll 1$. Near either singularity it is

$$zy'' + y' - \frac{m^2}{4z}y = 0$$

This has solutions $y = z^{+m/2}$ and $z^{-m/2}$, the former being well behaved if, as we may assume without loss of generality, $m \geqslant 0$.

We are therefore led to make the change of variable $y = v(1 - x^2)^{m/2}$ in (1-69), thus factoring out the behavior at both singularities simultaneously. The equation becomes

$$(1 - x^2)v'' - 2(m + 1)xv' + [n(n + 1) - m(m + 1)]v = 0 \tag{1-70}$$

---

[7] For a precise definition of $H_n(x)$ see Eq. (7-103).

with a straightforward series solution about $x = 0$. The recursion relation for the coefficients is

$$\frac{c_{r+2}}{c_r} = \frac{(r + m)(r + m + 1) - n(n + 1)}{(r + 1)(r + 2)}$$

$$= \frac{(r + m - n)(r + m + n + 1)}{(r + 1)(r + 2)} \tag{1-71}$$

Again we obtain even and odd series solutions for $v(x)$, both of which behave like $(1 - x^2)^{-m}$ near $x = \pm 1$ if they do not terminate.[8]   A bounded solution exists only if n and m are such that one series terminates after some term x^r. From (1-71), the condition is

$$(n - m) = r = \text{an integer} \geqslant 0 \tag{1-72}$$

Usually in physical applications n and m are both integers.   It may be verified that y is then simply a constant times

$$(1 - x^2)^{m/2} \left(\frac{d}{dx} \right)^m P_n(x) \tag{1-73}$$

which is called an *associated Legendre function*.   These functions will be discussed in more detail in Chapter 7.

1-3 MISCELLANEOUS APPROXIMATE METHODS

One can often get a qualitative picture of the solutions of a differential equation by means of a graphical approach.

For example, consider the first-order equation

$$\frac{dy}{dx} = e^{-2xy} \tag{1-74}$$

We may draw little lines in the xy-plane, each indicating the slope of the solution passing through that point, as illustrated in Figure 1–2. The approximate shape of the solutions is obvious.

As a second example, consider the second-order, nonlinear equation

$$y'' = x - y^2 \tag{1-75}$$

---

[8] See Problem 1-32.

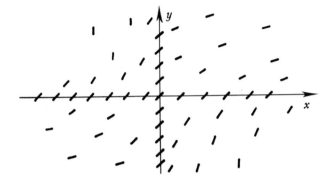

Figure 1-2   Slopes of the solutions of (1-74)

If we draw the parabola $x = y^2$, as in Figure 1-3, we see that y'' is positive inside the parabola, and negative outside of it.   If we limit ourselves to solutions starting near the origin and going toward the right, we can identify several types which are sketched in Figure 1-3:

1. Solutions oscillating about the upper branch of the parabola.
2. A critical nonoscillating solution approaching the upper branch of the parabola from above.
3. Solutions going to $y = -\infty$.
4. A critical nonoscillating solution approaching the lower branch of the parabola from above.

The power-series solution (1-45) we found earlier [with $y(0) = 0$, $y'(0) = 1$] is of type 1.   An approximation to solution 2 is given by expression (1-80).

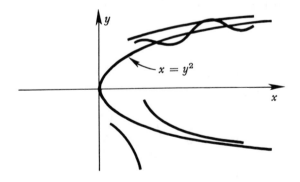

Figure 1-3   Several types of solution for the equation $y'' = x - y^2$

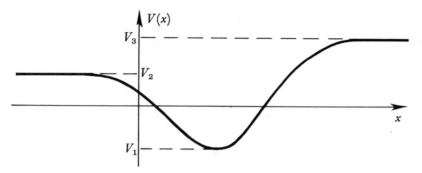

Figure 1-4   A potential $V(x)$ for the Schrödinger equation

As a third example, consider the one-dimensional Schrödinger equation for a particle of mass m in a potential $V(x)$:

$$\frac{d^2\psi}{dx^2} = -\frac{2m}{\hbar^2}[E - V(x)]\psi \tag{1-76}$$

If $E > V(x)$, then $\psi''/\psi < 0$, and ψ curves *toward* the x-axis; that is, ψ has an oscillatory or "sinusoidal" character. If $E < V(x)$, then $\psi''/\psi > 0$, and ψ curves *away from* the x-axis; that is, ψ has an "exponential" character. If we impose the boundary condition that ψ remain finite everywhere, then a solution with unbounded exponential behavior is unacceptable.

Suppose $V(x)$ looks as shown in Figure 1-4.

If $E > V_3$, all solutions are oscillatory everywhere and all are acceptable, that is, none blow up at infinity.

If $V_2 < E < V_3$, "most" solutions blow up at $x \to +\infty$. However, if we start just right, we can find a solution that falls off exponentially as $x \to +\infty$. This defines the acceptable solution uniquely (except in amplitude) and, in particular, fixes the phase in the left-hand region where the solution is oscillatory. Such a solution is illustrated qualitatively in Figure 1-5.

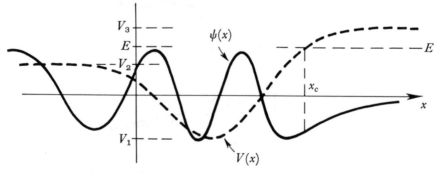

Figure 1-5   A physically acceptable solution of the Schrödinger equation for $V_2 < E < V_3$

If $V_1 < E < V_2$, things are really difficult. ψ behaves "exponentially" at both ends. If we adjust it so that it does not blow up at the left, it is almost sure to do so at the right. Only for certain values of E can satisfactory solutions be found. These are the *eigenvalues*; compare (1-68).

If $E < V_1$, no satisfactory solutions exist.

Another method, which can often provide useful approximate solutions, consists of experimenting with the equation, that is, dropping small terms, iterating, and so forth.

EXAMPLE

$$\frac{dy}{dx} = e^{-2xy} \tag{1-77}$$

Let's look for a solution near $x \to \infty$. Assume $y > 0$. Thus, we try

$$\frac{dy}{dx} \approx 0 \to y \approx a \qquad (= \text{constant})$$

Now put this guess back into (1-77)

$$\frac{dy}{dx} \approx e^{-2ax}$$

Therefore,

$$y \approx a - \frac{e^{-2ax}}{2a}$$

Iterate once more.

$$\frac{dy}{dx} \approx \exp\left[-2x\left(a - \frac{e^{-2ax}}{2a}\right)\right]$$

$$\approx e^{-2ax}\left(1 + \frac{x}{a}e^{-2ax} + \cdots\right)$$

$$y \approx a - \frac{e^{-2ax}}{2a} - \frac{1}{4a^2}\left(x + \frac{1}{4a}\right)e^{-4ax} + \cdots \qquad \text{etc.} \tag{1-78}$$

If the procedure seems to be converging, it is probably giving us a solution. Sometimes, though, it does not converge.

EXAMPLE

$$\frac{d^2y}{dx^2} = x - y^2 \tag{1-79}$$

(See the graphical discussion on pp. 22–23.)   Try $y \approx \sqrt{x}$.   Substituting into the right side of (1-79) gives $y'' \approx 0$, which in turn implies $y = ax + b$, with a and b constant.   This does not look anything like our first try.   The mistake, of course, was substituting our first guess into the "large" term y^2, rather than into the "small" term y''.   Doing it the other way gives

$$y^2 = x - y'' \approx x - \left(\sqrt{x}\right)'' = x(1 + \tfrac{1}{4}x^{-5/2})$$

$$y \approx \sqrt{x} + \tfrac{1}{8}x^{-2}$$

Iterating once more gives

$$y = \sqrt{x} + \tfrac{1}{8}x^{-2} - \tfrac{49}{128}x^{-9/2} + \cdots \tag{1-80}$$

This is the solution (2) enumerated on p. 23.

Alternatively, we may try a different method.   Write

$$y = \sqrt{x} + \eta(x) \qquad |\eta| \ll |\sqrt{x}| \tag{1-81}$$

Substituting into the differential equation (1-79) gives

$$-\tfrac{1}{4}x^{-3/2} + \eta'' = -2\eta\sqrt{x} - \eta^2 \tag{1-82}$$

We neglect the last term, but the equation is still complicated.   We may try further neglecting *either* the first or second term.   If we assume $|\eta''| \ll |2\eta\sqrt{x}|$, we find

$$\eta \approx \tfrac{1}{8}x^{-2}$$

and we obtain our previous solution (1-80).   We should, of course, check the consistency of the solution and the approximation made in finding it; namely, that

$$(\tfrac{1}{8}x^{-2})'' \ll 2(\tfrac{1}{8}x^{-2})\sqrt{x}$$

This is satisfied for large x.

We can obtain another solution by neglecting the $x^{-3/2}$ term of Eq. (1-82) instead of η''.   That is, if we assume

$$|\eta''| \gg |-\tfrac{1}{4}x^{-3/2}|$$

we obtain the differential equation

$$\eta'' + 2\sqrt{x}\,\eta = 0 \tag{1-83}$$

An approximate solution of this equation may be found by the WKB method, which we discuss in the next section.   The result is

$$\eta \approx \frac{A}{x^{1/8}} \sin\left(\frac{4\sqrt{2}}{5}x^{5/4} + \delta\right) \tag{1-84}$$

which gives an oscillatory solution of type 1 listed on p. 23. Again, we should go back and verify the consistency of the assumption

$$|\eta''| \gg |\tfrac{1}{4}x^{-3/2}|$$

Why can we not neglect the first term on the *right* of (1-82)?

1-4   THE WKB METHOD

The WKB method provides approximate solutions of differential equations of the form

$$\frac{d^2y}{dx^2} + f(x)y = 0 \qquad (1\text{-}85)$$

provided $f(x)$ satisfies certain restrictions discussed below, which may be summarized in the phrase "$f(x)$ is slowly varying." Recall that any linear homogeneous second-order equation may be put in this form by the transformation (1-41). The one-dimensional Schrödinger equation is of this form and the method was developed for quantum-mechanical applications by Wentzel, by Kramers,[9] and by Brillouin, whence the name. The method had been given previously by Jeffreys.[10]

The solutions of Eq. (1-85) with $f(x)$ constant suggest the substitution

$$y = e^{i\phi(x)} \qquad (1\text{-}86)$$

The differential equation becomes

$$-(\phi')^2 + i\phi'' + f = 0 \qquad (1\text{-}87)$$

If we assume ϕ'' small, a first approximation is

$$\phi' = \pm\sqrt{f} \qquad \phi(x) = \pm \int \sqrt{f(x)}\, dx \qquad (1\text{-}88)$$

The condition of validity (that ϕ'' be "small") is

$$|\phi''| \approx \frac{1}{2}\left|\frac{f'}{\sqrt{f}}\right| \ll |f| \qquad (1\text{-}89)$$

From (1-86) and (1-88) we see that $1/\sqrt{f}$ is roughly $1/(2\pi)$ times one "wavelength" or one "exponential length" of the solution y. Thus the condition of validity of our approximation is simply the intuitively reasonable one that the change in $f(x)$ in one wavelength should be small compared to $|f|$.

---

[9] See, for example, Kramers (K5).
[10] See Jeffreys and Jeffreys (J4).

A second approximation is easily found by iteration.   From (1-88),

$$\phi'' \approx \pm\tfrac{1}{2}f^{-1/2}f'$$

Substituting this estimate for the small term ϕ'' into (1-87), we obtain

$$(\phi')^2 \approx f \pm \frac{i}{2}\frac{f'}{\sqrt{f}}$$

$$\phi' \approx \pm\sqrt{f} + \frac{i}{4}\frac{f'}{f}$$

$$\phi(x) \approx \pm \int \sqrt{f(x)}\,dx + \frac{i}{4}\ln f$$

The two choices of sign give two (approximate) solutions which may be combined to give the general solution

$$y(x) \approx \frac{1}{(f(x))^{1/4}}\left\{ c_+ \exp\left[i\int \sqrt{f(x)}\,dx\right] + c_- \exp\left[-i\int \sqrt{f(x)}\,dx\right]\right\} \quad (1\text{-}90)$$

where c_+ and c_- are arbitrary constants.

We have thus found an approximation to the general solution of the original equation (1-85) in any region where the condition of validity (1-89) holds.   The method fails if $f(x)$ changes too rapidly or if $f(x)$ passes through zero.   The latter is a serious difficulty since we often wish to join an

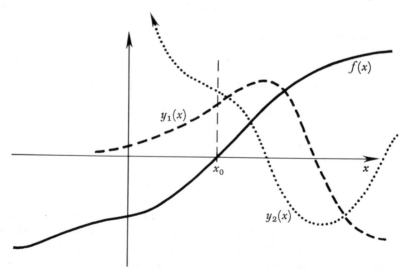

Figure 1–6   Graph of $f(x)$ and two exact solutions of the equation (1-85); one of them, $y_1(x)$, being the special solution which decreases to zero on the left

oscillatory solution in a region where $f(x) > 0$ to an "exponential" one in a region where $f(x) < 0$ [see, for example, the discussion of Eq. (1-76)].

We shall investigate this problem in some detail[11] in order to derive the so-called *connection formulas* relating the constants c_+ and c_- of the WKB solutions on either side of a point where $f(x) = 0$.

Suppose $f(x)$ passes through zero at x_0 and is positive on the right as shown in Figure 1-6. Suppose further that $f(x)$ satisfies the condition of validity (1-89) in regions both to the left and right of x_0 so that any specific solution $y(x)$ may be approximated in these regions by

$x \ll x_0, f(x) < 0$:

$$y(x) \approx \frac{a}{\sqrt[4]{-f(x)}} \exp\left[+\int_x^{x_0} \sqrt{-f(x)}\, dx\right]$$

$$+ \frac{b}{\sqrt[4]{-f(x)}} \exp\left[-\int_x^{x_0} \sqrt{-f(x)}\, dx\right] \quad (1\text{-}91)$$

$x \gg x_0, f(x) > 0$:

$$y(x) \approx \frac{c}{\sqrt[4]{f(x)}} \exp\left[+i\int_{x_0}^x \sqrt{f(x)}\, dx\right]$$

$$+ \frac{d}{\sqrt[4]{f(x)}} \exp\left[-i\int_{x_0}^x \sqrt{f(x)}\, dx\right] \quad (1\text{-}92)$$

where the symbols $\sqrt{}$ and $\sqrt[4]{}$ mean positive real roots throughout. If (x) is real, a solution which is real in the left region will also be real on the right. This "reality condition" states that if a and b are real, $d = c^*$.

Our problem is to "connect" the approximations on either side of x_0 so that they refer to the same exact solution, that is, to find c and d if we know a and b and vice versa. To make this connection, we need to use an approximate solution which is valid along the entire length of some path connecting the regions of x on either side of x_0, where the WKB approximations are valid. One procedure,[12] followed by Kramers and by Jeffreys, is to use a solution valid on the real axis through x_0 (see Problem 7-36 and also the discussion at the end of Section 4–5).

A second procedure, used by Zwann and by Kemble,[13] is to avoid the real axis near x_0 and use a path encircling x_0 in the complex plane, along

---

[11] If we were using fine print in this book, the material from here to Eq. (1-122) would be so displayed. The connection formulas (1-113) and (1-122) are, however, important and should be understood even by those who wish to skip the intervening derivation.

[12] See Schiff (S2) Section 34, for a discussion of this approach, and numerous references.

[13] See Kemble (K1) Section 21.

which the WKB approximations themselves remain valid.   We shall follow this second procedure because it provides not only the connection formulas, but also the means to estimate errors in the WKB approximations.   Also, the techniques employed are more generally instructive.

The question of errors is important because we wish to use the approximate solutions over a large range of x covering many "wavelengths" or "exponential lengths."   Thus, one might worry that errors would gradually accumulate and that the approximate solution would get badly "out of phase," for example, in the region where it is oscillatory.

Define *WKB functions* associated with Eq. (1-85) as

$$W_{\pm}(x) = [f(x)]^{-1/4} \exp\left[\pm i \int_{x_0}^{x} \sqrt{f(x)}\, dx\right] \tag{1-93}$$

We must consider these to be functions of the complex variable x, and we should draw suitable branch cuts in the x plane to avoid ambiguities from the roots of $f(x)$ (see the Appendix, Section A–1).   These functions (1-93) satisfy (exactly) a differential equation which we can find by differentiation:

$$W'_{\pm} = \left(\pm i\sqrt{f(x)} - \frac{1}{4}\frac{f'}{f}\right)W_{\pm}$$

$$W''_{\pm} + \left[f(x) + \frac{1}{4}\frac{f''}{f} - \frac{5}{16}\left(\frac{f'}{f}\right)^2\right]W_{\pm} = 0 \tag{1-94}$$

If we define

$$g(x) = \frac{1}{4}\frac{f''}{f} - \frac{5}{16}\left(\frac{f'}{f}\right)^2 \tag{1-95}$$

then $W_{\pm}(x)$ are exact solutions of

$$W''_{\pm} + [f(x) + g(x)]W_{\pm} = 0 \tag{1-96}$$

and approximate solutions of

$$y'' + f(x)y = 0 \tag{1-97}$$

provided $|g(x)| \ll |f(x)|$.   Equation (1-97) is regular at x_0, whereas Eq. (1-96) has a singularity there.   Correspondingly, $y(x)$ is regular at x_0 but the $W_{\pm}$ are singular there.

We now define functions $\alpha_{\pm}(x)$ by

$$y(x) = \alpha_{+}(x)W_{+}(x) + \alpha_{-}(x)W_{-}(x) \tag{1-98}$$

$$y'(x) = \alpha_{+}(x)W'_{+}(x) + \alpha_{-}(x)W'_{-}(x) \tag{1-99}$$

where $y(x)$ is a solution of (1-97).   The $\alpha_{\pm}$ are just osculating parameters as discussed on p. 9 in connection with the variation of parameters method.

The WKB approximation corresponds simply to setting α_+ and α_- equal to constants.

Solving (1-98) and (1-99) for $\alpha_\pm(x)$,

$$\alpha_+ = \frac{yW'_- - y'W_-}{W_+ W'_- - W'_+ W_-} \qquad \alpha_- = -\frac{yW'_+ - y'W_+}{W_+ W'_- - W'_+ W_-}$$

where the denominators are the *Wronskian* of W_+ and W_-. The Wronskian is a constant, as may easily be shown from the differential equation (1-96). From the explicit forms (1-93), it is

$$W_+ W'_- - W'_+ W_- = -2i$$

so that

$$\alpha_+ = \frac{i}{2}(yW'_- - y'W_-) \qquad \alpha_- = -\frac{i}{2}(yW'_+ - y'W_+) \qquad (1\text{-}100)$$

Differentiating, and using the differential equations (1-96) and (1-97) to eliminate y'' and $W''_\pm$, we obtain

$$\frac{d\alpha_\pm}{dx} = \pm\frac{i}{2}(yW''_\mp - y''W_\mp) = \mp\frac{i}{2}g(x)yW_\mp \qquad (1\text{-}101)$$

or, remembering (1-93) and (1-98),

$$\frac{d\alpha_\pm}{dx} = \mp\frac{i}{2}\frac{g(x)}{[f(x)]^{1/2}}\left\{\alpha_\pm + \alpha_\mp \exp\left[\mp 2i\int_{x_0}^x \sqrt{f(x)}\,dx\right]\right\} \qquad (1\text{-}102)$$

One or the other of these expressions (1-101) or (1-102) may be used to estimate the error which may accumulate in the WKB approximation over a long range of x.

EXAMPLE

The WKB functions associated with the equation

$$y'' + xy = 0 \qquad (1\text{-}103)$$

for $x \gg 0$ are

$$W_\pm(x) = x^{-1/4}\exp\left(\pm i\int_0^x x^{1/2}\,dx\right) = x^{-1/4}\exp\left(\pm i\tfrac{2}{3}x^{3/2}\right) \qquad (1\text{-}104)$$

These satisfy exactly the equation (1-96):

$$W'' + (x - \tfrac{5}{16}x^{-2})W = 0$$

That is,

$$f(x) = x \qquad g(x) = -\tfrac{5}{16}x^{-2}$$

We may write the solution of (1-103) in the form (1-98)

$$y(x) = \alpha_+(x)W_+(x) + \alpha_-(x)W_-(x)$$

At large x, a general solution of (1-103) is accurately described by the WKB approximation:

$$y(x) \approx Ax^{-1/4} \cos(\tfrac{2}{3}x^{3/2} + \delta) \qquad \text{as } x \to \infty \qquad (1\text{-}105)$$

so that $\alpha_+ \to (A/2)e^{i\delta}$, $\alpha_- \to (A/2)e^{-i\delta}$ for $x \to \infty$. We wish to investigate the error in this WKB solution when it is carried in to small values of x, this error being measured by the deviation of $\alpha_+(x)$ and $\alpha_-(x)$ from the constant values above. If the $\alpha_\pm(x)$ do not change much, we have approximately, from (1-102),

$$\frac{d\alpha_\pm}{dx} \approx \mp \frac{i}{2}\left(-\frac{5}{16}x^{-2}\right)x^{-1/2}\left[\frac{A}{2}e^{\pm i\delta} + \frac{A}{2}e^{\mp i\delta}\exp\left(\mp 2i\frac{2}{3}x^{3/2}\right)\right]$$

and $\Delta\alpha_\pm$, the changes in $\alpha_\pm$ from x_1 to infinity, are given by

$$\frac{\Delta\alpha_\pm}{A/2} \approx \frac{2}{A}\int_{x_1}^{\infty}\frac{d\alpha_\pm}{dx}\,dx$$

$$= \pm i\frac{5}{32}e^{\pm i\delta}\left[\frac{2}{3}x_1^{-3/2} + e^{\mp 2i\delta}\int_{x_1}^{\infty}x^{-5/2}\exp\left(\mp i\frac{4}{3}x^{3/2}\right)dx\right] \quad (1\text{-}106)$$

The complicated second term in the bracket is less important than the first (why?), and we see that the relative error $|\Delta\alpha_\pm|/|\alpha_\pm|$ is only of the order of 10 to 20 percent even for a value of x as low as $x_1 = 1$, and at larger x_1 the error becomes very small. In particular, no significant error in phase accumulates even after an arbitrarily large number of "wavelengths."

We now return to the problem of estimating the variation in the functions $\alpha_\pm(x)$ in the general case. Integrating Eq. (1-102) along a path Γ in the complex plane from x_1 to x_2,

$$\Delta\alpha_\pm = \int_\Gamma \frac{d\alpha_\pm}{dx}\,dx = \mp\frac{i}{2}\int_\Gamma \frac{g(x)}{\sqrt{f(x)}}\alpha_\pm\,dx \mp \frac{i}{2}\int_\Gamma \frac{g(x)}{\sqrt{f(x)}}\alpha_\mp\left(\frac{W_\mp}{W_\pm}\right)dx \quad (1\text{-}107)$$

We shall consider only paths such that

$$\int_\Gamma \left|\frac{g(x)}{\sqrt{f(x)}}\right|ds \ll 1 \qquad (1\text{-}108)$$

where $ds = |dx|$ is the element of length along Γ. Kemble calls such a path a *good path*.

Along a good path, the first term on the right side of (1-107) does not contribute a significant change $\Delta\alpha_\pm$. The second term can be significant only if $|W_\mp/W_\pm| \gg 1$. That is, along a good path α_+ can change only in a region where "its" function W_+ is small in magnitude compared to W_-.

Even there, α_+ cannot change if α_- is zero or nearly zero. Similar statements hold for α_- if we interchange $+$ and $-$ everywhere.

In the regions where $|W_-/W_+| \gg 1$ or $|W_+/W_-| \gg 1$, the coefficient of the *small* function may vary markedly, a behavior known as *Stokes' phenomenon*.[14] (This variation has no appreciable effect on the function $y(x)$ in this region, so that there is no inconsistency with the fact that the WKB approximation, with constant α_+ and α_-, is supposed to be valid!)

The above results will now be used to derive the WKB connection formulas. We recall the definitions (1-93) and draw a branch line from x_0 to $+\infty$ along the real axis. We take roots $[f(x)]^{1/4}$ and $[f(x)]^{1/2}$ to be the positive real roots on the top side of this branch cut. Then, for $x < x_0$,

$$W_\pm(x) = e^{-i(\pi/4)}[-f(x)]^{-1/4} \exp\left[\pm \int_x^{x_0} \sqrt{-f(x)}\, dx\right] \qquad x < x_0 \quad (1\text{-}109)$$

Referring to (1-91) and (1-92), we have

$$
\begin{array}{lll}
\alpha_+ \to a e^{i(\pi/4)} & \alpha_- \to b e^{i(\pi/4)} & x \ll x_0 \\
\alpha_+ \to c & \alpha_- \to d & x \gg x_0
\end{array}
\qquad (1\text{-}110)
$$

In order to locate the "Stokes' regions", where $|W_+/W_-| \gg 1$ or $|W_-/W_+| \gg 1$, we note[15] that near x_0, $f(x) \approx K(x - x_0)$, and the boundaries of these regions will be qualitatively similar to those for $f(x) = K(x - x_0)$. These boundaries are where $\int_{x_0}^x \sqrt{f(x)}\, dx$ is real since then $|W_+| = |W_-|$. For $f(x) = K(x - x_0)$, if we put $x - x_0 = re^{i\theta}$,

$$\int_{x_0}^x \sqrt{f(x)}\, dx = K^{1/2}\tfrac{2}{3}(x - x_0)^{3/2} = \tfrac{2}{3}K^{1/2}r^{3/2}e^{i(3\theta/2)} \qquad (1\text{-}111)$$

This is real in directions $\theta = 0$, $\tfrac{2}{3}\pi$, $\tfrac{4}{3}\pi$, and it is easy to see from (1-93) which of W_+ or W_- dominates in each region between these boundaries.

A map of the Stokes' regions is given in Figure 1-7, and a good path Γ is shown connecting the regions $x \ll x_0$ and $x \gg x_0$ on the real axis, where the WKB solutions are to be connected.

It is now easy to find the connection formula for the special solution $y_1(x)$ which has a decreasing exponential behavior to the left of x_0, as shown in Figure 1-6. This solution has $a = 0$, in the notation of (1-91), or, from (1-110), $\alpha_+(x) = 0$ at the start of path Γ. Thus, $\alpha_-(x)$ remains constant along Γ even though $|W_+/W_-| \gg 1$ in region 2. α_+ remains essentially zero. In region 1, $|W_-/W_+| \gg 1$ so that α_- remains constant although α_+ may

---

[14] For a rather amusing description of Stokes' discovery of this phenomenon, as well as one of the earliest discussions of the connection problem, see R. E. Langer (L4).

[15] The reader will note a tacit assumption: we assume that a radius for the path Γ can be chosen which is small enough for the assumption $f(x) \approx K(x - x_0)$ to be valid, and at the same time large enough that Γ is a "good" path.

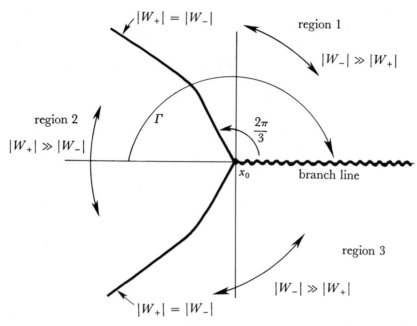

Figure 1–7 **Complex x-plane showing the Stokes' regions around the point x_0 where $f(x)$ vanishes**

(and does) change. Since α_- is constant all along Γ, we find from (1-110)

$$d = be^{i(\pi/4)} \tag{1-112}$$

The reality condition gives $c = d^*$ if b is real.

The connection formula for the solution $y_1(x)$ with decreasing exponential behavior for $x < x_0$ is thus (normalizing $b = 1$)

$$\frac{1}{\sqrt[4]{-f(x)}} \exp\left[-\int_x^{x_0} \sqrt{-f(x)}\,dx \right] \to 2\,\frac{1}{\sqrt[4]{f(x)}} \cos\left[\int_{x_0}^x \sqrt{f(x)}\,dx - \frac{\pi}{4} \right] \tag{1-113}$$

Reversal of the arrow is delicate. We know only that the growing exponential term on the left will be absent for some phase near $-\pi/4$ on the right.

Next, consider a more general solution $y_2(x)$ such that $a \neq 0$. Since the basic equation is linear, the coefficients in expressions (1-91) and (1-92) must be linearly related:

$$c = Aa + Bb$$
$$d = Ca + Db \tag{1-114}$$

We have found in (1-112) that

$$B = e^{-i(\pi/4)} \quad \text{and} \quad D = e^{i(\pi/4)} \tag{1-115}$$

The reality condition $d = c^*$, if a and b are real, says

$$C = A^* \quad \text{and} \quad D = B^* \tag{1-116}$$

the second of which we already know.

Another general relation may be obtained from the fact that the Wronskian $y_1 y_2' - y_2 y_1'$, of two independent solutions of our equation $y'' + f(x)y = 0$, is a constant. (To see this, just differentiate the Wronskian.) Write

$$y_1(x) = \alpha_+^{(1)} W_+ + \alpha_-^{(1)} W_- \qquad y_2(x) = \alpha_+^{(2)} W_+ + \alpha_-^{(2)} W_- \tag{1-117}$$

Remembering that $y' = \alpha_+ W_+' + \alpha_- W_-'$ and the Wronskian

$$W_+ W_-' - W_+' W_- = -2i$$

one finds

$$\alpha_+^{(1)} \alpha_-^{(2)} - \alpha_+^{(2)} \alpha_-^{(1)} = \text{constant} = K \tag{1-118}$$

Evaluating this constant at $x \ll x_0$ [from (1-110)],

$$K = e^{i(\pi/2)}(a_1 b_2 - a_2 b_1)$$

where a_1 and b_1 are the coefficients a and b in the WKB approximation (1-91) for $y_1(x)$, and a_2 and b_2 are the coefficients for $y_2(x)$. Similarly, on the other side, where $x \gg x_0$,

$$K = c_1 d_2 - c_2 d_1 = (AD - BC)(a_1 b_2 - a_2 b_1)$$

Thus

$$e^{i(\pi/2)} = i = AD - BC = AB^* - A^*B = 2i \operatorname{Im} AB^* \tag{1-119}$$

But $B = e^{-i(\pi/4)}$, so $Ae^{i(\pi/4)} = R + i/2$ where R is an undetermined real constant. Collecting our results,

$$\begin{aligned} A &= Re^{-i(\pi/4)} + \tfrac{1}{2} e^{i(\pi/4)} & B &= e^{-i(\pi/4)} \\ C &= Re^{i(\pi/4)} + \tfrac{1}{2} e^{-i(\pi/4)} & D &= e^{i(\pi/4)} \end{aligned} \tag{1-120}$$

and because R is undetermined, we still cannot write c and d in terms of a and b. However, this is as far as we can proceed; R is indeterminate. Another way of expressing this indeterminacy is to observe that the asymptotic form of a solution like $y_2(x)$ in Figure 1–6 for $x < x_0$ does not fix $y_2(x)$ for $x > x_0$. What about the reverse problem of writing a and b in terms of *c* and *d*? Equations (1-114), with the constants A, B, C, D given by (1-120), are easily inverted with the result

$$a = e^{-i(\pi/4)}c + e^{i(\pi/4)}d$$

$$b = [\tfrac{1}{2}e^{i(\pi/4)} - Re^{-i(\pi/4)}]c + [\tfrac{1}{2}e^{-i(\pi/4)} - Re^{i(\pi/4)}]d \qquad (1\text{-}121)$$

Thus we can find a but not b, which might be expected since $\alpha_-(x)$ fluctuates wildly in region 2 of Figure 1–7 except in the special case $\alpha_+ = 0$. We might also expect this result from the fact that, when $a \neq 0$, the value of b is immaterial since the decreasing exponential is negligible compared to the increasing one. Omitting b for this reason, we can write a second formula connecting the two expressions (1-91) and (1-92) for $y(x)$. It is conventional to choose a real solution by setting $c = \tfrac{1}{2}e^{i\phi}$ and $d = \tfrac{1}{2}e^{-i\phi}$. Then

$$\sin\left(\phi + \frac{\pi}{4}\right) \frac{1}{\sqrt[4]{-f(x)}} \exp\left[\int_x^{x_0} \sqrt{-f(x)}\, dx\right]$$

$$\leftarrow \frac{1}{\sqrt[4]{f(x)}} \cos\left[\int_{x_0}^x \sqrt{f(x)}\, dx + \phi\right] \qquad (1\text{-}122)$$

provided ϕ is not too near $-\pi/4$. The arrow cannot be reversed, as we have seen, since the phase ϕ cannot be determined if we only know how much large exponential is present.

It is very easy to remember the connection formulas (1-113) and (1-122) by drawing qualitative pictures like Figure 1–6. Consider first $y_1(x)$ with the decreasing exponential to the left of x_0. To the right of x_0, $y_1(x)$ looks like a cosine wave for which the phase is between $-\pi/2$ and 0 when $x = x_0$. We remember the phase is $-\pi/4$. Furthermore, the amplitude of the cosine wave on the right "looks" larger than that of the exponential on the left. We remember it is twice as large. These comments reproduce the result (1-113). A similar mnemonic for remembering (1-122) results from considering a solution $y_2(x)$ which is 90° out of phase with $y_1(x)$ in the right-hand region. By this same procedure, it is a straightforward matter to write down connection formulas similar to (1-113) and (1-122) for the case where the slope of $f(x)$ is negative at x_0 instead of positive.

We shall conclude by giving an example of the use of the WKB method in quantum mechanics. Consider the Schrödinger equation for a particle in a potential well

$$\frac{d^2\psi}{dx^2} + \frac{2m}{\hbar^2}[E - V(x)]\psi = 0 \qquad (1\text{-}123)$$

with $V(x)$ as shown in Figure 1–8.

$$f(x) = \frac{2m}{\hbar^2}[E - V(x)] \qquad \text{is} \quad \begin{cases} \text{positive for } a < x < b \\ \text{negative for } x < a,\ x > b \end{cases}$$

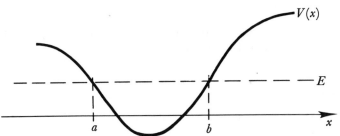

Figure 1–8   A typical one–dimensional potential well

If $\psi(x)$ is to be bounded for $x < a$, then in $a < x < b$ our connection formula (1-113) tells us

$$\psi(x) \approx \frac{A}{(E - V)^{1/4}} \cos \left(\int_a^x \sqrt{\frac{2m}{\hbar^2} (E - V)} \, dx - \frac{\pi}{4} \right) \qquad (1\text{-}124)$$

where A is an arbitrary constant. If $\psi(x)$ is to be bounded for $x > b$, then in $a < x < b$ similarly

$$\psi(x) \approx \frac{B}{(E - V)^{1/4}} \cos \left(\int_x^b \sqrt{\frac{2m}{\hbar^2} (E - V)} \, dx - \frac{\pi}{4} \right) \qquad (1\text{-}125)$$

where B is an arbitrary constant. These two expressions must be the same, which gives the condition

$$\int_a^b \sqrt{2m(E - V)} \, dx = (n + \tfrac{1}{2})\pi\hbar \qquad (1\text{-}126)$$

This result is very similar to the Bohr–Sommerfeld quantization condition of pre-1925 quantum mechanics.

REFERENCES

There are many good books on the subject of ordinary differential equations. Two small but remarkably complete ones are Burkill (B9) and Ince (I1). The latter is a condensed version of the treatise (I2) by the same author. Another treatise, excellent and well worth looking through, even if it is slightly old-fashioned, is Forsyth (F5).

Modern texts on this subject tend to place more emphasis on numerical solutions, integral transform methods, and special techniques for nonlinear equations, than do the older texts mentioned above. Several good examples are Birkhoff and Rota (B3); Golomb and Shanks (G5); and Rabenstein (R1).

The WKB method and its applications to quantum mechanics are discussed in any good book on quantum theory, such as Schiff (S2) Section 34;

Merzbacher (M4) Chapter 7; or Landau and Lifshitz (L2) Chapter VII. Note that the last-named book refers to the WKB method as the "quasi-classical case."

PROBLEMS

Find the general solutions of Problems 1-1 to 1-20:

1-1 $x^2y' + y^2 = xyy'$

1-2 $y' = \dfrac{x\sqrt{1+y^2}}{y\sqrt{1+x^2}}$

1-3 $y' = \dfrac{a^2}{(x+y)^2}$

1-4 $y' + y\cos x = \frac{1}{2}\sin 2x$

1-5 $(1 - x^2)y' - xy = xy^2$

1-6 $2x^3y' = 1 + \sqrt{1 + 4x^2y}$

1-7 $y'' + y'^2 + 1 = 0$

1-8 $y'' = e^y$

1-9 $x(1 - x)y'' + 4y' + 2y = 0$

1-10 $(1 - x)y^2\,dx - x^3\,dy = 0$

1-11 $xy' + y + x^4y^4e^x = 0$

1-12 $(1 + x^2)y' + y = \tan^{-1} x$

1-13 $x^2y'^2 - 2(xy - 4)y' + y^2 = 0$ (general solution *and* singular solution)

1-14 $yy'' - y'^2 - 6xy^2 = 0$

1-15 $x^4yy'' + x^4y'^2 + 3x^3yy' - 1 = 0$

1-16 $x^2y'' - 2y = x$

1-17 $y''' - 2y'' - y' + 2y = \sin x$

1-18 $y^{iv} + 2y'' + y = \cos x$

1-19 $y'' + 3y' + 2y = \exp[e^x]$

1-20 $a^2y''^2 = (1 + y'^2)^3$

1-21 The differential equation obeyed by the charge q on a capacitor C connected in series with a resistance R to a voltage

$$V = V_0\left(\frac{t}{\tau}\right)^2 e^{-t/\tau}$$

is

$$R\frac{dq}{dt} + \frac{1}{C}q = V_0\left(\frac{t}{\tau}\right)^2 e^{-t/\tau}$$

Find $q(t)$ if $q(0) = 0$.

1-22 In the activation of an indium foil by a constant slow neutron flux, the number N of radioactive atoms obeys the equation

$$\frac{dN}{dt} = \lambda N_s - \lambda N$$

where N_s is the constant number after "saturation." Find $N(t)$ if $N(0) = 0$.

1-23 Find the general solution of

$$A(x)y''(x) + A'(x)y'(x) + \frac{y(x)}{A(x)} = 0$$

where $A(x)$ is a known function and $y(x)$ is unknown.

1-24 Find the general solution of

$$xy'' + 2y' + n^2xy = \sin \omega x$$

Hint: Eliminate the first derivative term.

1-25 Note that $y = x$ would be a solution of

$$(1 - x)y'' + xy' - y = (1 - x)^2$$

if the right side were zero.   Use this fact to obtain the general solution of the equation as given.

1-26 Consider the differential equation

$$y'' + p(x)y' + q(x)y = 0$$

on the interval $a \leqslant x \leqslant b$. Suppose we know two solutions, $y_1(x)$ and $y_2(x)$, such that

$$y_1(a) = 0 \qquad y_2(a) \neq 0$$
$$y_1(b) \neq 0 \qquad y_2(b) = 0$$

Give the solution of the equation

$$y'' + p(x)y' + q(x)y = f(x)$$

which obeys the conditions $y(a) = y(b) = 0$, in the form

$$y(x) = \int_a^b G(x, x')f(x') \, dx'$$

where $G(x, x')$, the so-called Green's function, involves only the solutions y_1 and y_2 and assumes different functional forms for $x' < x$ and $x' > x$.

Illustrate by solving

$$y'' + k^2 y = f(x)$$
$$y(a) = y(b) = 0$$

1-27 Find the general solution of the differential equation

$$xy'' + \frac{3}{x}y = 1 + x^3$$

in real form (no i's in answer).

1-28 Consider the equation

$$\frac{d^2y}{dx^2} + \frac{2}{x}\frac{dy}{dx} + \left\{K + \frac{2}{x} - \frac{l(l+1)}{x^2}\right\}y = 0 \qquad 0 \leqslant x \leqslant \infty$$

where l = nonnegative integer

Find all values of the constant K which can give a solution which is finite on the entire range of x (including ∞). An equation like this arises in solving the Schrödinger equation for the hydrogen atom.

Hint: Let $y = v/x$, then "factor out" the behavior at infinity.

1-29 For what values of the constant K does the differential equation

$$y'' - \left(\frac{1}{4} + \frac{K}{x}\right)y = 0 \qquad (0 < x < \infty)$$

have a nontrivial solution vanishing at $x = 0$ and $x = \infty$?

1-30 Find the values of the constant k for which the equation

$$xy'' - 2xy' + (k - 3x)y = 0$$

has a solution bounded on the range $0 \leqslant x < \infty$.

1-31 A solution of the differential equation

$$xy'' + 2y' + (E - x)y = 0$$

is desired such that $y(0) = 1$, $y(\infty) = 0$. For what values of E is this possible?

1-32 By considering Eq. (1-71) for $r \to \infty$, verify that $v(x)$ indeed behaves like $(1 - x^2)^{-m}$ as $x \to \pm 1$.

1-33 Bessel's equation for $m = 0$ is

$$x^2y'' + xy' + x^2y = 0$$

We have found one solution

$$J_0(x) = 1 - \frac{x^2}{4} + \frac{x^4}{64} - + \cdots$$

Show that a second solution exists of the form

$$J_0(x) \ln x + Ax^2 + Bx^4 + Cx^6 + \cdots$$

and find the first three coefficients A, B, C.

1-34 Consider the differential equation

$$xy'' + (2 - x)y' - 2y = 0$$

Give two solutions, one regular at the origin and having the value 1 there, the other of the form

$$\frac{1}{x} + A(x) \ln x + B(x)$$

where $A(x)$ and $B(x)$ are regular at the origin.   Three terms of each series will suffice.

1-35 Find an approximate expression (say three terms) for large x for the solution of

$$y'' = \frac{1}{(x^2 + y^2)^2}$$

which approaches the x axis as $x \to +\infty$.

1-36 Consider the differential equation

$$y'' - xy + y^3 = 0$$

for large positive x.
(a) Find an oscillating solution with two arbitrary constants.
(b) Find a (nontrivial) particular nonoscillating solution.

1-37 Find an (approximate) oscillating solution of

$$y'' = (y - x)^2 - e^{2(y-x)}$$

You may let α denote the real number such that

$$\alpha^2 = e^{2\alpha} \qquad (\alpha \approx -0.57)$$

1-38 Consider the differential equation

$$\frac{dy}{dx} = e^{y/x}$$

(a) Suppose $y(1) = 0$. Give a series expansion for $y(x)$ which is valid for x near 1.   Neglect terms of order $(x - 1)^4$.
(b) Suppose $y(x_0) = +\infty$ ($x_0 > 0$). Give an approximate expression for $y(x)$ which is useful for x slightly less than x_0.

1-39 Use the WKB method to find approximate negative values of the constant E for which the equation

$$\frac{d^2y}{dx^2} + [E - V(x)]y = 0$$

has a solution which is finite for all x between $x = -\infty$ and $x = +\infty$, *inclusive*. $V(x)$ is the function shown below.

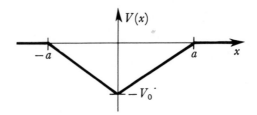

1-40 Find a good approximation, for x large and positive, to the solution of the equation

$$y'' - \frac{3}{x} y' + \left(\frac{15}{4x^2} + x^{1/2}\right) y = 0$$

Hint: Remove first derivative term.

1-41 Obtain an approximate formula for the Bessel function $J_m(x)$ by the WKB method and give the limiting form of this expression for large $x(x \gg m)$. Do not worry about getting the constant in front correct. You may assume $m \gg \frac{1}{2}$.

1-42 Consider the differential equation $y'' + xy = 0$.

(a) If $y \sim \dfrac{1}{x^{1/4}} \cos \dfrac{2}{3} x^{3/2}$ as $x \to +\infty, y \sim$? as $x \to -\infty$.

(b) If $y \sim \dfrac{1}{(-x)^{1/4}} \exp\left[-\dfrac{2}{3}(-x)^{3/2}\right]$ as $x \to -\infty$,

$y \sim$? as $x \to +\infty$.

(c) If $y \sim \dfrac{1}{(-x)^{1/4}} \exp\left[+\dfrac{2}{3}(-x)^{3/2}\right]$ as $x \to -\infty$,

$y \sim$? as $x \to +\infty$.

Note: The answer to *one* of (a), (b), (c) is not defined. Be sure to indicate which one is not defined, as well as giving correct answers for the other two.

1-43 Consider the solution of the differential equation

$$\frac{d^2y}{dx^2} + x^2 y = 0$$

which has zero value and unit slope at $x = 10$.

(a) Give (approximately) the location of the next zero of $y(x)$ greater than 10.

(b) Give (approximately) the value of $y(x)$ at its first maximum for $x > 10$.

1-44 Consider a solution $y_1(x)$ of the differential equation

$$y'' + x^2 y = 0$$

such that $y_1(x)$ has a zero at $x = 5$.   Give approximately the location of the 25th zero beyond the one at $x = 5$, and *estimate the error in your result.*

TWO

INFINITE SERIES

In this chapter we recall some tests for the convergence of series and present a number of methods for obtaining the sum of a series in closed form. Numerical and approximate methods are not discussed here, but in Chapter 13.

2–1 CONVERGENCE

An infinite series

$$\sum_{n=1}^{\infty} a_n = a_1 + a_2 + a_3 + \cdots$$

is said to *converge* to the sum S provided the sequence of partial sums has the limit S; that is, provided

$$\lim_{N \to \infty} \left(\sum_{n=1}^{N} a_n \right) = S$$

If the series $\sum_{n=1}^{\infty} a_n$ converges to the sum S, we will simply write this as an equality:

$$\sum_{n=1}^{\infty} a_n = S$$

The series $\sum_{n=1}^{\infty} a_n$ is said to converge *absolutely* if the related series $\sum_{n=1}^{\infty} |a_n|$ converges. Absolute convergence implies convergence, but not vice versa; for example, the series

$$1 - \tfrac{1}{2} + \tfrac{1}{3} - + \cdots$$

44

converges (to the sum ln 2), but it does not converge absolutely, because

$$1 + \tfrac{1}{2} + \tfrac{1}{3} + \cdots$$

does not converge (that is, it *diverges*).

It should be emphasized that the numbers a_n may be *complex* numbers. We must, of course, give the symbol $|a_n|$ of the preceding paragraph its usual meaning when a_n is complex:

$$|a_n| = \sqrt{(Re\ a_n)^2 + (Im\ a_n)^2}$$

That is, $|a_n|$ denotes the absolute value (or modulus) of a_n.

The simplest means for determining the convergence of an infinite series is to compare it with a series which is known to converge or diverge. For example, the geometric series

$$\frac{1}{1-x} = 1 + x + x^2 + x^3 + \cdots \tag{2-1}$$

converges for $|x| < 1$ and diverges for $|x| > 1$. This leads to the *ratio test*: If the ratio $|a_{n+1}/a_n|$ of successive terms in the infinite series

$$a_1 + a_2 + a_3 + \cdots$$

has as a limit a number less than one as $n \to \infty$, the series converges, and, in fact, converges absolutely. If the limit is greater than one, the series diverges. If the limit is one (or if there is no limit), we must investigate further.

A second criterion is comparison with an infinite integral. The series

$$f(1) + f(2) + f(3) + \cdots$$

converges or diverges with the infinite integral

$$\int^{\infty} f(x)\ dx$$

provided $f(x)$ is monotonically decreasing. For example, consider the series for the *Riemann zeta function*

$$\zeta(s) = 1 + \frac{1}{2^s} + \frac{1}{3^s} + \frac{1}{4^s} + \cdots \tag{2-2}$$

The ratio of successive terms is

$$\frac{a_{n+1}}{a_n} = \left(\frac{n}{n+1}\right)^s = \left(1 + \frac{1}{n}\right)^{-s} \underset{n \to \infty}{\sim} 1 - \frac{s}{n} + \cdots \tag{2-3}$$

The ratio approaches one as $n \to \infty$. Thus, comparison with the geometric series fails. However,

$$\int \frac{dx}{x^s} = \frac{-1}{s-1} \frac{1}{x^{s-1}} \tag{2-4}$$

so that the criterion for convergence of the zeta-function series is $s > 1$. This enables us to sharpen our ratio test; if

$$\left| \frac{a_{n+1}}{a_n} \right| \to 1 - \frac{s}{n} \tag{2-5}$$

with s greater than 1, the series converges (absolutely).

EXAMPLE

Consider the *hypergeometric series*

$$F(a, b; c; x) = 1 + \frac{ab}{c} \frac{x}{1!} + \frac{a(a+1)b(b+1)}{c(c+1)} \frac{x^2}{2!} + \cdots \tag{2-6}$$

The ratio of successive terms is

$$\frac{a_{n+1}}{a_n} = \frac{(a+n)(b+n)}{(c+n)(n+1)} x$$

$$= \frac{\left(1 + \dfrac{a}{n}\right)\left(1 + \dfrac{b}{n}\right)}{\left(1 + \dfrac{c}{n}\right)\left(1 + \dfrac{1}{n}\right)} x$$

$$= \left(1 + \frac{a+b-c-1}{n} + \cdots\right) x$$

Thus the series converges if $|x| < 1$, or, when $|x| = 1$, if [1]

$$a + b - c < 0 \tag{2-7}$$

We can sharpen our ratio test even more by considering a very slowly converging series such as

$$\sum_{n=2}^{\infty} \frac{1}{n(\ln n)^s} = \frac{1}{2(\ln 2)^s} + \frac{1}{3(\ln 3)^s} + \cdots \tag{2-8}$$

---

[1] If we allow a, b, c to be *complex* numbers, condition (2-7) becomes
$$\text{Re}\,(a + b - c) < 0$$

By comparing with the integral

$$\int \frac{dx}{x(\ln x)^s} = \frac{-1}{s-1} \frac{1}{(\ln x)^{s-1}} \tag{2-9}$$

we see that the series (2-8) converges provided $s > 1$.   The ratio of successive terms in (2-8) is

$$\frac{a_{n+1}}{a_n} = \frac{n}{n+1} \left[\frac{\ln n}{\ln (n+1)} \right]^s$$

$$= \left(1 - \frac{1}{n} + \cdots\right) \left[\frac{\ln n}{\ln n + \ln\left(1 + \frac{1}{n}\right)} \right]^s$$

$$= \left(1 - \frac{1}{n} + \cdots\right) \left[1 + \frac{1}{n \ln n} + \cdots \right]^{-s}$$

$$= 1 - \frac{1}{n} - \frac{s}{n \ln n} + \cdots \tag{2-10}$$

Thus a series for which $|a_{n+1}/a_n|$ is of the form (2-10) converges if $s > 1$ and diverges if $s < 1$.

EXAMPLE

Consider the series solution (1-51) of Legendre's equation

$$1 - n(n+1)\frac{x^2}{2!} + n(n+1)(n-2)(n+3)\frac{x^4}{4!} - + \cdots$$

The ratio of successive terms is

$$\frac{a_i}{a_{i-1}} = -\frac{(n - 2i + 4)(n + 2i - 3)}{(2i - 3)(2i - 2)} x^2$$

For large i, this ratio is

$$\frac{a_i}{a_{i-1}} \sim \left[1 - \frac{1}{i} + 0\left(\frac{1}{i^2}\right) \right] x^2$$

Thus, if $x^2 = 1$, the series diverges.

There are, of course, other convergence criteria; for example, if the signs of the a_n alternate and a_n approaches zero monotonically, then the series $\sum_n a_n$ converges (but not necessarily absolutely).   The reader interested in a more rigorous and more complete treatment of convergence, absolute convergence, and so forth, should consult any of the standard references, such as those listed at the end of this chapter.

2–2 FAMILIAR SERIES

The reader should be familiar with at least the following simple series:

$$(1 + x)^n = 1 + nx + n(n - 1)\frac{x^2}{2!} + \cdots = \sum_{\alpha=0}^{\infty} \frac{n!}{(n - \alpha)!\,\alpha!} x^\alpha \qquad (2\text{-}11)$$

$$e^x = 1 + x + \frac{x^2}{2!} + \frac{x^3}{3!} + \cdots = \sum_{\alpha=0}^{\infty} \frac{x^\alpha}{\alpha!} \qquad (2\text{-}12)$$

Using the Euler relation $e^{ix} = \cos x + i \sin x$, we deduce from (2-12)

$$\sin x = x - \frac{x^3}{3!} + \frac{x^5}{5!} - + \cdots \qquad (2\text{-}13)$$

$$\cos x = 1 - \frac{x^2}{2!} + \frac{x^4}{4!} - + \cdots \qquad (2\text{-}14)$$

Term-by-term integration of the series for $(1 + x)^{-1}$ and $(1 + x^2)^{-1}$ yields

$$\ln(1 + x) = x - \frac{x^2}{2} + \frac{x^3}{3} - \frac{x^4}{4} + - \cdots \qquad (2\text{-}15)$$

$$\tan^{-1} x = x - \frac{x^3}{3} + \frac{x^5}{5} - + \cdots \qquad (2\text{-}16)$$

The above series are simple enough to be easily remembered. Other elementary functions such as tan x, ctn x, sec x, and so on, have series expansions which are probably less familiar. For some, the coefficients are expressed in terms of the so-called *Bernoulli numbers* or the *Euler numbers* (which may be defined, in fact, by these coefficients). See Problem 2-12 for sec x.

We shall introduce the Bernoulli numbers by an expansion of the function $(e^x - 1)^{-1}$. Multiplying by x to eliminate the singularity at $x = 0$, we can write

$$\frac{x}{e^x - 1} = c_0 + c_1 x + c_2 x^2 + \cdots \qquad (2\text{-}17)$$

Therefore,

$$x = (c_0 + c_1 x + c_2 x^2 + \cdots)\left(x + \frac{x^2}{2!} + \frac{x^3}{3!} + \cdots\right)$$

and, dividing by x,

$$1 = (c_0 + c_1 x + c_2 x^2 + \cdots)\left(1 + \frac{x}{2!} + \frac{x^2}{3!} + \cdots\right)$$

Let $c_n = B_n/n!$. Then

$$1 = \left(B_0 + \frac{B_1 x}{1!} + \frac{B_2 x^2}{2!} + \cdots\right)\left(1 + \frac{x}{2!} + \frac{x^2}{3!} + \cdots\right) \qquad (2\text{-}18)$$

Equating powers of x in (2-18),

$$1 = B_0$$

$$0 = \frac{B_0}{2!} + \frac{B_1}{1!}$$

$$0 = \frac{B_0}{3!} + \frac{B_1}{2!1!} + \frac{B_2}{1!2!} \qquad (2\text{-}19)$$

etc.

Except for the first one, Eqs. (2-19) may be written symbolically in the concise form $(B + 1)^n - B^n = 0$, with the understanding that B^s really means B_s. The first few of these *Bernoulli numbers* are

$$B_0 = 1 \qquad B_2 = \tfrac{1}{6} \qquad B_4 = -\tfrac{1}{30} \qquad B_6 = \tfrac{1}{42} \qquad B_8 = -\tfrac{1}{30} \cdots$$
$$B_1 = -\tfrac{1}{2} \qquad B_3 = B_5 = B_7 = \cdots = 0 \qquad (2\text{-}20)$$

Notations vary, so that care must be exercised when looking up formulas. For example, the notations of Dwight (D8) and Pierce and Foster (P1) differ from ours.

As an example of the usefulness of Bernoulli numbers as coefficients in power series of familiar functions, consider

$$\operatorname{ctn} x = i \frac{e^{ix} + e^{-ix}}{e^{ix} - e^{-ix}} \qquad (2\text{-}21)$$

Let $ix = y/2$. Then

$$\operatorname{ctn} x = i \frac{e^{y/2} + e^{-y/2}}{e^{y/2} - e^{-y/2}}$$

$$= i\left(1 + \frac{2}{e^y - 1}\right)$$

$$= \frac{2i}{y}\left(\frac{y}{2} + \frac{y}{e^y - 1}\right)$$

$$= \frac{2i}{y} \sum_{n\,\text{even}} \frac{B_n y^n}{n!}$$

Replacing y by $2ix$ gives

$$\text{ctn } x = \frac{1}{x} \sum_{n \text{ even}} (-1)^{n/2} \frac{B_n (2x)^n}{n!} \tag{2-22}$$

The relation $\tan x = \text{ctn } x - 2 \text{ ctn } 2x$ enables us to deduce the power series for $\tan x$.

2–3 TRANSFORMATION OF SERIES

Various devices may be used to reduce a given unfamiliar series to a known one.

Differentiation and integration are often useful.

EXAMPLE

$$f(x) = 1 + 2x + 3x^2 + 4x^3 + \cdots \tag{2-23}$$

Integrating term by term,

$$\int_0^x f(x)\, dx = x + x^2 + x^3 + \cdots = \frac{x}{1-x}$$

Then, differentiating,

$$f(x) = \frac{d}{dx}\left(\frac{x}{1-x}\right) = \frac{1}{(1-x)^2}$$

[We might have recognized this immediately from the binomial series (2-11).]

EXAMPLE

$$f(x) = \frac{1}{1 \cdot 2} + \frac{x}{2 \cdot 3} + \frac{x^2}{3 \cdot 4} + \frac{x^3}{4 \cdot 5} + \cdots \tag{2-24}$$

$$x^2 f(x) = \frac{x^2}{1 \cdot 2} + \frac{x^3}{2 \cdot 3} + \frac{x^4}{3 \cdot 4} + \cdots$$

Differentiating twice,

$$(x^2 f)'' = 1 + x + x^2 + x^3 + \cdots = \frac{1}{1-x}$$

From this, two integrations give

$$f(x) = \frac{1}{x} + \frac{1-x}{x^2} \ln (1-x)$$

The constants of integration which arise in this procedure must be evaluated by knowing the series at certain values of x; for example, it is obvious from (2-24) that $f(0) = \frac{1}{2}$.

Complex variables sometimes provide useful transformations.

EXAMPLE

$$f(\theta) = 1 + a \cos \theta + a^2 \cos 2\theta + \cdots$$
$$= \mathrm{Re}\,(1 + ae^{i\theta} + a^2 e^{2i\theta} + \cdots) \tag{2-25}$$

This is just a simple geometric series.   Therefore,

$$f(\theta) = \mathrm{Re}\, \frac{1}{1 - ae^{i\theta}} = \frac{1 - a \cos \theta}{1 - 2a \cos \theta + a^2}$$

The trick of differentiation or integration may be employed even if the series does not contain a variable.

EXAMPLE

$$S = \frac{1}{2!} + \frac{2}{3!} + \frac{3}{4!} + \cdots \tag{2-26}$$

Define

$$f(x) = \frac{x^2}{2!} + \frac{2x^3}{3!} + \frac{3x^4}{4!} + \cdots$$

Then $S = f(1)$.

$$f'(x) = x + x^2 + \frac{x^3}{2!} + \frac{x^4}{3!} + \cdots = xe^x$$

$$f(x) = \int_0^x xe^x \, dx = xe^x - e^x + 1$$

$$S = f(1) = 1$$

EXAMPLE

$$S = 1 + m + \frac{m(m-1)}{2!} + \frac{m(m-1)(m-2)}{3!} + \cdots = \sum_n \frac{m!}{n!(m-n)!} \tag{2-27}$$

Let

$$f(x) = \sum_n \frac{m!}{n!(m-n)!} x^n \qquad S = f(1)$$

But

$$f(x) = (1 + x)^m$$

Therefore $S = 2^m$.

We shall now consider the deceptively simple series [compare (2-2)]

$$S = \sum_{n=1}^{\infty} \frac{1}{n^2} = 1 + \frac{1}{4} + \frac{1}{9} + \frac{1}{16} + \cdots = \zeta(2) \qquad (2\text{-}28)$$

Previous examples suggest defining

$$f(x) = x + \frac{x^2}{4} + \frac{x^3}{9} + \frac{x^4}{16} + \cdots \qquad S = f(1) \qquad (2\text{-}29)$$

$$f'(x) = 1 + \frac{x}{2} + \frac{x^2}{3} + \cdots = -\frac{1}{x} \ln(1 - x)$$

Thus

$$f(x) = -\int_0^x \frac{\ln(1-x)\,dx}{x} \qquad \text{and} \qquad S = -\int_0^1 \frac{\ln(1-x)\,dx}{x} \qquad (2\text{-}30)$$

Unfortunately, this integral is just as difficult as the series was.

The series (2-28) can be expressed in terms of a Bernoulli number. To find this relation, we begin by writing the Fourier series for $\cos kx$. (We shall discuss Fourier series in more detail in Chapter 4.)

$$\cos kx = \frac{A_0}{2} + \sum_{n=1}^{\infty} (A_n \cos nx + B_n \sin nx) \qquad (2\text{-}31)$$

$\cos kx$ is an even function, so there are no sine terms ($B_n = 0$). The coefficients A_n are given by

$$A_n = \frac{2}{\pi} \int_0^\pi \cos nx \cos kx \, dx = (-1)^n \frac{2k \sin k\pi}{\pi(k^2 - n^2)} \qquad (2\text{-}32)$$

Thus

$$\cos kx = \frac{2k \sin k\pi}{\pi} \left(\frac{1}{2k^2} - \frac{\cos x}{k^2 - 1} + \frac{\cos 2x}{k^2 - 4} - \frac{\cos 3x}{k^2 - 9} + \cdots \right) \qquad (2\text{-}33)$$

Setting $x = \pi$ in (2-33) gives

$$k\pi \operatorname{ctn} k\pi = 1 + 2k^2 \left(\frac{1}{k^2 - 1} + \frac{1}{k^2 - 4} + \frac{1}{k^2 - 9} + \cdots \right) \qquad (2\text{-}34)$$

This is the *partial fraction representation* of the cotangent. For those who know the relevant mathematics, this partial fraction representation could have been written down immediately by using the Mittag–Lefler theorem[2] from complex variable theory.

In any case, by expanding (2-34) as a power series in k, we obtain

$$k\pi \operatorname{ctn} k\pi = 1 - 2k^2\left(1 + \frac{1}{2^2} + \frac{1}{3^2} + \cdots\right)$$

$$- 2k^4\left(1 + \frac{1}{2^4} + \frac{1}{3^4} + \cdots\right) - \cdots$$

$$= 1 - 2\sum_{n=1}^{\infty} \zeta(2n)k^{2n} \tag{2-35}$$

where $\zeta(2n)$ is defined in (2-2). Comparing (2-35) with our power series (2-22) for the cotangent in terms of Bernoulli numbers, we find

$$\zeta(2n) = \frac{(-1)^{n+1} B_{2n}(2\pi)^{2n}}{2(2n)!} \tag{2-36}$$

Thus

$$1 + \frac{1}{4} + \frac{1}{9} + \cdots = \zeta(2) = \frac{B_2 4\pi^2}{4} = \frac{\pi^2}{6} \tag{2-37}$$

$$1 + \frac{1}{16} + \frac{1}{81} + \cdots = \zeta(4) = -\frac{B_4 16\pi^4}{48} = \frac{\pi^4}{90} \tag{2-38}$$

etc.

We shall conclude this section by discussing a transformation which is useful both for analytical and numerical summing of series. Let

$$g(z) = \sum_{n=0}^{\infty} b_n z^n \tag{2-39}$$

be a series whose sum we know, and suppose we want to sum the series

$$f(z) = \sum_{n=0}^{\infty} c_n b_n z^n \tag{2-40}$$

[2] See Copson (C8) Section 6.8, or Whittaker and Watson (W5) Section 7.4. The theorem, roughly speaking, says that a function $f(z)$ which is analytic everywhere in the finite z-plane except for simple poles at $z = a_1, a_2, \ldots$ with residues $b_1, b_2, \ldots,$ and which (except near the poles) remains finite as $|z|$ approaches infinity, may be written in the form

$$f(z) = f(0) + \sum_n b_n\left(\frac{1}{z - a_n} + \frac{1}{a_n}\right)$$

(2-34) follows immediately if we take for $f(z)$ the function $\pi z \operatorname{ctn} \pi z$.

We eliminate the b_n from (2-40):

$$
\begin{aligned}
f(z) &= c_0 b_0 + c_1 b_1 z + c_2 b_2 z^2 + c_3 b_3 z^3 + \cdots \\
&= c_0 g(z) + (c_1 - c_0) b_1 z + (c_2 - c_0) b_2 z^2 + (c_3 - c_0) b_3 z^3 + \cdots \\
&= c_0 g(z) + (c_1 - c_0) z g'(z) + (c_2 - 2c_1 + c_0) b_2 z^2 \\
&\quad + (c_3 - 3c_1 + 2c_0) b_3 z^3 + \cdots \\
&= c_0 g(z) + (c_1 - c_0) z g'(z) + (c_2 - 2c_1 + c_0) \frac{z^2 g''(z)}{2!} \\
&\quad + (c_3 - 3c_2 + 3c_1 - c_0) \frac{z^3 g'''(z)}{3!} + \cdots
\end{aligned}
\tag{2-41}
$$

The successive coefficients are just the leading differences in a difference table of the coefficients c_n:

$$
\begin{array}{cccc}
c_0 & & & \\
 & c_1 - c_0 & & \\
c_1 & & c_2 - 2c_1 + c_0 & \\
 & c_2 - c_1 & & c_3 - 3c_2 + 3c_1 - c_0 \\
c_2 & & c_3 - 2c_2 + c_1 & \\
 & c_3 - c_2 & & \cdots \\
c_3 & & \cdots & \\
 & \cdots & & \cdots \\
\cdots & & \cdots & \\
 & \cdots & & \cdots
\end{array}
$$

The most common example of this transformation is *Euler's transformation*. Set

$$
g(z) = \frac{1}{1 + z} = 1 - z + z^2 - z^3 + - \cdots
\tag{2-42}
$$

Thus

$$
b_n = (-1)^n
$$

$$
z g'(z) = \frac{-z}{(1 + z)^2}
$$

$$
\frac{z^2 g''(z)}{2!} = \frac{z^2}{(1 + z)^3}
$$

etc.

Thus, if $f(z) = \sum_{n=0}^{\infty} (-1)^n c_n z^n$, the Euler transformation tells us that

$$f(z) = \frac{1}{1+z}\left[c_0 - (c_1 - c_0)\left(\frac{z}{1+z}\right)\right.$$
$$\left. + (c_2 - 2c_1 + c_0)\left(\frac{z}{1+z}\right)^2 - + \cdots\right] \qquad (2\text{-}43)$$

For example, consider the series

$$f(z) = 1 - \tfrac{1}{2}z + \tfrac{1}{3}z^2 - \tfrac{1}{4}z^3 + - \cdots \qquad (2\text{-}44)$$

[We already know $f(z) = (1/z)\ln(1+z)$, of course.] This is a rather poorly converging series; in fact, it diverges for $|z| > 1$. From the difference table

$$
\begin{array}{ccccccc}
1 \\
 & -\tfrac{1}{2} \\
\tfrac{1}{2} & & \tfrac{1}{3} \\
 & -\tfrac{1}{6} & & -\tfrac{1}{4} \\
\tfrac{1}{3} & & \tfrac{1}{12} & & \tfrac{1}{5} \\
 & -\tfrac{1}{12} & & -\tfrac{1}{20} \\
\tfrac{1}{4} & & \tfrac{1}{30} \\
 & -\tfrac{1}{20} \\
\tfrac{1}{5}
\end{array}
$$

we see that

$$f(z) = \frac{1}{1+z}\left[1 + \frac{1}{2}\left(\frac{z}{1+z}\right) + \frac{1}{3}\left(\frac{z}{1+z}\right)^2 + \cdots\right] \qquad (2\text{-}45)$$

This new series converges much better for z near 1, for example. In fact, it converges for all positive z.

This is just one example of the tricks that are available to turn one series into another, more useful, one. The important practical problem of how to find the numerical value of a series will be treated in Chapter 13.

REFERENCES

Infinite series are discussed quite thoroughly in numerous books on complex variable theory. See, for example, Whittaker and Watson (W5) Chapter II and III, or Apostol (A5) Chapters 12 and 13.

Several fairly elementary references are Hyslop (H13); Green (G7); and Stanaitis (S11). The monograph by Hirschman (H10) is also very readable, but it is at a somewhat more advanced level than the three preceding references.

Two standard treatises, with somewhat of a nineteenth-century flavor, are those by Bromwich (B8) and Knopp (K3).

Davis (D2) presents in a systematic manner many devices for summing series. Many elementary series, including those leading to Bernoulli numbers and Euler numbers, are given near the beginning of Dwight (D8).

PROBLEMS

Find the sums of the series 2-1 through 2-7:

2-1 $1 + \dfrac{1}{4} - \dfrac{1}{16} - \dfrac{1}{64} + \dfrac{1}{256} + \dfrac{1}{1024} - - + + \cdots$

2-2 $\dfrac{1}{1 \cdot 3} + \dfrac{1}{2 \cdot 4} + \dfrac{1}{3 \cdot 5} + \dfrac{1}{4 \cdot 6} + \cdots$

2-3 $1 - \dfrac{1}{5 \cdot 3^2} - \dfrac{1}{7 \cdot 3^3} + \dfrac{1}{11 \cdot 3^5} + \dfrac{1}{13 \cdot 3^6} - - + + \cdots$

2-4 $\dfrac{1}{0!} + \dfrac{2}{1!} + \dfrac{3}{2!} + \cdots$

2-5 $1 + \dfrac{1}{9} + \dfrac{1}{25} + \dfrac{1}{49} + \cdots$

2-6 $1 + \dfrac{1}{9^2} + \dfrac{1}{25^2} + \dfrac{1}{49^2} + \cdots$

2-7 $1 - \dfrac{1}{4^2} + \dfrac{1}{9^2} - \dfrac{1}{16^2} + - \cdots$

2-8 Evaluate in closed form the sum

$$f(\theta) = \sin \theta + \tfrac{1}{3} \sin 2\theta + \tfrac{1}{5} \sin 3\theta + \tfrac{1}{7} \sin 4\theta + \cdots$$

(you may assume $0 < \theta < \pi$ for definiteness).

Do the following series converge or not?

2-9 $\dfrac{\frac{1}{2} \cdot \frac{1}{2}}{9 \cdot 7 \cdot 25 \cdot 1!} + \dfrac{\frac{1}{2} \cdot \frac{3}{2} \cdot \frac{3}{2}}{11 \cdot 9 \cdot 49 \cdot 2!} + \dfrac{\frac{1}{2} \cdot \frac{3}{2} \cdot \frac{5}{2} \cdot \frac{5}{2}}{13 \cdot 11 \cdot 81 \cdot 3!} + \cdots$

2-10 $\dfrac{(1 \cdot 3)^2}{1 \cdot 1 \cdot 1^2} + \dfrac{(1 \cdot 3 \cdot 5)^2}{4 \cdot 2 \cdot (1 \cdot 2)^2} + \dfrac{(1 \cdot 3 \cdot 5 \cdot 7)^2}{16 \cdot 3 \cdot (1 \cdot 2 \cdot 3)^2}$

$$+ \dfrac{(1 \cdot 3 \cdot 5 \cdot 7 \cdot 9)^2}{64 \cdot 4 \cdot (1 \cdot 2 \cdot 3 \cdot 4)^2} + \cdots$$

2-11 Evaluate the series

$$f(x) = \sum_{n=0}^{\infty} \frac{(-1)^{n+1} n^2 x^{2n-1}}{(2n-1)!}$$

$$= x - \frac{4x^3}{3!} + \frac{9x^5}{5!} - + \cdots$$

in closed form, by comparing with

$$\sin x = x - \frac{x^3}{3!} + \frac{x^5}{5!} - + \cdots$$

2-12 Consider the so-called Euler numbers, defined by

$$\sec z = \sum_{n=0}^{\infty} (-1)^n \frac{E_{2n}}{(2n)!} z^{2n}$$

(*a*) Show that

$$(E+1)^k + (E-1)^k = 0 \qquad (E^n \to E_n) \qquad (k \text{ even})$$

Give E_2, E_4, E_6, E_8.

(*b*) Using the partial fraction expansion of the secant

$$k\pi \sec k\pi = 4k\left(\frac{1}{1-4k^2} - \frac{3}{9-4k^2} + \frac{5}{25-4k^2} - + \cdots\right)$$

evaluate:

(1) $1 - \dfrac{1}{3^3} + \dfrac{1}{5^3} - \dfrac{1}{7^3} + - \cdots$

(2) $1 - \dfrac{1}{3^{2n+1}} + \dfrac{1}{5^{2n+1}} - + \cdots$

THREE

EVALUATION OF INTEGRALS

In the first three sections of this chapter we discuss various techniques for the analytic evaluation of definite integrals. These techniques include differentiation or integration with respect to a parameter, the exploitation of symmetries, and evaluation by contour integration. Tabulated integrals such as the gamma function, beta function, exponential integral, elliptic integrals, and so forth, are described in Section 3–4. Approximate expansions, especially asymptotic expansions, are treated in Section 3–5, and the saddle-point method is discussed in Section 3–6.

3–1 ELEMENTARY METHODS

We first review some useful elementary techniques for doing integrals. The simplest device is changing variables. For example, the standard trick for evaluating

$$\int_0^\infty e^{-t^2}\, dt = \tfrac{1}{2}\sqrt{\pi} \tag{3-1}$$

by using polar coordinates is presumably familiar. From this result, by setting $t = u\sqrt{a}$ we find

$$\int_0^\infty e^{-au^2}\, du = \frac{1}{2}\sqrt{\frac{\pi}{a}} \tag{3-2}$$

or by setting $t = u^2$, we may deduce

$$\int_0^\infty e^{-u^4} u \, du = \tfrac{1}{4}\sqrt{\pi} \tag{3-3}$$

How about $\int_0^\infty e^{-t^4} \, dt$? This is not so easy. Let

$$I_\alpha = \int_0^\infty e^{-t^\alpha} \, dt \tag{3-4}$$

Make the change of variable

$$t^\alpha = u \qquad t = u^{1/\alpha} \qquad dt = \frac{1}{\alpha} u^{(1/\alpha)-1} \, du$$

Then

$$I_\alpha = \frac{1}{\alpha} \int_0^\infty e^{-u} u^{(1/\alpha)-1} \, du$$

This does not look any easier, but one defines the *gamma function* $\Gamma(z)$ by

$$\Gamma(z) = \int_0^\infty e^{-u} u^{z-1} \, du \tag{3-5}$$

so that

$$I_\alpha = \frac{1}{\alpha} \Gamma\!\left(\frac{1}{\alpha}\right) \tag{3-6}$$

We shall say more about the gamma function in Section 3-4; note that (3-1) gives $\Gamma(\tfrac{1}{2}) = \sqrt{\pi}$.

Another useful technique is to introduce complex variables.

EXAMPLE

$$I = \int_0^\infty e^{-ax} \cos \lambda x \, dx$$

$$= \operatorname{Re} \int_0^\infty e^{-ax} e^{i\lambda x} \, dx$$

$$= \operatorname{Re} \frac{1}{a - i\lambda} \tag{3-7}$$

therefore,[1]

$$I = \frac{a}{a^2 + \lambda^2}$$

[1] Note that our method seems to require that λ be real, but the results are in fact correct for λ anywhere in the strip $|\operatorname{Im} \lambda| < a$. How did this happen?

This method gives us another integral at the same time, from the imaginary part,

$$\int_0^\infty e^{-ax} \sin \lambda x \, dx = \frac{\lambda}{a^2 + \lambda^2} \tag{3-8}$$

The method of integration by parts is very useful, and is presumably familiar by now.

Another useful trick is differentiation or integration with respect to a parameter.

EXAMPLE

$$I = \int_0^\infty e^{-ax} \cos \lambda x \, x \, dx \tag{3-9}$$

Let

$$I(a) = \int_0^\infty e^{-ax} \cos \lambda x \, dx = \frac{a}{a^2 + \lambda^2}$$

Then

$$I = -\frac{d}{da} I(a) = \frac{a^2 - \lambda^2}{(a^2 + \lambda^2)^2}$$

We have assumed that the order of differentiation and integration can be reversed. For necessary and sufficient conditions see mathematics books, such as Whittaker and Watson (W5) Chapter IV, or Apostol (A5) Chapter 9. In physical applications it will nearly always work.

EXAMPLE

$$I = \int_0^\infty \frac{\sin x}{x} \, dx \tag{3-10}$$

Let

$$I(\alpha) = \int_0^\infty \frac{e^{-\alpha x} \sin x}{x} \, dx \qquad \text{so that } I = I(0)$$

$$\frac{dI(\alpha)}{d\alpha} = -\int_0^\infty e^{-\alpha x} \sin x \, dx = \frac{-1}{\alpha^2 + 1}$$

$$I(\alpha) = -\int \frac{d\alpha}{\alpha^2 + 1} = C - \tan^{-1} \alpha$$

But $I(\infty) = 0$.   Therefore $C = \pi/2$.

$$I(\alpha) = \frac{\pi}{2} - \tan^{-1} \alpha \qquad \text{and} \qquad I = I(0) = \frac{\pi}{2} \tag{3-11}$$

Sometimes one can combine several derivatives of an integral to form a differential equation.

EXAMPLE

$$I(\alpha) = \int_0^\infty \frac{e^{-\alpha x}}{1 + x^2}\, dx \tag{3-12}$$

$$I''(\alpha) + I(\alpha) = \int_0^\infty e^{-\alpha x}\, dx = \frac{1}{\alpha}$$

This equation is easily solved by the variation of parameters method discussed in Section 1-1.   The result is

$$I(\alpha) = -\cos \alpha \int^\alpha \frac{\sin t}{t}\, dt + \sin \alpha \int^\alpha \frac{\cos t}{t}\, dt$$

But $I(\alpha)$ and all its derivatives vanish at $\alpha = \infty$.   Thus

$$I(\alpha) = \sin \alpha \int_\infty^\alpha \frac{\cos t}{t}\, dt - \cos \alpha \int_\infty^\alpha \frac{\sin t}{t}\, dt$$

The *cosine-integral* and *sine-integral* functions are defined by

$$\text{Ci } x = \int_\infty^x \frac{\cos t}{t}\, dt \qquad \text{Si } x = \int_0^x \frac{\sin t}{t}\, dt \tag{3-13}$$

Thus $I(\alpha) = \sin \alpha \text{ Ci } \alpha + \cos \alpha(\pi/2 - \text{Si } \alpha)$.   The functions Si α and Ci α are tabulated in Jahnke *et al.* (J3) and Abramowitz and Stegun (A1).

Finally, the useful integrals $\int_0^\infty e^{-\alpha x^2} x^n\, dx$ $(n = 0, 1, 2, \ldots)$ can be obtained from the first two,

$$\int_0^\infty e^{-\alpha x^2}\, dx = \frac{1}{2}\sqrt{\frac{\pi}{a}} \qquad \int_0^\infty e^{-\alpha x^2} x\, dx = \frac{1}{2a} \tag{3-14}$$

by repeated differentiation.

We observe in passing that if a parameter in an integral also appears in the limit(s), the differentiation with respect to that parameter proceeds according to the following rule:

$$\frac{d}{d\alpha}\left[\int_{a(\alpha)}^{b(\alpha)} f(x, \alpha)\, dx\right] = \int_{a(\alpha)}^{b(\alpha)} \frac{\partial f(x, \alpha)}{\partial \alpha}\, dx + \frac{db}{d\alpha} f(b(\alpha), \alpha) - \frac{da}{d\alpha} f(a(\alpha), \alpha) \tag{3-15}$$

3–2 USE OF SYMMETRY ARGUMENTS

The evaluation of some integrals may be greatly simplified by exploiting the symmetries present in the problem. We shall illustrate the principles involved by means of some integrals over solid angle in three dimensions.

EXAMPLE

Consider the integral

$$I_1(\mathbf{k}) = \int \frac{d\Omega}{1 + \mathbf{k} \cdot \hat{\mathbf{r}}} = \int_0^{2\pi} d\phi \int_{-1}^{+1} d(\cos\theta) \frac{1}{1 + \mathbf{k} \cdot \hat{\mathbf{r}}} \tag{3-16}$$

where $\hat{\mathbf{r}}$ is the unit radius vector and (θ, ϕ) are conventional spherical polar coordinates:

$$\hat{\mathbf{r}}_x = \sin\theta \cos\phi \qquad \hat{\mathbf{r}}_y = \sin\theta \sin\phi \qquad \hat{\mathbf{r}}_z = \cos\theta$$

Since the orientation of our coordinate system is arbitrary, we may choose the z axis along $\mathbf{k}$ and obtain (we assume $k < 1$)

$$I_1(\mathbf{k}) = \int_0^{2\pi} d\phi \int_{-1}^{+1} \frac{d(\cos\theta)}{1 + k\cos\theta} = \frac{2\pi}{k} \ln\left(\frac{1+k}{1-k}\right) \tag{3-17}$$

The integrals

$$I_m(\mathbf{k}) = \int \frac{d\Omega}{(1 + \mathbf{k} \cdot \hat{\mathbf{r}})^m} \tag{3-18}$$

may be obtained from I_1 by differentiation (replace the 1 in the denominator by α, differentiate with respect to α, and set α equal to 1 again). For example,

$$I_2(\mathbf{k}) = \int \frac{d\Omega}{(1 + \mathbf{k} \cdot \hat{\mathbf{r}})^2} = \frac{4\pi}{1 - k^2} \tag{3-19}$$

Another example is

$$I_1(\mathbf{k}, \mathbf{a}) = \int \frac{\mathbf{a} \cdot \hat{\mathbf{r}} \, d\Omega}{1 + \mathbf{k} \cdot \hat{\mathbf{r}}} \tag{3-20}$$

This integral is complicated by the fact that *two* directions are given by the two vectors $\mathbf{a}$ and $\mathbf{k}$, and we cannot choose our polar axis along both of them. However, the direction $\mathbf{a}$ is trivial, in that it may be "factored out." For, consider

$$\mathbf{J}(\mathbf{k}) = \int \frac{\hat{\mathbf{r}} \, d\Omega}{1 + \mathbf{k} \cdot \hat{\mathbf{r}}} \tag{3-21}$$

Clearly

$$I_1(\mathbf{k}, \mathbf{a}) = \mathbf{a} \cdot \mathbf{J}$$

Now $\mathbf{J}(\mathbf{k})$ is a vector and must point in the direction $\mathbf{k}$, since no other direction is specified in the definition (3-21) of $\mathbf{J}$. Therefore,

$$\mathbf{J}(\mathbf{k}) = A\mathbf{k} \qquad (3\text{-}22)$$

To evaluate the scalar A, we dot $\mathbf{k}$ into both sides of (3-22) and obtain

$$A = \frac{1}{k^2} \mathbf{k} \cdot \mathbf{J}(\mathbf{k}) = \frac{1}{k^2} \int \frac{\mathbf{k} \cdot \hat{\mathbf{r}}}{1 + \mathbf{k} \cdot \hat{\mathbf{r}}} \, d\Omega$$

$$= \frac{1}{k^2} \int d\Omega \left(1 - \frac{1}{1 + \mathbf{k} \cdot \hat{\mathbf{r}}} \right)$$

$$= \frac{4\pi}{k^2} \left(1 - \frac{1}{2k} \ln \frac{1 + k}{1 - k} \right)$$

Thus our original integral (3-20) is

$$I_1(\mathbf{k}, \mathbf{a}) = A\mathbf{a} \cdot \mathbf{k} = \frac{4\pi}{k^2} \mathbf{a} \cdot \mathbf{k} \left[1 - \frac{1}{2k} \ln \frac{1 + k}{1 - k} \right] \qquad (3\text{-}23)$$

What about the integral

$$I_2(\mathbf{k}, \mathbf{a}) = \int \frac{\mathbf{a} \cdot \hat{\mathbf{r}} \, d\Omega}{(1 + \mathbf{k} \cdot \hat{\mathbf{r}})^2} \, ?$$

We could obtain $I_2(\mathbf{k}, \mathbf{a})$ from $I_1(\mathbf{k}, \mathbf{a})$ by replacing the 1 by α and differentiating as before, but a simpler method is to differentiate $I_1(\mathbf{k})$ with respect to $\mathbf{k}$. On the one hand,

$$\frac{\partial I_1(\mathbf{k})}{\partial \mathbf{k}} = \frac{\partial}{\partial \mathbf{k}} \int \frac{d\Omega}{1 + \mathbf{k} \cdot \hat{\mathbf{r}}} = -\int \frac{d\Omega \, \hat{\mathbf{r}}}{(1 + \mathbf{k} \cdot \hat{\mathbf{r}})^2} \qquad (3\text{-}24)$$

On the other hand, using (3-17),

$$\frac{\partial I_1(\mathbf{k})}{\partial \mathbf{k}} = \frac{\mathbf{k}}{k} \frac{\partial I_1(\mathbf{k})}{\partial k} = \frac{2\pi \mathbf{k}}{k^2} \left[\frac{2}{1 - k^2} - \frac{1}{k} \ln \left(\frac{1 + k}{1 - k} \right) \right] \qquad (3\text{-}25)$$

Comparing (3-24) and (3-25) gives

$$\int \frac{d\Omega \, \hat{\mathbf{r}}}{(1 + \mathbf{k} \cdot \hat{\mathbf{r}})^2} = \frac{2\pi \mathbf{k}}{k^2} \left(\frac{-2}{1 - k^2} + \frac{1}{k} \ln \frac{1 + k}{1 - k} \right)$$

and therefore

$$\int \frac{\mathbf{a} \cdot \hat{\mathbf{r}} \, d\Omega}{(1 + \mathbf{k} \cdot \hat{\mathbf{r}})^2} = \mathbf{a} \cdot \int \frac{\hat{\mathbf{r}} \, d\Omega}{(1 + \mathbf{k} \cdot \hat{\mathbf{r}})^2} = \frac{2\pi \mathbf{k} \cdot \mathbf{a}}{k^2} \left[\frac{-2}{1 - k^2} + \frac{1}{k} \ln \left(\frac{1 + k}{1 - k} \right) \right] \qquad (3\text{-}26)$$

Other examples of the use of symmetry arguments are the evaluation of integrals such as

$$\Phi_1(\mathbf{a}, \mathbf{b}) = \int d\Omega \, \hat{\mathbf{r}} \cdot \mathbf{a} \, \hat{\mathbf{r}} \cdot \mathbf{b}$$

$$\Phi_2(\mathbf{a}, \mathbf{b}, \mathbf{c}, \mathbf{d}) = \int d\Omega \, \hat{\mathbf{r}} \cdot \mathbf{a} \, \hat{\mathbf{r}} \cdot \mathbf{b} \, \hat{\mathbf{r}} \cdot \mathbf{c} \, \hat{\mathbf{r}} \cdot \mathbf{d} \tag{3-27}$$

etc.

To evaluate Φ_1, we observe that it is a scalar which is linear in both **a** and **b**. The only possibility is that $\Phi_1 = A\mathbf{a} \cdot \mathbf{b}$, where A is a number. To find A, let **a** and **b** both equal $\hat{\mathbf{z}}$. Then

$$\Phi_1(\hat{\mathbf{z}}, \hat{\mathbf{z}}) = A = \int d\Omega (\hat{\mathbf{r}} \cdot \hat{\mathbf{z}})^2 = \int d\Omega \, \cos^2 \theta = \frac{4\pi}{3}$$

Therefore $\Phi_1(\mathbf{a}, \mathbf{b}) = (4\pi/3)\mathbf{a} \cdot \mathbf{b}$.

The evaluation of Φ_2 proceeds similarly. Since Φ_2 is a scalar, linear in the four vectors **a**, **b**, **c**, **d**, as well as being invariant under any interchange of these vectors, it must have the form

$$\Phi_2 = B(\mathbf{a} \cdot \mathbf{b} \, \mathbf{c} \cdot \mathbf{d} + \mathbf{a} \cdot \mathbf{c} \, \mathbf{b} \cdot \mathbf{d} + \mathbf{a} \cdot \mathbf{d} \, \mathbf{b} \cdot \mathbf{c})$$

where B is a number. B is found by setting $\mathbf{a} = \mathbf{b} = \mathbf{c} = \mathbf{d} = \hat{\mathbf{z}}$, so that

$$\Phi_2(\hat{\mathbf{z}}, \hat{\mathbf{z}}, \hat{\mathbf{z}}, \hat{\mathbf{z}}) = 3B = \int d\Omega (\hat{\mathbf{r}} \cdot \hat{\mathbf{z}})^4 = \int d\Omega \, \cos^4 \theta = \frac{4\pi}{5}$$

Therefore $B = 4\pi/15$, and

$$\Phi_2(\mathbf{a}, \mathbf{b}, \mathbf{c}, \mathbf{d}) = \frac{4\pi}{15} (\mathbf{a} \cdot \mathbf{b} \, \mathbf{d} + \mathbf{c} \cdot \mathbf{a} \cdot \mathbf{c} \, \mathbf{b} \cdot \mathbf{d} + \mathbf{a} \cdot \mathbf{d} \, \mathbf{b} \cdot \mathbf{c})$$

As a final example of the sort of device one can use to simplify integrals, we mention the identity

$$\frac{1}{ab} = \int_0^1 \frac{du}{[au + b(1 - u)]^2} \tag{3-28}$$

which Feynman has used to simplify the evaluation of integrals arising in quantum field theory.[2] As an application of (3-28), we evaluate the integral

$$\psi(\mathbf{k}, \mathbf{l}) = \int \frac{d\Omega}{(1 + \mathbf{k} \cdot \hat{\mathbf{r}})(1 + \mathbf{l} \cdot \hat{\mathbf{r}})} \tag{3-29}$$

[2] See the Appendix of R. P. Feynman (F3).

Use of (3-28) converts (3-29) to

$$\psi(\mathbf{k}, \mathbf{l}) = \int_0^1 du \int \frac{d\Omega}{\{1 + \hat{\mathbf{r}} \cdot [ku + \mathbf{l}(1 - u)]\}^2} \qquad (3\text{-}30)$$

The solid-angle integral in (3-30) is just $I_2[ku + \mathbf{l}(1 - u)]$, as given by (3-19).   Thus

$$\psi(\mathbf{k}, \mathbf{l}) = 4\pi \int_0^1 \frac{du}{1 - [ku + \mathbf{l}(1 - u)]^2}$$

This is an elementary integral, although rather tedious; the answer has the interesting form

$$\psi(\mathbf{k}, \mathbf{l}) = \frac{4\pi}{\sqrt{A^2 - B^2}} \cosh^{-1} \frac{A}{B} \qquad (3\text{-}31)$$

where

$$A = 1 - \mathbf{k} \cdot \mathbf{l}$$

$$B = \sqrt{(1 - k^2)(1 - l^2)}$$

The proof of (3-31) is left as an exercise (Problem 3-37).

3-3   CONTOUR INTEGRATION

One of the most powerful means for evaluating definite integrals is provided by the theorem of residues from the theory of functions of a complex variable.   We shall illustrate this method of contour integration by a number of examples in this section.   Before reading this material, the student who does not know the theory of functions of a complex variable reasonably well should review (or learn) certain parts of this theory.   These parts are presented in the Appendix of this book to serve as an aid in the review (or as a guide to the study).

The theorem of residues [Appendix, Eq. (A-15)] tells us that if a function $f(z)$ is regular in the region bounded by a closed path C, except for a finite number of poles and isolated essential singularities in the interior of C, then the integral of $f(z)$ along the contour C is

$$\int_C f(z)\, dz = 2\pi i \sum \text{residues}$$

where $\sum$ residues means the sum of the residues at all the poles and essential singularities inside C.

The residues at poles and isolated essential singularities may be found as follows.

If $f(z)$ has a simple pole (pole or order one) at $z = z_0$, the residue is

$$a_{-1} = [(z - z_0)f(z)]_{z=z_0} \tag{3-32}$$

If $f(z)$ is written in the form $f(z) = q(z)/p(z)$, where $q(z)$ is regular and $p(z)$ has a simple zero at z_0, the residue of $f(z)$ at z_0 may be computed from

$$a_{-1} = \frac{q}{p'}\bigg|_{z=z_0} \tag{3-33}$$

If z_0 is a pole of order n, the residue is

$$a_{-1} = \frac{1}{(n-1)!}\left\{\left(\frac{d}{dz}\right)^{n-1}[(z - z_0)^n f(z)]\right\}_{z=z_0} \tag{3-34}$$

If z_0 is an isolated essential singularity, the residue is found from the Laurent expansion (Appendix, Section A–2, item 7).

We illustrate the method of contour integration by some examples.

EXAMPLE

$$I = \int_0^\infty \frac{dx}{1 + x^2} \tag{3-35}$$

Consider $\oint dz/(1 + z^2)$ along the contour of Figure 3–1. Along the real axis the integral is $2I$. Along the large semicircle in the upper-half plane we get zero, since

$$z = Re^{i\theta} \qquad dz = iRe^{i\theta}\, d\theta \qquad \frac{1}{1 + z^2} \approx \frac{e^{-2i\theta}}{R^2}$$

$$\int \frac{dz}{1 + z^2} \approx \frac{i}{R}\int e^{-i\theta}\, d\theta \to 0 \text{ as } R \to \infty$$

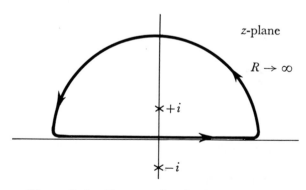

Figure 3–1   Contour for the integral (3-35)

The residue of $1/(1 + z^2) = 1/(z + i)(z - i)$ at $z = i$ is $1/(2i)$. Thus

$$2I = 2\pi i\left(\frac{1}{2i}\right) = \pi \qquad I = \frac{\pi}{2}$$

Note that an important part of the problem may be choosing the "return path" so that the contribution from it is simple (preferably zero).

EXAMPLE

Consider a resistance R and inductance L connected in series with a voltage $V(t)$ (Figure 3-2). Suppose $V(t)$ is a voltage impulse, that is, a very high pulse lasting for a very short time. As we shall see in Chapter 4, we can write to a good approximation

$$V(t) = \frac{A}{2\pi} \int_{-\infty}^{\infty} e^{i\omega t} \, d\omega$$

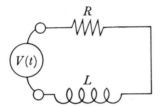

Figure 3-2   Series R-L circuit

where A is the area under the curve $V(t)$.

The current due to a voltage $e^{i\omega t}$ is $e^{i\omega t}/(R + i\omega L)$. Thus the current due to our voltage pulse is

$$I(t) = \frac{A}{2\pi} \int_{-\infty}^{\infty} \frac{e^{i\omega t} \, d\omega}{R + i\omega L} \tag{3-36}$$

Let us evaluate this integral.

If $t < 0$, the integrand is exponentially small for Im $\omega \to -\infty$, so that we may complete the contour by a large semicircle in the *lower*-half ω-plane, along which the integral vanishes.[3] The contour encloses no singularities, so that $I(t) = 0$.

If $t > 0$, we must complete the contour by a large semicircle in the *upper*-half plane. Then

$$I(t) = 2\pi i\left(\frac{A}{2\pi}\right)\frac{e^{-Rt/L}}{iL} = \frac{A}{L} e^{-Rt/L}$$

---

[3] A rigorous justification of this procedure is provided by *Jordan's lemma*; see Copson (C8) p. 137 for example.

EXAMPLE

$$I = \int_0^\infty \frac{dx}{1 + x^3} \tag{3-37}$$

The integrand is not even, so we cannot extend it to $-\infty$. Consider the integral

$$\int \frac{\ln z \, dz}{1 + z^3}$$

The integrand is many-valued; we may cut the plane as shown in Figure 3-3, and define $\ln z$ real ($=\ln x$) just above the cut. Then $\ln z = \ln x + 2\pi i$ below the cut, and integrating along the indicated contour,

$$\oint \frac{\ln z \, dz}{1 + z^3} = -2\pi i I$$

On the other hand, using the method of residues,

$$\oint \frac{\ln z \, dz}{1 + z^3} = -\frac{4\pi^2 i \sqrt{3}}{9}$$

Thus $I = (2\pi \sqrt{3})/9$.

When integrating around a branch point, as in this example, it is necessary to show that the integral on a vanishingly small circle around the branch point is zero. In this example, this part goes like $r \ln r$, which approaches zero as $r \to 0$.

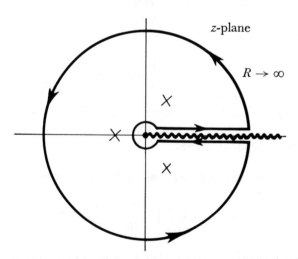

Figure 3–3   Contour for the integral (3-37)

A more straightforward method for evaluating the integral (3-37) consists of evaluating the integral

$$J = \oint \frac{dz}{1 + z^3}$$

along a closed contour consisting of (i) the real axis from 0 to $+\infty$, (ii) one-third of a large circle at $|z| = \infty$, and (iii) a return to the origin along the line $\arg z = 2\pi/3$. Evaluating J along these pieces, we find

$$J = (1 - e^{2\pi i/3})I$$

On the other hand, the integrand of J has a simple pole at $z = e^{\pi i/3}$; the residue theorem gives

$$J = \frac{2\pi i}{3e^{2\pi i/3}}$$

Therefore

$$I = \frac{2\pi i}{3} \frac{e^{-2\pi i/3}}{1 - e^{2\pi i/3}} = \frac{\pi}{3 \sin \frac{\pi}{3}} = \frac{2\pi}{3\sqrt{3}}$$

as before.

EXAMPLE

$$I = \int_0^\pi \frac{d\theta}{a + b \cos \theta} \qquad a > b > 0 \qquad \text{(3-38)}$$

The integrand is even, so

$$2I = \int_0^{2\pi} \frac{d\theta}{a + b \cos \theta}$$

If we integrate along the unit circle, as in Figure 3-4,

$$z = e^{i\theta} \qquad dz = ie^{i\theta}\, d\theta$$

$$\cos \theta = \frac{e^{i\theta} + e^{-i\theta}}{2} = \frac{1}{2}\left(z + \frac{1}{z}\right) \qquad \text{(3-39)}$$

Then

$$2I = \int_c \frac{dz/(iz)}{a + (b/2)[z + (1/z)]}$$

$$= \frac{2}{i} \int_c \frac{dz}{bz^2 + 2az + b}$$

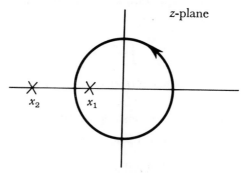

Figure 3–4   Contour for the integral (3-38)

The integrand has two poles, at the roots of the denominator. The product of the roots is $b/b = 1$; one is outside and one inside the unit circle. They are at

$$x_{1,2} = -\frac{a}{b} \pm \sqrt{\frac{a^2}{b^2} - 1}$$

Then

$$2I = \frac{2}{i}\, 2\pi i\left(\text{residue at } x_1 = -\frac{a}{b} + \sqrt{\frac{a^2}{b^2} - 1}\right)$$

$$I = 2\pi\left(\frac{1}{2bx_1 + 2a}\right) = \frac{\pi}{\sqrt{a^2 - b^2}}$$

EXAMPLE

$$I = \int_0^\infty \frac{\sqrt{x}\, dx}{1 + x^2} \qquad\qquad 3\text{-}40$$

Consider the integral

$$\oint \frac{\sqrt{z}\, dz}{1 + z^2}$$

along the contour of Figure 3–5. We choose $\sqrt{z}$ positive on top of the cut. Then

$$\oint \frac{\sqrt{z}\, dz}{1 + z^2} = 2I$$

But, using residues,

$$\oint \frac{\sqrt{z}\, dz}{1 + z^2} = \pi\sqrt{2}$$

Therefore $I = \pi/\sqrt{2}$.

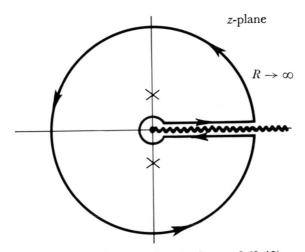

Figure 3–5   Contour for the integral (3-40)

EXAMPLE

$$I = \int_{-\infty}^{\infty} \frac{e^{ax}}{e^x + 1}\, dx \qquad (0 < a < 1) \tag{3-41}$$

We clearly wish to consider

$$\int \frac{e^{az}\, dz}{e^z + 1}$$

along some contour.   One choice is shown in Figure 3–6, involving the familiar large semicircle.   Note that we must give a a small imaginary part in

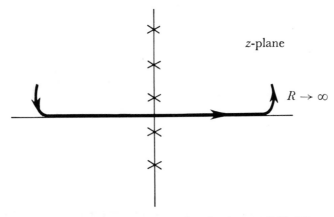

Figure 3–6   Possible contour for the integral (3-41)

order to be certain that the integral along the semicircle vanishes.   Then

$$I = 2\pi i \sum \text{residues}$$

There is an infinite number of poles.

At $z = i\pi$, residue is $-e^{i\pi a}$

At $z = 3i\pi$, residue is $-e^{3i\pi a}$, etc.

Thus

$$I = -2\pi i \frac{e^{i\pi a}}{1 - e^{2i\pi a}} = \frac{\pi}{\sin \pi a}$$

Now we can let Im $a \to 0$.

Alternatively, we could use the contour shown in Figure 3–7.   Along the real axis, we get I.   Along Im $z = 2\pi i$, we get

$$-\int_{-\infty}^{\infty} \frac{e^{a(x + 2\pi i)}}{e^{x + 2\pi i} + 1} \, dx = -e^{2\pi i a} I$$

Thus

$$(1 - e^{2\pi i a})I = 2\pi i \, (\text{residue at } z = \pi i)$$

$$= -2\pi i e^{i\pi a}$$

$$I = \frac{\pi}{\sin \pi a} \qquad \text{as before}$$

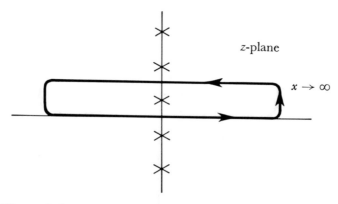

Figure 3–7   Another contour for the integral (3-41)

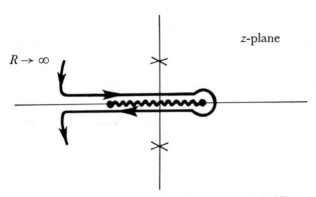

z-plane

$R \to \infty$

Figure 3–8   Contour for the integral (3-42)

EXAMPLE

We consider

$$I = \int_{-1}^{+1} \frac{dx}{\sqrt{1 - x^2}(1 + x^2)} \tag{3-42}$$

We consider

$$\oint \frac{dz}{\sqrt{1 - z^2}(1 + z^2)}$$

along the contour of Figure 3–8.   On the top side of the cut we get I, and we get another I from the bottom side.   Therefore,

$$2I = 2\pi i \left[\frac{1}{2i\sqrt{2}} + \frac{1}{2i\sqrt{2}} \right]$$

$$= \pi\sqrt{2}$$

$$I = \frac{\pi}{\sqrt{2}}$$

EXAMPLE

Consider the integral

$$I = \oint \frac{f(z)\, dz}{\sin \pi z} \tag{3-43}$$

around the contour of Figure 3–9, where $f(z)$ has several isolated singularities (indicated by crosses in the figure) and goes to zero at least as fast as $|z|^{-1}$ as $|z| \to \infty$.

On the one hand, the integral I may be evaluated by summing the residues at the zeros of $\sin \pi z$, indicated by dots in Figure 3–9.   The result is

$$I = 2\pi i \sum_{n=-\infty}^{\infty} \frac{1}{\pi} (-)^n f(n) \tag{3-44}$$

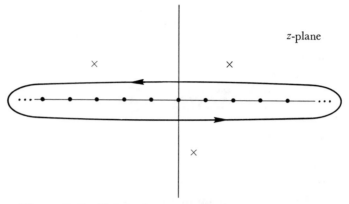

Figure 3–9   Original contour for the integral (3-43)

On the other hand, the integral along the contour of Figure 3–9 is clearly the same as that along the contour of Figure 3–10 since the integral around the circle at $|z| = \infty$ vanishes.   The singularities enclosed by the contour of Figure 3–10 are now the singularities of $f(z)$; if the locations and residues are denoted by z_k and R_k, respectively, then

$$I = -2\pi i \sum_k \frac{R_k}{\sin \pi z_k}$$

Comparison with (3-44) gives the summation formula

$$\sum_{n=-\infty}^{\infty} (-)^n f(n) = -\pi \sum_k \frac{R_k}{\sin \pi z_k} \qquad (3\text{-}45)$$

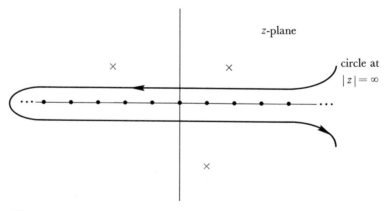

Figure 3–10   Deformed contour for the integral (3-43)

This device, which converts an infinite sum into a contour integral which is subsequently deformed, is known as a *Sommerfeld–Watson transformation.*[4]
    As an example, consider the series

$$S(x) = \frac{\sin x}{a^2 + 1} - \frac{2 \sin 2x}{a^2 + 4} + \frac{3 \sin 3x}{a^2 + 9} - + \cdots$$

$$= \sum_{n=1}^{\infty} (-)^{n+1} \frac{n \sin nx}{a^2 + n^2}$$

$$= -\frac{1}{2} \sum_{n=-\infty}^{\infty} (-)^n \frac{n \sin nx}{a^2 + n^2}$$

This is just the left side of equation (3-45), if we set

$$f(z) = -\frac{1}{2} \frac{z \sin xz}{a^2 + z^2}$$

(Note that we must require $|x| < \pi$ in order that the large circle in Figure 3–10 make no contribution to the integral.)   Now the locations and residues of the poles of $f(z)$ are

$$z_1 = ia \qquad\qquad z_2 = -ia$$

$$R_1 = -\frac{i}{4} \sinh ax \qquad R_2 = \frac{i}{4} \sinh ax$$

Therefore, we obtain from formula (3-45)

$$S(x) = -\pi \left[\frac{-\dfrac{i}{4} \sinh ax}{i \sinh a\pi} + \frac{\dfrac{i}{4} \sinh ax}{-i \sinh a\pi} \right] = \frac{\pi \sinh ax}{2 \sinh a\pi}$$

As we remarked above, this result is valid only if $|x| < \pi$. For x outside this range, we simply observe that $S(x)$, from its definition, is periodic in x with period 2π.

3–4   TABULATED INTEGRALS

We first mention (again) the *gamma function,* defined for Re $z > 0$ by the integral

$$\Gamma(z) = \int_0^{\infty} x^{z-1} e^{-x} \, dx \qquad\qquad (3\text{-}46)$$

---

[4] See Sommerfeld (S9) Appendix to Chapter VI, or G. N. Watson (W1).

Integration by parts gives

$$\Gamma(z) = (z - 1)\Gamma(z - 1) \qquad (\text{Re } z > 1) \tag{3-47}$$

Thus, from $\Gamma(1) = 1$, we obtain

$$\Gamma(2) = 1, \; \Gamma(3) = 2, \; \Gamma(4) = 6, \ldots, \; \Gamma(n) = (n - 1)\, ! \tag{3-48}$$

The gamma function may be analytically continued into the entire complex plane by means of the recursion relation (3-47), except for simple poles at $z = 0, -1, -2, \ldots$.

A related integral of some interest is encountered if we consider

$$\Gamma(r)\Gamma(s) = \int_0^\infty x^{r-1} e^{-x}\, dx \int_0^\infty y^{s-1} e^{-y}\, dy$$

Let $x + y = u$.

$$\Gamma(r)\Gamma(s) = \int_0^\infty du \int_0^u dx\; x^{r-1}(u - x)^{s-1} e^{-u}$$

Let $x = ut$.

$$\Gamma(r)\Gamma(s) = \int_0^\infty e^{-u} u^{r+s-1}\, du \int_0^1 dt\; t^{r-1}(1 - t)^{s-1}$$

$$= \Gamma(r + s)B(r, s)$$

where $B(r, s)$ is the *beta function*

$$B(r, s) = \frac{\Gamma(r)\Gamma(s)}{\Gamma(r + s)} = \int_0^1 x^{r-1}(1 - x)^{s-1}\, dx \tag{3-49}$$

This integral representation for $B(r, s)$ is obviously valid only for Re $r > 0$, Re $s > 0$; the first equality in (3-49) defines $B(r, s)$ for all r, s in the complex plane.

An interesting special case of this relation is

$$\Gamma(z)\Gamma(1 - z) = \Gamma(1)B(z, 1 - z)$$

$$= \int_0^1 x^{z-1}(1 - x)^{-z}\, dx$$

Let $x = t/(1 + t)$.   Then

$$\Gamma(z)\Gamma(1 - z) = \int_0^\infty \frac{t^{z-1}\, dt}{1 + t}$$

This integral can be done by using the contour shown in Figure 3–11.[5]   The result is

$$\Gamma(z)\Gamma(1 - z) = \frac{\pi}{\sin \pi z} \tag{3-50}$$

[5] Alternatively, we may make a simple change of variable in Example (3-41).

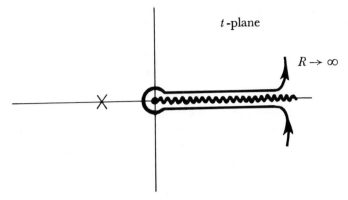

Figure 3–11   Contour for the integral $\int_0^\infty [t^{z-1}/(1+t)]\,dt$

Our derivation has only been valid for $0 < \mathrm{Re}\ z < 1$, but both sides of (3-50) are analytic functions of z (except at $z = 0, \pm 1, \pm 2, \ldots$), so we can extend the result into the entire plane.

Another integral which has been given a name is the so-called *exponential integral*,

$$Ei(x) = \int_{-\infty}^{x} \frac{e^t\,dt}{t} \tag{3-51}$$

This definition is conventionally supplemented by cutting the x-plane along the positive real axis.   Thus $Ei(x)$ is well defined for negative real x, but for positive real x we must distinguish between $Ei(x + i\varepsilon)$ and $Ei(x - i\varepsilon)$.

Two functions related to $Ei(x)$ are the *sine* and *cosine integrals*:

$$\mathrm{Si}\ x = \int_0^x \frac{\sin t\,dt}{t} \tag{3-52}$$

$$\mathrm{Ci}\ x = \int_\infty^x \frac{\cos t\,dt}{t} \tag{3-53}$$

The *error function* is defined as

$$\mathrm{erf}\ x = \frac{2}{\sqrt{\pi}} \int_0^x e^{-t^2}\,dt \tag{3-54}$$

The associated trigonometric integrals are known as *Fresnel integrals*[6]:

$$C(x) = \int_0^x \cos\left(\frac{\pi t^2}{2}\right) dt \qquad S(x) = \int_0^x \sin\left(\frac{\pi t^2}{2}\right) dt \tag{3-55}$$

---

[6] Conventions differ; our definitions (3-55) agree with Magnus *et al.* (M1) but not with Erdelyi *et al.* (E5).

Another class of integral which is tabulated in many places is the class of *elliptic integrals*. These are integrals of the form

$$\int dx \; \frac{A(x) + B(x)\sqrt{S(x)}}{C(x) + D(x)\sqrt{S(x)}} \tag{3-56}$$

where A, B, C, D are polynomials and S is a polynomial of the third or fourth degree (not a perfect square, of course!).

We shall not go into the general theory of such integrals, but just make a few remarks, which the student can verify.[7] In the first place,

$$\frac{A + B\sqrt{S}}{C + D\sqrt{S}} = E + \frac{F}{\sqrt{S}} \tag{3-57}$$

where E and F are rational (ratios of polynomials). By decomposing F in partial fractions, we see that the only nonelementary integrals we need are

$$J_n = \int \frac{x^n \, dx}{\sqrt{S}} \qquad H_n = \int \frac{dx}{(x - c)^n \sqrt{S}} \tag{3-58}$$

But we can evaluate all the J_n from J_0, J_1, and J_2, and all the H_n from H_1, J_0, J_1, and J_2. Thus we only need

$$J_0 = \int \frac{dx}{\sqrt{S}} \quad J_1 = \int \frac{x \, dx}{\sqrt{S}} \quad J_2 = \int \frac{x^2 \, dx}{\sqrt{S}} \quad H_1 = \int \frac{dx}{(x - c)\sqrt{S}} \tag{3-59}$$

There are various standard forms for S; we shall consider only Legendre's, namely, $S = (1 - x^2)(1 - k^2 x^2)$. Then

$$J_0 = \int \frac{dx}{\sqrt{(1 - x^2)(1 - k^2 x^2)}}$$

The integral

$$F = \int_0^x \frac{dx}{\sqrt{(1 - x^2)(1 - k^2 x^2)}} \tag{3-60}$$

is called the *Legendre elliptic integral of the first kind*. Generally, one sets $x = \sin \phi$, and defines

$$F(\phi, k) = \int_0^\phi \frac{d\phi}{\sqrt{1 - k^2 \sin^2 \phi}} \tag{3-61}$$

J_1 is an "elementary" integral, if we set $x^2 = u$.

---

[7] See, for example, Abramowitz and Stegun (A1) Chapter 17 for a more complete and useful discussion.

Instead of J_2, one takes for a standard integral the *Legendre elliptic integral of the second kind*,

$$E = \int_0^x \frac{\sqrt{1 - k^2 x^2} \, dx}{\sqrt{1 - x^2}} \tag{3-62}$$

Again one generally sets $x = \sin \phi$, and tabulates

$$E(\phi, k) = \int_0^\phi \sqrt{1 - k^2 \sin^2 \phi} \, d\phi \tag{3-63}$$

The integral H_1 is hard. One defines the *Legendre elliptic integral of the third kind*,

$$\Pi(\phi, n, k) = \int_0^x \frac{dx}{(1 + nx^2)\sqrt{(1 - x^2)(1 - k^2 x^2)}} \tag{3-64}$$

$$= \int_0^\phi \frac{d\phi}{(1 + n \sin^2 \phi)\sqrt{1 - k^2 \sin^2 \phi}} \tag{3-65}$$

When $\phi = \pi/2$, we have the *complete elliptic integrals*,

$$K(k) = F\left(\frac{\pi}{2}, k\right) = \int_0^{\pi/2} \frac{d\phi}{\sqrt{1 - k^2 \sin^2 \phi}} \tag{3-66}$$

$$E(k) = E\left(\frac{\pi}{2}, k\right) = \int_0^{\pi/2} \sqrt{1 - k^2 \sin^2 \phi} \, d\phi \tag{3-67}$$

$$\Pi(n, k) = \Pi\left(\frac{\pi}{2}, n, k\right) = \int_0^{\pi/2} \frac{d\phi}{(1 + n \sin^2 \phi)\sqrt{1 - k^2 \sin^2 \phi}} \tag{3-68}$$

The formulas for reducing S to a standard form are given in various places, for example, Magnus *et al.* (M1) Section 10-1 or Abramowitz and Stegun (A1) Chapter 17. Elliptic integrals occur, among other places, in

1. The motion of the simple pendulum
2. Finding inductances of coils
3. Finding lengths of conic sections
4. Calculating the solid angles of circles seen obliquely

A large number of integrals are related to tabulated functions; these *integral representations* are sometimes very useful. For example,

$$\int_0^1 \frac{\cos tx \, dt}{\sqrt{1 - t^2}} = \frac{\pi}{2} J_0(x) \tag{3-69}$$

$$\int_0^\theta \frac{\cos(n + \frac{1}{2})\phi}{\sqrt{\cos \phi - \cos \theta}} \, d\phi = \frac{\pi}{\sqrt{2}} P_n(\cos \theta) \tag{3-70}$$

$J_0(x)$ is a suitably normalized solution of Bessel's equation, with $m = 0$, and is known as a Bessel function; P_n is a Legendre polynomial. We shall study such integral representations of the "special functions" in greater detail in Chapter 7.

3–5 APPROXIMATE EXPANSIONS

One can often obtain a useful expression for an integral by expanding the integrand in some sort of series.

EXAMPLE

$$\text{erf } x = \frac{2}{\sqrt{\pi}} \int_0^x e^{-t^2}\, dt$$

$$= \frac{2}{\sqrt{\pi}} \int_0^x \left(1 - t^2 + \frac{t^4}{2!} - \frac{t^6}{3!} + - \cdots \right) dt$$

$$= \frac{2}{\sqrt{\pi}} \left(x - \frac{x^3}{3} + \frac{x^5}{5 \cdot 2!} - \frac{x^7}{7 \cdot 3!} + - \cdots \right) \qquad (3\text{-}71)$$

This series converges for all x, but is only useful for small x, $(x \lesssim 1)$.

Integrations by parts are often useful.

EXAMPLE

Suppose we want erf x for large x.   As $x \to \infty$, erf $x \to 1$.   Let us compute the difference from 1.

$$1 - \text{erf } x = \frac{2}{\sqrt{\pi}} \int_x^\infty e^{-t^2}\, dt$$

We perform a sequence of integrations by parts.

$$\int_x^\infty e^{-t^2}\, dt = \frac{e^{-x^2}}{2x} - \int_x^\infty \frac{e^{-t^2}\, dt}{2t^2}$$

$$= \frac{e^{-x^2}}{2x} - \frac{e^{-x^2}}{4x^3} + \int_x^\infty \frac{3}{4} \frac{e^{-t^2}}{t^4}\, dt$$

etc.

The result after n integrations by parts is ($n > 1$):

$$\text{erf } x = 1 - \frac{2}{\sqrt{\pi}} e^{-x^2} \left[\frac{1}{2x} - \frac{1}{2^2 x^3} + \frac{1 \cdot 3}{2^3 x^5} - \frac{1 \cdot 3 \cdot 5}{2^4 x^7} + - \cdots \right.$$

$$\left. + (-1)^{n-1} \frac{1 \cdot 3 \cdot 5 \cdots (2n-3)}{2^n x^{2n-1}} \right]$$

$$+ (-1)^n \frac{1 \cdot 3 \cdot 5 \cdots (2n-1)}{2^n} \frac{2}{\sqrt{\pi}} \int_x^\infty \frac{e^{-t^2}}{t^{2n}} dt \qquad (3\text{-}72)$$

The terms in the brackets do not form the beginning of a convergent infinite series. The series does not converge for any x, since the individual terms eventually increase as n increases. Nevertheless, this expression with a finite number of terms is very useful for large x.

The expression (3-72) is exact if we include the "remainder," that is, the last term, containing the integral. This remainder alternates in sign as n increases, which means that the error after n terms in the series is smaller in magnitude than the next term. Thus the accuracy of the approximate expression in the brackets is highest if we stop one term before the smallest.

The series in brackets in (3-72) is an example of an *asymptotic series* if continued indefinitely.[8] The precise definition of an asymptotic series is the following:

$$S(z) = c_0 + \frac{c_1}{z} + \frac{c_2}{z^2} + \cdots \qquad (3\text{-}73)$$

is an asymptotic series expansion of $f(z)$ [written $f(z) \sim S(z)$, where $\sim$ reads "is asymptotically equal to"] provided, for any n, the error involved in terminating the series with the term $c_n z^{-n}$ goes to zero faster than z^{-n} as $|z|$ goes to ∞ for some range of arg z, that is,

$$\lim_{|z| \to \infty} z^n [f(z) - S_n(z)] = 0 \qquad (3\text{-}74)$$

for arg z in the given interval; $S_n(z)$ means $c_0 + c_1/z + \cdots + c_n/z^n$.

[A *convergent* series approaches $f(z)$ as $n \to \infty$ for given z, whereas an *asymptotic* series approaches $f(z)$ as $z \to \infty$ for given n.]

From the definition, it is easy to show that asymptotic series may be added, multiplied, and integrated to obtain asymptotic series for the sum, product, and integral of the corresponding functions. Also, the asymptotic expansion of a given function $f(z)$ is unique, but the reverse is not true. An asymptotic series does not specify a function $f(z)$ uniquely [see Whittaker and Watson (W5) Chapter VIII or Jeffreys and Jeffreys (J4) Chapter 17].

---

[8] The idea is probably due to Poincaré; see reference (P2) Chapter VIII.

We have not proved that our series in (3-72) is an asymptotic series, but that is not hard; we leave it to the interested reader.

Another *example*:

$$Ei(-x) = \int_{-\infty}^{-x} \frac{e^t}{t}\, dt = \int_{\infty}^{x} e^{-t}\frac{dt}{t} \qquad (x > 0)$$

$$= -\frac{e^{-x}}{x} - \int_{\infty}^{x} \frac{e^{-t}}{t^2}\, dt$$

$$= -\frac{e^{-x}}{x} + \frac{e^{-x}}{x^2} + 2\int_{\infty}^{x} \frac{e^{-t}}{t^3}\, dt$$

Continuing in this way, we obtain

$$-Ei(-x) = \frac{e^{-x}}{x}\left[1 - \frac{1}{x} + \frac{2!}{x^2} - \frac{3!}{x^3} + - \cdots + \frac{(-1)^n n!}{x^n}\right]$$

$$+ (-1)^n(n + 1)! \int_{\infty}^{x} \frac{e^{-t}}{t^{n+2}}\, dt \quad (3\text{-}75)$$

We can use this result in two ways:

1. As an asymptotic series for $Ei(-x)$:

$$-Ei(-x) \sim \frac{e^{-x}}{x}\left(1 - \frac{1}{x} + \frac{2!}{x^2} - \frac{3!}{x^3} + - \cdots\right) \qquad (3\text{-}76)$$

2. As an exact expression for computing certain integrals, assuming we have available a table of $Ei(-x)$. From (3-75)

$$\int_{x}^{\infty} \frac{e^{-t}\, dt}{t^n} = \frac{(-1)^n}{(n-1)!}\, Ei(-x) + \frac{(-1)^n}{(n-1)!}\frac{e^{-x}}{x}$$

$$\times \left[1 - \frac{1}{x} + \frac{2!}{x^2} - + \cdots + (-1)^n \frac{(n-2)!}{x^{n-2}}\right] \quad (3\text{-}77)$$

3–6 SADDLE-POINT METHODS

Other important methods for approximating integrals are known as *saddle-point methods,* for reasons which will soon be clear. The most important of these is the *method of steepest descent.* We shall illustrate the method by looking for an approximation to $\Gamma(x + 1)$ for x large, positive, and real.

$$\Gamma(x + 1) = \int_{0}^{\infty} t^x e^{-t}\, dt \qquad (3\text{-}78)$$

For large x, the integrand looks as shown in Figure 3–12. To find the maximum, we have

$$0 = \frac{d}{dt}(e^{-t}t^x) = e^{-t}[-t^x + xt^{x-1}]$$

$$t = x$$

We now approximate the integrand in a particular way, by writing it in an exponential form $e^{f(t)}$ and using a Taylor's series approximation for $f(t)$ near its maximum. (A Taylor's expansion of the integrand itself, for example, would not be useful if only a few terms were kept.) The integrand of (3-78) is

$$e^{f(t)} = t^x e^{-t} = e^{x \ln t - t}$$

so that

$$f(t) = x \ln t - t$$

$$f'(t) = \frac{x}{t} - 1 \qquad f' = 0 \text{ for } t = x$$

$$f''(t) = -\frac{x}{t^2}$$

Then, expanding about $t = x$,

$$\Gamma(x + 1) \approx \int_0^\infty \exp\left[x \ln x - x - \frac{1}{2x}(t - x)^2\right] dt$$

$$\approx e^{x \ln x - x} \int_{-\infty}^\infty \exp\left[-\frac{1}{2x}(t - x)^2\right] dt$$

(If $x \gg 1$, we make a very small error by extending the integral to $-\infty$.) Doing the t integral gives

$$\Gamma(x + 1) \approx \sqrt{2\pi x}\, x^x e^{-x}$$

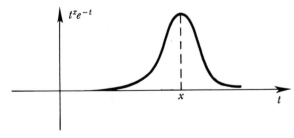

Figure 3–12   Graph of $t^x e^{-t}$ for large x

This is the first term of *Stirling's formula*, which is the asymptotic expansion of $n! = \Gamma(n + 1)$:

$$n! \sim \sqrt{2\pi n} \left(\frac{n}{e}\right)^n \left(1 + \frac{1}{12n} + \frac{1}{288n^2} + \cdots\right) \tag{3-79}$$

We shall return to this asymptotic series at the end of this chapter.

The method of steepest descents is applicable, in general, to integrals of the form

$$I(\alpha) = \int_C e^{\alpha f(z)} \, dz \qquad (\alpha \text{ large and positive}) \tag{3-80}$$

where C is a path in the complex plane such that the ends of the path do not contribute significantly to the integral. [This method usually gives the first term in an asymptotic expansion of $I(\alpha)$, valid for large α.]

If $f(z) = u + iv$, we would expect most of the contribution to $I(\alpha)$ to come from parts of the contour where u is largest. The idea of the method of steepest descents is to deform the contour C so that the region of large u is compressed into as short a space as possible.

To see how to deform the contour it is necessary to examine the behavior of the functions u and v in more or less detail for the example at hand. However, some general features apply to any regular function $f(z)$. Neither u nor v can have a maximum or minimum except at a singularity, because $\nabla^2 u = 0$ and $\nabla^2 v = 0$.[9]   For example, if

$$\frac{\partial^2 u}{\partial x^2} < 0 \qquad \text{then} \qquad \frac{\partial^2 u}{\partial y^2} > 0$$

so that a "flat spot" of the surface $u(x, y)$, where

$$\frac{\partial u}{\partial x} = \frac{\partial u}{\partial y} = 0 \tag{3-81}$$

must be a "saddle point," where the surface looks like a saddle or a mountain pass, as shown in Figure 3–13.

By the Cauchy–Riemann equations (A-8), we see that (3-81) implies $\partial v/\partial y = 0$ and $\partial v/\partial x = 0$, so that $f'(z) = 0$. Thus a saddle point of the function $u(x, y)$ is also a saddle point of $v(x, y)$ as well as a point where $f'(z) = 0$.

Near the saddle point z_0,

$$f(z) \approx f(z_0) + \tfrac{1}{2} f''(z_0)(z - z_0)^2$$

---

[9] These follow immediately from the Cauchy–Riemann equations, equations (A-8) of the Appendix.

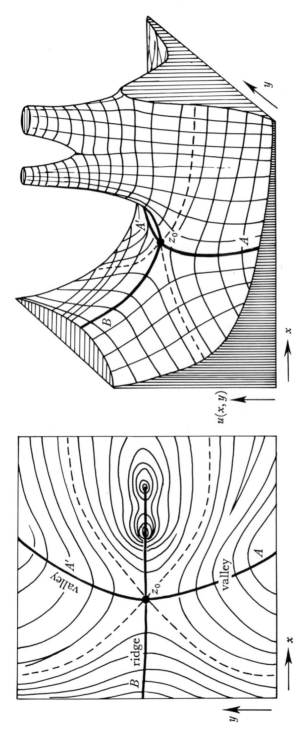

Figure 3-13 Topography of the surface $u = \mathrm{Re}\,f(z)$ near the saddle point z_0, for a typical function $f(z)$. The heavy solid curves follow the centers of the ridges and valleys from the saddle point, and the dashed curves follow level contours, $u = u(x_0, y_0) = \text{constant}$. The curve AA' is the path of steepest descent

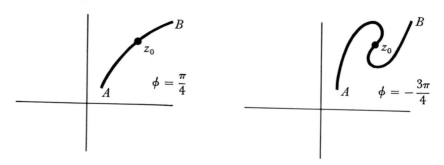

Figure 3–14   Alternative possibilities for steepest descent contours

Let $f''(z_0) = \rho e^{i\theta}$, $z - z_0 = s e^{i\phi}$.   Then

$$u \approx u(x_0, y_0) + \tfrac{1}{2}\rho s^2 \cos(\theta + 2\phi)$$

$$v \approx v(x_0, y_0) + \tfrac{1}{2}\rho s^2 \sin(\theta + 2\phi) \tag{3-82}$$

We see that on the surface $u(x, y)$ paths of steepest descent from the saddle point into the valleys start out in the directions where $\cos(\theta + 2\phi) = -1$.   In these directions, $v = v(x_0, y_0)$ is constant.   As we go further from z_0, the steepest path follows the direction $-\text{grad } u$, which is normal to the contours of equal u.   Thus the path follows a curve $v(x, y) = \text{constant} = v(x_0, y_0)$, so that the factor $e^{i\alpha v}$ in the original integrand of (3-80) will not produce destructive oscillations.   We therefore deform our original contour C so as to go over the pass on this path of steepest descent.   Using the approximation (3-82) in the integral (3-80) gives

$$I(\alpha) \approx e^{\alpha f(z_0)} \int_{-\infty}^{\infty} e^{-(\alpha/2)\rho s^2} e^{i\phi}\, ds$$

$$I(\alpha) \approx \sqrt{\frac{2\pi}{\alpha\rho}}\, e^{\alpha f(z_0)} e^{i\phi} \tag{3-83}$$

where ϕ has one of the values $-\theta/2 \pm \pi/2$, depending on which direction we travel over the pass.   (ϕ is just the inclination of the path at the saddle point.)   For example, if $\theta = \pi/2$ we have the two alternatives of Figure 3–14. The first is much more likely to be correct, but to be sure we should examine the "mountain range" we are passing through.   If it looks anything like that shown in part (a) of Figure 3–15, then $\phi = \pi/4$ is indeed correct.   On the other hand, if the range is like that shown in part (b) of Figure 3–15, the second alternative, $\phi = -3\pi/4$, is the right one.

As an example of the method of steepest descents with a more general contour, we shall work out the asymptotic approximation to $\Gamma(z + 1)$ for

complex z, without making any simplifications based on our previous work
with $\Gamma(x + 1)$.

$$\Gamma(z) = \int_0^\infty e^{-t}t^{z-1}\, dt$$

$$\Gamma(z + 1) = \int_0^\infty e^{-t+z\ln t}\, dt \tag{3-84}$$

Let

$$z = \alpha e^{i\beta} \tag{3-85}$$

Then

$$\Gamma(z + 1) = \int_0^\infty \exp\left[\alpha\left(\ln t - \frac{t}{z}\right)e^{i\beta}\right] dt$$

and

$$f(t) = \left(\ln t - \frac{t}{z}\right)e^{i\beta}$$

$$f'(t) = \left(\frac{1}{t} - \frac{1}{z}\right)e^{i\beta} \qquad t_0 = z$$

$$f''(t) = \frac{-1}{t^2}\, e^{i\beta}$$

$$f(t_0) = (\ln z - 1)e^{i\beta} \qquad f''(t_0) = -\frac{e^{i\beta}}{z^2} = \frac{-1}{\alpha^2}\, e^{-i\beta}$$

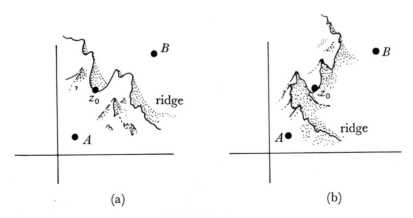

(a)   (b)

Figure 3–15 Alternative "mountain ranges" for the two cases of
Figure 3–14

Thus $\rho = 1/\alpha^2$, $\theta = \pi - \beta$.   What is ϕ?   Use of the condition $\cos(\theta + 2\phi) = -1$ gives

$$\phi = \frac{\beta}{2} \quad \text{or} \quad \frac{\beta}{2} - \pi$$

In our previous example, with z real, $\beta = 0$ and $\phi = 0$, so the choice $\phi = \beta/2$ would clearly seem most reasonable.   Since it is sometimes necessary to examine the topography of the surface $u = \text{Re}\, f(t)$ in more detail, however, we show the ridges and valleys for the present example, with $\beta = \pi/4$, in Figure 3–16.   This figure confirms the choice $\phi = \beta/2$.

The steepest descent approximation for $\Gamma(z + 1)$ may now be written down from (3-80):

$$\Gamma(z + 1) \sim \sqrt{2\pi\alpha}\, e^{z \ln z - z} e^{i\beta/2}$$

$$\Gamma(z + 1) \sim \sqrt{2\pi}\, z^{z+1/2} e^{-z} \tag{3-86}$$

The original integral representation for $\Gamma(z)$ is only valid for $\text{Re}\, z > 0$.

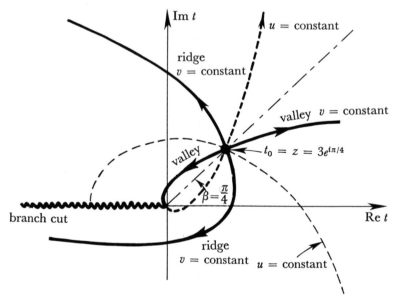

Figure 3–16   Topography of the surface $u = \text{Re}\, f(t) = \text{Re}\,(\ln t - t/z)e^{i\beta}$ for the case $z = 3e^{i(\pi/4)}$. It is clear that the original path of integration along the real t-axis should be deformed so as to go over the saddle point, $t_0 = z$, on the path of steepest descents (valleys) in a direction specified by the angle $\phi = \beta/2$

However, the result (3-86) is actually valid for all $|z| \to \infty$ provided we stay away from the negative real axis (see Whittaker and Watson (W5) Section 12.33).

The result (3-86) is just the first term of an asymptotic series for the gamma function, which we shall now find.   First write

$$\Gamma(z) \approx \sqrt{2\pi}\,(z-1)^{z-1/2}e^{-(z-1)}$$

$$\approx \sqrt{2\pi}\,z^{z-1/2}\left(1-\frac{1}{z}\right)^{z-1/2}e^{-(z-1)}$$

$$\approx \sqrt{2\pi}\,z^{z-1/2}e^{-z}$$

Let us now set

$$\Gamma(z) \sim \sqrt{2\pi}\,z^{z-1/2}e^{-z}\left(1+\frac{A}{z}+\frac{B}{z^2}+\cdots\right) \qquad (3\text{-}87)$$

The constants $A, B, \ldots$ may be found from the recursion relation for the gamma function:

$$\Gamma(z+1) = z\Gamma(z)$$

For, from (3-87),

$$\Gamma(z+1) = \sqrt{2\pi}\,(z+1)^{z+1/2}e^{-(z+1)}\left[1+\frac{A}{z+1}+\frac{B}{(z+1)^2}+\cdots\right]$$

$$= \sqrt{2\pi}\,\exp\left[\left(z+\frac{1}{2}\right)\log(z+1)-(z+1)\right]$$

$$\times\left[1+\frac{A}{z+1}+\frac{B}{(z+1)^2}+\cdots\right]$$

$$= \sqrt{2\pi}\,\exp\left[\left(z+\frac{1}{2}\right)\log z-z+\left(\frac{1}{12z^2}-\frac{1}{12z^3}+\frac{3}{40z^4}-+\cdots\right)\right]$$

$$\times\left[1+\frac{A}{z+1}+\frac{B}{(z+1)^2}+\cdots\right]$$

$$= \sqrt{2\pi}\,z^{z+1/2}e^{-z}\left(1+\frac{1}{12z^2}-\frac{1}{12z^3}+\frac{113}{1440z^4}-+\cdots\right)$$

$$\times\left[1+\frac{A}{z}+\frac{B-A}{z^2}+\frac{C-2B+A}{z^3}+\cdots\right]$$

On the other hand, (3-87) gives immediately

$$z\Gamma(z) = \sqrt{2\pi}\, z^{z+1/2} e^{-z}\left(1 + \frac{A}{z} + \frac{B}{z^2} + \cdots\right)$$

Equating corresponding terms in the two series we find

$$A = 1/12, \; B = 1/288, \ldots$$

This result agrees with the formula (3-79) previously quoted for $z = n$. We have not proved the validity of this heuristic procedure, but it gives the right answer.

One last procedure for approximately evaluating integrals should be mentioned. This is the so-called *method of stationary phase*; it is concerned with integrals of the form

$$I = \int_C e^{i\alpha f(z)}\, dz \tag{3-88}$$

where α is large and positive, and $f(z)$ is real along the contour C. Unless $f'(z) = 0$, the contributions to I from the neighborhood of z will largely cancel because of the rapidly oscillating character of $e^{i\alpha f(z)}$. Thus we look for points along the contour where $f'(z) = 0$, and use the result

$$\int_{-\infty}^{\infty} e^{i\alpha u^2}\, du = \sqrt{\frac{\pi}{\alpha}}\, e^{i(\pi/4)}$$

This method is clearly very closely related to the method of steepest descent. Note that if the integrands for the two methods are written in the same form, $e^{\alpha f(z)}$, the paths of integration are along a curve $\operatorname{Im} f(z) = $ constant for the method of steepest descent, and along a curve $\operatorname{Re} f(z) = $ constant for the method of stationary phase.

REFERENCES

Convenient collections of useful integrals are to be found in Dwight (D8) and Pierce and Foster (P1). Grobner and Hofreiter (G8) and Gradshteyn and Ryzhik (G6) give more extensive lists. The possibility of finding a given integral in a table of integral transforms, such as those of Erdelyi *et al.* (E6), should not be overlooked.

Many useful tables and graphs of functions defined in this and later chapters may be found in Jahnke *et al.* (J3), and Abramowitz and Stegun (A1). Additional properties of these functions are collected in Magnus *et al.* (M1) and Erdelyi *et al.* (E5). In particular, the reader who is faced with the problem of actually evaluating some elliptic integral will find convenient transformations and reduction formulas in Magnus *et al.* (M1) Chapter X;

Abramowitz and Stegun (A1) Section 17; Jahnke *et al.* (J3) Chapter V; Milne-Thomson(M7) pp. 26–38; and Hancock (H6).   Also, the lovely monograph by Artin (A6) on the gamma function should not be overlooked.

Evaluation of integrals by contour integration is treated in many books; among them are Copson (C8) Chapter VI; Whittaker and Watson (W5) Chapter III; and Morse and Feshbach (M9) Chapter 4.

Asymptotic expansions are treated by Jeffreys and Jeffreys (J4) Chapter 17; Whittaker and Watson (W5) Chapter VIII; and Smith (S5) Chapter 8. Two smaller books, devoted specifically to asymptotic expansions and related questions, are Erdelyi (E3) and de Bruijn (D4).   Saddle-point methods are discussed in the last-named reference, as well as by Jeffreys and Jeffreys (J4) Chapter 17; Smith (S5) Chapter 8; and Morse and Feshbach (M9) Chapter 4.

PROBLEMS

Evaluate the integrals in Problems 3-1 to 3-9:

3-1 $\displaystyle\int_0^\infty \frac{e^{-ay} - e^{-by}}{y}\, dy$

3-2 $\displaystyle\int_0^\infty \sin bx\, dx$ (apply a convergence factor; do integral; remove the convergence factor)

3-3 $\displaystyle\int_0^\infty \frac{\cos ax\, dx}{1 + x^2}$

3-4 $\displaystyle\int_0^\infty \frac{\cos ax\, dx}{(1 + x^2)^2}$

3-5 $\displaystyle\int d^3x\, e^{i\mathbf{a}\cdot\mathbf{x}} e^{-br^2}$ (the symbol d^3x stands for $dx\, dy\, dz$, or in general a volume element in three dimensions)

3-6 $\displaystyle\int d^3x\, \mathbf{x}\, e^{i\mathbf{a}\cdot\mathbf{x}} e^{-br^2}$

3-7 $\displaystyle\int_0^1 \frac{dx}{x} \ln\frac{1 + x}{1 - x}$

Hint: Expand the integrand in a power series.

3-8 $\displaystyle\int_0^\infty \frac{dx}{\cosh x}$

Hint: Expand the integrand in a series which is useful near $x = \infty$.

3-9 $\displaystyle\int_{-\infty}^\infty \frac{dx}{(1 + x^2)^2}$

Do the integrals 3-10 to 3-24 by contour integration:

3-10 $\displaystyle\int_0^\infty \frac{dx}{1 + x^4}$

3-11 $\displaystyle\int_{-\infty}^\infty \frac{e^{i\omega t}\, d\omega}{\omega^2 - \omega_0^2}$ (put poles slightly *above* real axis)

3-12 $\displaystyle\int_{-\infty}^\infty \frac{x^2\, dx}{(a^2 + x^2)^2}$

3-13 $\displaystyle\int \frac{d^3x}{(a^2 + r^2)^3}$

3-14 $\displaystyle\int_{-1}^{+1} \frac{dx}{\sqrt{1 - x^2}\,(a + bx)}$ $(a > b > 0)$

3-15 $\displaystyle\int_0^\infty \frac{x\, dx}{1 + x^5}$

3-16 $\displaystyle\int_0^{2\pi} \frac{\sin^2\theta\, d\theta}{a + b\cos\theta}$ $(a > |b|)$

3-17 $\displaystyle\int_{-\infty}^\infty \frac{\sinh ax}{\sinh \pi x}\, dx$

3-18 $\displaystyle\int_0^\infty \frac{(\ln x)^2}{1 + x^2}\, dx$

3-19 $\displaystyle\int_0^\infty \frac{dx}{1 + x^2 + x^4}$

3-20 $\displaystyle\int_0^\infty \frac{dx}{(a + bx^2)^3}$ $(a > 0, b > 0)$

3-21 $\displaystyle\int_0^\infty \frac{x^2\, dx}{(a^2 + x^2)^3}$

3-22 $\displaystyle\int_0^\infty \frac{\sin x\, dx}{x(a^2 + x^2)}$

3-23 $\displaystyle\int_0^{2\pi} \frac{d\theta}{(a + b\cos\theta)^2}$ $(a > b > 0)$

3-24 $\displaystyle\int_0^\infty \frac{\ln x \, dx}{(x+1)^2}$

3-25 Evaluate $\displaystyle\int_0^\infty e^{-x^2} \operatorname{Ci}(ax)\, dx$

3-26 Evaluate $\displaystyle\int_0^\infty e^{-ax} \operatorname{erf} x \, dx$

3-27 Evaluate (a) $\Gamma(\tfrac{1}{2})$ (d) $B(\tfrac{1}{2}, -\tfrac{1}{2})$
      (b) $\Gamma(\tfrac{5}{2})$ (e) $\Gamma(\tfrac{1}{3})\,\Gamma(-\tfrac{1}{3})$
      (c) $B(1, 3)$

3-28 Evaluate $\displaystyle\int_0^\infty e^{-sx}[-Ei(-x)]\, dx$

3-29 Evaluate $\displaystyle\int_{-2}^{2} (4 - x^2)^{1/6}\, dx$ in terms of beta and gamma functions.

3-30 Consider the integral $F(z) = \int_C dt(-t)^{z-1}e^{-t}$ where the t-plane is cut along the positive real axis, $(-t)^{z-1}$ is defined to equal $\exp[(z-1)\ln(-t)]$ with $\ln(-t)$ real on the negative real t-axis, and the path of integration C comes in from $t = +\infty$ below the cut, goes around the origin and returns to $t = +\infty$ above the cut. This integral defines the gamma function $\Gamma(z)$ throughout the complex plane [unlike the definition (3-46)]; more precisely, $F(z) = $ (something) $\times \Gamma(z)$. Evaluate (something).

3-31 (a) Consider the contour integral $\oint f(z)\operatorname{ctn}\pi z\, dz$ around a suitable large contour, and obtain thereby a formula for the sum

$$\sum_{n=-\infty}^{\infty} f(n)$$

(b) Evaluate

$$g(a) = \sum_{n=-\infty}^{\infty} \frac{1}{n^2 + a^2}$$

3-32 The absorption mean free path λ for neutrons in a certain material is measured by an absorption experiment as follows: A thin foil detector in the shape of a disk of radius b is irradiated at a distance a from a point source of neutrons, and its activity A_0 is measured. An absorber of thickness T is then placed between the source and foil as shown, and the activity A produced in the same length of time is measured. Find λ if $a = b = 12$ cm, $T = 1$ cm, and $A/A_0 = 0.25$.

Remarks:
1. Source emits neutrons isotropically.
2. Neglect scattering of neutrons.

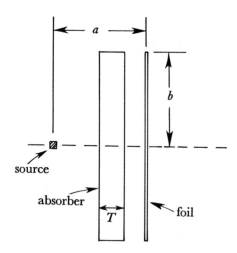

3. In traveling a distance d through absorber, neutron intensity is reduced by a factor $e^{-d/\lambda}$.

4. Activity produced by neutron in foil is proportional to distance traveled by neutron in foil.

3-33 Evaluate $\oint \Gamma(z)e^{\alpha z}\, dz$ around the contour $|z| = \frac{5}{2}$ once in a positive sense.

3-34 Obtain two expansions of Si x, one useful for small x and one useful for large x.

3-35 Evaluate $I(x) = \int_0^\infty dt\, e^{xt - e^t}$ approximately for large positive x.

3-36 Energy in a star is produced by nuclear reactions. The number of collisions with CM kinetic energy in the interval from E to $E + dE$ is

$$Ne^{-E/kT}E\, dE$$

per unit time, where k is Boltzmann's constant, T is the temperature, and N is a constant. The probability that a collision with CM kinetic energy E will result in a nuclear reaction is

$$Me^{-\alpha/\sqrt{E}}$$

where M and α are constants. Find an approximate expression for the total number of nuclear reactions per unit time, assuming

$$\left(\frac{kT}{\alpha^2}\right)^{1/3} \ll 1$$

3-37 Verify the result (3-31).

3-38 Find an asymptotic approximation for large x (real and positive) to the function

$$f_n(x) = \int_C e^{-ix\,\sin t + int}\,dt$$

where C is the contour shown below.

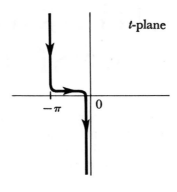

FOUR

INTEGRAL TRANSFORMS

In this chapter we discuss Fourier series, Fourier transforms, and Laplace transforms. Some other transform pairs are listed in Section 4–4, but not discussed. The use of integral transforms in solving problems is briefly illustrated by a few examples in Section 4–5. Further applications of integral transforms to the solution of partial differential equations and integral equations will be made later, in Chapters 8 and 11.

4–1 FOURIER SERIES

We begin by considering a function $f(\theta)$ defined for $0 \leqslant \theta < 2\pi$. We seek an expansion of the form

$$f(\theta) = \frac{A_0}{2} + \sum_{n=1}^{\infty} (A_n \cos n\theta + B_n \sin n\theta) \tag{4-1}$$

The coefficients may be found by multiplying both sides of (4-1) by $\cos n\theta$ (or $\sin n\theta$) and integrating from 0 to 2π:

$$A_n = \frac{1}{\pi} \int_0^{2\pi} f(\theta) \cos n\theta \, d\theta$$

$$\tag{4-2}$$

$$B_n = \frac{1}{\pi} \int_0^{2\pi} f(\theta) \sin n\theta \, d\theta$$

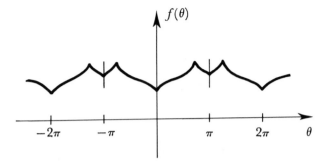

Figure 4–1   An even function of period 2π

The series (4-1), known as the *Fourier series* of $f(\theta)$, converges at all points to $\frac{1}{2}[f(\theta+) + f(\theta-)]$ provided $f(\theta)$ is of bounded variation[1] in $0 \leqslant \theta < 2\pi$. We have not worried about rigor in the above discussion, and refer the interested student to a mathematical reference book such as Whittaker and Watson (W5) Chapter IX or Apostol (A5) Chapter 15.

A Fourier series is periodic so that it repeats $f(\theta)$ in $2\pi \leqslant \theta < 4\pi$, $-2\pi \leqslant \theta < 0$, etc. We need not have started with the interval from 0 to 2π; any interval of length 2π would do. Often the interval from $-\pi$ to π is more convenient.

One should be able to recognize types of functions whose Fourier series obviously have certain terms missing.

1. Even functions: $f(-\theta) = f(\theta)$, or $f(2\pi - \theta) = f(\theta)$, etc. Only cosine terms occur, that is, $B_n = 0$. An example is shown in Figure 4–1.

2. Odd functions: $f(-\theta) = -f(\theta)$. Only sine terms occur, that is, $A_n = 0$. An example is shown in Figure 4–2.

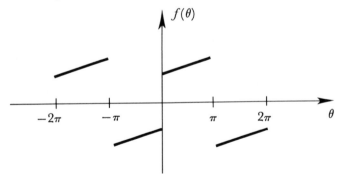

Figure 4–2   An odd function of period 2π

[1] For a function $f(x)$ defined on some interval, choose a set of n points x_i from that interval and evaluate V (the "variation") $= |f(x_1) - f(x_2)| + |f(x_2) - f(x_3)| + \cdots + |f(x_{n-1}) - f(x_n)|$. If there is a constant B, such that $V < B$ for all choices of x_i as $n \to \infty$, we say that $f(x)$ is of *bounded variation* on that interval.

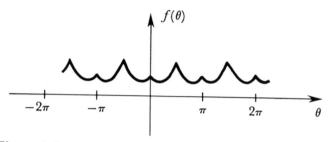

Figure 4-3   An even function symmetric about $\pi/2$

3. Even functions symmetric about $\pi/2 : f((\pi/2) + \theta) = f((\pi/2) - \theta)$. Only cosine terms with *even n* occur, that is, $B_n = 0$, $A_{2n+1} = 0$.   Figure 4-3 shows an example.

Obviously there are many such possibilities. We shall now consider a few examples of Fourier series.

EXAMPLE

$$f(\theta) = \begin{cases} +1 & 0 < \theta < \pi \\ -1 & \pi < \theta < 2\pi \end{cases} \qquad (4\text{-}3)$$

This is an odd funtion, so there are no cosine terms.   It is symmetric about $\pi/2$, which means that there are no even sine terms.   For odd n,

$$B_n = \frac{1}{\pi} \int_0^{2\pi} f(\theta) \sin n\theta \, d\theta = \frac{4}{\pi} \int_0^{\pi/2} f(\theta) \sin n\theta \, d\theta = \frac{4}{n\pi}$$

Therefore

$$f(\theta) = \frac{4}{\pi} \left(\sin \theta + \frac{\sin 3\theta}{3} + \frac{\sin 5\theta}{5} + \cdots \right) \qquad (4\text{-}4)$$

This series exhibits the nonuniformity of the convergence of a Fourier series near a discontinuity.   Successive approximations are illustrated in Figure 4-4.   Note the overshoot, which is called *Gibbs' phenomenon*.   In

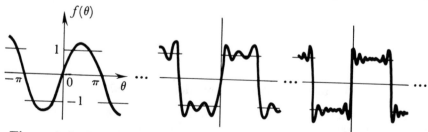

Figure 4-4   Partial sums of the Fourier series (4-4), with increasing number of terms

the limit of infinitely many terms, the overshoot remains finite—approximately 0.18 (see Problem 4-2).

If we set $\theta = \pi/2$ in the Fourier series (4-4), we obtain

$$\frac{\pi}{4} = 1 - \frac{1}{3} + \frac{1}{5} - \frac{1}{7} + - \cdots \tag{4-5}$$

a series known as *Gregory's series*. It may also be derived from the series

$$\tan^{-1} x = x - \frac{x^3}{3} + \frac{x^5}{5} - + \cdots$$

EXAMPLE

$$f(\theta) = \cos k\theta \qquad (-\pi < \theta < \pi)$$

The Fourier series for $\cos k\theta$ was already mentioned in connection with Bernoulli numbers (Section 2–3).

Since $f(\theta)$ is even, only cosine terms are present.

$$A_n = \frac{2}{\pi} \int_0^\pi \cos k\theta \cos n\theta \, d\theta$$

$$= \frac{(-1)^n 2k \sin k\pi}{\pi(k^2 - n^2)}$$

Thus

$$\cos k\theta = \frac{2k \sin k\pi}{\pi} \left(\frac{1}{2k^2} - \frac{\cos \theta}{k^2 - 1} + \frac{\cos 2\theta}{k^2 - 4} - + \cdots \right) \tag{4-6}$$

We can, of course, generalize Fourier series to represent functions which are periodic with some period L other than 2π. Let

$$x = \frac{\theta L}{2\pi}$$

Then an interval of length 2π in the variable θ becomes an interval of length L in the variable x. Our formulas become

$$f(x) = \frac{A_0}{2} + \sum_{n=1}^\infty \left(A_n \cos \frac{2\pi n x}{L} + B_n \sin \frac{2\pi n x}{L} \right)$$

$$A_n = \frac{2}{L} \int_0^L f(x) \cos \frac{2\pi n x}{L} \, dx \tag{4-7}$$

$$B_n = \frac{2}{L} \int_0^L f(x) \sin \frac{2\pi n x}{L} \, dx$$

All our remarks about evenness, oddness, and so on may be carried over to this new interval.

One must learn to be clear about what the fundamental interval L is for any particular problem. Suppose a function $f(x)$ is given to us in the interval $0 < x < a$. We may expand $f(x)$ in a Fourier series of period a:

$$f(x) = \frac{A_0}{2} + \sum_{n=1}^{\infty} \left(A_n \cos \frac{2\pi n x}{a} + B_n \sin \frac{2\pi n x}{a} \right) \tag{4-8}$$

Clearly we need both the cosines and sines for an arbitrary $f(x)$. In other words, sines or cosines (with period a) alone are *incomplete*.

On the other hand, we can expand *in sines alone* by defining $f(x)$ on $-a < x < 0$ by $f(-x) = -f(x)$. Now take $2a$ to be the period of our Fourier series, so that

$$f(x) = \sum_{n=1}^{\infty} B_n \sin \frac{n\pi x}{a} \tag{4-9}$$

Thus, so to speak, we have thrown away the cosines but doubled the number of sine terms, and we still have a complete set of functions. We could, of course, also expand $f(x)$ in a series containing only *cosines* of period $2a$ by. defining $f(-x) = +f(x)$ for $0 < x < a$.

It is often convenient to write Fourier series in complex form:

$$f(\theta) = \sum_{n=-\infty}^{\infty} a_n e^{in\theta} \tag{4-10}$$

or, more generally,

$$f(x) = \sum_{n=-\infty}^{\infty} a_n e^{2\pi i n x / L} \tag{4-11}$$

To evaluate the coefficients a_n in (4-11), multiply both sides by $e^{-2\pi i m x / L}$ and integrate from 0 to L (or any other interval of length L). Then we observe that

$$\int_0^L e^{-2\pi i m x / L} e^{2\pi i n x / L} \, dx = L \, \delta_{mn} \tag{4-12}$$

where δ_{mn} is called the *Kronecker delta* and is defined by

$$\delta_{mn} = \begin{cases} 0 & \text{if } m \neq n \\ 1 & \text{if } m = n \end{cases} \tag{4-13}$$

Therefore

$$\frac{1}{L} \int_0^L f(x) e^{-2\pi i m x / L} \, dx = \sum_{n=-\infty}^{\infty} a_n \delta_{mn} = a_m$$

Be sure to notice the change in sign in the exponential:

$$f(x) = \sum_{n=-\infty}^{\infty} a_n e^{2\pi inx/L} \qquad a_n = \frac{1}{L}\int_0^L f(x)e^{-2\pi inx/L}\,dx \qquad (4\text{-}14)$$

The quantity $(1/L)\int_0^L |f(x)|^2\,dx$, namely, the average absolute square of $f(x)$, is often of interest in physical applications. Expanding $f(x)$ in a complex Fourier series (4-11), we obtain[2]

$$\frac{1}{L}\int_0^L |f(x)|^2\,dx = \frac{1}{L}\int_0^L dx\left(\sum_n a_n e^{2\pi inx/L}\right)\left(\sum_m a_m^* e^{-2\pi imx/L}\right)$$

$$= \sum_{mn} a_n a_m^* \delta_{mn}$$

$$= \sum_n |a_n|^2 \qquad (4\text{-}15)$$

This result (4-15) shows that each Fourier component of $f(x)$ makes its separate contribution to the integral $\int_0^L |f(x)|^2\,dx$ independently of the other Fourier components. There are no *interference* terms of the form $a_n^* a_m$. [Compare the comments made below in connection with the example (4-26).]

4-2   FOURIER TRANSFORMS

We begin with the complex Fourier series of a function $f(x)$:

$$f(x) = \sum_n a_n e^{2\pi inx/L} \qquad a_n = \frac{1}{L}\int_{-L/2}^{L/2} f(x)e^{-2\pi inx/L}\,dx$$

We wish to consider the case $L \to \infty$. Then the sum may be converted into an integral as follows: define

$$\frac{2\pi n}{L} = y \qquad \text{and} \qquad La_n = g(y)$$

Then, since n increases in steps of unity in the sum,

$$\sum_{n=-\infty}^{\infty} F_n = \int_{-\infty}^{\infty} F_n\,dn = \frac{L}{2\pi}\int_{-\infty}^{\infty} F(y)\,dy$$

---

[2] Notice the general rule:

$$\sum_m (\text{anything})_m\,\delta_{mn} = (\text{anything})_n$$

provided the range of summation over m includes the value n; otherwise, of course, the sum is zero.

where $F(y) = F_n$. Our formulas thus become

$$f(x) = \frac{1}{2\pi} \int_{-\infty}^{\infty} g(y)e^{ixy}\, dy \qquad g(y) = \int_{-\infty}^{\infty} f(x)e^{-ixy}\, dx \qquad (4\text{-}16)$$

$g(y)$ is called the *Fourier transform* of $f(x)$, or vice versa. The position of the 2π is quite arbitrary; often one defines things more symmetrically,

$$f(x) = \sqrt{\frac{1}{2\pi}} \int_{-\infty}^{\infty} g(y)e^{ixy}\, dy \qquad g(y) = \sqrt{\frac{1}{2\pi}} \int_{-\infty}^{\infty} f(x)e^{-ixy}\, dx \quad (4\text{-}17)$$

We may combine the expressions (4-16) [or (4-17)] to obtain

$$f(x) = \frac{1}{2\pi} \int_{-\infty}^{\infty} dy\, e^{ixy} \int_{-\infty}^{\infty} f(x')e^{-ix'y}\, dx'$$

$$= \int_{-\infty}^{\infty} dx' f(x') \frac{1}{2\pi} \int_{-\infty}^{\infty} e^{i(x-x')y}\, dy \qquad (4\text{-}18)$$

The fact that (4-18) holds for *any* function $f(x)$ tells us something remarkable about the integral

$$\frac{1}{2\pi} \int_{-\infty}^{\infty} e^{i(x-x')y}\, dy$$

considered as a function of x'. It vanishes everywhere but at $x' = x$, and its integral with respect to x' over any interval including x is unity. That is, we may think of this function as having an infinitely high, infinitely narrow peak at $x' = x$.

It is conventional to define the (Dirac) *delta function*[3] $\delta(x)$ by

$$\delta(x) = 0 \qquad x \neq 0$$

$$\int_{-a}^{+b} \delta(x)\, dx = 1 \qquad a, b > 0 \qquad (4\text{-}19)$$

These equations imply

$$\int f(y)\, \delta(x - y)\, dy = f(x) \qquad (4\text{-}20)$$

for any function $f(x)$, provided the range of integration includes the point x.

Comparison of (4-18) and (4-20) now shows that

$$\delta(x) = \frac{1}{2\pi} \int_{-\infty}^{\infty} e^{ixy}\, dy \qquad (4\text{-}21)$$

which is an integral representation of the δ-function.

[3] See Dirac (D6) Section 15.

We may use the delta function to evaluate the important integral

$$\int_{-\infty}^{\infty} |f(x)|^2 \, dx$$

in terms of the Fourier transform (4-16) of $f(x)$, as follows:

$$\int_{-\infty}^{\infty} |f(x)|^2 \, dx = \int_{-\infty}^{\infty} dx \, \frac{1}{2\pi} \int_{-\infty}^{\infty} g^*(y) e^{-ixy} \, dy \, \frac{1}{2\pi} \int_{-\infty}^{\infty} g(y') e^{ixy'} \, dy'$$

$$= \frac{1}{2\pi} \int_{-\infty}^{\infty} dy \, g^*(y) \int_{-\infty}^{\infty} dy' g(y') \frac{1}{2\pi} \int_{-\infty}^{\infty} dx e^{i(y'-y)x}$$

$$= \frac{1}{2\pi} \int_{-\infty}^{\infty} dy \, g^*(y) \int_{-\infty}^{\infty} dy' g(y') \, \delta(y' - y)$$

$$= \frac{1}{2\pi} \int_{-\infty}^{\infty} dy \, g^*(y) \, g(y) \qquad (4\text{-}22)$$

This is called *Parseval's theorem*. It is useful in understanding the physical interpretation of the transform function $g(y)$ when the physical significance of $f(x)$ is known [see the discussion following (4-26)].

Suppose $f(x)$ is an even function. Then

$$g(y) = \int_0^{\infty} f(x) e^{-ixy} \, dx + \int_{-\infty}^0 f(x) e^{-ixy} \, dx$$

$$= \int_0^{\infty} f(x)(e^{ixy} + e^{-ixy}) \, dx$$

$$= 2 \int_0^{\infty} f(x) \cos xy \, dx \qquad (4\text{-}23)$$

Now note that $g(y)$ is even. Therefore

$$f(x) = \frac{1}{\pi} \int_0^{\infty} g(y) \cos xy \, dy \qquad (4\text{-}24)$$

$f(x)$ and $g(y)$, which now need only be defined for positive x and y, are called *Fourier cosine transforms* of each other.

By considering the Fourier transform of an odd function, we similarly obtain the relations between *Fourier sine transforms*

$$f(x) = \frac{1}{\pi} \int_0^{\infty} g(y) \sin xy \, dy \qquad g(y) = 2 \int_0^{\infty} f(x) \sin xy \, dx \qquad (4\text{-}25)$$

We may symmetrize by putting $\sqrt{2/\pi}$ before each integral in (4-23), (4-24), and (4-25) if we wish.

Extensive tables of Fourier transforms are given by Erdelyi *et al.* (E6).

EXAMPLE

$$f(t) = \begin{cases} 0 & (t < 0) \\ e^{-t/T} \sin \omega_0 t & (t > 0) \end{cases} \qquad (4\text{-}26)$$

This function is shown in Figure 4–5. $f(t)$ might represent the displacement of a damped harmonic oscillator, or the electric field in a radiated wave, or the current in an antenna, for example.

The Fourier transform of $f(t)$ is

$$g(\omega) = \int_{-\infty}^{\infty} f(t) e^{-i\omega t} \, dt$$

$$= \int_{0}^{\infty} e^{-t/T} e^{-i\omega t} \sin \omega_0 t \, dt$$

$$= \frac{1}{2} \left(\frac{1}{\omega + \omega_0 - \dfrac{i}{T}} - \frac{1}{\omega - \omega_0 - \dfrac{i}{T}} \right)$$

We may interpret the physical meaning of $g(\omega)$ with the help of Parseval's theorem (4-22). For example, if $f(t)$ is a radiated electric field, the radiated power is proportional to $|f(t)|^2$ and the total energy radiated is proportional to $\int_{0}^{\infty} |f(t)|^2 \, dt$. This is equal to

$$\frac{1}{2\pi} \int_{-\infty}^{\infty} |g(\omega)|^2 \, d\omega$$

by Parseval's theorem. Then $|g(\omega)|^2$ may be interpreted as the energy radiated per unit frequency interval (times some constant).

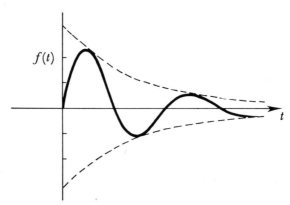

Figure 4–5   A damped sine wave

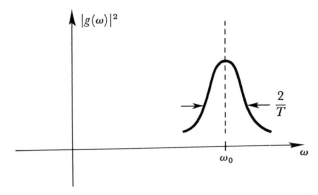

Figure 4–6   Energy spectrum for the damped oscillation of Figure 4–5

Let us assume that T is rather large ($\omega_0 T \gg 1$). Then our "frequency spectrum" $g(\omega)$ is sharply peaked near $\omega = \pm\omega_0$. For example, near $\omega = \omega_0$,

$$g(\omega) \approx -\frac{1}{2} \frac{1}{\omega - \omega_0 - \dfrac{i}{T}}$$

$$|g(\omega)| \approx \frac{1}{2} \frac{1}{\sqrt{(\omega - \omega_0)^2 + \dfrac{1}{T^2}}}$$

When $\omega = \omega_0 \pm 1/T$, the "amplitude" $g(\omega)$ is down by a factor $\sqrt{1/2}$, and the radiated energy $|g(\omega)|^2$ is down by a factor $1/2$. In other words, the width Γ at half (power) maximum is given by $\Gamma \approx 2/T$. The energy spectrum $|g(\omega)|^2$ is sketched in Figure 4–6.

This result is a typical *uncertainty principle* and is very closely related to the Heisenberg uncertainty principle in quantum mechanics. The length of time (T) during which something oscillates is inversely proportional to the width Γ, which is a measure of the "uncertainty" in frequency.

Fourier transforms may easily be generalized to more than one dimension. For example, in three-dimensional space we have the transform pair

$$\phi(\mathbf{k}) = \int d^3x \, f(\mathbf{x}) e^{-i\mathbf{k}\cdot\mathbf{x}}$$

$$f(\mathbf{x}) = \int \frac{d^3k}{(2\pi)^3} \, \phi(\mathbf{k}) e^{i\mathbf{k}\cdot\mathbf{x}} \tag{4-27}$$

From these we may deduce, as before, the integral representation

$$\delta(\mathbf{x}) = \int \frac{d^3k}{(2\pi)^3} \, e^{i\mathbf{k}\cdot\mathbf{x}} \tag{4-28}$$

where the three-dimensional delta function is defined by

$$\delta(\mathbf{x}) = 0 \qquad \mathbf{x} \neq 0$$

$$\int d^3x\, \delta(\mathbf{x}) = 1 \qquad \begin{array}{l}\text{provided origin is inside region of} \\ \text{integration}\end{array} \qquad (4\text{-}29)$$

$$\int d^3x f(\mathbf{x})\, \delta(\mathbf{x} - \mathbf{y}) = f(\mathbf{y})$$

Such transform pairs are of interest in quantum mechanics. If $f(\mathbf{x})$ is the wave function of a particle, the Fourier transform $\phi(\mathbf{k})$ is the so-called "wave function in momentum space." $|f(\mathbf{x})|^2$ and $|\phi(\mathbf{k})|^2$ are the probability distributions for the position and momentum, respectively.

EXAMPLE

$$f(\mathbf{x}) = \left(\frac{2}{\pi a^2}\right)^{3/4} e^{-r^2/a^2} = N e^{-r^2/a^2} \qquad (r = |\mathbf{x}|) \qquad (4\text{-}30)$$

This is a wave function which gives a Gaussian[4] probability distribution, $|f(\mathbf{x})|^2$, centered at $r = 0$, and normalized so that $\int d^3x\, |f(\mathbf{x})|^2 = 1$. The Fourier transform is

$$\phi(\mathbf{k}) = N \int d^3x\, e^{-r^2/a^2} e^{-i\mathbf{k}\cdot\mathbf{x}}$$

Introduce polar coordinates, with the z axis along $\mathbf{k}$. Let $\cos\theta = \alpha$. Then

$$\phi(\mathbf{k}) = 2\pi N \int_0^\infty r^2\, dr \int_{-1}^{+1} d\alpha\, e^{-r^2/a^2} e^{-ikr\alpha}$$

$$= \frac{4\pi}{k} N \int_0^\infty r\, dr\, e^{-r^2/a^2} \sin kr$$

$$= \frac{4\pi}{k} \frac{N}{2i} \int_{-\infty}^\infty r\, dr\, e^{-r^2/a^2} e^{ikr}$$

$$= \frac{2\pi}{ik} N e^{-k^2a^2/4} \int_{-\infty}^\infty r\, dr \exp\left[-\frac{1}{a^2}\left(r - \frac{ika^2}{2}\right)^2 \right]$$

$$= \frac{2\pi}{ik} N e^{-k^2a^2/4} \int_{-\infty}^\infty \left(y + \frac{ika^2}{2}\right) dy\, e^{-y^2/a^2}$$

$$= \frac{2\pi}{ik} N e^{-k^2a^2/4} \frac{ika^2}{2} a\sqrt{\pi}$$

---

[4] Functions of the form $A e^{-Bx^2}$ (A, B constant) are often called *Gaussian* functions because of their occurrence in the least-squares method of data analysis, which originated with Gauss. See also the footnote on p. 378.

Finally, remembering $N = (2/\pi a^2)^{3/4}$, we obtain

$$\phi(\mathbf{k}) = (2\pi a^2)^{3/4} e^{-k^2 a^2/4} \qquad (4\text{-}31)$$

Note that the Fourier transform of a Gaussian is another Gaussian.   The narrower the distribution in $\mathbf{x}$ (that is, the smaller the value of a), the broader is the distribution in $\mathbf{k}$.   The widths Δx and Δk are roughly inverse to each other,

$$\Delta x \Delta k \approx 1$$

The prescription $\mathbf{p} = \hbar \mathbf{k}$ from quantum mechanics ($\mathbf{p} =$ momentum) converts his into another quantum mechanical uncertainty relation

$$\Delta x \Delta p \approx \hbar$$

The probability distribution $|\phi(\mathbf{k})|^2$ found above is a Gaussian centered at $k = 0$.   We might expect that a similar distribution centered at an arbitrary k_0 could be obtained, starting with the same probability distribution in position $|f(\mathbf{r})|^2$.   This may be done by simply multiplying $f(\mathbf{x})$ by a factor $e^{i k_0 \cdot \mathbf{x}}$, as the reader may easily verify.

4–3   LAPLACE TRANSFORMS

We are often interested in functions whose Fourier transforms do not exist.   For example, the simple functions $f(x) = A = $ constant, and $f(x) = x^2$, have Fourier transform integrals which do not converge.   For many functions, the trouble at $x \to +\infty$ may be fixed by multiplication with a factor e^{-cx} if c is real and larger than some minimum value α.   This factor may make the behavior at $x \to -\infty$ worse, but we are often interested in a function only for positive x.   Thus we can take care of the behavior for negative x by a second factor, the (Heaviside) step function:

$$H(x) = \begin{cases} 0 & x < 0 \\ 1 & x > 0 \end{cases} \qquad (4\text{-}32)$$

[Note that $(d/dt)H(t) = \delta(t)$, as can be shown by integrating the δ function.]
The function $f(x)e^{-cx}H(x)$ now has the Fourier transform:

$$g(y) = \int_{-\infty}^{\infty} f(x)e^{-cx}H(x)e^{-ixy}\,dx = \int_{0}^{\infty} f(x)e^{-cx}e^{-ixy}\,dx$$

The inverse transform is

$$f(x)e^{-cx}H(x) = \frac{1}{2\pi}\int_{-\infty}^{\infty} g(y)e^{ixy}\,dy$$

It is conventional to introduce a new variable,

$$s = c + iy \tag{4-33}$$

and to define $F(s) \equiv g(y)$. The above two integrals become

$$F(s) = \int_0^\infty f(x)e^{-sx}\, dx \tag{4-34}$$

$$f(x)H(x) = \frac{1}{2\pi i} \int_C F(s)e^{sx}\, ds \tag{4-35}$$

where the path of integration C is upward along the straight line, Re $s = c =$ constant (Figure 4–7).

$F(s)$, as given by (4-34), is called the *Laplace transform* of $f(x)$. The integral exists only in the "right-half" s-plane, Re $s > \alpha$, where α is the minimum value for c mentioned above. In this region, $F(s)$ is analytic; $F(s)$ may usually be defined in the left-half plane by analytic continuation.

The second integral (4-35) is called the Laplace *inversion integral*. Note that for $x > 0$ it gives $f(x)$ {or more precisely $\frac{1}{2}[f(x+) + f(x-)]$} but for $x < 0$ it automatically gives zero; when $x < 0$ the contour C may be closed by the addition of a large semicircle on the *right*, where $F(s)$ is analytic.

We shall hereafter omit the $H(x)$ in (4-35), and write simply

$$f(x) = \frac{1}{2\pi i} \int_{c-i\infty}^{c+i\infty} F(s)e^{sx}\, ds \tag{4-36}$$

with the understanding that all functions $f(x)$ which are to be Laplace transformed vanish for negative arguments.

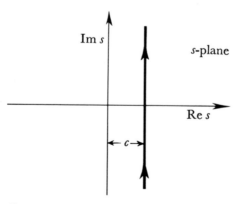

Figure 4–7   Contour for the Laplace inversion integral (4-35)

EXAMPLE

$$f(x) = 1$$

$$F(s) = \int_0^\infty f(x)e^{-sx}\, dx = \int_0^\infty e^{-sx}\, dx = \frac{1}{s}$$

(4-37)

where the integral exists for Re $s > 0$ (that is, $\alpha = 0$).   Note that $F(s)$ has a singularity (a simple pole in this example) on the limiting line, Re $s = 0$.
We may verify the inversion formula:

$$f(x) = \frac{1}{2\pi i} \int_{c-i\infty}^{c+i\infty} F(s)e^{sx}\, ds = \frac{1}{2\pi i} \int_{c-i\infty}^{c+i\infty} \frac{e^{sx}}{s}\, ds$$

where $c > 0$.

If $x > 0$, we complete the contour by a large semicircle to the *left*, and

$$f(x) = 1$$

If $x < 0$, we complete the contour to the *right*, and

$$f(x) = 0$$

A short table of Laplace transforms appears at the end of this chapter. More complete tables may be found in Erdelyi *et al.* (E6); Magnus *et al.* (M1); and elsewhere.

4–4   OTHER TRANSFORM PAIRS

Fourier–Bessel transforms:
(or Hankel transform)

$$g(k) = \int_0^\infty f(x)J_m(kx)\, x\, dx$$

(4-38)

$$f(x) = \int_0^\infty g(k)J_m(kx)\, k\, dk$$

(4-39)

Mellin transform:

$$\phi(z) = \int_0^\infty t^{z-1}f(t)\, dt$$

(4-40)

$$f(t) = \frac{1}{2\pi i} \int_{-i\infty}^{i\infty} t^{-z}\phi(z)\, dz$$

(4-41)

Hilbert transform:

$$g(y) = \frac{1}{\pi} P \int_{-\infty}^{\infty} \frac{f(x)\, dx}{x - y}$$

(4-42)

$$f(x) = \frac{1}{\pi} P \int_{-\infty}^{\infty} \frac{g(y)\, dy}{y - x}$$

(4-43)

[P denotes that the (Cauchy) *principal value* of the integral is to be taken; see Appendix, Section A–2, number 8.]

4–5 APPLICATIONS OF INTEGRAL TRANSFORMS

We shall first summarize some basic properties of Fourier and Laplace transforms:

1. Both transforms are *linear*, that is, the transform of $\alpha f(x) + \beta g(x)$ equals α times the transform of $f(x)$ plus β times the transform of $g(x)$.

2. *Derivatives*:

$$\int_0^\infty e^{-sx} f'(x)\, dx = e^{-sx} f(x) \Big|_0^\infty + s \int_0^\infty e^{-sx} f(x)\, dx$$

We shall write $\mathscr{L}[f(x), s]$ or $\mathscr{L}[f(x)]$ for the Laplace transform of $f(x)$, and similarly $\mathscr{F}[f(x)]$ for the Fourier transform. Then our relation above may be written

$$\mathscr{L}[f'(x)] = s\mathscr{L}[f(x)] - f(0) \tag{4-44}$$

Similarly,

$$\mathscr{L}[f''(x)] = s^2 \mathscr{L}[f(x)] - sf(0) - f'(0) \tag{4-45}$$

and so forth (note that 0 really means $0+$, the limit as zero is approached from the *positive* side). For Fourier transforms the integrated parts vanish, and

$$\mathscr{F}[f'(x)] = iy\mathscr{F}[f(x)] \qquad \text{etc.} \tag{4-46}$$

3. *Integrals*:

$$\mathscr{L}\left[\int_0^x f(t)\, dt\right] = \int_0^\infty dx\, e^{-sx} \int_0^x f(t)\, dt$$

$$= \int_0^\infty dt \int_t^\infty dx\, f(t)\, e^{-sx}$$

$$= \frac{1}{s} \int_0^\infty dt\, f(t)\, e^{-st}$$

Thus

$$\mathscr{L}\left[\int_0^x f(t)\, dt\right] = \frac{1}{s}\mathscr{L}[f(t)] \tag{4-47}$$

For Fourier transforms, things are not quite so simple. Suppose $g(x) = \int f(x)\, dx$ is an indefinite integral of $f(x)$. Then, from (4-46),

$$\mathscr{F}[f(x)] = iy\mathscr{F}[g(x)]$$

However, we cannot immediately conclude that

$$\mathscr{F}[g(x)] = \frac{\mathscr{F}[f(x)]}{iy}$$

Why not?   Consider the equation $xf(x) = g(x)$.   Can we conclude that

$$f(x) = \frac{g(x)}{x}$$

No, not at $x = 0$.   The general result is

$$f(x) = \frac{g(x)}{x} + C\,\delta(x)$$

where C is an arbitrary constant.   Thus

$$\mathscr{F}\left[\int f(x)\,dx\right] = \frac{\mathscr{F}[f(x)]}{iy} + C\,\delta(y) \tag{4-48}$$

This arbitrariness is also obvious from the fact that $\int f(x)\,dx$ is uncertain to within an arbitrary additive constant C, and

$$\mathscr{F}[C] = 2\pi C\,\delta(y)$$

4. *Translation*:

$$\mathscr{F}[f(x+a)] = \int_{-\infty}^{\infty} f(x+a)\,e^{-ixy}\,dx$$

$$= \int_{-\infty}^{\infty} f(x)\,e^{-iy(x-a)}\,dx$$

Therefore

$$\mathscr{F}[f(x+a)] = e^{iay}\mathscr{F}[f(x)] \tag{4-49}$$

For Laplace transforms, we must be a little more careful.   Consider the cases $a > 0$, $a < 0$ separately.

For $a > 0, f(x+a)$ is shown in Figure 4–8b.

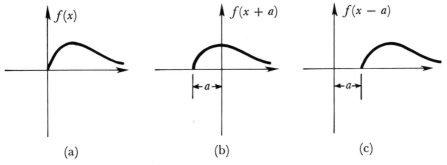

Figure 4–8 $f(x)$ translated to the left, $f(x+a)$, and to the right, $f(x-a)$

Since the Laplace transform ignores $f(x)$ for $x < 0$ (in fact, assumes it is zero), we must chop off some of our function, and

$$\mathscr{L}[f(x + a)] = \int_0^\infty f(x + a)\, e^{-sx}\, dx = \int_a^\infty f(x)\, e^{-s(x-a)}\, dx$$

so that

$$\mathscr{L}[f(x + a)] = e^{as}\left\{\mathscr{L}[f(x)] - \int_0^a f(x)\, e^{-sx}\, dx\right\} \qquad a > 0 \qquad (4\text{-}50)$$

On the other hand, $f(x - a)$ is shown in Figure 4-8c, and

$$\mathscr{L}[f(x - a)] = \int_0^\infty f(x - a)\, e^{-sx}\, dx$$

$$= \int_{-a}^\infty f(x)\, e^{-s(x+a)}\, dx$$

Therefore,

$$\mathscr{L}[f(x - a)] = e^{-as}\mathscr{L}[f(x)] \qquad a > 0 \qquad (4\text{-}51)$$

5. *Multiplication by an exponential*:
The following two formulas are easily verified:

$$\mathscr{F}[e^{\alpha x}f(x); y] = \mathscr{F}[f(x); y + i\alpha] \qquad (4\text{-}52)$$

$$\mathscr{L}[e^{\alpha x}f(x); s] = \mathscr{L}[f(x); s - \alpha] \qquad (4\text{-}53)$$

6. *Multiplication by a power of x*:
If

$$g(y) = \int_{-\infty}^\infty f(x)\, e^{-ixy}\, dx$$

then

$$g'(y) = -i \int_{-\infty}^\infty x f(x)\, e^{-ixy}\, dx$$

Thus

$$\mathscr{F}[xf(x)] = i\frac{d}{dy}\mathscr{F}[f(x)] \qquad (4\text{-}54)$$

An analogous result is easily shown to hold for Laplace transforms:

$$\mathscr{L}[xf(x)] = -\frac{d}{ds}\mathscr{L}[f(x)] \qquad (4\text{-}55)$$

7. *Convolution theorems*:

Let $f_1(x)$, $f_2(x)$ be two arbitrary functions.   We define their *convolution* (*faltung* in German) to be

$$g(x) = \int_{-\infty}^{\infty} f_1(y) f_2(x - y) \, dy \qquad (4\text{-}56)$$

What is the Fourier transform of such a convolution?   A straightforward change of variable shows that

$$\mathscr{F}[g(x)] = \mathscr{F}[f_1(x)]\mathscr{F}[f_2(x)] \times \text{constant} \qquad (4\text{-}57)$$

The value of the constant in (4-57) depends on our convention for Fourier transforms, that is, whether we use (4-16) or (4-17) and, if (4-16), which of $f(x)$ and $g(y)$ is considered the original function and which the Fourier transform.   The student is urged to evaluate the constant in (4-57) for at least one convention.   In any case, the important result is that, to within some constant, the Fourier transform of a convolution is the product of the Fourier transforms of the "factors" of the convolution.

An analogous result holds for Laplace transforms; if

$$g(x) = \int_{0}^{x} dt \, f_1(t) \, f_2(x - t) \qquad (4\text{-}58)$$

then

$$\mathscr{L}[g(x)] = \mathscr{L}[f_1(x)]\mathscr{L}[f_2(x)]$$

An interesting converse relation holds for Laplace transforms.   Suppose

$$\mathscr{L}[f_1] = g_1(s) \qquad \text{and} \qquad \mathscr{L}[f_2] = g_2(s)$$

where the Laplace integrals for $g_1(s)$ and $g_2(s)$ exist for $\operatorname{Re} s > \alpha_1$ and $\operatorname{Re} s > \alpha_2$, respectively.   Then the Laplace transform of the product $f_1 f_2$ is given by

$$\mathscr{L}[f_1 f_2] = \frac{1}{2\pi i} \int_{c-i\infty}^{c+i\infty} g_1(z) g_2(s - z) \, dz \qquad (4\text{-}59)$$

where the path of integration is along the line $\operatorname{Re} z = c$, with $(\operatorname{Re} s - \alpha_2) > c > \alpha_1$.   This result may be obtained by substituting the Laplace inversion integral for $f_1(z)$ in the integral $\mathscr{L}[f_1 f_2]$.

The corresponding relation for Fourier transforms is[5]

$$\mathscr{F}[f_1 f_2] = \frac{1}{2\pi} \int_{-\infty}^{\infty} g_1(z) g_2(y - z) \, dz \qquad (4\text{-}60)$$

[5] Again, the factor $1/2\pi$ depends on one's convention for Fourier transforms; see remarks after Eq. (4-57).

We now consider several examples of the application of integral transforms. The general procedure is to "transform" the problem, or restate it in terms of the transform function, and hope that in this form it is easier to solve.

EXAMPLE

Find the current in the circuit of Figure 4–9 if the switch is closed at time $t = 0$, and the initial charge on the condenser is Q_0.

$$RI + L\frac{dI}{dt} + \frac{Q}{C} = E_0 \tag{4-61}$$

Strictly speaking, we should write $E_0 H(t)$, where $H(t)$ is the step function (4-32), but in dealing with Laplace transforms we assume everything vanishes for $t < 0$ anyway.

Since $dQ/dt = I$, $Q(t) = Q_0 + \int_0^t I(t')\,dt'$, and (4-61) may be written

$$RI + L\frac{dI}{dt} + \frac{1}{C}\left[Q_0 + \int_0^t I(t')\,dt'\right] = E_0$$

Now take the Laplace transform of both sides.   Let $\mathscr{L}[I(t)] = i(s)$:

$$Ri(s) + L[si(s) - I(0)] + \frac{1}{C}\left[\frac{Q_0}{s} + \frac{i(s)}{s}\right] = \frac{E_0}{s}$$

Solving for $i(s)$ gives (by simple algebra!)

$$i(s) = \frac{E_0 - \dfrac{Q_0}{C}}{L} \frac{1}{(s+a)^2 + b^2}$$

where

$$a = \frac{R}{2L} \qquad b = \sqrt{\frac{1}{LC} - \frac{R^2}{4L^2}} \qquad \text{[and } I(0) = 0\text{]}$$

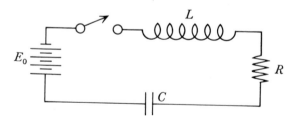

Figure 4–9   Series *RLC* circuit

But

$$\mathcal{L}[e^{-at}\sin bt] = \frac{b}{(s+a)^2 + b^2}$$

Therefore

$$I(t) = \frac{E_0 - \dfrac{Q_0}{C}}{L}\, \frac{e^{-at}\sin bt}{b} \tag{4-62}$$

EXAMPLE

Consider the coupled pendulums shown in Figure 4–10.   Assume the

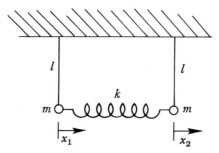

Figure 4–10   Coupled pendulums

initial conditions $x_1 = x_2 = 0$, $\dot{x}_1 = v$, $\dot{x}_2 = 0$ at $t = 0$.   Newton's equations
are

$$m\ddot{x}_1 = -\frac{mg}{l}x_1 + k(x_2 - x_1)$$

$$m\ddot{x}_2 = -\frac{mg}{l}x_2 + k(x_1 - x_2) \tag{4-63}$$

Let $\mathcal{L}[x_i(t)] = F_i(s)$.   Then the Laplace transforms of our two differential
equations (4-63) are

$$m(s^2 F_1 - v) = -\frac{mg}{l}F_1 + k(F_2 - F_1)$$

$$ms^2 F_2 = -\frac{mg}{l}F_2 + k(F_1 - F_2)$$

We must now solve these simultaneous algebraic equations for F_1 and F_2. We find

$$F_1(s) = \frac{v\left(s^2 + \dfrac{g}{l} + \dfrac{k}{m}\right)}{\left(s^2 + \dfrac{g}{l} + 2\dfrac{k}{m}\right)\left(s^2 + \dfrac{g}{l}\right)}$$

$$= \frac{v}{2}\left(\frac{1}{s^2 + \dfrac{g}{l} + 2\dfrac{k}{m}} + \frac{1}{s_2 + \dfrac{g}{l}}\right)$$

Therefore,

$$x_1(t) = \frac{v}{2}\left(\frac{\sin\sqrt{\dfrac{g}{l} + 2\dfrac{k}{m}}\,t}{\sqrt{\dfrac{g}{l} + 2\dfrac{k}{m}}} + \frac{\sin\sqrt{\dfrac{g}{l}}\,t}{\sqrt{\dfrac{g}{l}}}\right)$$

Similarly, we can work out $x_2(t)$. Note that $\sqrt{g/l + 2(k/m)}$ and $\sqrt{g/l}$ are the (angular) frequencies of the two normal modes.

We shall conclude this chapter by carrying out an alternative derivation of the WKB connection formula (1-113) given in Chapter 1, relating the exponentially small solution on one side of a turning point to an oscillatory solution on the other side. Consider the differential equation

$$\frac{d^2y}{dx^2} + xy = 0 \tag{4-64}$$

This equation has a solution $y(x)$ which looks something like the sketch in Figure 4–11. Let

$$g(\omega) = \int_{-\infty}^{\infty} y(x)e^{-i\omega x}\, dx$$

Then the Fourier transform of our differential equation (4-64) is

$$-\omega^2 g(\omega) + i\frac{dg}{d\omega} = 0$$

which gives, inverting the transform,

$$g(\omega) = Ae^{-i(\omega^3/3)} \qquad (A = \text{arbitrary const.})$$

Thus

$$y(x) = A\int_{-\infty}^{\infty} \frac{d\omega}{2\pi} \exp\left[i\left(\omega x - \frac{\omega^3}{3}\right)\right] \tag{4-65}$$

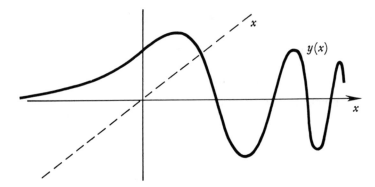

Figure 4–11   Sketch of the solution of (4-64), which has a decreasing exponential character for $x < 0$

We shall drop the irrelevant constant $A/2\pi$.

Unfortunately, this integral is nontrivial.   Partly for this reason it has a name, being known as an *Airy integral*.   For the present application of finding the WKB connection formula, we only need the asymptotic forms for large $|x|$, which may be found by the saddle-point method of Section 3–6. The two cases $x \to \pm\infty$ must be considered separately.

1. Let $x \to +\infty$.   The saddle points occur on the real axis, and we must go over both of them.   Using the notation introduced in Section 3–6.

$$f(\omega) = i\left(\omega - \frac{\omega^3}{3x}\right) \qquad f'(\omega) = i\left(1 - \frac{\omega^2}{x}\right) \qquad f''(\omega) = -\frac{2i\omega}{x}$$

$$f'(\omega_0) = 0 \Rightarrow \omega_0 = \pm\sqrt{x}$$

$$\rho e^{i\theta} = \mp\frac{2i}{\sqrt{x}} \qquad \rho = \frac{2}{\sqrt{x}} \qquad \theta = \mp\frac{\pi}{2}$$

$$\phi = \mp\frac{\pi}{4}$$

$$f(\omega_0) = \pm\tfrac{2}{3}i\sqrt{x}$$

Therefore,

$$y(x) \sim \sum_{\pm} \sqrt{\frac{2\pi\sqrt{x}}{2x}} \exp\left[\pm\tfrac{2}{3}ix^{3/2}\right] \exp\left[\mp\frac{i\pi}{4}\right]$$

$$y(x) \sim \frac{2\sqrt{\pi}}{x^{1/4}} \cos\left(\tfrac{2}{3}x^{3/2} - \frac{\pi}{4}\right) \qquad\qquad (4\text{-}66)$$

2. Let $x \to -\infty$.   Now the large positive parameter is $-x$, and

$$f(\omega) = -i\left(\omega - \frac{\omega^3}{3x}\right) \qquad f'(\omega) = -i\left(1 - \frac{\omega^2}{x}\right) \qquad f''(\omega) = \frac{2i\omega}{x}$$

$$f'(\omega_0) = 0 \Rightarrow \omega_0 = \pm i\sqrt{-x}$$

The saddle points now occur on the imaginary axis.

$$\rho e^{i\theta} = \frac{2i}{x}\left(\pm i\sqrt{-x}\right) = \frac{\pm 2}{\sqrt{-x}}$$

$$\rho = \frac{2}{\sqrt{-x}} \qquad \theta = \binom{0}{\pi}$$

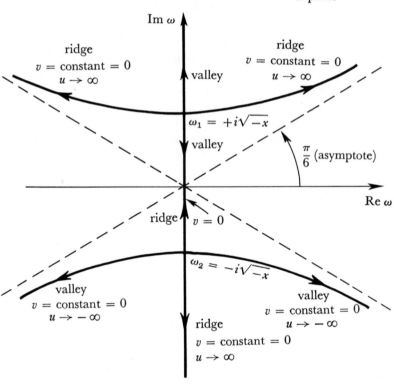

Figure 4–12   **Topography of the surface $u = \mathrm{Re}\, f(\omega)$ in the ω-plane near the saddle points** $\omega_{1,2} = \pm i\sqrt{-x}$, **for the function**

$$f(\omega) = -i[\omega + \omega^3/3(-x)]$$

If $\omega = re^{i\theta}$, $u = \mathrm{Re}\, f(\omega) = r \sin\theta + [r^3/3(-x)] \sin 3\theta$, **and** $v = \mathrm{Im}\, f(\omega) = -r \cos\theta - [r^3/3(-x)] \cos 3\theta$

The topography of the surface $u = \mathrm{Re}\, f(\omega)$ is indicated in Figure 4–12.   We go through the saddle point at $\omega_2 = -i\sqrt{-x}$, with $\phi = 0$.   Then

$$y(x) \sim \sqrt{\frac{2\pi\sqrt{-x}}{2(-x)}}\, \exp\left[-(-x)\frac{2}{3}\sqrt{-x}\right]$$

$$y(x) \sim \frac{\sqrt{\pi}}{(-x)^{1/4}}\, \exp\left[-\frac{2}{3}(-x)^{3/2}\right] \tag{4-67}$$

We see that our two asymptotic formulas (4-66) and (4-67) are in fact the respective WKB solutions, and that they are indeed related by the connection formula (1-113), derived previously in Chapter 1.

In Table 4–1 we list a few of the most frequently occurring Laplace transforms; the results enumerated as 1, 2, $\cdots$, 7 at the beginning of Section 4–5 may be used to extend this list.

Table 4–1   Laplace Transforms

$f(x)$	$F(s) = \int_0^\infty e^{-sx} f(x)\, dx$
1	$\dfrac{1}{s}$
$\delta(x - x_0)$ $(x_0 > 0)$	e^{-sx_0}
$\sin \lambda x$	$\dfrac{\lambda}{s^2 + \lambda^2}$
$\cos \lambda x$	$\dfrac{s}{s^2 + \lambda^2}$
x^n	$\dfrac{n!}{s^{n+1}}$
$e^{-\lambda x}$	$\dfrac{1}{s + \lambda}$

REFERENCES

Rigorous discussions of Fourier series may be found in Apostol (A5) and in Whittaker and Watson (W5) as well as in other mathematical texts.

Fourier integrals are treated by Morse and Feshbach (M9) and by Titchmarsh (T3).   Do not be deceived by the title of the last-named reference; it is not for the casual reader.

In recent years, various new formalisms have been developed, among whose advantages is rigorous treatment of delta functions and similar functions.   Readable discussions of some of this new material are given in

Lighthill (L8); Mikusinski (M5); and Erdelyi (E4).   The first of these three is especially recommended.

The reference work of Erdelyi *et al.* (E6) contains useful tables of Laplace and Fourier transforms as well as several other integral transforms.

Many examples of the application of Laplace transforms to physical problems can be found in Carslaw and Jaeger (C2).

PROBLEMS

4-1   Expand the function $f(x)$ shown below, in a Fourier series.

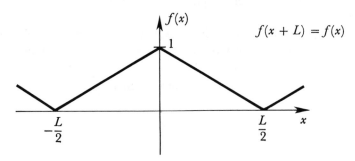

4-2   (*a*) Consider the Fourier series for

$$f(\theta) = \begin{cases} +1 & 0 < \theta < \pi \\ -1 & \pi < \theta < 2\pi \end{cases}$$

$f(\theta + 2\pi) = f(\theta)$

Just to the right of $\theta = 0$, the sum of the first n terms looks as shown below.

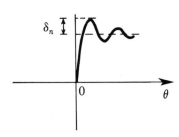

Find δ_n, the "overshoot" of the first maximum.

(*b*) Show that

$$\lim_{n \to \infty} \delta_n \approx 0.18$$

(Evaluate it precisely.)   This is known as Gibbs' phenomenon.

4-3   A function $f(x)$ equals e^{-x} for $0 < x < 1$.
(a) Expand $f(x)$ as a Fourier series of the form $\sum_n B_n \sin n\pi x$.
(b) Expand $f(x)$ as a Fourier series of period 1.

4-4   A linear system is driven by a periodic input $f(t)$, such that $f(t + T) = f(t)$. The response of the system is such that a sinusoidal input of angular frequency ω is multiplied by $(\omega_0/\omega)^2$, unless $\omega = 0$, in which case no output occurs.   The output can be written in the form

$$g(t) = \frac{1}{T} \int_0^T G(t - t') f(t')\, dt'$$

Find the function $G(t)$.

4-5   Of what function is

$$\cos\theta + \frac{\cos 3\theta}{9} + \frac{\cos 5\theta}{25} + \cdots$$

the Fourier series?   Work it out; don't look it up!

4-6   Show that

$$\int_a^b f(x)\, \delta[g(x)]\, dx = \frac{f(x_0)}{|g'(x_0)|}$$

provided $g(x) = 0$ has a single root x_0 in the interval $a < x < b$.

4-7   Evaluate $\int_0^\pi dx \int_1^2 dy\, \delta(\sin x)\, \delta(x^2 - y^2)$.

4-8   Find the Fourier transform of the wave function for a $2p$ electron in hydrogen:

$$\psi(\mathbf{x}) = \frac{1}{\sqrt{32\pi a_0^5}}\, z e^{-r/2a_0}$$

where $a_0 = $ radius of first Bohr orbit and z is a rectangular coordinate.

4-9   A linear system has a response $G(\omega)e^{-i\omega t}$ to the input signal $e^{-i\omega t}$ (ω arbitrary).   If the input $f(t)$ has the particular form

$$f(t) = \begin{cases} 0 & (t < 0) \\ e^{-\lambda t} & (t > 0) \end{cases}$$

where λ is some fixed constant, the output is observed to be

$$F(t) = \begin{cases} 0 & (t < 0) \\ (1 - e^{-\alpha t})e^{-\lambda t} & (t > 0) \end{cases}$$

where α is another fixed constant.
(a) Find $G(\omega)$.
(b) Find the reponse of the system to the input $f(t) = A\, \delta(t)$.

4-10 Find the Laplace transform $\mathscr{L}[f(x)]$ of the function sketched below.

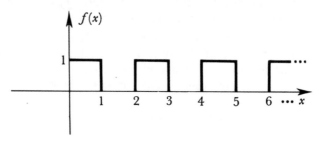

4-11 A function $f(x)$ has the series expansion

$$f(x) = \sum_{n=0}^{\infty} \frac{c_n x^n}{n!}$$

Write the function $g(y) = \sum_{n=0}^{\infty} c_n y^n$ in closed form in terms of $f(x)$.

4-12 By using the integral representation

$$J_0(x) = \frac{1}{2\pi} \int_0^{2\pi} \cos(x \cos \theta) \, d\theta$$

find the Laplace transform of $J_0(x)$.

4-13 Of what function is

$$\frac{1}{(s^2 + 1)(s - 1)}$$

the Laplace transform?

4-14 Three radioactive nuclei decay successively in series, so that the numbers $N_i(t)$ of the three types obey the equations

$$\frac{dN_1}{dt} = -\lambda_1 N_1$$

$$\frac{dN_2}{dt} = \lambda_1 N_1 - \lambda_2 N_2$$

$$\frac{dN_3}{dt} = \lambda_2 N_2 - \lambda_3 N_3$$

If initially $N_1 = N$, $N_2 = 0$, $N_3 = n$, find $N_3(t)$ by using Laplace transforms.

FIVE

FURTHER APPLICATIONS OF COMPLEX VARIABLES

In Chapter 3 we made use of the powerful methods provided by the theory of functions of a complex variable for the evaluation of integrals, and in Chapter 4 contour integrals played an important role in the study of integral transforms. A number of other applications of the theory of complex variables are important in physics. In the first section of this chapter, we give a rather brief treatment of conformal transformations and their applications. The second section contains an elementary description of dispersion relations.

5–1 CONFORMAL TRANSFORMATIONS

Three properties of the mappings or transformations,

$$W = W(z) \qquad z = x + iy \qquad W = U + iV$$

produced by analytic functions make them particularly useful for the solution of two-dimensional problems in electrostatics, magnetostatics, heat flow, hydrodynamics, elasticity, and other fields. (By a two-dimensional problem we do not mean that the physical space of the problem is two-dimensional, but only that there is no dependence on one Cartesian coordinate, say z.)

First, the Cauchy–Riemann differential equations [Appendix, Eqs. (A-8)] imply

$$\frac{\partial^2 U}{\partial x^2} + \frac{\partial^2 U}{\partial y^2} = 0 \qquad \frac{\partial^2 V}{\partial x^2} + \frac{\partial^2 V}{\partial y^2} = 0 \qquad (5\text{-}1)$$

so that U and V are solutions of the two-dimensional Laplace equation and can represent physical quantities which satisfy this equation.

Second, the Cauchy–Riemann equations imply

$$\frac{\partial U}{\partial x} \frac{\partial V}{\partial x} + \frac{\partial U}{\partial y} \frac{\partial V}{\partial y} = (\text{grad } U) \cdot (\text{grad } V) = 0 \qquad (5\text{-}2)$$

so that the vectors grad U and grad V are normal to each other. These vectors are normal to the curves $U = \text{constant}$ and $V = \text{constant}$, respectively, so that the latter curves are likewise mutually perpendicular.

Third, the transformation is *conformal*, meaning that the angle of inter- section of two curves is not changed by the mapping. This follows from the analytic property:

$$W(z) - W(z_0) \approx W'(z_0)(z - z_0) \qquad z \text{ near } z_0 \qquad (5\text{-}3)$$

The mapping is not conformal at a singular point of the transformation, that is, a point where the function $W(z)$ or its inverse $z(W)$ is not analytic.

In applications of conformal transformations, one looks for a transforma- tion which gives an equivalent problem with a simpler boundary configuration.

EXAMPLE

Find a solution of $\nabla^2 \phi(x, y) = 0$ in the first quadrant, which obeys the boundary condition $\phi = 0$ along the axes $x = 0$, $y = 0$. Let $z = x + iy$ and $\zeta = z^2$. Then the first quadrant of the z-plane becomes the upper half ζ-plane, with the boundaries of our problem transformed into the real ζ-axis, as shown in Figure 5–1. It is trivial to find an analytic function of ζ whose imaginary part vanishes on the real axis, namely, $f(\zeta) = c\zeta$, where c is an arbitrary real constant. This is also an analytic function of z, whose imagi- nary part is thus a solution to our original problem

$$\phi = \text{Im } cz^2 = 2cxy \qquad (5\text{-}4)$$

As it stands, this example is rather academic, although the solution ϕ might represent the electrostatic potential near the inside corner of a bent

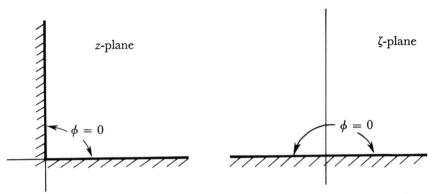

Figure 5-1   Application of the conformal transformation $\zeta = z^2$

conducting sheet.   However, we can apply the result to a problem of considerable practical importance in the design of "optical" systems for focusing charged particles, including the alternating gradient synchrotron.   The alternating gradient focusing systems[1] make use of magnetic (or electric) fields having the property

$$B_x = Cy \qquad B_y = Cx \qquad (C = \text{constant}) \qquad (5\text{-}5)$$

If our solution (5-4) for ϕ is interpreted as a magnetostatic potential, the resulting field $B = -\nabla\phi$ is of just the form (5-5).   To realize the field in practice, one places a pole piece along one of the hyperbolic equipotential surfaces and puts it at the proper potential.   By symmetry, we can place similar pole pieces in the other three quadrants with potentials of alternating sign.   The result is a *quadrupole magnet* such as is shown in Figure 5-2. Such a magnet is *focusing* for particles having orbits displaced in the $\pm y$ directions and *defocusing* for orbits displaced in the $\pm x$ directions.[2]   Two such magnets in succession, rotated 90° with respect to each other about their common axis, can give net focusing in both directions.

Returning to conformal transformations, if $V = \text{Im} f(z)$ represents the function of physical interest obeying Laplace's equation, what is the significance of U?   The qualitative significance of U should be obvious from the fact that the curves $U = \text{constant}$ and $V = \text{constant}$ are mutually orthogonal.   For example, if V is the electrostatic potential, then the field lines, or lines of force, will lie along curves $U = \text{constant}$.   Furthermore, the Cauchy–Riemann equations imply

$$U(Q) - U(P) = -\int_P^Q (\nabla V)_n \, ds \qquad (5\text{-}6)$$

---

[1] E. Courant *et al.* (C11).

[2] We are assuming the particles to be *positively* charged and moving *into* the plane of the figure.   The student should be able to verify the focusing and defocusing properties just by using the right-hand rule for forces on moving charges in magnetic fields.

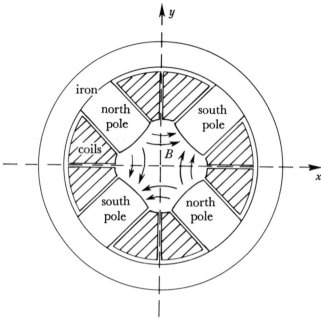

Figure 5–2   The cross section of a quadrupole focusing magnet with hyperbolic pole pieces for producing a field $B_x = Cy$, $B_y = Cx$. The particles move normal to the plane of the paper

where P and Q are any two points, and the integral is along any path connecting P and Q. ds is the element of path length and $(\nabla V)_n$ is the component of the gradient normal to the path, the positive direction being to the right as one proceeds from P to Q.   Similarly,

$$V(Q) - V(P) = \int_P^Q (\nabla U)_n \, ds \qquad (5\text{-}7)$$

Now if V is the electrostatic potential, $-(\nabla V)_n = E_n =$ the component of electric field normal to the path.   If the path lies along the surface of a conductor (that is, along the intersection of the conductor surface with the x, y plane), then E_n is the electric field normal to the conducting surface, which is proportional to the surface charge density by Gauss's law in electrostatics.   Specifically, in mks units, the surface charge density is $\sigma = \varepsilon_0 E_n$, while in cgs units $\sigma = E_n/4\pi$.

By integrating σ along the conducting surface, we may find the total charge in this region (per unit distance in the z-direction).   If we define

$C(P, Q)$ to be the surface charge between P and Q per unit length normal to the x, y plane, then we have shown (in mks units)

$$C(P, Q) = -\varepsilon_0 \int_P^Q (\nabla V)_n \, ds = \varepsilon_0 [U(Q) - U(P)] \qquad (5\text{-}8)$$

If one of the functions U, V represents the potential, the other function is conventionally called the *stream function*.

EXAMPLE

Use the conformal transformation $\zeta = a \sin z$ to find the potential produced by an infinitely long conducting strip of width $2a$ and charge λ per unit length. The mapping $\zeta = a \sin z$ is shown in Figure 5-3.

Consider the charged strip to be infinitely long in the direction normal to the ζ-plane, and to occupy the real axis between $-a$ and a in that plane, as in Figure 5-3. The electrostatic potential is constant along the strip, of course, and from symmetry we know that the lines $O'D'$, $B'E'$, and $A'C'$ all lie along lines of force, or field lines. Then we may restate our electrostatics problem as a mathematical exercise: find an analytic function $f(\zeta)$ whose imaginary part (the electrostatic potential) is constant for ζ on the real axis between $-a$ and a, and whose real part (the stream function) is constant for ζ on the real axis outside this strip, and also on the positive and negative imaginary axes (that is, these are lines of force).

This problem is not easy to solve by inspection, but the conformal transformation $\zeta = a \sin z$ makes it trivial, when we look at the z-plane of Figure

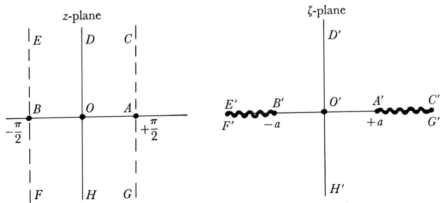

Figure 5-3   The mapping $\zeta = a \sin z$. Note that the ζ-plane is "cut" along the real axis for $|\operatorname{Re} \zeta| > a$. The vertical strip $-\pi/2 \leqslant \operatorname{Re} z \leqslant \pi/2$ maps onto the entire ζ-plane, as do infinitely many other vertical strips of width π in the z-plane

5–3.   We now want an analytic function $F(z)[=f(\zeta)]$ whose imaginary part is constant on BA, and whose real part is constant on BE, OD, and AC.   The solution is obvious, namely $F(z) = Kz$, where K is a real, but otherwise arbitrary, constant.   In terms of the variable ζ, we have

$$f(\zeta) = F(z) = Kz = K \sin^{-1} \frac{\zeta}{a} \tag{5-9}$$

Now we evaluate the constant K, using the fact that the total charge per unit length on the strip is λ.   We have already seen [Eq. (5-8)] that the total charge per unit length on the *top* of the strip, between A' and B' in Figure 5–3, is just ε_0 times the increase in the real part of $f(\zeta)$ as we go from A' to B'.   From (5-9), this increase is

$$K \sin^{-1}(-1) - K \sin^{-1}(1) = -K\pi$$

Therefore the total charge per unit length on the top of the strip is $-K\pi\varepsilon_0$. But, by symmetry, half of the total charge is on the top side of the strip; the other half, of course, is on the bottom.   Therefore

$$-K\pi\varepsilon_0 = \frac{1}{2}\lambda \Rightarrow K = -\frac{\lambda}{2\pi\varepsilon_0}$$

With this value of K,

$$f(\zeta) = -\frac{\lambda}{2\pi\varepsilon_0} \sin^{-1} \frac{\zeta}{a}$$

and the electrostatic potential is

$$V = \operatorname{Im} f(\zeta) = -\frac{\lambda}{2\pi\varepsilon_0} \operatorname{Im} \sin^{-1} \frac{\zeta}{a}$$

What do the equipotentials look like?   These are lines of constant $\operatorname{Im} f(\zeta)$, or, from (5-9), lines of constant $\operatorname{Im} z$.   If we write $z = x + iy$ and $\zeta = \xi + i\eta$, our conformal transformation becomes

$$\xi + i\eta = a \sin (x + iy)$$

or

$$\xi = a \sin x \cosh y$$
$$\eta = a \cos x \sinh y$$

The equipotentials are lines of constant y; eliminating x from these two equations gives

$$\frac{\xi^2}{a^2 \cosh^2 y} + \frac{\eta^2}{a^2 \sinh^2 y} = 1$$

or, since

$$V = \operatorname{Im} f(\zeta) = \operatorname{Im} F(z) = Ky = -\frac{\lambda}{2\pi\varepsilon_0} y$$

the equation for equipotentials may be written

$$\frac{\xi^2}{a^2 \cosh^2\left(\dfrac{2\pi\varepsilon_0 V}{\lambda}\right)} + \frac{\eta^2}{a^2 \sinh^2\left(\dfrac{2\pi\varepsilon_0 V}{\lambda}\right)} = 1$$

The equipotentials are ellipses.

5-2 DISPERSION RELATIONS

Cauchy's integral formula [Eq. (A-10) of the Appendix] expresses the remarkable result that the values of an analytic function $f(z)$ along a closed curve within its region of regularity determine the values of $f(z)$ everywhere inside this curve.   Namely,

$$f(z) = \frac{1}{2\pi i} \int_C \frac{f(\zeta)}{(\zeta - z)} \, d\zeta$$

where C is any closed contour within and on which $f(z)$ is regular and z is any point inside C.

When applied in a manner to be described below to certain functions having physical significance on a portion of the real axis, Cauchy's formula leads to the so-called *dispersion relations*.   This name comes from the fact that relations of this sort were first derived by Kronig and by Kramers in the theory of optical and X-ray dispersion.   In optics, a dispersion relation is an integral relationship between the refractive part and the absorptive part of the refractive index at different frequencies.   Since these are the real and imaginary parts, respectively, of the (complex) refractive index we apply the term dispersion relation to any such integral relationship between the real and imaginary parts of a function of a complex variable.

Dispersion relations now have extensive applications in the theory of elementary particle interactions, where they are applied to scattering amplitudes.   This application is closely related to the original one since the refractive index of light is closely related to its scattering amplitude.   Other applications are found in electrical circuit analysis.   For example, see Seshu and Balabanian (S4), Chapter 7.

Consider first the elementary situation where $f(z)$ has no singularities

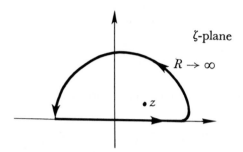

Figure 5–4   Contour used to derive the dispersion relations (5-12)

in the upper-half plane and $f(z) \to 0$ when $|z| \to \infty$.[3]   We then apply Cauchy's formula with a contour consisting of the real axis and a large upper semi-circle, as in Figure 5–4.   The integral on the semicircle vanishes because, as $R \to \infty$,

$$\left| \int_{\text{semicircle}} \frac{f(\zeta)}{(\zeta - z)} \, d\zeta \right| \leqslant |f(\zeta)|_{\max} \int \frac{Re^{i\theta} i d\theta}{Re^{i\theta}} = \pi i \, |f(\zeta)|_{\max} \to 0$$

where $|f(\zeta)|_{\max}$ is the maximum magnitude of $f(\zeta)$ on the semicircle, which vanishes for $R \to \infty$.   Thus

$$f(z) = \frac{1}{2\pi i} \int_{\text{real axis}} \frac{f(\zeta)}{\zeta - z} \, d\zeta \qquad \text{Im } z > 0 \tag{5-10}$$

Suppose further that the function of physical interest is $f(z)$ on the real axis, or, in case there is a branch line, the limit as z approaches the real axis from above.   That is, calling this function $F(x)$,

$$F(x) = \lim_{\varepsilon \to 0} f(x + i\varepsilon) \tag{5-11}$$

where ε is positive.   Then it follows from (5-10) that

$$2\pi i F(x) = \lim_{\varepsilon \to 0} \int_{-\infty}^{\infty} \frac{f(x')}{x' - x - i\varepsilon} \, dx' = P \int_{-\infty}^{\infty} \frac{F(x')}{x' - x} \, dx' + \pi i F(x)$$

where P designates the Cauchy principal value [see Appendix, Section A–2, item 8].

---

[3] We may briefly sketch the physical basis of this assumption.   The "response" $\phi(t)$ of a linear system to an "input" $g(t)$ may be written in the form $\phi(t) = \int dt' \, K(t - t')g(t')$, where $K(t)$ is the Green's function (Section 9–4) of the system.   The requirement of *causality*, namely, that no response shall precede its input, means that $K(t) = 0$ for $t < 0$.   The functions $f(z)$ being discussed are Fourier transforms (Section 4–2) of such Green's functions, and the condition $K(t) = 0$ for $t < 0$ implies that the Fourier transform $f(z) = \int dt \, e^{itz} K(t)$ of $K(t)$ has no singularities for Im $z > 0$, and that $f(z) \to 0$ for $|z| \to \infty$ in the upper-half z-plane (cf. item 2, p. 246).

Thus

$$F(x) = \frac{1}{\pi i} P \int_{-\infty}^{\infty} \frac{F(x')}{x' - x} dx'$$

and, equating real and imaginary parts,

$$\text{Re } F(x) = \frac{1}{\pi} P \int_{-\infty}^{\infty} \frac{\text{Im } F(x')}{x' - x} dx'$$

$$\text{Im } F(x) = -\frac{1}{\pi} P \int_{-\infty}^{\infty} \frac{\text{Re } F(x')}{x' - x} dx'$$

(5-12)

These are dispersion relations for the function of physical interest $F(x)$

In most applications, these relations are not in a form suitable for comparison with experiment. The reason is that the variable x usually has physical significance only for $x > 0$ or even $x > x_1 \geq 0$, so that the integrals above extend over a large "nonphysical" region. For example, x may be the energy of a scattering particle, which cannot be less than the rest mass. In this situation, some symmetry of $f(z)$ may be used to relate the integral over a nonphysical region to that over a physical region.

A simple case occurs if the physical range of x is $x \geq 0$ and if $f(z)$ has "even" or "odd" symmetry, defined by $f(-z) = \pm f^*(z^*)$, respectively. For the "even" symmetry,

$$F(x) = \frac{1}{\pi i} P \int_{\infty}^{0} \frac{F^*(x')}{x' + x} dx' + \frac{1}{\pi i} P \int_{0}^{\infty} \frac{F(x')}{x' - x} dx'$$

$$\text{Re } F(x) = \frac{2}{\pi} P \int_{0}^{\infty} \frac{x' \text{ Im } F(x')}{(x')^2 - x^2} dx'$$

(5-13)

$$\text{Im } F(x) = -\frac{2}{\pi} P \int_{0}^{\infty} \frac{x \text{ Re } F(x')}{(x')^2 - x^2} dx'$$

(5-14)

Similar results hold for the "odd" symmetry case. The optical dispersion relations have this form, where x is the optical frequency and $F(x)$ is the refractive index.

Thus far, we have focused attention on dispersion relations in their historical form. A related point is the way in which certain analytic functions may be specified uniquely by describing their singularities and behavior at infinity. Along a branch line, the information required is simply the discontinuity across the branch line, that is, the difference in the values of the function on opposite sides of the cut.

Some of the conditions under which such a representation is possible will become clear in the examples below.

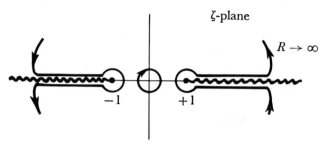

Figure 5–5   Contour used in the example below

EXAMPLE

Suppose we have the following information about a function $f(z)$:

1. $f(z)$ is analytic everywhere except for a pole of residue 1 at $z = 0$, and branch lines from $+1$ to $+\infty$ and -1 to $-\infty$ along the real axis.

2. $f(z) \to 0$ as $|z| \to \infty$.

3. $f(z)$ is real on the real axis from -1 to $+1$.

It follows from the *Schwarz reflection principle* (p. 482) that $f(z^*) = f^*(z)$.

Now apply the Cauchy formula along the contour shown in Figure 5–5. As in our previous example, the integral around the large circle vanishes as $R \to \infty$. We shall assume that near the branch points ± 1, $f(z)$ behaves in such a way that the integrals on the two little circles around these points also vanish. Then

$$2\pi i f(z) = 2\pi i \left(\frac{1}{z}\right) + \lim_{\varepsilon \to 0}\left[\int_1^\infty \frac{f(x' + i\varepsilon)\,dx'}{x' + i\varepsilon - z} - \int_1^\infty \frac{f(x' - i\varepsilon)\,dx'}{x' - i\varepsilon - z}\right.$$
$$\left. + \int_{-\infty}^{-1} \frac{f(x' + i\varepsilon)\,dx'}{x' + i\varepsilon - z} - \int_{-\infty}^{-1} \frac{f(x' - i\varepsilon)\,dx'}{x' - i\varepsilon - z}\right]$$

Combining the first two integrals and the last two, we see that only the discontinuity of $f(z)$ across the cuts is involved, as mentioned before. The discontinuity is supplied in this example by the reflection symmetry $f(x' - i\varepsilon) = f^*(x' + i\varepsilon)$, from which

$$f(x' + i\varepsilon) - f(x' - i\varepsilon) = 2i \operatorname{Im} f(x' + i\varepsilon)$$

Thus

$$f(z) = \frac{1}{z} + \frac{1}{\pi}\int_1^\infty \frac{\operatorname{Im} F(x')}{x' - z}\,dx' + \frac{1}{\pi}\int_{-\infty}^{-1} \frac{\operatorname{Im} F(x')}{x' - z}\,dx' \qquad (5\text{-}15)$$

where $F(x) = \lim_{\varepsilon \to 0} f(x + i\varepsilon)$ as before. This completes the representation of $f(z)$ in terms of its prescribed singularities.

If $F(x)$ is the function of physical interest, we can write a dispersion relation for it by going to the limit $z = x + i\varepsilon \to x$.

$$F(x) = \frac{1}{x} + \frac{1}{\pi} P \int_1^\infty \frac{\operatorname{Im} F(x')}{x' - x} dx' + \frac{1}{\pi} P \int_{-\infty}^{-1} \frac{\operatorname{Im} F(x')}{x' - x} dx' + \frac{1}{\pi} \pi i \operatorname{Im} F(x)$$

If $|x| > 1$, the last term comes from the singularity in one of the integrals. If $|x| < 1$, $\operatorname{Im} F(x) = 0$ by assumption, and we make no error by including the last term. In any case, we obtain

$$\operatorname{Re} F(x) = \frac{1}{x} + \frac{1}{\pi} P \int_{-\infty}^{-1} \frac{\operatorname{Im} F(x')}{x' - x} dx' + \frac{1}{\pi} P \int_1^\infty \frac{\operatorname{Im} F(x')}{x' - x} dx \quad (5\text{-}16)$$

We do not obtain by this procedure the " other " dispersion relation giving $\operatorname{Im} F(x)$ in terms of an integral over $\operatorname{Re} F(x)$.

If the physical region of the variable x is $x > 1$, we still need a symmetry principle to relate $\operatorname{Im} F(x)$ for $x < -1$ to $\operatorname{Im} F(x)$ for $x > 1$. This symmetry relation must be found in the physics of the problem.

In obtaining the representation (5-15), we made use of the special contour chosen. However, the dispersion relation (5-16) could also have been obtained by using Cauchy's formula along the contour shown in Figure 5–6, which gives

$$2\pi i f(z) = P \int_{-\infty}^\infty \frac{f(x')}{x' - z} dx' - \pi i \left(\frac{1}{-z} \right)$$

Let $z = x + i\varepsilon$ and $\varepsilon \to 0$. Then

$$2F(x) = \frac{1}{x} + \frac{1}{\pi i} \left[P \int_{-\infty}^\infty \frac{F(x')}{x' - x} dx' + \pi i F(x) \right]$$

$$F(x) = \frac{1}{x} + \frac{1}{\pi i} P \int_{-\infty}^\infty \frac{F(x')}{x' - x} dx' \quad (5\text{-}17)$$

This result contains (5-16) as its real part.

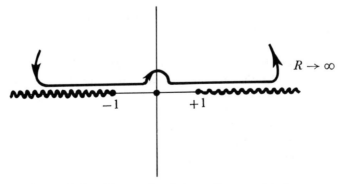

$R \to \infty$

-1 $+1$

Figure 5–6 Alternative integration contour

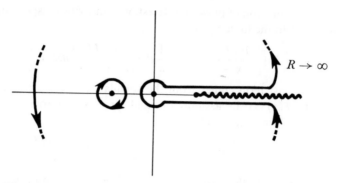

Figure 5–7   Integration contour for example below

EXAMPLE

Find a function $f(z)$ which has the following properties:

1. $f(z)$ is analytic except for a branch line from $z = 0$ to $+\infty$ along the real axis, and a simple pole of residue 1 at $z = -1$.

2. $f(z) \to 0$ as $|z| \to \infty$.

3. $f(z)$ is real on the negative real axis.

4. For $x > 0$, $\operatorname{Im} f(x + i\varepsilon) = 1/(1 + x^2)$.

We apply Cauchy's formula along the contour shown in Figure 5–7. The result may be written down from our work in the previous example.

$$f(z) = \frac{1}{z+1} + \frac{1}{\pi} \int_0^\infty \frac{dx'}{(1 + x'^2)(x' - z)}$$

The integral is straightforward and yields the result

$$f(z) = \frac{1}{z+1} - \frac{1}{\pi} \frac{\ln z}{(1 + z^2)} - \frac{1}{(1 + z^2)} \left(\frac{z}{2} - i \right) \tag{5-18}$$

A little care is required with the multivalued function $\ln z$. The answer given in (5-18) assumes that $\ln z$ is real on the top side of the cut.

We conclude this section with a description of a modification which may be made in order to obtain dispersion relations for a function which does not satisfy the condition $f(z) \to 0$ as $|z| \to \infty$.

If $(1/z)f(z) \to 0$ as $|z| \to \infty$ and $(1/z)f(z)$ also satisfies other conditions such as those in the above examples, then a dispersion relation for $f(z)$ may be obtained from one for $(1/z)f(z)$.

A convergence factor $1/(z - x_0)$ will serve just as well as $1/z$, of course. The pole introduced by this factor at $z = x_0$ [or a pole already present in $f(z)$] may be removed by subtracting a pole with the same residue at the same point. For example, a dispersion relation for the function $[f(z) - f(x_0)]/(z - x_0)$ corresponding to (5-12) is

$$\text{Re}\left\{\frac{1}{x - x_0}[F(x) - F(x_0)]\right\} = \frac{1}{\pi} P \int_{-\infty}^{\infty} \frac{\text{Im } F(x') - \text{Im } F(x_0)}{(x' - x_0)(x' - x)} dx' \quad (5\text{-}19)$$

Generally, $\text{Im } F(x_0) = 0$, in which case

$$\text{Re } F(x) = \text{Re } F(x_0) + \frac{(x - x_0)}{\pi} P \int_{-\infty}^{\infty} \frac{\text{Im } F(x')}{(x' - x_0)(x' - x)} dx' \quad (5\text{-}20)$$

This is called a *subtracted* dispersion relation. The subtraction procedure improves the convergence of the dispersion integral, as is easily seen in the above example.

REFERENCES

For further applications of conformal transformations to two-dimensional problems, see Smythe (S6), Chapter 4; Irving and Mullineux (I3), Chapter VIII; Smith (S5), Chapter 9; Morse and Feshbach (M9), Section 4.7; and Wilf (W7), Chapter 6.

An elementary treatment of dispersion relations is given by E. Corinaldesi (C9); the connection with causality is discussed by J. S. Toll (T6); the quantum theory case, by Gell-Mann *et al.* (G2).

PROBLEMS

5-1 Use the method of conformal transformations to find the electrostatic potential arising from a series of coplanar charged strips, each of width $2a$, separated by spaces of width $2b$, and each with charge λ per unit length.

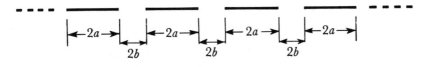

5-2 Line charges $\pm\lambda$ are placed at the points $(0, \pm d)$, respectively. If $\lambda \to \infty$, $d \to 0$ in such a way that $p = 2\lambda d = $ constant, find the function $W(z)$ whose real part gives the electrostatic potential in the xy plane.

5-3 Consider a line charge λ at $(0, a)$ between the two grounded planes $y = 0$ and $y = b$ $(0 < a < b)$. Find the potential everywhere between the planes.
Hint: The transformation $\zeta = e^{z\pi/b}$ is useful.

5-4 Derive (5-8) carefully; in particular be sure you know which *sign* the charge has.

5-5 Consider an array of many parallel semi-infinite conducting plates, alternately charged to potentials $+V_0$ and $-V_0$ as shown in the figure.

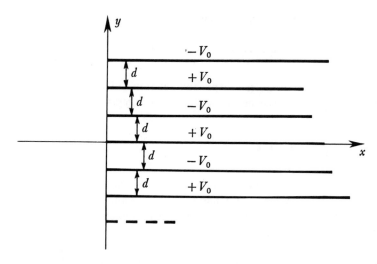

(*a*) Show that the conformal transformation

$$\zeta = \frac{2V_0}{\pi} \sin^{-1} e^{\pi z/d}$$

reduces the problem to a trivial one in the ζ-plane.

(*b*) Show that, to a first approximation, the capacitance of such a parallel-plate capacitor is increased, because of edge effects, by an amount which is equivalent to the addition of a strip of width $(d/\pi) \ln 2$ to each plate.

5-6 (*a*) Consider the problem of finding the potential distribution in the upper-half plane, if the potential along the x-axis is fixed thus:

$$\phi(y = 0) = \begin{cases} 0 & x < -a \\ V_0 & -a < x < a \\ 0 & x > a \end{cases}$$

Show that this is the same problem as finding the potential due to two line charges, if we exchange the roles of potential and stream function. The solution is

$$\phi = \frac{V_0}{\pi} \left(\tan^{-1} \frac{y}{x-a} - \tan^{-1} \frac{y}{x+a} \right)$$

(b) With the aid of the solution to part a, and the conformal transformation

$$\zeta = \cosh \frac{\pi z}{2d}$$

find the temperature distribution in the semi-infinite metal plate shown below, if the three faces are maintained at the indicated temperatures. Note that $\nabla^2 T = 0$ inside the metal.

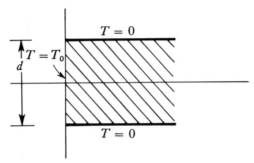

5-7 The Schwartz Transformation

(a) Consider a mapping $\zeta = \zeta(z)$ such that

$$\frac{d\zeta}{dz} = A(z - a_1)^{s_1}(z - a_2)^{s_2}$$

where a_1, a_2, s_1, s_2 are arbitrary real constants, while A is an arbitrary complex constant. Show that, as z moves along the real axis, ζ also moves in a straight line, except for sudden changes of direction when $z = a_1$ or $z = a_2$. As z passes (infinitesimally above) a_1 from left to right, show that ζ turns to the right through an angle πs_1, and similarly by πs_2 when z passes above a_2 (see the figure below).

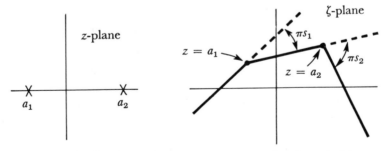

Thus the upper half z-plane maps into a region bounded by an arbitrary polygon in the ζ-plane. This method can clearly be generalized to an arbitrary number of corners.

(b) Apply the method of (a) to find the mapping $\zeta(z)$ which carries the upper-half z-plane into the interior of the rectangular trough which is shaded in the figure below.

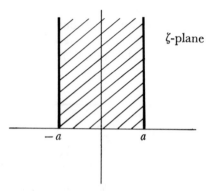

Compare your result with Figure 5–3 and the associated example.

5-8   Verify that the solution of the second example in Section 5–1 approaches the potential of a line charge λ at large distances, that is, for large values of $|\zeta|$.

5-9   A function $f(z)$ has the following properties:
(a) $f(z)$ is analytic except for: (1) a branch line from 0 to $+\infty$ along the real axis; (2) a simple pole of residue 2 at $z = -2$.
(b) $f(z) \to 0$ as $|z| \to \infty$.
(c) $f(z)$ is real on the negative real axis.
(d) For $x > 0$, $\operatorname{Im} f(x + i\varepsilon) = x/(1 + x^2)$.   Find the function

$$F(x) = f(x + i\varepsilon).$$

5-10   Consider the subtracted dispersion relation (5-20).   Suppose that in fact $|F(x)| \to 0$ as $|x| \to \infty$.
(a) Derive from (5-20) the "sum rule"

$$\operatorname{Re} F(0) = \frac{1}{\pi} P \int_{-\infty}^{\infty} \frac{\operatorname{Im} F(x')\, dx'}{x'}$$

(b) How can this sum rule be obtained from the unsubtracted dispersion relation (5-12)?
(c) Suppose now that $F(x)$ approaches zero at infinity so rapidly that $|x\, F(x)| \to 0$ as $|x| \to \infty$.   Derive a further sum rule from (5-12), by analogy with part (a).

Such relations are called "superconvergence relations," because of the rapid convergence at infinity which is imposed on $F(x)$.   See, for example, V. de Alfaro *et al.* (D3).   (The name "superconvergence relation" was coined after this reference appeared.)

SIX

VECTORS AND MATRICES

In this chapter we discuss linear vector spaces and linear operators and the representation of both vectors and operators by means of components in specified coordinate systems. We discuss the algebra and other properties of matrices which represent linear operators, and we also discuss matrices which serve the quite different purpose of describing transformations of coordinates.

Eigenvalue problems involving matrices are treated in Section 6–5, with emphasis on Hermitian matrices because of their importance in physics. The related problem of diagonalizing matrices is discussed in Section 6–6. We conclude the chapter with a descriptive treatment of linear vector spaces having an infinite number of dimensions.

6–1   LINEAR VECTOR SPACES

The use of ordinary three-dimensional vectors to represent physical quantities such as position, velocity, forces, and so forth, is quite familiar. We abstract from the well-known properties of such vectors the following definition of a linear vector space.

A linear vector space is a set of objects (vectors) **a, b, c,** ... which is closed under two operations:

1. Addition, which is commutative and associative:

$$\mathbf{a} + \mathbf{b} = \mathbf{c} = \mathbf{b} + \mathbf{a}$$

$$(\mathbf{a} + \mathbf{b}) + \mathbf{c} = \mathbf{a} + (\mathbf{b} + \mathbf{c}) \tag{6-1}$$

2. Multiplication by a scalar (any complex number), which is distributive and associative, that is,

$$\lambda(\mathbf{a} + \mathbf{b}) = \lambda\mathbf{a} + \lambda\mathbf{b}$$

$$\lambda(\mu\mathbf{a}) = (\lambda\mu)\mathbf{a} \tag{6-2}$$

$$(\lambda + \mu)\mathbf{a} = \lambda\mathbf{a} + \mu\mathbf{a}$$

In addition, we assume a null vector **0** exists such that for all **a**

$$\mathbf{a} + \mathbf{0} = \mathbf{a} \tag{6-3}$$

that multiplication by the scalar 1 leaves every vector unchanged,

$$1\mathbf{a} = \mathbf{a}$$

and finally, that for every **a**, a vector $-\mathbf{a}$ exists such that

$$\mathbf{a} + (-\mathbf{a}) = \mathbf{0} \tag{6-4}$$

A set of vectors **a, b,** ..., **u** is said to be *linearly independent* provided no equation

$$\lambda\mathbf{a} + \mu\mathbf{b} + \cdots + \sigma\mathbf{u} = \mathbf{0} \tag{6-5}$$

holds except the trivial one with $\lambda = \mu = \cdots = \sigma = 0$.

If in a particular vector space there exist n linearly independent vectors but no set of $n + 1$ linearly independent ones, the space is said to be *n-dimensional*. We shall mainly consider finite-dimensional spaces in this chapter, but we shall briefly discuss infinite-dimensional ones at the end.

Let $\mathbf{e}_1, \mathbf{e}_2, \ldots, \mathbf{e}_n$ be a set of n linearly independent vectors in an n-dimensional vector space. Then if **x** is an arbitrary vector in the space, there exists a relation

$$\lambda\mathbf{e}_1 + \mu\mathbf{e}_2 + \cdots + \sigma\mathbf{e}_n + \tau\mathbf{x} = 0$$

with not all the constants equal to zero, and in particular $\tau \neq 0$. Thus **x** can be written as a linear combination of the $\mathbf{e}_i$:

$$\mathbf{x} = \sum_{i=1}^{n} x_i \mathbf{e}_i \tag{6-6}$$

The vectors e_i are said to form a *basis*, or *coordinate system*, and the numbers x_i are the *components* of x in this system.   The e_i are called *base vectors*. The fact that an arbitrary vector x can be written as a linear combination of the e_i is often expressed thus: The set of base vectors e_i is *complete*.   The idea of completeness is very important, and will occur several times later in this text (for example, Section 9–2).

6–2   LINEAR OPERATORS

Next we consider a *linear vector function* of a vector, that is, a rule which associates with every vector x a vector $\phi(x)$, in a linear way,

$$\phi(\lambda a + \mu b) = \lambda \phi(a) + \mu \phi(b) \tag{6-7}$$

It is sufficient to know the n vectors $\phi(e_i)$, which may be conveniently described in terms of the basis e_i, that is,

$$\phi(e_i) = \sum_j A_{ji} e_j \tag{6-8}$$

where A_{ji} is thus the jth component of the vector $\phi(e_i)$.
If we now consider an arbitrary vector x, and call $\phi(x) = y$, we have

$$y = \phi\left(\sum_i x_i e_i\right) = \sum_i x_i \sum_j A_{ji} e_j$$

Thus the components of x and y are related by

$$y_j = \sum_i A_{ji} x_i \tag{6-9}$$

We may also describe the above relations in another way, saying that the association of y with x is accomplished by a *linear operator* $\mathscr{A}$ operating on x.   Symbolically,

$$y = \mathscr{A}x \tag{6-10}$$

Then the numbers A_{ij} are the *components* of the linear operator $\mathscr{A}$ (or the vector function ϕ) in the coordinate system e_i.   Specifically, from (6-8), A_{ij} is the ith component of the vector $\mathscr{A}e_j$.
Just as with vectors, a linear operator often has a physical meaning which is independent of a specific coordinate system, and may be described without reference to a specific system.

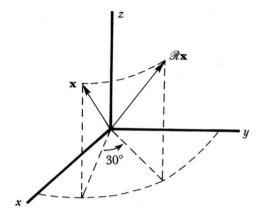

Figure 6–1   The result of a simple rotation operator

EXAMPLE

We may describe a simple rotation operator $\mathscr{R}$ in words by the rule: $\mathscr{R}\mathbf{x}$ is the vector obtained by rotating $\mathbf{x}$ by 30° about a vertical axis in the positive (right-hand) sense. This operation is illustrated in Figure 6–1. The coordinate system in the figure is not necessary for the definition of the operator, since "vertical" has a meaning independent of coordinate systems. However, we could also define $\mathscr{R}$ by giving its components in the system shown.

The sum and product of linear operators and the product of an operator and a scalar number may be defined by the relations

$$(\mathscr{A} + \mathscr{B})\mathbf{x} = \mathscr{A}\mathbf{x} + \mathscr{B}\mathbf{x} \tag{6-11}$$

$$(\mathscr{A}\mathscr{B})\mathbf{x} = \mathscr{A}(\mathscr{B}\mathbf{x}) \tag{6-12}$$

$$(\lambda\mathscr{A})\mathbf{x} = \lambda(\mathscr{A}\mathbf{x}) \tag{6-13}$$

In general $\mathscr{A}\mathscr{B} \neq \mathscr{B}\mathscr{A}$, but, if these are equal, $\mathscr{A}$ and $\mathscr{B}$ are said to *commute*. The null and identity operators 0 and 1 have obvious meanings, namely,

$$0\mathbf{x} = \mathbf{0} \quad \text{and} \quad 1\mathbf{x} = \mathbf{x} \tag{6-14}$$

for every vector $\mathbf{x}$ in our space. Two operators $\mathscr{A}$ and $\mathscr{B}$ are equal if $\mathscr{A}\mathbf{x} = \mathscr{B}\mathbf{x}$ for every vector $\mathbf{x}$. Finally, if an operator $\mathscr{A}^{-1}$ exists with the properties

$$\mathscr{A}\mathscr{A}^{-1} = \mathscr{A}^{-1}\mathscr{A} = 1 \tag{6-15}$$

$\mathscr{A}^{-1}$ is called the *inverse* of $\mathscr{A}$. Operators which have an inverse are said to be *nonsingular*.

In the above discussion, specifically in Eq. (6-8), we have assumed that the vector $\phi(x) = \mathscr{A}x$ is in the same space as the vector x. This need not be so, as illustrated by the following elementary example.

EXAMPLE

Consider the space of three-dimensional position vectors x and a conventional xyz Cartesian coordinate system such as shown in Figure 6–1. We define a *projection operator* $\mathscr{P}$ such that $\mathscr{P}x$ is the projection of x on the xy-plane, that is, $\mathscr{P}x$ has the same x- and y-coordinates as x but zero z-component. In fact, the space of the vectors $\mathscr{P}x$ is only two-dimensional and is thus different from the space of the vectors x.[1] It is also clear that this operator $\mathscr{P}$ does not have an inverse.

The only change required in the above analysis if $\phi(x)$ is in a different space from x is to express $\phi(e_i)$ in terms of a basis f_i in the ϕ space, so that (6-8) becomes

$$\phi(e_i) = \sum_j A_{ji} f_j \qquad (6\text{-}16)$$

Then the components A_{ji} of the operator $\mathscr{A}$ refer to the two bases e_i and f_j. It is clear, furthermore, that the two spaces may have different numbers of dimensions, as in the above example. If so, the operator $\mathscr{A}$ cannot have an inverse, as is perhaps obvious from the example and as will become clear from the discussion of matrices which follows.

6–3 MATRICES

The numbers A_{ij} introduced above can be displayed as a *matrix*:

$$A = \begin{pmatrix} A_{11} & A_{12} & \cdots & A_{1n} \\ A_{21} & A_{22} & \cdots & A_{2n} \\ & & \cdot & \\ A_{m1} & A_{m2} & \cdots & A_{mn} \end{pmatrix} \qquad (6\text{-}17)$$

The numbers A_{ij} are called the *elements* of the matrix.

In general, a matrix is such a rectangular array of numbers which obeys certain rules of algebra. We can easily find the rules which are required if matrices are to represent linear operators *linearly*[2] in our vector spaces.

---

[1] The two-dimensional space of vectors $\mathscr{P}x$ may, of course, be thought of as a subspace of the original space of vectors x. This need not be true in general.

[2] That is, if the operators $\mathscr{A}$ and $\mathscr{B}$ are represented by the matrices A and B, respectively, we would like the operator $\lambda\mathscr{A} + \mu\mathscr{B}$ to be represented by the matrix $\lambda A + \mu B$.

For example, the j-components in a given coordinate system of the operator equations (6-11), (6-12), and (6-13) are

(6-11): $\displaystyle\sum_i (A + B)_{ji}\, x_i = \sum_i A_{ji}\, x_i + \sum_i B_{ji}\, x_i$

(6-12): $\displaystyle\sum_i (AB)_{ji}\, x_i = \sum_l A_{jl}(Bx)_l = \sum_{li} A_{jl}\, B_{li}\, x_i$

(6-13): $\displaystyle\sum_i (\lambda A)_{ji}\, x_i = \lambda \sum_i A_{ji}\, x_i$

Since x is arbitrary, these give immediately the rules for adding and multiplying matrices, and for multiplication by a constant:

$$(A + B)_{ji} = A_{ji} + B_{ji} \qquad (6\text{-}18)$$

$$(AB)_{ji} = \sum_l A_{jl}\, B_{li} \qquad (6\text{-}19)$$

$$(\lambda A)_{ji} = \lambda A_{ji} \qquad (6\text{-}20)$$

The least trivial of these is (6-19), about which we note the following:

1. Matrix multiplication is not commutative; AB need not equal BA.

2. The ji element of AB is the sum of products of elements from the jth row of A and the ith column of B.

3. The number of columns in A must equal the number of rows in B if the definition (6-19) of the product AB is to make sense.

The elements of the null matrix 0 are all zero, and the elements of the identity 1 are δ_{ij}, the Kronecker delta defined in Eq. (4-13).   That is, 1 is *diagonal* with diagonal elements unity.

For an inverse matrix A^{-1}, we have the conditions

$$A^{-1}A = AA^{-1} = 1 \qquad (6\text{-}21)$$

corresponding to the operator equations of the same form.   These can only be satisfied if A is a square matrix with nonzero determinant, in which case[3]

$$(A^{-1})_{ij} = \frac{\text{cofactor of } A_{ji}}{\det A} \qquad (6\text{-}22)$$

Then

$$(AA^{-1})_{ij} = \sum_k A_{ik}(A^{-1})_{kj} = \sum_k A_{ik}\frac{\text{cofactor of } A_{jk}}{\det A}$$

$$= \delta_{ij}$$

[3] The *cofactor* of the element A_{ij} in the (square) matrix A equals $(-)^{i+j}$ times the determinant of the matrix which A becomes if its ith row and jth column are deleted.   If every element in any one row (or column) of A is multiplied by its cofactor, and the products added together, the result is just the determinant of A.

and similarly for the product $A^{-1}A$, so that the conditions (6-21) are indeed satisfied.

A number of matrices closely related to a given matrix A are given in Table 6-1.

Some further definitions are the following:

A is *real* if $A^* = A$

symmetric if $\tilde{A} = A$

antisymmetric if $\tilde{A} = -A$

Hermitian if $A^\dagger = A$

orthogonal if $A^{-1} = \tilde{A}$

unitary if $A^{-1} = A^\dagger$

diagonal if $A_{ij} = 0$ for $i \neq j$

idempotent if $A^2 = A$

The *trace* or *spur* of a square matrix A is the sum of its diagonal elements:

$$\mathrm{Tr}\, A = \sum_i A_{ii}$$

We list below a few relations involving products of square matrices, which are frequently useful and may be easily verified by the student:

$$(ABC)^{-1} = C^{-1}B^{-1}A^{-1} \qquad \text{(provided all inverses exist)}$$

$$\widetilde{(ABC)} = \tilde{C}\tilde{B}\tilde{A} \tag{6-23}$$

$$\mathrm{Tr}\,(AB) = \mathrm{Tr}\,(BA)$$

$$\det(AB) = (\det A)(\det B) = \det(BA)$$

Table 6-1   Matrices Related to a Matrix A

Matrix	Components	Example
A	A_{ij}	$\begin{pmatrix} 1 & i \\ 1+i & 2 \end{pmatrix}$
Transpose $\tilde{A}$	$(\tilde{A})_{ij} = A_{ji}$	$\begin{pmatrix} 1 & 1+i \\ i & 2 \end{pmatrix}$
Complex conjugate A^*	$(A^*)_{ij} = (A_{ij})^*$	$\begin{pmatrix} 1 & -i \\ 1-i & 2 \end{pmatrix}$
Hermitian conjugate or *adjoint* $A^\dagger = (\tilde{A})^*$	$(A^\dagger)_{ij} = (\tilde{A}_{ji})^*$	$\begin{pmatrix} 1 & 1-i \\ -i & 2 \end{pmatrix}$
Inverse A^{-1} (square matrices with determinant $\neq 0$ only)	$(A^{-1})_{ij} = \dfrac{\text{cofactor of } A_{ji}}{\det A}$	$\dfrac{1}{(3-i)}\begin{pmatrix} 2 & -i \\ -1-i & 1 \end{pmatrix}$

Finally, we point out that matrices may sometimes be subdivided into submatrices or *blocks* in such a way as to simplify certain algebraic relations (and work).

EXAMPLE

If we subdivide A and B as follows,

$$A = \left(\begin{array}{cc|ccc} a_{11} & a_{12} & a_{13} & a_{14} & a_{15} \\ a_{21} & a_{22} & a_{23} & a_{24} & a_{25} \\ \hline a_{31} & a_{32} & a_{33} & a_{34} & a_{35} \\ a_{41} & a_{42} & a_{43} & a_{44} & a_{45} \\ a_{51} & a_{52} & a_{53} & a_{54} & a_{55} \end{array}\right) = \begin{pmatrix} A_{11} & A_{12} \\ A_{21} & A_{22} \end{pmatrix} \qquad (6\text{-}24)$$

$$B = \left(\begin{array}{ccc|cc} b_{11} & b_{12} & b_{13} & b_{14} & b_{15} \\ b_{21} & b_{22} & b_{23} & b_{24} & b_{25} \\ \hline b_{31} & b_{32} & b_{33} & b_{34} & b_{35} \\ b_{41} & b_{42} & b_{43} & b_{44} & b_{45} \\ b_{51} & b_{52} & b_{53} & b_{54} & b_{55} \end{array}\right) = \begin{pmatrix} B_{11} & B_{12} \\ B_{21} & B_{22} \end{pmatrix} \qquad (6\text{-}25)$$

then A and B have the form of 2×2 *block matrices* whose elements A_{ij}, B_{ij} are themselves matrices. We can easily see that the correct product AB results if the block matrices are multiplied according to the usual rule,

$$AB = \begin{pmatrix} A_{11} & A_{12} \\ A_{21} & A_{22} \end{pmatrix}\begin{pmatrix} B_{11} & B_{12} \\ B_{21} & B_{22} \end{pmatrix} = \begin{pmatrix} A_{11}B_{11} + A_{12}B_{21} & A_{11}B_{12} + A_{12}B_{22} \\ A_{21}B_{11} + A_{22}B_{21} & A_{21}B_{12} + A_{22}B_{22} \end{pmatrix}$$

$$(6\text{-}26)$$

provided all the matrix products in the last matrix make sense. This will be true provided the original division of the columns in the first matrix is the same as the division of the rows in the second. Thus, the above division is not suitable for working out the product BA in terms of the 2×2 block matrices. (Is there a different subdivision of B which will work for both products AB and BA?)

We now return to the discussion of vectors and linear operators.

6–4 COORDINATE TRANSFORMATIONS

As we have seen, a linear operator is represented by a matrix in a given coordinate system. The right side of Eq. (6-9)

$$y_j = \sum_i A_{ji} x_i$$

is a special case of a matrix product Ax, in which the right-hand matrix x (and thus the product y) has a single column.   Such matrices are called *column vectors*.

We now ask how the components of vectors and linear operators transform when we change the coordinate system.   Consider a basis e'_j defined in terms of the unprimed system by

$$e'_j = \sum_i \gamma_{ij} e_i \qquad (6\text{-}27)$$

The coefficient γ_{ij} is the ith component of e'_j in the unprimed system (note the convention adopted for the order of the indices).

The n^2 coefficients γ_{ij} form the elements of a *transformation matrix* γ, which effects the transformation from one system to the other.   Consider an arbitrary vector x with components x_i and x'_j in the two systems.   Then

$$x = \sum_i x_i e_i$$

$$= \sum_j x'_j e'_j = \sum_j x'_j \sum_i \gamma_{ij} e_i$$

Therefore,

$$x_i = \sum_j \gamma_{ij} x'_j \qquad (6\text{-}28)$$

which is equivalent to the matrix equation

$$x = \gamma x' \qquad (6\text{-}29)$$

The linear independence of the base vectors e'_j assures us that the matrix γ is nonsingular so that it has an inverse.   Multiplying (6-29) by γ^{-1},

$$x' = \gamma^{-1} x \qquad (6\text{-}30)$$

Notice that the components x_i transform differently than the vectors e_i [compare (6-27) with (6-30)].   Things that transform like the x_i's are said to transform *contragrediently* to things that transform like the e_i's.   Things that transform in the same way are said to transform *cogrediently* to each other.

Next we find the transformation law for the components of linear operators by writing the operator equation

$$y = \mathscr{A}x$$

as matrix equations in the two coordinate systems:

$$y = Ax \qquad y' = A'x'$$

Expressing x and y in the first equation with the aid of (6-29) gives

$$\gamma y' = A\gamma x'$$

or

$$y' = \gamma^{-1} A\gamma x'$$

Thus, the desired transformation is

$$A' = \gamma^{-1}A\gamma \qquad A = \gamma A'\gamma^{-1} \tag{6-31}$$

This is an example of a *similarity transformation* which is defined to be a transformation of square matrices of the form

$$A' = S^{-1}AS \tag{6-32}$$

Any algebraic matrix equation remains unchanged under a similarity transformation; for example, the equation

$$ABC + \lambda D = 0$$

implies

$$S^{-1}A(SS^{-1})B(SS^{-1})CS + S^{-1}\lambda DS = 0$$

or

$$A'B'C' + \lambda D' = 0$$

Coordinate transformations in which the transformation matrix γ is orthogonal or unitary play an especially important role in physics for reasons which will become clear below. First, we consider a *linear scalar function* of a vector. This is a concept related to that of a linear vector function which was used above to introduce matrices. A linear scalar function is a rule which, for every vector $\mathbf{x}$, defines a scalar $\phi(\mathbf{x})$ in a linear way:

$$\phi(\lambda\mathbf{x} + \mu\mathbf{y}) = \lambda\phi(\mathbf{x}) + \mu\phi(\mathbf{y}) \tag{6-33}$$

If we have a basis $\mathbf{e}_i$, the linear scalar function may be conveniently specified by the n numbers $\alpha_i = \phi(\mathbf{e}_i)$, which we shall call the *components* of the scalar function ϕ. Then

$$\phi(\mathbf{x}) = \phi\left(\sum_i x_i\mathbf{e}_i\right) = \sum_i \alpha_i x_i \tag{6-34}$$

This expression fits the rule for matrix multiplication if the components α_i form a *row vector*, that is, a matrix having a single row. Then (6-34) is the matrix equation

$$\phi(\mathbf{x}) = \alpha x \tag{6-35}$$

If we write the components in another (primed) coordinate system, the scalar number $\phi(\mathbf{x})$ must remain the same, that is,

$$\phi(\mathbf{x}) = \alpha'x' = \alpha'\gamma^{-1}x$$

Thus the requirement of invariance determines the law of transformation of the α_i:

$$\alpha = \alpha'\gamma^{-1} \qquad \alpha' = \alpha\gamma \tag{6-36}$$

or

$$\alpha'_j = \sum_i \alpha_i \gamma_{ij}$$

Comparison with (6-27) and (6-30) shows that the components α_i transform contragrediently to the components x_i of a vector, and cogrediently to the base vectors e_i.

The linear scalar functions may be thought of as forming a vector space which is distinct from the space of the original vectors, but has the same dimensionality. It is sometimes called the *dual space*.

We now introduce the idea of a scalar product, and thereby a *metric*, in our original vector space. For every pair of vectors **a** and **b**, we define a scalar product (or *inner product*),

$$\mathbf{a} \cdot \mathbf{b} \qquad \text{or} \qquad (\mathbf{a}, \mathbf{b})$$

as a scalar function of the two vectors **a**, **b** with the following properties:

(1) $\mathbf{a} \cdot \mathbf{b} = (\mathbf{b} \cdot \mathbf{a})^*$ (6-37)

(2) $\mathbf{a} \cdot (\lambda \mathbf{b} + \mu \mathbf{c}) = \lambda \mathbf{a} \cdot \mathbf{b} + \mu \mathbf{a} \cdot \mathbf{c}$ (6-38)

(3) $\mathbf{a} \cdot \mathbf{a} \geqslant 0$ and $\mathbf{a} \cdot \mathbf{a} = 0$ implies $\mathbf{a} = 0$ (6-39)

Two vectors whose scalar product is zero are said to be *orthogonal*, and the *length* of a vector **a** is defined to be $|\mathbf{a} \cdot \mathbf{a}|^{\frac{1}{2}}$.

Notice that properties (6-37) and (6-38) imply that the scalar product $\mathbf{a} \cdot \mathbf{b}$ is *antilinear* (or *conjugate linear*) in its first argument, that is,

$$(\lambda \mathbf{a} + \mu \mathbf{b}) \cdot \mathbf{c} = \lambda^* \mathbf{a} \cdot \mathbf{c} + \mu^* \mathbf{b} \cdot \mathbf{c}$$ (6-40)

A conventional way to define a scalar product obeying the axioms (6-37), (6-38), and (6-39) is to define

$$\mathbf{a} \cdot \mathbf{b} = \sum_i a_i^* b_i$$ (6-41)

where a_i and b_i are the components of **a** and **b** *in some particular coordinate system*. The base vectors e_i of this coordinate system clearly satisfy the condition

$$\mathbf{e}_i \cdot \mathbf{e}_j = \delta_{ij}$$ (6-42)

that is, the base vectors are *orthonormal* (orthogonal and normalized to unit length). Conversely, if our base vectors are orthonormal [that is (6-42) holds], it follows immediately that the scalar product takes the form (6-41).

Suppose that we have a vector space in which a metric is defined, and a coordinate system with orthonormal base vectors e_i. If we wish to introduce a new basis $\mathbf{e}'_i$, what condition must we impose on the transformation matrix γ in order that the new base vectors be also orthonormal? We require

$$\delta_{ij} = \mathbf{e}'_i \cdot \mathbf{e}'_j = \left(\sum_k \gamma_{ki}\,\mathbf{e}_k\right) \cdot \left(\sum_l \gamma_{lj}\,\mathbf{e}_l\right)$$

$$= \sum_k \gamma^*_{ki}\gamma_{kj} = (\gamma^\dagger\gamma)_{ij}$$

Thus, $\gamma^\dagger\gamma = 1$ and γ must be unitary in order that the form (6-41) of our scalar product remain unchanged. In classical physics, γ is usually real as well as unitary, that is, orthogonal.

Other important properties of unitary transformations have to do with transformations of matrices rather than vectors. A similarity transformation by a unitary matrix leaves a unitary matrix unitary and a Hermitian matrix Hermitian. These properties may be easily demonstrated by the student (see Problem 6-3).

Finally, we point out that a scalar product enables us to establish a one-to-one correspondence between vectors and linear scalar functions, that is, between our original vector space and its dual. We associate with each vector $\mathbf{a}$ the linear scalar function $\phi_\mathbf{a}(\mathbf{x})$ defined by

$$\phi_\mathbf{a}(\mathbf{x}) = \mathbf{a} \cdot \mathbf{x}$$

In the coordinate system used above in (6-41) and (6-42), $\phi_\mathbf{a}$ has the components

$$\phi_\mathbf{a}(\mathbf{e}_i) = a^*_i$$

6-5 EIGENVALUE PROBLEMS

When an operator $\mathscr{A}$ acts on a vector $\mathbf{x}$, the resulting vector $\mathscr{A}\mathbf{x}$ is, in general, distinct from $\mathbf{x}$. However, there may exist certain (nonzero) vectors for which $\mathscr{A}\mathbf{x}$ is just $\mathbf{x}$ multiplied by a constant λ. That is,

$$\mathscr{A}\mathbf{x} = \lambda\mathbf{x} \qquad\qquad (6\text{-}43)$$

Such a vector is called an *eigenvector* of the operator $\mathscr{A}$, and the constant λ is called an *eigenvalue*. The eigenvector is said to "belong" to the eigenvalue.

EXAMPLE

Consider the rotation operator $\mathscr{R}$ illustrated in Figure 6–1. Any vector lying along the axis of rotation (z-axis) is an eigenvector of $\mathscr{R}$ belonging to the eigenvalue 1. (Are there any other eigenvectors?)

In a given coordinate system, the i-component of Eq. (6-43) is

$$\sum_j A_{ij} x_j = \lambda x_i \qquad i = 1, 2, \ldots, n \qquad (6\text{-}44)$$

or, in matrix notation,

$$Ax = \lambda x \qquad (6\text{-}45)$$

The problem of finding the eigenvalues λ for which the system of linear equations (6-44) has a nontrivial solution is a very important one.

EXAMPLE

If A is the matrix

$$A = \begin{pmatrix} 1 & 2 & 3 \\ 4 & 5 & 6 \\ 7 & 8 & 9 \end{pmatrix}$$

the system of equations (6-44) is

$$(1 - \lambda)x_1 + 2x_2 + 3x_3 = 0$$
$$4x_1 + (5 - \lambda)x_2 + 6x_3 = 0$$
$$7x_1 + 8x_2 + (9 - \lambda)x_3 = 0$$

If these equations are to have a nontrivial solution, the determinant of the coefficients must vanish:

$$\begin{vmatrix} 1 - \lambda & 2 & 3 \\ 4 & 5 - \lambda & 6 \\ 7 & 8 & 9 - \lambda \end{vmatrix} = 0$$

This gives a third-order polynomial in λ whose three roots are the eigenvalues λ_i.

In general, the eigenvalues of a matrix A are determined by the equation

$$\det (A - \lambda 1) = 0 \qquad (\text{secular equation}) \qquad (6\text{-}46)$$

If A is a $n \times n$ matrix, there will be n roots λ, not necessarily all different.

We shall now prove two important theorems concerning the eigenvalues and eigenvectors of Hermitian matrices. Let H be a Hermitian matrix, λ_1 and λ_2 two of its eigenvalues, and x_1 and x_2 two eigenvectors, belonging to λ_1 and λ_2, respectively. That is,

$$Hx_1 = \lambda_1 x_1 \qquad Hx_2 = \lambda_2 x_2 \qquad (6\text{-}47)$$

Take the scalar product[4] of x_2 with the first equation and x_1 with the second.

$$x_2^\dagger H x_1 = \lambda_1 x_2^\dagger x_1$$
$$x_1^\dagger H x_2 = \lambda_2 x_1^\dagger x_2 \qquad (6\text{-}48)$$

The left sides are complex conjugates since[5]

$$(x_2^\dagger H x_1)^* = \sum_{ij} (x_{i2}^* H_{ij} x_{j1})^*$$
$$= \sum_{ij} x_{i2} H_{ij}^* x_{j1}^* = \sum_{ij} x_{j1}^* H_{ji} x_{i2}$$

that is,

$$(x_2^\dagger H x_1)^* = x_1^\dagger H x_2 \qquad (6\text{-}49)$$

Therefore, from (6-48),

$$(\lambda_1 - \lambda_2^*) x_2^\dagger x_1 = 0 \qquad (6\text{-}50)$$

Suppose first that $\lambda_1 = \lambda_2$, $x_1 = x_2 \neq 0$. Then $x_2^\dagger x_1 = x_1^\dagger x_1 > 0$, so that $\lambda_1 = \lambda_1^*$. *The eigenvalues of a Hermitian matrix are real.*

Alternatively, suppose $\lambda_1 \neq \lambda_2$. Then $x_2^\dagger x_1 = 0$. *Eigenvectors of a Hermitian matrix belonging to different eigenvalues are orthogonal.*

Several eigenvectors may belong to the same eigenvalue; such an eigenvalue is said to be *degenerate*. The eigenvectors in question are also often called degenerate. We have seen above that eigenvectors of a Hermitian matrix belonging to different eigenvalues are orthogonal. What about eigenvectors belonging to the same (degenerate) eigenvalue? This question, which is a special case of the more general problem involving Hermitian *operators* and their eigenvectors (compare Section 9–2), is resolved as follows:

Suppose the three independent eigenvectors $\mathbf{x}_1$, $\mathbf{x}_2$, and $\mathbf{x}_3$ belong to the eigenvalue λ. Clearly, any linear combination of these is also an eigenvector. Let

$$\mathbf{y}_1 = \mathbf{x}_1$$
$$\mathbf{y}_2^* = \mathbf{x}_2 + \alpha \mathbf{y}_1$$

We wish to choose α so that $\mathbf{y}_1$ and $\mathbf{y}_2$ are orthogonal. Therefore

$$\mathbf{y}_1 \cdot \mathbf{y}_2 = 0 = \mathbf{y}_1 \cdot \mathbf{x}_2 + \alpha \mathbf{y}_1 \cdot \mathbf{y}_1$$

$$\alpha = -\frac{\mathbf{y}_1 \cdot \mathbf{x}_2}{\mathbf{y}_1 \cdot \mathbf{y}_1}$$

---

[4] We are taking, as the scalar product of two column vectors u and v, the expression

$$u^\dagger v = \sum_i u_i^* v_i$$

[5] Any notation for the components of a set of vectors $\mathbf{x}_1, \mathbf{x}_2, \ldots, \mathbf{x}_n$, which are distinguished by subscripts, is likely to be confusing. Our convention is that x_{ij} is the i-component of $\mathbf{x}_j$ [compare Eqs. (6-8) and (6-27)].

Now set

$$\mathbf{y}_3 = \mathbf{x}_3 + \beta\mathbf{y}_1 + \gamma\mathbf{y}_2$$

We wish to choose β and γ so that $\mathbf{y}_3$ is orthogonal to both $\mathbf{y}_1$ and $\mathbf{y}_2$.   Thus

$$0 = \mathbf{y}_1 \cdot \mathbf{x}_3 + \beta\mathbf{y}_1 \cdot \mathbf{y}_1 \Rightarrow \beta = -\frac{\mathbf{y}_1 \cdot \mathbf{x}_3}{\mathbf{y}_1 \cdot \mathbf{y}_1}$$

$$0 = \mathbf{y}_2 \cdot \mathbf{x}_3 + \gamma\mathbf{y}_2 \cdot \mathbf{y}_2 \Rightarrow \gamma = -\frac{\mathbf{y}_2 \cdot \mathbf{x}_3}{\mathbf{y}_2 \cdot \mathbf{y}_2}$$

Thus we have constructed three mutually orthogonal eigenvectors $\mathbf{y}_i$. This procedure, known as the *Gram–Schmidt orthogonalization procedure*, clearly can be extended to an arbitrary number of degenerate eigenvectors. Thus, we can arrange things so that all n eigenvectors of a Hermitian matrix are mutually orthogonal (provided all n eigenvectors are linearly independent, which is true as we shall show later, in Section 6–6).

The eigenvalues and eigenvectors of a linear operator are independent of the particular coordinate system used in finding them, as is clear from Eq. (6-43).   It is also clear from the matrix equation (6-45)

$$Ax = \lambda x$$

which implies

$$\gamma^{-1}A\gamma\gamma^{-1}x = \lambda\gamma^{-1}x$$

or

$$A'x' = \lambda x'$$

Thus, if x is an eigenvector of A, its transform $x' = \gamma^{-1}x$ is an eigenvector of the transformed matrix A', and the eigenvalues are the same.

Suppose we make a transformation to a coordinate system in which the base vectors $\mathbf{e}'_i$ are eigenvectors of a linear operator $\mathscr{A}$ (assuming these form a linearly independent set),

$$\mathscr{A}\mathbf{e}'_i = \lambda_i\mathbf{e}'_i \tag{6-51}$$

In this system, the matrix element A'_{ij} is the ith component of the vector $\mathscr{A}\mathbf{e}'_j$, which, from (6-51), is zero or λ_j according to $i \neq j$ or $i = j$, respectively. That is,

$$A'_{ij} = \lambda_j\,\delta_{ij} \tag{6-52}$$

Thus the matrix A' is diagonal, and the diagonal elements are the eigenvalues!

Because of its importance, we write the specific form of the transformation matrix γ in this case.   Recall [Eq. (6-27)] that γ_{ij} is the ith component of the new e'_j in the old (unprimed) system.   That is, the jth column of γ consists of the components of the eigenvector e'_j .

$$\gamma = \begin{pmatrix} e'_1 & e'_2 & & e'_n \\ \downarrow & \downarrow & \cdots & \downarrow \end{pmatrix} \tag{6-53}$$

If the transformation is to be unitary, the eigenvectors must be mutually orthogonal (made so by the Gram–Schmidt procedure if necessary) and they must be normalized.

We have seen above that the eigenvalues of a matrix remain unchanged when the matrix undergoes a similarity transformation.   They are *invariants* of the matrix.   Other invariants are the trace and the determinant:

$$\text{Tr } A' = \text{Tr } S^{-1}AS = \text{Tr } ASS^{-1} = \text{Tr } A \tag{6-54}$$

$$\det A' = \det (S^{-1}AS) = \det (S^{-1})(\det A)(\det S) = \det A \tag{6-55}$$

These invariants are not independent of the eigenvalues.   In fact, we may evaluate them in the system in which A is diagonal, with the result

$$\text{Tr } A = \sum_i \lambda_i \tag{6-56}$$

$$\det A = \prod_i \lambda_i \tag{6-57}$$

The eigenvalues are the only independent invariants.

We conclude this section with an example from physics of an eigenvalue problem involving matrices.

EXAMPLE

Consider the problem of molecular vibrations, or, in general, "small vibrations" of a (classical) mechanical system. For example, in Figure 6–2, are shown schematically a water molecule and a simple vibrating system consisting of three masses connected by springs.

The configuration or *state* of such a system may be conveniently described by an n-dimensional vector x where n is the number of degrees of freedom,

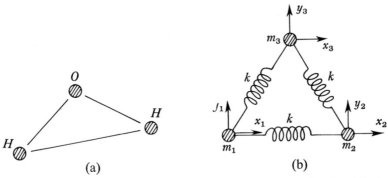

Figure 6–2   **Simple vibrating systems: (a) a water molecule; (b) three masses connected by springs**

that is, three times the number of masses.   A simple coordinate system consists of the base vectors:

$\mathbf{e}_1$: mass 1 displaced a unit distance in the x-direction

$\mathbf{e}_2$: mass 1 displaced a unit distance in the y-direction

$\mathbf{e}_3$: mass 1 displaced a unit distance in the z-direction

$\mathbf{e}_4$: mass 2 displaced a unit distance in the x-direction

etc.

In terms of components in this coordinate system, the kinetic energy is

$$T = \tfrac{1}{2} \sum_i m_i \dot{x}_i^2 = \tfrac{1}{2}\tilde{x}M\dot{x} \qquad (6\text{-}58)$$

where m_i is the mass associated with x_i (that is $m_1 = m_2 = m_3 =$ mass of particle 1; $m_4 = m_5 = m_6 =$ mass of particle 2, etc.), $\dot{x}_i = dx_i/dt$, and M is the diagonal matrix $M_{ij} = m_i \, \delta_{ij}$.

For vibrations of small amplitude, we expand the potential energy in a Taylor's series

$$V = V_0 + \sum_i \left(\frac{\partial V}{\partial x_i}\right)_0 x_i + \sum_{ij} \frac{1}{2}\left(\frac{\partial^2 V}{\partial x_i \, \partial x_j}\right)_0 x_i x_j + \cdots \qquad (6\text{-}59)$$

The force $(\delta V/\delta x_i)_0$ at equilibrium must be zero, and we can choose $V_0 = 0$, so neglecting third-order terms the potential energy has the form

$$V = \tfrac{1}{2} \sum_{ij} V_{ij} x_i x_j = \tfrac{1}{2}\tilde{x}Vx \qquad (6\text{-}60)$$

where V is a real symmetric matrix; $V_{ij} = [\delta^2 V/(\delta x_i \, \delta x_j)]_0$.

The expression for kinetic energy is further simplified if we transform to a new coordinate system with basis

$$\mathbf{e}_i' = \frac{1}{\sqrt{m_i}} \, \mathbf{e}_i \qquad (6\text{-}61)$$

The transformation matrix

$$\gamma = \begin{pmatrix} \dfrac{1}{\sqrt{m_1}} & 0 & 0 & \cdots & 0 \\ 0 & \dfrac{1}{\sqrt{m_2}} & 0 & \cdots & 0 \\ \cdot & \cdot & \cdot & \cdots & \cdot \\ 0 & 0 & 0 & \cdots & \dfrac{1}{\sqrt{m_n}} \end{pmatrix} \qquad (6\text{-}62)$$

is *not* unitary so that, for example, the form of a scalar product will not remain invariant. But this is just what we desire in order to obtain the expression (6-64) below for T. We have

$$x = \gamma x' \qquad \tilde{x} = \tilde{x}'\tilde{\gamma} \qquad (6\text{-}63)$$

so that

$$T = \tfrac{1}{2}\tilde{\dot{x}}'\tilde{\gamma}M\gamma\dot{x}' = \tfrac{1}{2}\tilde{\dot{x}}'\dot{x}'$$

$$V = \tfrac{1}{2}\tilde{x}'\tilde{\gamma}V\gamma x' = \tfrac{1}{2}\tilde{x}'Ax' \qquad (6\text{-}64)$$

where we define

$$A = \tilde{\gamma}V\gamma \qquad (6\text{-}65)$$

(A transformation of this form, $\tilde{S}VS$, is called a *congruent transformation*.)

Now A is real and symmetric (and thus Hermitian) so we know it has n real eigenvalues λ_k and n real eigenvectors[6] $\mathbf{u}_k$ which can be made orthonormal:

$$A\mathbf{u}_k = \lambda_k \mathbf{u}_k \qquad (6\text{-}66)$$

If we make a unitary[7] transformation from the $\mathbf{e}'_j$ system to the $\mathbf{u}_k$ system and call the coordinates of the vector $\mathbf{x}$ in the latter system q_k, the result is that the potential energy matrix is also diagonalized:

$$T = \tfrac{1}{2}\tilde{\dot{q}}\dot{q} = \tfrac{1}{2}\sum_k \dot{q}_k^2$$

$$V = \tfrac{1}{2}\tilde{q}\Lambda q = \tfrac{1}{2}\sum_k \lambda_k q_k^2 \qquad (6\text{-}67)$$

Λ is the diagonal matrix $\Lambda_{kl} = \lambda_k \delta_{kl}$. We know that all the eigenvalues λ_k are real and that $\lambda_k \geqslant 0$ since V can never be negative if the equilibrium is stable.

---

[6] The linear independence of these eigenvectors is proved in the next section.
[7] Since the $\mathbf{u}_k$ are real, our transformation is orthogonal as well as unitary.

Lagrange's equations of motion in terms of the generalized coordinates q_k are

$$\frac{d}{dt}\frac{\partial T}{\partial \dot{q}_k} + \frac{\partial V}{\partial q_k} = 0 \tag{6-68}$$

or

$$\ddot{q}_k + \lambda_k q_k = 0 \tag{6-69}$$

The solutions are

(1) $\qquad \lambda_k > 0: q_k = a_k \sin(\omega_k t + \delta_k) \qquad \omega_k = \sqrt{\lambda_k}$ $\qquad$ (6-70)

(2) $\qquad \lambda_k = 0: q_k = b_k t + c_k$

The frequencies ω_k are called the *normal frequencies* and the q_k are the *normal coordinates*. If only one $q_k \neq 0$, the system is vibrating in a *normal mode*, in which all the original coordinates x_i vary harmonically with time with a common frequency ω_k.

Some of the zero frequencies, $\lambda_k = 0$, correspond to translations and rotations of the system as a whole. We know that the eigenvectors describing the true vibrational modes are orthogonal to these (and to each other). One can show that the conditions of orthogonality with the translational and rotational modes translate into the physical statements that in a vibrational mode the center of mass is stationary and the angular momentum is zero.

Finally, we transform back to the original (unprimed) coordinate system.

$$x' = U'q$$
$$x = \gamma x' = \gamma U'q \tag{6-71}$$

where γ is the matrix (6-62) and U' is the *unitary* transformation matrix from the $\mathbf{e}'_i$ to the $\mathbf{u}_k$ system; $U'_{ik} = u'_{ik} = \mathbf{e}'_i \cdot \mathbf{u}_k =$ the ith coordinate in the $\mathbf{e}'_i$ system of the eigenvector $\mathbf{u}_k$. Then each Cartesian component x_i is expressed as a sum over normal modes:

$$x_i = \sum_{jk} \gamma_{ij} U'_{jk} q_k$$

$$= \sum_k \frac{1}{\sqrt{m_i}} u'_{ik} q_k = \sum_k \frac{u'_{ik}}{\sqrt{m_i}} a_k \sin(\omega_k t + \delta_k)$$

$$= \sum_k u_{ik} a_k \sin(\omega_k t + \delta_k) \tag{6-72}$$

where u_{ik} is the ith component of $\mathbf{u}_k$ in the unprimed system.

The orthogonality and normalization relations for the eigenvectors $\mathbf{u}_k$, when expressed in terms of the elementary (unprimed) system, are

$$\delta_{kl} = \mathbf{u}_k \cdot \mathbf{u}_l = \sum_j u'_{jk} u'_{jl} = \sum_j m_j u_{jk} u_{jl} \tag{6-73}$$

6–6 DIAGONALIZATION OF MATRICES

We have seen above that a square matrix can be diagonalized by a unitary matrix formed from its eigenvectors, provided the eigenvectors form a complete orthonormal set. We now show that this is true for a Hermitian matrix H.

There is at least one eigenvalue and eigenvector

$$Hu_1 = \lambda_1 u_1 \qquad (6\text{-}74)$$

because the secular equation (6-46) has at least one solution. Choose an orthonormal coordinate system e_i' in which the first member $e_1' = \mathbf{u}_1$. (This can be done because there exist sets of n linearly independent vectors which include $\mathbf{u}_1$, and these can be rearranged in orthogonal combinations by the Gram–Schmidt procedure if necessary.) Transforming H to this system, the elements in the first column are

$$H_{i1}' = \mathbf{e}_i' \cdot H\mathbf{u}_1 = \lambda_1\, \delta_{i1} \qquad (6\text{-}75)$$

Then, the fact that H' is Hermitian determines the first row, and we have the following form for H':

$$H' = \begin{pmatrix} \lambda_1 & 0 & 0 & \cdots & 0 \\ \hline 0 & & & & \\ 0 & & & G & \\ \vdots & & & & \\ 0 & & & & \end{pmatrix} \qquad (6\text{-}76)$$

where G is a Hermitian matrix of dimensionality $n - 1$, in the vector subspace normal to $\mathbf{u}_1$. We now repeat the same process with G and continue in this way until H is completely diagonalized. We have found in the process n independent eigenvectors of H, which form a unitary transformation matrix which diagonalizes H.

Now suppose we have two Hermitian matrices H_1 and H_2. Can they be diagonalized "simultaneously" by the same unitary transformation? That is, can we find a unitary matrix U such that

$$D_1 = U^{-1}H_1 U \qquad \text{and} \qquad D_2 = U^{-1}H_2 U \qquad (6\text{-}77)$$

are both diagonal?

Since diagonal matrices clearly commute with each other,

$$0 = D_1 D_2 - D_2 D_1 = U^{-1}(H_1 H_2 - H_2 H_1)U \qquad (6\text{-}78)$$

$$H_1 H_2 - H_2 H_1 = 0$$

Therefore, a *necessary* condition is that H_1 and H_2 commute.

This condition is also *sufficient*.   For suppose H_1 and H_2 commute.   Let

$$U^{-1}H_1U = D \qquad \text{(diagonal)}$$

$$U^{-1}H_2 U = M \qquad \text{(maybe not diagonal)}$$

Now D and M commute

$$DM = MD$$

The ij element of this equation is

$$D_{ii} M_{ij} = M_{ij} D_{jj} \tag{6-79}$$

Thus, if $D_{ii} \neq D_{jj}$, $M_{ij} = 0$.   This does not mean that M is diagonal, however, because H_1 might have some degenerate eigenvalues, that is, several of the elements of D may be equal.   Suppose, for example, the first three are equal.   Then

$$D = \begin{pmatrix} \lambda & 0 & 0 & & & \\ 0 & \lambda & 0 & & 0 & \\ 0 & 0 & \lambda & & & \\ \hline & & & \lambda_4 & & 0 \\ & 0 & & & \cdot & \\ & & & 0 & & \cdot \end{pmatrix} \qquad M = \begin{pmatrix} a & b & c & & \\ d & e & f & & 0 \\ g & h & i & & \\ \hline & 0 & & & N \end{pmatrix} \tag{6-80}$$

That is, M is in block diagonal form.   The submatrix in the upper left-hand corner of M is Hermitian and can be diagonalized by a unitary transformation which involves only the first three rows and columns.   This corner of D is just a multiple of the unit matrix and is therefore unaffected.   Repeating this operation, clearly both M and D can be simultaneously diagonalized.

This leads to the general condition that a matrix be diagonalizable by means of a unitary transformation.   Consider an arbitrary matrix M.   We can write

$$M = A + iB \tag{6-81}$$

where A and B are both Hermitian, by choosing

$$A = \frac{M + M^\dagger}{2} \qquad B = \frac{M - M^\dagger}{2i} \tag{6-82}$$

(This is just like splitting a complex number into its real and imaginary parts.)   Now A and B can be diagonalized separately, but in order that M may be diagonalized, we must be able to diagonalize A and B simultaneously.   The requirement for this is that A and B commute, which is equivalent to requiring that M and $M^\dagger$ commute. A matrix which commutes with its Hermitian conjugate is said to be *normal*; a matrix M can be diagonalized by a unitary transformation if, and only if, M is normal. Note that unitary matrices as well as Hermitian matrices satisfy this condition.

6–7 SPACES OF INFINITE DIMENSIONALITY

In this section we given a descriptive treatment of how some of the relations discussed above can be extended to a vector space with an infinite number of dimensions. As a heuristic introduction, consider the problem of vibrations of a "loaded string," that is, a stretched string of length L under tension T with a number n of equal masses fastened to it at equal intervals. This system is shown in Figure 6–3.

As in the example at the end of Section 6–5, we represent the configuration, or "state," of the system by an n-dimensional vector **y**. An elementary basis can be formed with the vectors $\mathbf{e}_l$, where $\mathbf{e}_l$ is the configuration with the lth mass displaced a distance $1/\sqrt{m_l}$ and all others in their equilibrium positions. (This is the *primed* coordinate system of Section 6–5.) The coordinate y_l of **y** in this elementary system is simply the displacement of the lth mass in units $1/\sqrt{m_l}$. One can show that the eigenvectors $\mathbf{u}_k$ corresponding to the normal modes have components (in the elementary system)

$$u_{lk} = (\mathbf{u}_k)_l = \sqrt{\frac{2a}{L}} \sin \frac{lk\pi a}{L} \tag{6-83}$$

where a is the spacing between masses; $(n+1)a = L$; and the constant $\sqrt{2a/L}$ is for normalization:

$$\mathbf{u}_k \cdot \mathbf{u}_k = 1 = \sum_l \left(\frac{2a}{L}\right) \sin^2 \frac{lk\pi a}{L} \tag{6-84}$$

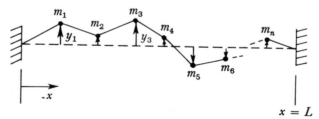

Figure 6–3   A vibrating "loaded string."   All masses are equal, $m_l = m$

An arbitrary configuration of the system at any given time may be expressed in terms of normal coordinates by (6-71):

$$y_l = \sum_k u_{lk} q_k = \sum_k u_{lk} a_k \sin(\omega_k t + \delta_k) \tag{6-85}$$

where the frequencies are found to be

$$\omega_k = \sqrt{\frac{T}{ma}} \, 2 \sin \frac{k\pi a}{2L} \tag{6-86}$$

Suppose we now let the number of masses n become infinite, holding the length L fixed. Instead of designating a given mass by an integer l, it then becomes convenient to use the position variable $x = la$, which becomes a continuous variable in the limiting process. A number of other changes in notation occur.

1. The components of y in the elementary coordinate system are

$$y_l \to y(x)$$

$$\mathbf{y} = \sum_l y_l \mathbf{e}_l \to \mathbf{y} = \int y(x) \mathbf{e}_x \, dx \tag{6-87}$$

2. The elementary coordinate system becomes a *point coordinate system*, that is, a base vector $\mathbf{e}_b$ represents a δ-function displacement at only one point b. In fact, the components of a base vector in the elementary system are

$$(\mathbf{e}_l)_m = \delta_{lm} \to (\mathbf{e}_b)_x = \delta(x - b) \tag{6-88}$$

3. The transformation matrix to go from the elementary system to the eigenvector coordinate system has elements such that the kth "column" is made up of the components of $\mathbf{u}_k$ in the elementary system, that is,

$$U_{lk} = (\mathbf{u}_k)_l = \sqrt{\frac{2a}{L}} \sin \frac{lk\pi a}{L} \to u_k(x) = \sqrt{\frac{2}{L}} \sin \frac{k\pi x}{L} \tag{6-89}$$

4. The normalization-orthogonality relations for the eigenfunctions, evaluated in the elementary system, are

$$\mathbf{u}_k \cdot \mathbf{u}_l = \sum_{j=1}^{n} \frac{2}{(n+1)} \sin \frac{jk\pi}{(n+1)} \sin \frac{jl\pi}{(n+1)} = \delta_{kl}$$

$$\to \int_0^L u_k(x) u_l(x) \, dx = \int_0^L \left(\frac{2}{L}\right) \sin \frac{k\pi x}{L} \sin \frac{l\pi x}{L} \, dx = \delta_{kl} \tag{6-90}$$

5. The expansion (6-85) of an elementary component in terms of normal coordinates becomes

$$y_l = \sum_k u_{lk} q_k \to y(x) = \sum_k u_k(x) q_k$$

$$= \sum_k \sqrt{\frac{2}{L}} q_k \sin \frac{k\pi x}{L} \tag{6-91}$$

The last expression is just a Fourier (sine) series. q_k is the k-component of $\mathbf{y}$ in the $\mathbf{u}_k$ coordinate system.

6. The inverse relation is

$$q_k = \sum_l U_{kl}^{-1} y_l = \sum_l u_{lk}^* y_l$$

$$\rightarrow q_k = \int u_k^*(x) y(x)\, dx$$

$$= \sqrt{\frac{2}{L}} \int_0^L y(x) \sin \frac{k\pi x}{L}\, dx \tag{6-92}$$

which is the Fourier inversion integral.

Two kinds of infinite dimensional coordinate systems have been used above. One has a denumerably infinite number of base vectors, the eigenvectors $\mathbf{u}_k$, and these are normalizable in the usual way, $\mathbf{u}_k \cdot \mathbf{u}_l = \delta_{kl}$. The other system is nondenumerably infinite, and the normalization takes the form

$$\mathbf{e}_a \cdot \mathbf{e}_b = \int \delta(x - a)\, \delta(x - b)\, dx = \delta(a - b) \tag{6-93}$$

We may now make some generalizations based on the above simple example. We see that ordinary functions (satisfying certain mathematical conditions of good behavior which we shall not go into) form a linear vector space of infinite dimensionality. Such a vector space with a scalar product is called a *Hilbert space*.

Some further analogies with relations in a finite dimensional space follow.

7. A convenient definition for the scalar product, evaluated in the point coordinate system, is

$$\boldsymbol{\phi} \cdot \boldsymbol{\psi} = \int_a^b \phi^*(x) \psi(x)\, dx \tag{6-94}$$

8. An orthonormal set $\boldsymbol{\phi}_k$ is complete if an "arbitrary" vector $\boldsymbol{\psi}$ can be expanded in terms of the $\boldsymbol{\phi}_k$, which thus form a basis

$$\boldsymbol{\psi} = \sum_k a_k \boldsymbol{\phi}_k \tag{6-95}$$

or, taking the x-component

$$\psi(x) = \sum_k a_k \phi_k(x) \tag{6-96}$$

a_k is the k-component of $\boldsymbol{\psi}$ in the $\boldsymbol{\phi}_k$ system.

$$a_k = \boldsymbol{\phi}_k \cdot \boldsymbol{\psi} = \int_a^b \phi_k^*(x) \psi(x)\, dx \tag{6-97}$$

9. Linear operators.   Suppose $\mathcal{K}$ is a linear operator such that symbolically

$$\psi = \mathcal{K} \phi$$

In components, we have

$$\psi_i = \sum_j K_{ij} \phi_j \rightarrow \psi(x) = \int_a^b K(x, x')\phi(x')\, dx' \qquad (6\text{-}98)$$

The last expression is reminiscent of the integral transforms of Chapter 4; $K(x, x')$ is called the *kernel* of the transform.

10. The eigenvalue equation

$$\mathcal{K} \psi = \lambda \psi$$

becomes

$$\int_a^b K(x, x')\psi(x')\, dx' = \lambda \psi(x) \qquad (6\text{-}99)$$

This is a homogeneous integral equation which we shall meet again in Chapter 11.

We now terminate our discussion of infinite dimensional spaces. The ideas introduced here will reappear in Chapter 8 and elsewhere in this book.

REFERENCES

Matrices are discussed in most texts on mathematical methods in physics, such as Margenau and Murphy (M2) Chapter 10; Goertzel and Tralli (G3) Part I; Sokolnikoff and Redheffer (S8) Chapter 4; Hildebrand (H9) Chapter 1; Friedman (F8) Chapters 1 and 2; and Dennery and Krzywicki (D5) Chapter II. The last-named book is a particularly pleasant blend of rigor and informality.

In addition, there are of course many purely mathematical approaches to the subject of matrices and linear algebra. Three particularly fine texts are Halmos (H2); Gel'fand (G1); and Jacobson (J2). In addition, we mention Frazer *et al.* (F6); Aitken (A2); Turnbull (T9); and Schwartz (S3).

PROBLEMS

6-1   Show that
   (a) $\widetilde{(AB)} = \tilde{B}\tilde{A}$
   (b) $(AB)^\dagger = B^\dagger A^\dagger$
   (c) $A(BC) = (AB)C$
   (d) $\text{Tr } ABC = \text{Tr } BCA$
      where A, B, C are matrices.

6-2 Let U be a unitary matrix and let x_1, x_2 be two eigenvectors of U belonging to the eigenvalues λ_1, λ_2, respectively. Show that
(a) $|\lambda_1| = |\lambda_2| = 1$
(b) If $\lambda_1 \neq \lambda_2$, $x_1^\dagger x_2 = 0$

6-3 Suppose the matrices A and B are Hermitian and the matrices C and D are unitary. Prove that
(a) $C^{-1}AC$ is Hermitian
(b) $C^{-1}DC$ is unitary
(c) $i(AB - BA)$ is Hermitian

6-4 Find the eigenvalues and normalized eigenvectors of the matrix

$$\begin{pmatrix} 1 & 2 & 3 \\ 4 & 5 & 6 \\ 7 & 8 & 9 \end{pmatrix}$$

Express your answers numerically (3 significant figures).

6-5 Using a particular coordinate system, a linear transformation in an abstract vector space is represented by the matrix

$$\begin{pmatrix} 2 & 1 & 0 \\ 1 & 2 & 0 \\ 0 & 0 & 5 \end{pmatrix}$$

and a particular (abstract) vector by the column vector

$$\begin{pmatrix} 1 \\ 2 \\ 3 \end{pmatrix}$$

Give the matrix and column vector in a new coordinate system, in terms of which the old base vectors are represented by

$$\mathbf{e}_1 = \begin{pmatrix} 1 \\ 1 \\ 0 \end{pmatrix} \qquad \mathbf{e}_2 = \begin{pmatrix} 1 \\ -1 \\ 0 \end{pmatrix} \qquad \mathbf{e}_3 = \begin{pmatrix} 0 \\ 0 \\ 1 \end{pmatrix}$$

6-6 Transform the matrix A and vector $\mathbf{x}$ given below to a coordinate system in which A is diagonal.

$$A = \begin{pmatrix} 0 & -i & 0 & 0 & 0 \\ i & 0 & 0 & 0 & 0 \\ 0 & 0 & 3 & 0 & 0 \\ 0 & 0 & 0 & 1 & -i \\ 0 & 0 & 0 & i & 1 \end{pmatrix} \qquad x = \begin{pmatrix} 1 \\ a \\ i \\ b \\ -1 \end{pmatrix}$$

6-7 The inertia tensor of a rigid body may be thought of as a linear operator; the components of a particular inertia tensor in a particular Cartesian coordinate system form the matrix

$$
\begin{pmatrix}
\dfrac{5}{2} & \sqrt{\dfrac{3}{2}} & \sqrt{\dfrac{3}{4}} \\[2ex]
\sqrt{\dfrac{3}{2}} & \dfrac{7}{3} & \sqrt{\dfrac{1}{18}} \\[2ex]
\sqrt{\dfrac{3}{4}} & \sqrt{\dfrac{1}{18}} & \dfrac{13}{6}
\end{pmatrix}
$$

Find the components, in this same coordinate system, of three unit vectors (the so-called *principal axes* of the rigid body) such that the matrix, representing the inertia tensor in a coordinate system using these as base vectors, is diagonal. What are these diagonal elements (the *principal moments of inertia*)?

6-8 Find the normal modes and normal frequencies for the linear vibrations of the CO_2 molecule (that is, vibrations in the line of the molecule).

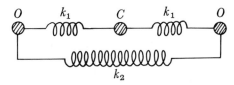

6-9 Prove the statements made in the text that, for a molecule, the orthogonality of a true vibrational mode with the translational and rotational modes means physically that in the vibration the center of mass remains stationary and the angular momentum is zero.

6-10 (a) Let **a** and **b** be any two vectors in a linear vector space, and define $\mathbf{c} = \mathbf{a} + \lambda\mathbf{b}$, where λ is a scalar. By requiring that $\mathbf{c} \cdot \mathbf{c} \geqslant 0$ for all λ, derive the *Cauchy–Schwartz inequality*

$$(\mathbf{a} \cdot \mathbf{a})(\mathbf{b} \cdot \mathbf{b}) \geqslant (\mathbf{a} \cdot \mathbf{b})^2$$

(When does equality hold?)

(b) In an infinite-dimensional space, questions of convergence arise and the expansion $\mathbf{x} = \sum x_i \mathbf{e}_i$ of an arbitrary vector $\mathbf{x}$ in terms of the base vectors $\mathbf{e}_1, \mathbf{e}_2, \ldots$ may lack meaning. A useful result can, however, be derived: Assume $\mathbf{e}_i \cdot \mathbf{e}_j = \delta_{ij}$, define $x_k = \mathbf{e}_k \cdot \mathbf{x}$, and define

$$\mathbf{x}^{(n)} = \sum_{i=1}^{n} x_i \mathbf{e}_i$$

Apply the Cauchy–Schwartz inequality (Part a) and derive the inequality

$$\sum_{i=1}^{n} |x_i|^2 \leq \mathbf{x} \cdot \mathbf{x}$$

This result, which is valid for any n no matter how large, is known as *Bessel's inequality*.

6-11 Consider a linear chain of N atoms, with atom N adjacent to atom 1 (cyclic boundary condition), as if the chain were bent into a circle. Each atom can be in any of three states, called A, B, and C, except that an atom in state A cannot be adjacent to an atom in state C. Find the entropy per atom of such a chain as $N \to \infty$. (Entropy $= k \ln W$, where W is the total number of allowed configurations.) *Hint*: Define the three-dimensional column vectors $v^{(j)}$ such that the three elements of $v^{(j)}$ are the total number of allowed configurations of the j-atom chain 1, 2, 3, ..., j, with atom j in state A, B, or C, respectively. For example,

$$v^{(1)} = \begin{pmatrix} 1 \\ 1 \\ 1 \end{pmatrix}, \qquad v^{(2)} = \begin{pmatrix} 2 \\ 3 \\ 2 \end{pmatrix}, \qquad v^{(3)} = \begin{pmatrix} 5 \\ 7 \\ 5 \end{pmatrix}, \dots$$

Clearly, $v^{(j+1)} = M v^{(j)}$, where

$$M = \begin{pmatrix} 1 & 1 & 0 \\ 1 & 1 & 1 \\ 0 & 1 & 1 \end{pmatrix}$$

is the so-called *transfer matrix* appropriate to this problem. Then

$$W = \text{tr } M^N \underset{N \to \infty}{\approx} (\text{largest eigenvalue of } M)^N$$

Show thereby that the entropy per atom is $k \ln (1 + \sqrt{2})$. Suppose atom N is *not* adjacent to atom 1. What is the entropy per atom?

Suppose the constraint forbidding an atom in state A to be adjacent to an atom in state C is removed. Now what is the entropy per atom?

SEVEN

SPECIAL
FUNCTIONS

The so-called "special functions of mathematical physics" are solutions of certain frequently occurring linear second-order differential equations. As examples of the elementary ones, we discuss Legendre functions, associated Legendre functions, and Bessel functions. These functions have many representations: as specified solutions of given differential equations, series, various integral representations, by recurrence relations, and by generating functions. Starting from one of these representations one can obtain the others. Thus, the particular paths followed in this chapter to get from one to the other are by no means unique but are illustrative.

Hypergeometric functions, which include many other functions as special cases, are treated in Sections 7–3 and 7–4. Mathieu functions are discussed in Section 7–5 in a way which is applicable to solutions of other equations having periodic coefficients. Finally, in Section 7–6 we present an introductory discussion of the properties of Jacobian elliptic functions.

7–1 LEGENDRE FUNCTIONS

Recall the Legendre differential equation (1-49)

$$(1 - x^2)\frac{d^2y}{dx^2} - 2x\frac{dy}{dx} + n(n + 1)y = 0 \tag{7-1}$$

In Chapter 1 we found two independent series solutions [Eq. (1-51)], such that if n is an integer, one or the other solution is simply a polynomial in x. Let us study these polynomials in somewhat greater detail.

A straightforward rewriting by means of factorials enables us to express the solution (1-51) in the form

$$
y = c_0 \sum_k (-)^k \frac{\left(\frac{n}{2}\right)!}{\left(\frac{n}{2} + k\right)!} \frac{\left(\frac{n}{2}\right)!}{\left(\frac{n}{2} - k\right)!} \frac{(n + 2k)!}{n!} \frac{x^{2k}}{(2k)!}
$$

$$
+ c_1 \sum_k (-)^k \frac{\left(\frac{n-1}{2}\right)!}{\left(\frac{n-1}{2} + k\right)!} \frac{\left(\frac{n-1}{2}\right)!}{\left(\frac{n-1}{2} - k\right)!} \frac{(n + 2k)!}{n!} \frac{x^{2k+1}}{(2k + 1)!}
$$

If $n = 0, 2, 4, \ldots$, the first series becomes a polynomial. We shall introduce the new summation index $r = n/2 - k$, and write this solution as

$$
c_0 \frac{\left[\left(\frac{n}{2}\right)!\right]^2}{n!} (-)^{n/2} \sum_r (-)^r \frac{(2n - 2r)!}{(n - r)! r! (n - 2r)!} x^{n - 2r}
$$

If, on the other hand, $n = 1, 3, 5, \ldots$, it is the second series that concerns us. In this series we define $r = [(n - 1)/2] - k$, and obtain the solution

$$
\frac{c_1}{2} \frac{\left[\left(\frac{n-1}{2}\right)!\right]^2}{n!} (-)^{\frac{n-1}{2}} \sum_r (-)^r \frac{(2n - 2r)!}{(n - r)! r! (n - 2r)!} x^{n - 2r}
$$

It is now clear that for any nonnegative integer n, even or odd, the polynomial

$$
P_n(x) = K_n \sum_r (-)^r \frac{(2n - 2r)!}{(n - r)! r! (n - 2r)!} x^{n - 2r} \qquad (K_n \text{ arbitrary})
$$

is a solution of Legendre's equation.

Before fixing the normalization constant K_n, we shall engage in some more factorial manipulation, and write

$$
P_n(x) = K_n \sum_r \frac{(-)^r}{(n - r)! r!} \left(\frac{d}{dx}\right)^n x^{2n - 2r}
$$

$$
= \frac{K_n}{n!} \left(\frac{d}{dx}\right)^n \sum_r (-)^r \frac{n!}{r! (n - r)!} x^{2n - 2r}
$$

$$
= \frac{K_n}{n!} \left(\frac{d}{dx}\right)^n (x^2 - 1)^n \qquad\qquad (7-2)
$$

It is conventional to impose the normalization $P_n(1) = 1$ for all n. The resulting value of K_n may be found by inspection, since $P_n(1)$ equals $K_n/n!$ times the nth derivative of the product

$$\overbrace{(x-1)(x-1)\cdots(x-1)}^{n\ \text{factors}} \qquad \overbrace{(x+1)(x+1)\cdots(x+1)}^{n\ \text{factors}}$$

evaluated at $x = 1$.   If a single factor $(x - 1)$ survives after the n differentiations, setting $x = 1$ will cause that term to vanish.   Thus the only non-vanishing contributions to $P_n(1)$ occur when each of the n differentiations turns one factor $(x - 1)$ into 1.   There are $n!$ ways this can happen, and setting $x = 1$ gives $(1 + 1)^n = 2^n$ each time; the result is

$$P_n(1) = \frac{K_n}{n!}\, n!\, 2^n = 2^n K_n$$

and we therefore choose

$$K_n = \frac{1}{2^n} \tag{7-3}$$

The resulting expression

$$P_n(x) = \frac{1}{2^n n!} \left(\frac{d}{dx}\right)^n (x^2 - 1)^n \tag{7-4}$$

is called *Rodrigues' formula* for the *Legendre polynomials*.   The first few are

$$\begin{aligned}
P_0(x) &= 1 & P_3(x) &= \tfrac{1}{2}(5x^3 - 3x) \\
P_1(x) &= x & P_4(x) &= \tfrac{1}{8}(35x^4 - 30x^2 + 3) \\
P_2(x) &= \tfrac{1}{2}(3x^2 - 1) & P_5(x) &= \tfrac{1}{8}(63x^5 - 70x^3 + 15x)
\end{aligned} \tag{7-5}$$

A contour integral representation follows immediately from Rodrigues' formula.   We make use of Cauchy's formula

$$f(z) = \frac{1}{2\pi i} \oint \frac{f(t)\, dt}{t - z}$$

provided $f(t)$ is regular within the contour, which encloses z once in a positive sense [see Appendix, Eq. (A–10)].   Differentiating n times with respect to z gives

$$\left(\frac{d}{dz}\right)^n f(z) = \frac{n!}{2\pi i} \oint \frac{f(t)\, dt}{(t - z)^{n+1}} \tag{7-6}$$

Therefore, from (7-4) we obtain

$$P_n(z) = \frac{1}{2^n} \frac{1}{2\pi i} \oint \frac{(t^2 - 1)^n}{(t - z)^{n+1}}\, dt \tag{7-7}$$

This is called *Schläfli's integral representation* for the $P_n(z)$.

We can deduce another integral representation as follows. Let the contour in Schläfli's integral (7-7) be a circle about z with a radius $|\sqrt{z^2 - 1}|$. That is,

$$t = z + \sqrt{z^2 - 1}\, e^{i\phi}$$

where ϕ goes from 0 to 2π. It then follows by straightforward algebra, that

$$t^2 - 1 = 2(t - z)(z + \sqrt{z^2 - 1}\cos\phi)$$

and

$$dt = i(t - z)\, d\phi$$

Therefore, by substituting into (7-7), we find

$$P_n(z) = \frac{1}{\pi}\int_0^\pi \left(z + \sqrt{z^2 - 1}\cos\phi\right)^n d\phi \tag{7-8}$$

This is called *Laplace's integral representation* for the Legendre polynomials. Next, we consider the series

$$\sum_{n=0}^{\infty} h^n P_n(z) = F(h, z) \tag{7-9}$$

Using Laplace's representation, we obtain

$$F(h, z) = \frac{1}{\pi}\int_0^\pi \sum_{n=0}^{\infty} h^n \left(z + \sqrt{z^2 - 1}\cos\phi\right)^n d\phi$$

$$= \frac{1}{\pi}\int_0^\pi \frac{d\phi}{1 - hz - h\sqrt{z^2 - 1}\cos\phi}$$

so that

$$F(h, z) = \frac{1}{\sqrt{1 - 2hz + h^2}} = \sum_{n=0}^{\infty} h^n P_n(z) \tag{7-10}$$

$F(h, z)$ is called the *generating function* for the Legendre polynomials.

Equation (7-10) gives a very useful representation for the *inverse distance* between two points in three-dimensional space. If the points are $\mathbf{x}$ and $\mathbf{x}'$, with $r' < r$, then

$$\frac{1}{|\mathbf{x} - \mathbf{x}'|} = \frac{1}{\sqrt{r^2 + r'^2 - 2rr'\cos\theta}}$$

$$= \frac{1}{r}\frac{1}{\sqrt{1 - 2\dfrac{r'}{r}\cos\theta + \left(\dfrac{r'}{r}\right)^2}}$$

$$= \sum_{l=0}^{\infty} \frac{(r')^l}{r^{l+1}} P_l(\cos\theta) \tag{7-11}$$

where θ is the angle between the vectors $\mathbf{x}$ and $\mathbf{x}'$.   If $r < r'$, r and r' must be exchanged[1] in formula (7-11); sometimes both formulas are combined by writing

$$\frac{1}{|\mathbf{x} - \mathbf{x}'|} = \sum_{l=0}^{\infty} \frac{r_<^l}{r_>^{l+1}} P_l(\cos \theta) \qquad (7\text{-}12)$$

where $r_<$ and $r_>$ denote the lesser and greater, respectively, of r and r'.

As an example of the usefulness of a generating function, we shall deduce the recursion relations among the Legendre polynomials:

$$F(h, z) = \frac{1}{\sqrt{1 - 2hz + h^2}}$$

$$\frac{\partial F}{\partial h} = \frac{z - h}{1 - 2hz + h^2} F$$

$$(1 - 2hz + h^2)\frac{\partial F}{\partial h} = (z - h)F$$

Equating coefficients of h^n on both sides gives

$$(n + 1)P_{n+1}(z) - 2znP_n(z) + (n - 1)P_{n-1}(z) = zP_n(z) - P_{n-1}(z)$$

or

$$(n + 1)P_{n+1}(z) - (2n + 1)zP_n(z) + nP_{n-1}(z) = 0 \qquad (7\text{-}13)$$

A second recursion relation is obtained by differentiating the generating function (7-10) with respect to z:

$$(1 - 2hz + h^2)\frac{\partial F}{\partial z} = hF$$

which gives

$$P_n'(z) - 2zP_{n-1}'(z) + P_{n-2}'(z) = P_{n-1}(z) \qquad (7\text{-}14)$$

From the two recursion relations (7-13) and (7-14), we can derive many others, for example,

$$P_{n+1}'(z) - zP_n'(z) = (n + 1)P_n(z)$$

$$zP_n'(z) - P_{n-1}'(z) = nP_n(z)$$

$$P_{n+1}'(z) - P_{n-1}'(z) = (2n + 1)P_n(z)$$

$$(z^2 - 1)P_n'(z) = nzP_n(z) - nP_{n-1}(z)$$

etc.

[1] Otherwise the series (7-11) fails to converge; compare Problem 7-7.

Another use of the generating function is in evaluating $P_n(z)$ at various points. For example, at $z = 1$, (7-10) gives

$$\sum_{n=0}^{\infty} h^n P_n(1) = \frac{1}{1-h} = 1 + h + h^2 + h^3 + \cdots$$

Therefore,

$$P_n(1) = 1 \tag{7-15}$$

At $z = 0$,

$$\sum_{n=0}^{\infty} h^n P_n(0) = (1 + h^2)^{-1/2} = 1 - \tfrac{1}{2}h^2 + (-\tfrac{1}{2})(-\tfrac{3}{2})\frac{h^4}{2!} + \cdots$$

Therefore,

$$P_n(0) = \begin{cases} 0 & \text{if } n \text{ is odd} \\ \dfrac{(n-1)!!(-1)^{n/2}}{2^{n/2}\left(\dfrac{n}{2}\right)!} & \text{if } n \text{ is even} \end{cases} \tag{7-16}$$

where we use the notation

$$n!! \ (n \text{ double factorial}) = n(n-2)(n-4)\cdots 5 \cdot 3 \cdot 1$$

We now discuss the orthogonality and normalization properties of the Legendre polynomials. For this purpose we use Rodrigues' expression (7-4). Consider the integral

$$I_{mn} = \int_{-1}^{+1} P_m(x)P_n(x)\,dx \qquad (m < n)$$

$$= \frac{1}{2^{m+n}}\frac{1}{m!\,n!}\int_{-1}^{+1}\left[\left(\frac{d}{dx}\right)^m (x^2-1)^m\right]\left[\left(\frac{d}{dx}\right)^n (x^2-1)^n\right]dx$$

We integrate by parts n times:

$$I_{mn} = \frac{-1}{2^{m+n}}\frac{1}{m!\,n!}\int_{-1}^{+1}\left[\left(\frac{d}{dx}\right)^{m+1}(x^2-1)^m\right]\left[\left(\frac{d}{dx}\right)^{n-1}(x^2-1)^n\right]dx$$

$$\cdot \ \cdot \ \cdot \ \cdot \ \cdot \ \cdot \ \cdot \ \cdot \ \cdot \ \cdot \ \cdot \ \cdot \ \cdot \ \cdot$$

$$= \frac{(-1)^n}{2^{m+n}\,m!\,n!}\int_{-1}^{+1}\left[\left(\frac{d}{dx}\right)^{m+n}(x^2-1)^m\right](x^2-1)^n\,dx$$

Therefore,

$$I_{mn} = 0 \quad \text{since} \quad \left(\frac{d}{dx}\right)^{m+n}(x^2-1)^m = 0 \quad \text{if} \quad m < n \tag{7-17}$$

Suppose $m = n$. Then

$$I_{nn} = \int_{-1}^{+1} [P_n(x)]^2 \, dx = \frac{1}{2^{2n}(n!)^2} \int_{-1}^{+1} \left[\left(\frac{d}{dx}\right)^n (x^2 - 1)^n \right]^2 dx$$

n integrations by parts now give

$$I_{nn} = \frac{(-1)^n}{2^{2n}(n!)^2} \int_{-1}^{+1} (x^2 - 1)^n \left(\frac{d}{dx}\right)^{2n} (x^2 - 1)^n \, dx$$

$$= \frac{(2n)!}{2^{2n}(n!)^2} \int_{-1}^{+1} (1 - x^2)^n \, dx$$

Let $x = 2u - 1$.

$$I_{nn} = \frac{2(2n)!}{(n!)^2} \int_{0}^{1} du(1 - u)^n u^n$$

$$= \frac{2(2n)!}{(n!)^2} B(n + 1, n + 1) \qquad [\text{compare (3-49)}]$$

Therefore,

$$I_{nn} = \frac{2}{2n + 1} \tag{7-18}$$

We have shown that

$$\int_{-1}^{+1} P_m(x) P_n(x) \, dx = \frac{2}{2n + 1} \delta_{mn} \tag{7-19}$$

Any "reasonable"[2] function $f(x)$ on the interval $-1 \leqslant x \leqslant 1$ can be expanded in a series of Legendre polynomials:

$$f(x) = \sum_{n=0}^{\infty} c_n P_n(x) \tag{7-20}$$

In the language of Chapter 6, the c_n may be thought of as the components of a vector f in a coordinate system with base vectors P_n. Multiplying both sides of (7-20) by $P_m(x)$ and integrating from -1 to $+1$ gives

$$c_m = \frac{2m + 1}{2} \int_{-1}^{+1} P_m(x) f(x) \, dx \tag{7-21}$$

---

[2] The condition that a function is required to be reasonable, or reasonably behaved, and so on, will occur from time to time in this text. We do not propose to be very precise about the exact conditions a function must satisfy to be a reasonable function. On the one hand, if $f(x)$ is continuous and has a finite number of maxima and minima, or if it has in addition a finite number of jump discontinuities, then the expansion (7-20) is valid, that is, the series on the right of (7-20) converges to $f(x)$. On the other hand, for far more general functions $f(x)$, the expansion (7-20) still converges "in the mean" (see Section 12-3).

What is the second linearly independent solution of Legendre's equation, besides $P_n(x)$? If we make the change of variable [compare the discussion following Eq. (1-40)]

$$y = vP_n(x) \qquad (7\text{-}22)$$

in Legendre's equation, we obtain the equation

$$(1 - x^2)v''P_n(x) + v'[2(1 - x^2)P_n'(x) - 2xP_n(x)] = 0$$

The solution of this equation is straightforward and gives

$$v = C \int \frac{dx}{(1 - x^2)[P_n(x)]^2} + C'$$

Thus we have found a second solution; the conventional definition is

$$Q_n(z) = -P_n(z) \int_\infty^z \frac{dz}{(z^2 - 1)[P_n(z)]^2} \qquad (7\text{-}23)$$

Notice that $Q_n(z)$ approaches zero like $z^{-(n+1)}$ as $z \to \infty$. It can be shown (see Copson (C8) for example) that

$$Q_n(z) = \frac{1}{2} P_n(z) \ln \left(\frac{z + 1}{z - 1} \right) + f_{n-1}(z) \qquad (7\text{-}24)$$

where $f_{n-1}(z)$ is a polynomial of degree $n - 1$ in z. Note that $Q_n(z)$ is a many-valued function; to make it single valued, it is conventional to cut the z-plane between -1 and $+1$ along the real axis and to define $Q(z)$ real for real $z > 1$. Then, if $-1 < x < +1$,

$$Q_n(x \pm i\varepsilon) = \frac{1}{2} P_n(x) \left[\ln \left(\frac{1 + x}{1 - x} \right) \mp i\pi \right] + f_{n-1}(x) \qquad (7\text{-}25)$$

If $-1 < x < +1$, one usually means by $Q_n(x)$ the arithmetic mean

$$Q_n(x) = \frac{1}{2} [Q_n(x + i\varepsilon) + Q_n(x - i\varepsilon)] = \frac{1}{2} P_n(x) \ln \left(\frac{1 + x}{1 - x} \right) + f_{n-1}(x) \quad (7\text{-}26)$$

The polynomial $f_{n-1}(x)$ is determined by the condition that $Q_n(z) \to 0$ like $z^{-(n+1)}$ as $z \to \infty$. For example,

$$Q_0(z) = \frac{1}{2} \ln \frac{z + 1}{z - 1} = \frac{1}{z} + \frac{1}{3z^3} + \frac{1}{5z^5} + \cdots$$

$$Q_1(z) = \frac{1}{2} z \ln \frac{z + 1}{z - 1} + f_0$$

$$= z \left(\frac{1}{z} + \frac{1}{3z^3} + \cdots \right) + f_0$$

Clearly, $f_0 = -1$, and

$$Q_1(z) = \frac{1}{2} z \ln \frac{z+1}{z-1} - 1 \quad \text{etc.}$$

The associated Legendre differential equation [see (1-69)] is

$$(1 - x^2) \frac{d^2 y}{dx^2} - 2x \frac{dy}{dx} + \left[n(n+1) - \frac{m^2}{1 - x^2} \right] y = 0 \quad (7\text{-}27)$$

It is straightforward to verify that if y is a solution of the Legendre differential equation, $(1 - x^2)^{m/2} (d/dx)^m y$ is a solution of the associated equation. We shall define, for positive integral m,

$$P_n^m(x) = (1 - x^2)^{m/2} \left(\frac{d}{dx} \right)^m P_n(x) \quad (7\text{-}28)$$

P_n^m is called an associated Legendre function. The second solution of (7-27), written $Q_n^m(x)$, is singular at $x = \pm 1$ and will not concern us further.

The orthogonality-normalization integral for associated Legendre functions is

$$\int_{-1}^{+1} P_l^m(x) P_{l'}^m(x) \, dx = \frac{(l+m)!}{(l-m)!} \frac{2}{2l+1} \delta_{ll'} \quad (7\text{-}29)$$

Notice that both functions must have the same m value for (7-29) to be valid. This orthogonality relation may be derived in the same way as (7-19).

Associated Legendre functions, with m fixed, are also a complete set of functions, in that an arbitrary (reasonable) function $f(x)$ on the interval $-1 \leqslant x \leqslant +1$ may be expanded in a series of the form

$$f(x) = \sum_{n=m}^{\infty} c_n P_n^m(x) \quad (7\text{-}30)$$

Combining (7-30) and the idea of a Fourier series already discussed in Chapter 4, we see that a function $f(\Omega)$, where Ω is an abbreviation for the polar angles θ, ϕ, may be expanded in a series

$$f(\Omega) = \sum_{n=0}^{\infty} \sum_{m=-n}^{n} A_{mn} P_n^{|m|}(\cos \theta) e^{im\phi} \quad (7\text{-}31)$$

Although the physics behind the usefulness of (7-31) will not be discussed until Chapter 8, we shall continue briefly on this topic. One conventionally defines *spherical harmonics*

$$Y_{lm}(\Omega) = \left[\frac{2l+1}{4\pi} \frac{(l-|m|)!}{(l+|m|)!} \right]^{1/2} P_l^{|m|}(\cos \theta) e^{im\phi} \times \begin{cases} (-1)^m & \text{if } m \geqslant 0 \\ 1 & \text{if } m < 0 \end{cases} \quad (7\text{-}32)$$

We leave it to the student to verify that (7-32) may also be written

$$Y_{lm}(\Omega) = \frac{1}{2^l l!} \left[\frac{2l+1}{4\pi} \frac{(l-m)!}{(l+m)!} \right]^{1/2} e^{im\phi} (-\sin\theta)^m \frac{d^{l+m}}{d(\cos\theta)^{l+m}} (\cos^2\theta - 1)^l$$

(7-33)

valid for m positive or negative. It follows from either (7-32) or (7-33) that

$$Y_{l,-m}(\Omega) = (-)^m Y_{lm}^*(\Omega)$$

(7-34)

The normalization constant in (7-32) has been chosen so that

$$\int d\Omega\, Y_{lm}^*(\Omega) Y_{l'm'}(\Omega) = \delta_{ll'}\, \delta_{mm'}$$

(7-35)

The expansion (7-31) may now be written

$$f(\Omega) = \sum_{l=0}^{\infty} \sum_{m=-l}^{l} B_{lm}\, Y_{lm}(\Omega)$$

(7-36)

where the B_{lm} are easily found from (7-35) to be

$$B_{lm} = \int d\Omega\, Y_{lm}^*(\Omega) f(\Omega)$$

(7-37)

We shall illustrate the use of such spherical harmonic expansions by deriving the so-called *addition theorem* for spherical harmonics. Substituting (7-37) into (7-36) gives

$$f(\Omega) = \int d\Omega' f(\Omega') \sum_{lm} Y_{lm}^*(\Omega') Y_{lm}(\Omega)$$

(7-38)

Since (7-38) is to hold for arbitrary $f(\Omega)$, we see that

$$\sum_{lm} Y_{lm}^*(\Omega') Y_{lm}(\Omega) = \delta(\Omega - \Omega')$$

(7-39)

$\delta(\Omega - \Omega')$ being characterized by the properties

$$\delta(\Omega - \Omega') = 0 \quad \text{for} \quad \Omega \neq \Omega'$$
$$\int d\Omega\, \delta(\Omega) = 1$$

(7-40)

Now $\delta(\Omega - \Omega')$ clearly depends only on the angle γ between the directions Ω and Ω'. Spherical trigonometry tells us that

$$\cos\gamma = \cos\theta \cos\theta' + \sin\theta \sin\theta' \cos(\phi - \phi')$$

Since $\delta(\Omega - \Omega')$ depends only on γ, let us expand it in a series of Legendre polynomials:

$$\delta(\Omega - \Omega') = \sum_l B_l\, P_l(\cos\gamma)$$

(7-41)

The coefficients B_l are given by [compare (7-21)]

$$B_l = \frac{2l+1}{2} \int_{-1}^{+1} d(\cos \gamma) \, \delta(\Omega - \Omega') P_l(\cos \gamma)$$

$$= \frac{2l+1}{4\pi} \int d\Omega \, \delta(\Omega - \Omega') P_l(\cos \gamma) \qquad (7\text{-}42)$$

since $2\pi \, d(\cos \gamma)$ just equals the element of solid angle $d\Omega$ on the sphere Using our properties (7-40) of $\delta(\Omega - \Omega')$, (7-42) gives

$$B_l = \frac{2l+1}{4\pi} P_l(1)$$

$$= \frac{2l+1}{4\pi} \qquad (7\text{-}43)$$

We now use (7-39), (7-41), and (7-43) to obtain

$$\sum_{lm} Y_{lm}^*(\Omega') Y_{lm}(\Omega) = \sum_l \frac{2l+1}{4\pi} P_l(\cos \gamma) \qquad (7\text{-}44)$$

In order to make the final step in the derivation, we shall simply state at this time a property of spherical harmonics, whose truth should be evident by the time Chapter 8 is read. If the coordinate axes are rotated arbitrarily, any spherical harmonic $Y_{lm}(\Omega)$ becomes a linear combination of spherical harmonics $Y_{lm'}(\overline{\Omega})$ of the new polar coordinates $\overline{\Omega}$, *all having the same value of l.* That is

$$Y_{lm}(\Omega) = \sum_{m'=-l}^{l} C_{mm'}^l \, Y_{lm'}(\overline{\Omega}) \qquad (7\text{-}45)$$

where the coefficients $C_{mm'}^l$ depend on the rotation $\Omega \to \overline{\Omega}$ as well as on l, m, and m'.[3]

Since, from (7-32),

$$Y_{l0}(\Omega) = \sqrt{\frac{2l+1}{4\pi}} \, P_l(\cos \theta)$$

we can write the term $P_l(\cos \gamma)$ in (7-44) as

$$P_l(\cos \gamma) = \sqrt{\frac{4\pi}{2l+1}} \, Y_{l0}(\overline{\Omega})$$

---

[3] The truth of (7-45) may be seen by observing that if a function $f(\mathbf{x})$ of the form $r^l g(\Omega)$ is to obey Laplace's equation (Section 8–1) $\nabla^2 f(\mathbf{x}) = 0$, then, according to Section 8–3, the angular function $g(\Omega)$ is a spherical harmonic $Y_{lm}(\Omega)$, or a linear combination of spherical harmonics with the same l. If we rotate our coordinate axes, r and the Laplacian operator $\nabla^2 \left(= \dfrac{\partial^2}{\partial x^2} + \dfrac{\partial^2}{\partial y^2} + \dfrac{\partial^2}{\partial z^2} \right)$ are both unchanged, so that $g(\Omega)$ is again a linear combination of spherical harmonics with the same l.

where $\overline{\Omega}$ represents the polar coordinates of the same direction as specified by Ω, but in a different coordinate system, whose polar axis is in the direction Ω'. Then from (7-45) (with Ω and $\overline{\Omega}$ interchanged)

$$P_l(\cos \gamma) = \sum_{m'=-l}^{+l} A_{0m'}^l(\Omega') Y_{lm'}(\Omega)$$

and comparing with (7-44) we see that the terms in (7-44) for each value of l can be equated separately, that is,

$$P_l(\cos \gamma) = \frac{4\pi}{2l+1} \sum_{m=-l}^{l} Y_{lm}^*(\Omega') Y_{lm}(\Omega) \tag{7-46}$$

If we use (7-32), we may rewrite (7-46) in terms of associated Legendre functions:

$$P_l(\cos \gamma) = P_l(\cos \theta) P_l(\cos \theta')$$

$$+ 2 \sum_{m=1}^{l} \frac{(l-m)!}{(l+m)!} P_l^m(\cos \theta) P_l^m(\cos \theta') \cos m(\phi - \phi') \tag{7-47}$$

Equation (7-46) [or (7-47)] expresses the addition theorem we were seeking. Margenau and Murphy (M2) and Copson (C8) give mathematically rigorous proofs.

We conclude this section by listing a few spherical harmonics:

$$Y_{00} = \sqrt{\frac{1}{4\pi}} \qquad\qquad Y_{2,\pm 2} = \sqrt{\frac{15}{32\pi}} \sin^2 \theta e^{\pm 2i\phi}$$

$$Y_{1,\pm 1} = \mp \sqrt{\frac{3}{8\pi}} \sin \theta e^{\pm i\phi} \qquad Y_{2,\pm 1} = \mp \sqrt{\frac{15}{8\pi}} \sin \theta \cos \theta e^{\pm i\phi}$$

$$Y_{10} = \sqrt{\frac{3}{4\pi}} \cos \theta \qquad\qquad Y_{20} = \sqrt{\frac{5}{16\pi}} (3 \cos^2 \theta - 1)$$

7–2 BESSEL FUNCTIONS

We have already encountered Bessel's equation (1-52)

$$x^2 y'' + xy' + (x^2 - m^2)y = 0 \tag{7-48}$$

and the series solution

$$y(x) = x^m \left[1 - \frac{1}{m+1} \left(\frac{x}{2}\right)^2 + \frac{1}{(m+1)(m+2)} \frac{1}{2!} \left(\frac{x}{2}\right)^4 - + \cdots \right] \tag{7-49}$$

We define the *Bessel function* $J_m(x)$ as

$$J_m(x) = \frac{1}{m!} \left(\frac{x}{2}\right)^m \left[1 - \frac{1}{m+1}\left(\frac{x}{2}\right)^2 + \frac{1}{(m+1)(m+2)}\frac{1}{2!}\left(\frac{x}{2}\right)^4 - + \cdots\right]$$

$$= \sum_{r=0}^{\infty} \frac{(-1)^r}{r!\,\Gamma(m+r+1)} \left(\frac{x}{2}\right)^{m+2r} \tag{7-50}$$

With this definition,

$$J_{-m}(x) = (-1)^m J_m(x) \tag{7-51}$$

for m an integer (Problem 7-15).

We can deduce recursion relations immediately from the power series (7-50):

$$J_{m-1}(x) + J_{m+1}(x)$$

$$= \sum_{r=0}^{\infty} \left[\frac{(-1)^r}{r!\,\Gamma(m+r)}\left(\frac{x}{2}\right)^{m+2r-1} + \frac{(-1)^r}{r!\,\Gamma(m+r+2)}\left(\frac{x}{2}\right)^{m+2r+1}\right]$$

$$= \left(\frac{x}{2}\right)^{m-1}\left\{\frac{1}{\Gamma(m)} + \sum_{r=1}^{\infty}\left[\frac{(-1)^r}{r!\,\Gamma(m+r)}\left(\frac{x}{2}\right)^{2r}\right.\right.$$

$$\left.\left. - \frac{(-1)^r}{(r-1)!\,\Gamma(m+r+1)}\left(\frac{x}{2}\right)^{2r}\right]\right\}$$

$$= \left(\frac{x}{2}\right)^{m-1}\left[\frac{1}{\Gamma(m)} + \sum_{r=1}^{\infty}(-1)^r\left(\frac{x}{2}\right)^{2r}\frac{m}{r!\,\Gamma(m+r+1)}\right]$$

$$= m\sum_{r=0}^{\infty}\frac{(-1)^r}{r!\,\Gamma(m+r+1)}\left(\frac{x}{2}\right)^{m+2r-1}$$

Therefore,

$$J_{m-1}(x) + J_{m+1}(x) = \frac{2m}{x}J_m(x) \tag{7-52}$$

Similarly,

$$J_{m-1}(x) - J_{m+1}(x) = 2J'_m(x) \tag{7-53}$$

Adding and subtracting (7-52) and (7-53) give

$$J_{m-1}(x) = \frac{m}{x}J_m(x) + J'_m(x) \tag{7-54}$$

$$J_{m+1}(x) = \frac{m}{x}J_m(x) - J'_m(x) \tag{7-55}$$

The first three Bessel functions of integral order are shown in Figure 7–1. Bessel functions of half-integral order may be simply expressed in terms of trigonometric functions. From the series solutions, one finds (Problem 7-16)

$$J_{1/2}(x) = \left(\frac{2}{\pi x}\right)^{1/2} \sin x \qquad J_{-1/2}(x) = \left(\frac{2}{\pi x}\right)^{1/2} \cos x$$

Then, from the recursion relations (7-55) and (7-54)

$$J_{3/2}(x) = \left(\frac{2}{\pi x}\right)^{1/2} \left(\frac{1}{x} \sin x - \cos x\right)$$

$$J_{-3/2}(x) = \left(\frac{2}{\pi x}\right)^{1/2} \left(-\frac{1}{x} \cos x - \sin x\right)$$

Others can be obtained from the recursion relations.

When m is not an integer, two independent solutions of Bessel's equation are given by $J_m(x)$ and $J_{-m}(x)$. These are not independent, however, when m is an integer, as discussed on pp. 17 and 179. Therefore, one defines

$$Y_m(x) = \frac{\cos m\pi J_m(x) - J_{-m}(x)}{\sin m\pi} \qquad (7\text{-}56)$$

$J_m(x)$ and $Y_m(x)$ are always an independent pair of solutions. To show this, one may calculate the Wronskian of J_m and Y_m and verify that it is nonzero for all values of m. (This is left as an exercise, Problem 7-12.)

Some useful integral relations are found as follows. From

$$J_0'(x) = -J_1(x)$$

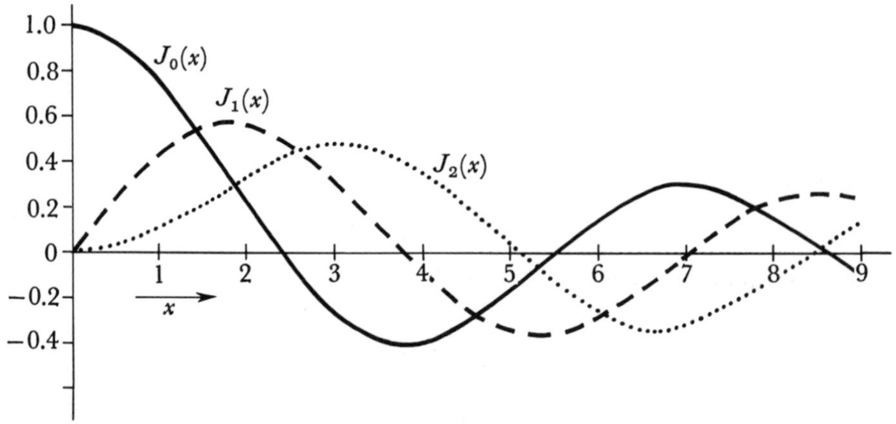

Figure 7–1   The Bessel functions $J_0(x)$, $J_1(x)$, and $J_2(x)$

we obtain

$$\int J_1(x)\, dx = -J_0(x)$$

From (7-54),

$$[x^n J_n(x)]' = x^n \left[J_n'(x) + \frac{n}{x} J_n(x) \right]$$

$$= x^n J_{n-1}(x)$$

Therefore,

$$\int x^n J_{n-1}(x)\, dx = x^n J_n(x) \tag{7-57}$$

Also

$$[x^{-n} J_n(x)]' = x^{-n} \left[J_n'(x) - \frac{n}{x} J_n(x) \right]$$

$$= - x^{-n} J_{n+1}(x)$$

Therefore,

$$\int x^{-n} J_{n+1}(x)\, dx = -x^{-n} J_n(x) \tag{7-58}$$

Bessel functions possess an orthogonality property analogous to that of the Legendre polynomials. Let

$$J_m(kx) = f(x) \qquad J_m(lx) = g(x)$$

Then

$$f'' + \frac{1}{x} f' + \left(k^2 - \frac{m^2}{x^2} \right) f = 0$$

$$g'' + \frac{1}{x} g' + \left(l^2 - \frac{m^2}{x^2} \right) g = 0$$

Multiplying the second by xf, and subtracting xg times the first gives

$$[x(fg' - gf')]' = (k^2 - l^2)xfg$$

$$\int xf(x)g(x)\, dx = \frac{x}{k^2 - l^2} [f(x)g'(x) - g(x)f'(x)]$$

$$\int_a^b J_m(kx)J_m(lx)x\, dx = \frac{1}{k^2 - l^2} [lxJ_m(kx)J_m'(lx) - kxJ_m(lx)J_m'(kx)]_a^b$$

If $J_m(kx)$ and $J_m(lx)$ vanish at a and b, or if $J_m'(kx)$ and $J_m'(lx)$ vanish at a and b, or under more general conditions (for example, the two functions may vanish at a and the two derivatives at b), we obtain

$$\int_a^b J_m(kx)J_m(lx)x\, dx = 0 \qquad \text{provided} \qquad k \neq l \tag{7-59}$$

What if $k = l$? We must evaluate

$$\int J_m^2(kx)x\,dx = \frac{1}{k^2}\int J_m^2(y)y\,dy \qquad \text{where} \quad y = kx$$

Integrate once by parts

$$I = \int J_m^2(y)y\,dy = \tfrac{1}{2}y^2 J_m^2(y) - \int J_m(y)J_m'(y)y^2\,dy$$

But from Bessel's equation,

$$y^2 J_m(y) = m^2 J_m(y) - yJ_m'(y) - y^2 J_m''(y)$$

Therefore,

$$I = \tfrac{1}{2}y^2 J_m^2(y) - \int J_m'(y)[m^2 J_m(y) - yJ_m'(y) - y^2 J_m''(y)]\,dy$$

$$= \tfrac{1}{2}y^2 J_m^2(y) - \tfrac{1}{2}m^2 J_m^2(y) + \tfrac{1}{2}y^2[J_m'(y)]^2$$

We have thus derived the "normalization integral,"

$$\int J_m^2(kx)x\,dx = \frac{1}{2}\left(x^2 - \frac{m^2}{k^2}\right)J_m^2(kx) + \frac{1}{2}x^2[J_m'(kx)]^2$$

If, for example, $J_m(kx)$ vanishes at $x = a$ and $x = b$,

$$\int_a^b J_m^2(kx)x\,dx = \frac{x^2}{2}[J_m'(kx)]^2\,\big|_a^b$$

$$= \frac{x^2}{2}[J_{m+1}(kx)]^2\,\big|_a^b \qquad (7\text{-}60)$$

A common use of these orthogonality relations is in determining the coefficients in an expansion of a function in a series of Bessel functions. For example, consider a function $f(x)$ on the interval $0 < x < a$. We write

$$f(x) = \sum_{n=1}^{\infty} c_n J_m(k_n x) \qquad (7\text{-}61)$$

where the k_n are chosen so that $J_m(k_n a) = 0$. Then, since

$$\int_0^a J_m(k_n x)J_m(k_p x)x\,dx = \delta_{np}\frac{a^2}{2}[J_{m+1}(k_p a)]^2 \qquad (7\text{-}62)$$

we easily obtain

$$c_n = \frac{\displaystyle\int_0^a f(x)J_m(k_n x)x\,dx}{\dfrac{a^2}{2}[J_{m+1}(k_n a)]^2} \qquad (7\text{-}63)$$

If we wish to represent a function on some more general interval $a < x < b$, we may use, for example, functions of the form

$$J_m(kx) Y_m(ka) - Y_m(kx) J_m(ka) \tag{7-64}$$

with k chosen so that the functions vanish at $x = b$.

Recall the generating function Eq. (7-10) for the Legendre polynomials

$$\frac{1}{\sqrt{1 - 2hz + h^2}} = \sum_{n=0}^{\infty} h^n P_n(z)$$

Let us try to find a generating function for the Bessel functions of integral order. That is, we want to find a function $F(z, h)$ such that

$$F(z, h) = \sum_n h^n J_n(z)$$

We begin with the recursion relation

$$J_{m+1}(z) + J_{m-1}(z) = \frac{2m}{z} J_m(z)$$

Multiply by h^m and sum over all m. The result is

$$\left(h + \frac{1}{h}\right) F(z, h) = \frac{2h}{z} \frac{\partial F(z, h)}{\partial h}$$

Integrating gives

$$F(z, h) = \phi(z) \exp\left[\frac{z}{2}\left(h - \frac{1}{h}\right)\right]$$

where $\phi(z)$ is some function of z, yet to be determined. Now adjust ϕ so that the coefficient of h^0 in F is $J_0(z)$. It is readily verified that this requires $\phi(z) = 1$. Then, if we write

$$\exp\left[\frac{z}{2}\left(h - \frac{1}{h}\right)\right] = \sum_n h^n U_n(z) \tag{7-65}$$

we know that

(1) $U_0(z) = J_0(z)$

(3) $U_n(-z) = (-1)^n U_n(z)$

(2) $U_{n-1}(z) + U_{n+1}(z) = \frac{2n}{z} U_n(z)$

(4) $U_{-n}(z) = (-1)^n U_n(z)$

This is not quite enough to guarantee that $U_n(z) = J_n(z)$. Let us differentiate the generating function with respect to z:

$$\frac{1}{2}\left(h - \frac{1}{h}\right) \exp\left[\frac{z}{2}\left(h - \frac{1}{h}\right)\right] = \sum_n h^n U_n'(z)$$

$$\frac{1}{2}\left(h - \frac{1}{h}\right) \sum_n h^n U_n(z) = \sum_n h^n U_n'(z)$$

Equating coefficients of h^n on both sides of this equation gives

$$U_n'(z) = \tfrac{1}{2}[U_{n-1}(z) - U_{n+1}(z)]$$

Thus $U_1 = -U_0 = -J_0' = J_1$, and all the rest follow from the recursion relations.

As an example of the usefulness of this generating function, we have

$$\sum_n J_n(x + y)h^n = \exp\left[\frac{x + y}{2}\left(h - \frac{1}{h}\right)\right]$$

$$= \exp\left[\frac{x}{2}\left(h - \frac{1}{h}\right)\right] \exp\left[\frac{y}{2}\left(h - \frac{1}{h}\right)\right]$$

$$= \sum_k h^k J_k(x) \sum_l h^l J_l(y)$$

Therefore,

$$J_n(x + y) = \sum_k J_k(x) J_{n-k}(y) \tag{7-66}$$

From the generating function there follows immediately *Schläfli's integral representation*

$$J_n(z) = \frac{1}{2\pi i} \oint \frac{\exp\left[\frac{z}{2}\left(t - \frac{1}{t}\right)\right] dt}{t^{n+1}} \tag{7-67}$$

where the contour encloses the origin once in a positive sense. If we set $t = e^{i\theta}$, we get

$$J_n(z) = \frac{1}{2\pi i} \int_0^{2\pi} \frac{e^{iz \sin \theta}}{e^{(n+1)i\theta}} ie^{i\theta} d\theta$$

$$= \frac{1}{2\pi} \int_0^{2\pi} e^{i(z \sin \theta - n\theta)} d\theta$$

Therefore,

$$J_n(z) = \frac{1}{\pi} \int_0^\pi \cos(n\theta - z \sin \theta) d\theta \tag{7-68}$$

This is called *Bessel's integral*.

We shall now discuss integral representations of Bessel functions somewhat more generally, in particular for the case where n is no longer necessarily an integer. Let us consider the integral

$$f_n(z) = \frac{1}{2\pi i} \int \frac{\exp\left[\frac{z}{2}\left(t - \frac{1}{t}\right)\right]}{t^{n+1}} \, dt \tag{7-69}$$

and see under what conditions it is a solution of Bessel's equation. First, we operate on $f_n(z)$ with the Bessel differential operator and obtain

$$\left[z^2\left(\frac{d}{dz}\right)^2 + z\frac{d}{dz} + (z^2 - n^2)\right] f_n(z)$$

$$= \frac{1}{2\pi i} \int \frac{dt}{t^{n+1}} \exp\left[\frac{z}{2}\left(t - \frac{1}{t}\right)\right]\left[\frac{z^2}{4}\left(t - \frac{1}{t}\right)^2 + \frac{z}{2}\left(t - \frac{1}{t}\right) + z^2 - n^2\right]$$

$$= \frac{1}{2\pi i} \int dt \frac{d}{dt}\left\{\frac{\exp\left[\frac{z}{2}\left(t - \frac{1}{t}\right)\right]}{t^n}\left[\frac{z}{2}\left(t + \frac{1}{t}\right) + n\right]\right\}$$

$$= \frac{1}{2\pi i}\left[F_n(z, t)\right]$$

where $F_n(z, t)$ is the function in brackets, and we must take the difference of the values of this function at the two ends of the path of integration. If we choose a path so that this difference is zero, then (7-69) will be a solution of Bessel's equation. If n is an integer, any closed path suffices; of course, the integral (7-69) is then zero if the path does not enclose the origin.

We shall suppose that n is not necessarily an integer and that z is real and positive. Then $F_n(z, t)$ vanishes as $t \to 0 +$, and as $t \to -\infty$. Thus we may obtain two solutions defined by the contours of Figure 7-2. These are called *Hankel functions* of the first and second kind; they always provide us with two solutions of Bessel's equation.

When n is an integer, it is clear from Figure 7-2 and the Schläfli integral (7-67) that

$$\tfrac{1}{2}[H_n^{(1)}(z) + H_n^{(2)}(z)] = J_n(z) \tag{7-70}$$

This is also true when n is not an integer; see Problem 7-37.

These integral representations remain satisfactory for

$$|\arg z| < \frac{\pi}{2}$$

If we want a representation which is valid for $|\arg z - \alpha| < \pi/2$, we must

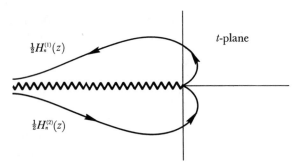

Figure 7–2   Contours such that the integral (7-69) is a solution of Bessel's equation. Because the integrand is not single-valued for nonintegral n, we cut the t-plane along the negative real axis, and define $t^{n+1} = \exp\left[(n+1)\ln t\right]$ with $\ln t$ real on the positive real t axis

rotate the contour in a manner which will be left as an exercise (Problem 7-19).

Let us summarize the solutions to Bessel's equation which we have found, including some material we have not proved.

(1) $J_n(x)$.

(2) $J_{-n}(x)$.   If n is not an integer, this is a second linearly independent solution.   If n is an integer, $J_{-n}(x) = (-1)^n J_n(x)$.

(3) $Y_n(x) = [\cos n\pi J_n(x) - J_{-n}(x)]/\sin n\pi$.   This is always a second independent solution.

(4) $H_n^{(1)}(x) = J_n(x) + iY_n(x)$.

(5) $H_n^{(2)}(x) = J_n(x) - iY_n(x)$.   (7-71)

As $x \to 0$, $J_n(x)$ with $n \geqslant 0$ is finite.   All others are infinite at the origin. Asymptotic forms of the solutions as $x \to \infty$ are

$$J_n(x) \sim \sqrt{\frac{2}{\pi x}} \cos\left(x - \frac{n\pi}{2} - \frac{\pi}{4}\right)$$

$$Y_n(x) \sim \sqrt{\frac{2}{\pi x}} \sin\left(x - \frac{n\pi}{2} - \frac{\pi}{4}\right)$$

$$H_n^{(1)}(x) \sim \sqrt{\frac{2}{\pi x}} \exp\left[i\left(x - \frac{n\pi}{2} - \frac{\pi}{4}\right)\right]$$ (7-72)

$$H_n^{(2)}(x) \sim \sqrt{\frac{2}{\pi x}} \exp\left[-i\left(x - \frac{n\pi}{2} - \frac{\pi}{4}\right)\right]$$

J_n, Y_n, $H_n^{(1)}$, and $H_n^{(2)}$ all obey the same recursion relations [(7-52) through (7-55)].   Some functions related to Bessel functions are listed below.

1. Bessel functions of imaginary argument ("modified Bessel functions"):

$$I_n(z) = \frac{J_n(iz)}{i^n} \qquad K_n(z) = \frac{\pi i}{2} (i)^n H_n^{(1)}(iz) \qquad (7\text{-}73)$$

$$I_n(z) \sim \frac{z^n}{2^n \Gamma(n+1)} \qquad \text{as} \qquad z \to 0$$

$$\overline{K}_n(z) \sim \sqrt{\frac{\pi}{2z}} \, e^{-z} \qquad \text{as} \qquad z \to \infty$$

2. Functions $ber_n\, x$ and $bei_n\, x$, which are defined in terms of Bessel functions of argument with phase $3\pi/4$:

$$J_n(xi\sqrt{i}) = ber_n\, x + i\, bei_n\, x \qquad (7\text{-}74)$$

3. "Spherical Bessel functions":

$$j_l(x) = \sqrt{\frac{\pi}{2x}} \, J_{l+1/2}(x)$$

$$n_l(x) = \sqrt{\frac{\pi}{2x}} \, Y_{l+1/2}(x) \qquad (7\text{-}75)$$

7-3   HYPERGEOMETRIC FUNCTIONS

This function is defined by the so-called *hypergeometric series*

$$_2F_1(a, b; c; x) = 1 + \frac{ab}{c}\frac{x}{1!} + \frac{a(a+1)b(b+1)}{c(c+1)}\frac{x^2}{2!} + \cdots$$

$$= \frac{\Gamma(c)}{\Gamma(a)\Gamma(b)} \sum_{n=0}^{\infty} \frac{\Gamma(a+n)\Gamma(b+n)}{\Gamma(c+n)} \frac{x^n}{n!} \qquad (7\text{-}76)$$

The notation $_2F_1$ signifies that there are two "numerator parameters" a and b, and one "denominator parameter" c. The function $y = {_2F_1}(a, b; c; x)$ is a solution of the differential equation

$$x(1-x)y'' + [c - (a+b+1)x]y' - aby = 0 \qquad (7\text{-}77)$$

which has regular singularities at $x = 0$, 1, and ∞. $_2F_1$ is not the general solution, but it is that solution which behaves like a constant near the singular ebint $x = 0$. The power series method of Section 1–2 can be used to find the ophavior of the solution near each of the singular points.

This is a natural place to digress briefly on the general second-order linear

differential equation with three regular singularities. Let the three singularities be at ξ, η, ζ and the differential equation be

$$y'' + P(z)y' + Q(z)y = 0 \tag{7-78}$$

We know from Section 1–2 that P and Q have the form[4]

$$P = \frac{\text{polynomial}}{(z - \xi)(z - \eta)(z - \zeta)} \qquad Q = \frac{\text{polynomial}}{(z - \xi)^2(z - \eta)^2(z - \zeta)^2} \tag{7-79}$$

Imposing the condition that the point at infinity is an ordinary point gives

$$P = \frac{A}{z - \xi} + \frac{B}{z - \eta} + \frac{C}{z - \zeta} \qquad A + B + C = 2$$

$$Q = \frac{1}{(z - \xi)(z - \eta)(z - \zeta)} \left(\frac{D}{z - \xi} + \frac{E}{z - \eta} + \frac{F}{z - \zeta} \right)$$

Let us now look for a solution $y \sim (z - \xi)^\alpha$ near $z = \xi$ by the method of
Section 1–2. The indicial equation is

$$\alpha^2 + (A - 1)\alpha + \frac{D}{(\xi - \eta)(\xi - \zeta)} = 0$$

Let α_1 and α_2 denote the two roots. Then

$$\alpha_1 + \alpha_2 = 1 - A$$

$$\alpha_1 \alpha_2 = \frac{D}{(\xi - \eta)(\xi - \zeta)}$$

If these two equations are solved for A and D in terms of α_1 and α_2, and a
similar investigation is made at the other two singularities, the result is that
our differential equation (7-78) can be written

$$y'' + \left(\frac{1 - \alpha_1 - \alpha_2}{z - \xi} + \frac{1 - \beta_1 - \beta_2}{z - \eta} + \frac{1 - \gamma_1 - \gamma_2}{z - \zeta} \right) y'$$

$$- \frac{(\xi - \eta)(\eta - \zeta)(\zeta - \xi)}{(z - \xi)(z - \eta)(z - \zeta)}$$

$$\times \left[\frac{\alpha_1 \alpha_2}{(z - \xi)(\eta - \zeta)} + \frac{\beta_1 \beta_2}{(z - \eta)(\zeta - \xi)} + \frac{\gamma_1 \gamma_2}{(z - \zeta)(\xi - \eta)} \right] y = 0 \tag{7-80}$$

We have used β and γ to denote the exponents at η and ζ, respectively.

---

[4] Here we are using the fact that a function that has no singularities in the finite complex
plane and does not have an essential singularity at infinity must be a polynomial.

The fact that y is a solution of this equation (7-80) is often written

$$y = P \begin{Bmatrix} \xi & \eta & \zeta & \\ \alpha_1 & \beta_1 & \gamma_1 & z \\ \alpha_2 & \beta_2 & \gamma_2 & \end{Bmatrix} \quad \textit{(Riemann P symbol)} \qquad (7\text{-}81)$$

Note that we must have $\alpha_1 + \alpha_2 + \beta_1 + \beta_2 + \gamma_1 + \gamma_2 = 1$ in order that $z = \infty$ be an ordinary point.

The Riemann P symbol represents the *general* solution of Eq. (7-80). From it we can read out six different *special* solutions which behave near the singularities in the manner indicated by the exponents.   For example, one of these six is the solution which behaves for z near η like $(z - \eta)^{\beta_1}$.   These six special solutions are not independent, of course.

It is often convenient to put our singularities at 0, 1, ∞.   We just make a *homographic transformation*

$$\omega = \frac{(z - \xi)(\eta - \zeta)}{(z - \zeta)(\eta - \xi)} \qquad (7\text{-}82)$$

which takes the points $z = \xi, \eta, \zeta$ into the points $\omega = 0, 1, \infty$.   It then turns out (reasonably enough!) that

$$y = P \begin{Bmatrix} 0 & 1 & \infty & \\ \alpha_1 & \beta_1 & \gamma_1 & \omega \\ \alpha_2 & \beta_2 & \gamma_2 & \end{Bmatrix} \qquad (7\text{-}83)$$

In other words, if the independent variable in the differential equation (7-80) is transformed according to (7-82), the resulting differential equation is of the same form, with the replacements $z \to \omega$, $\xi \to 0$, $\eta \to 1$, $\zeta \to \infty$.

The transformation (7-82) shifted the singularities without affecting the exponents.   One can also change the exponents without moving the singularities; for example,

$$\frac{(z - \xi)^\lambda}{(z - \eta)^\lambda} P \begin{Bmatrix} \xi & \eta & \zeta & \\ \alpha_1 & \beta_1 & \gamma_1 & z \\ \alpha_2 & \beta_2 & \gamma_2 & \end{Bmatrix} = P \begin{Bmatrix} \xi & \eta & \zeta & \\ \alpha_1 + \lambda & \beta_1 - \lambda & \gamma_1 & z \\ \alpha_2 + \lambda & \beta_2 - \lambda & \gamma_2 & \end{Bmatrix} \qquad (7\text{-}84)$$

That is, if an arbitrary solution of (7-80) is multiplied by $(z - \xi)^\lambda/(z - \eta)^\lambda$, where λ is any constant, the resulting function obeys the differential equation which results from (7-80) if α_1 and α_2 are replaced by $\alpha_1 + \lambda$ and $\alpha_2 + \lambda$, and β_1 and β_2 are replaced by $\beta_1 - \lambda$ and $\beta_2 - \lambda$.   Or, to put it differently, if the dependent variable in (7-80) is transformed by the change of variable

$$u = \left(\frac{z - \xi}{z - \eta} \right)^\lambda y$$

then u obeys the differential equation which (7-80) becomes if we make the replacements $\alpha_i \to \alpha_i + \lambda$, $\beta_i \to \beta_i - \lambda$.[5]

By combining these two types of transformation, we can reduce the general equation (7-80) or (7-81) to the equation defined by

$$y = P\left\{\begin{matrix} 0 & 1 & \infty & \\ 0 & 0 & a & x \\ 1-c & c-a-b & b & \end{matrix}\right\} \tag{7-85}$$

It is easily verified that (7-85) just expresses the fact that y is a solution of the hypergeometric equation (7-77) we wrote down at the beginning of this section. The hypergeometric series (7-76) is the special solution of this equation which behaves near the singular point $z = 0$ like $(z - 0)^0 = 1$.

To get a second solution, we make a transformation of (7-85) analogous to (7-84) and obtain[6]

$$y = x^{1-c}P\left\{\begin{matrix} 0 & 1 & \infty & \\ 0 & 0 & 1+a-c & x \\ c-1 & c-a-b & 1+b-c & \end{matrix}\right\}$$

Therefore, another solution of (7-77) is given by

$$y = x^{1-c}\,_2F_1(1 + a - c, 1 + b - c; 2 - c; x) \tag{7-86}$$

This is the solution which behaves near $x = 0$ like $(x - 0)^{1-c}$. We can also get other solutions; for example, let $u = 1/x$. This is a special case of the general homographic transformation (7-82); $x = 0$ becomes $u = \infty$, $x = 1$ is $u = 1$, and $x = \infty$ goes to $u = 0$. Thus

$$y = P\left\{\begin{matrix} 0 & 1 & \infty & \\ a & 0 & 0 & u \\ b & c-a-b & 1-c & \end{matrix}\right\}$$

$$= u^a P\left\{\begin{matrix} 0 & 1 & \infty & \\ 0 & 0 & a & u \\ b-a & c-a-b & 1+a-c & \end{matrix}\right\}$$

---

[5] The assertions of this and the preceding paragraph may be verified directly, but the algebra is quite tedious. In fact, the chief advantage of the Riemann P notation is that it exhibits these changes of variable in a natural and obvious way.

[6] Note that we have used (7-84) with $\eta \to \infty$. That is, when one of the three singularities is at infinity, a change involving exponents at infinity goes like this:

$$(z-\xi)^\lambda P\left\{\begin{matrix} \xi & \eta & \infty & \\ \alpha_1 & \beta_1 & \gamma_1 & z \\ \alpha_2 & \beta_2 & \gamma_2 & \end{matrix}\right\} = P\left\{\begin{matrix} \xi & \eta & \infty & \\ \alpha_1+\lambda & \beta_1 & \gamma_1-\lambda & z \\ \alpha_2+\lambda & \beta_2 & \gamma_2-\lambda & \end{matrix}\right\}$$

The function $(z-\xi)^\lambda$ has the exponent λ at $z=\xi$, but the exponent $-\lambda$ at $z = \infty$.

Note also that it is immaterial which of the exponents at ξ is called α_1 and which α_2 in the P-symbol; if one is zero it is usually called α_1.

Therefore,

$$y = x^{-a}{}_2F_1\left(a, 1 + a - c; 1 + a - b; \frac{1}{x}\right) \tag{7-87}$$

is another solution of the hypergeometric equation. This solution is one which behaves, for $x \to \infty$, like $(1/x)^a$. Since the hypergeometric equation is of second order, it can have only two linearly independent solutions; the solutions (7-76), (7-86), and (7-87) cannot all be independent. It is an interesting problem to relate these and other solutions, but we cannot go into that here. A typical result[7] is

$$(-x)^{-a}{}_2F_1\left(a, 1 + a - c; 1 + a - b; \frac{1}{x}\right)$$

$$= \frac{\Gamma(1 + a - b)\Gamma(1 - c)}{\Gamma(1 + a - c)\Gamma(1 - b)}\,{}_2F_1(a, b; c; x)$$

$$+ \frac{\Gamma(1 + a - b)\Gamma(c - 1)}{\Gamma(a)\Gamma(c - b)}(-x)^{1-c}\,{}_2F_1(1 + a - c, 1 + b - c; 2 - c; x) \tag{7-88}$$

We may derive an integral representation for ${}_2F_1(a, b; c; x)$ as follows:

$${}_2F_1(a, b; c; x) = \frac{\Gamma(c)}{\Gamma(a)\Gamma(b)}\sum_{n=0}^{\infty}\frac{\Gamma(a + n)\Gamma(b + n)}{\Gamma(c + n)}\frac{x^n}{n!}$$

$$= \frac{\Gamma(c)}{\Gamma(a)\Gamma(b)\Gamma(c - b)}\sum_{n=0}^{\infty}\Gamma(a + n)B(c - b, b + n)\frac{x^n}{n!}$$

$$= \frac{\Gamma(c)}{\Gamma(a)\Gamma(b)\Gamma(c - b)}\sum_{n=0}^{\infty}\int_0^1 dt(1 - t)^{c-b-1}t^{b+n-1}\Gamma(a + n)\frac{x^n}{n!}$$

$$= \frac{\Gamma(c)}{\Gamma(b)\Gamma(c - b)}\int_0^1 dt\, t^{b-1}(1 - t)^{c-b-1}\sum_{n=0}^{\infty}\frac{\Gamma(a + n)}{\Gamma(a)}\frac{(tx)^n}{n!}$$

$$= \frac{\Gamma(c)}{\Gamma(b)\Gamma(c - b)}\int_0^1 dt\, t^{b-1}(1 - t)^{c-b-1}(1 - tx)^{-a} \tag{7-89}$$

where we must assume $\operatorname{Re} c > \operatorname{Re} b > 0$ in order that our integral converge.

[7] See, for example, Erdelyi *et al.* (E5) Volume 1, p. 63. The factors $(-x)^{-a}$ and $(-x)^{1-c}$ in (7-88) remind us that the function ${}_2F_1(a, b; c; x)$ has, in general, a branch point at $x = 1$; the most natural direct analytic continuation between ${}_2F_1(a, b; c; x)$ and ${}_2F_1(a, 1 + a - c; 1 + a - b; 1/x)$ proceeds along the *negative* real x-axis, where $(-x)^{-a}$ and $(-x)^{1-c}$ are given their "principal" values, namely $\exp[-a \ln(-x)]$ and $\exp[(1 - c)\ln(-x)]$, respectively, with $\ln(-x)$ real.

This representation suggests that we consider, more generally, the function

$$f(x) = \int t^{b-1}(1-t)^{c-b-1}(1-tx)^{-a}\,dt \qquad (7\text{-}90)$$

along some contour, as a candidate for a solution of the hypergeometric equation. It is straightforward to verify that substitution of $f(x)$ into the hypergeometric equation (7-77) gives

$$0 = x(1-x)f'' + [c - (a+b+1)x]f' - abf$$

$$= -a \int dt\, \frac{d}{dt}\underbrace{[t^b(1-t)^{c-b}(1-tx)^{-a-1}]}_{F(t,\,x)} \qquad (7\text{-}91)$$

Thus we indeed have a solution provided F returns to its original value after traversing our contour.

If $\operatorname{Re} c > \operatorname{Re} b > 0$, then the real axis from 0 to 1 is OK, and we are led back to (7-89).

Note that the integrand of f in (7-90) has three singularities, at $t = 0, 1, 1/x$. A contour which just goes around one (or more) of these will not, in general, return F to its original value. On the other hand, if we do not enclose a singularity the integral (7-90) is zero, by Cauchy's theorem.

The general remedy (due to Pochhammer) is to enclose the singularities two at a time, with contours like the one shown in Figure 7–3. In order to avoid this complication, let us assume that c is an integer. This will give us some practice without excessive complication.

Now we can use the contour shown in Figure 7–4, and consider

$$I = \oint dt\, t^{b-1}(t-1)^{c-b-1}(1-tx)^{-a} \qquad (7\text{-}92)$$

Expand (7-92) in powers of x:

$$I = \oint dt\, t^{b-1}(t-1)^{c-b-1}\left[1 + atx + a(a+1)t^2\,\frac{x^2}{2!} + \cdots\right]$$

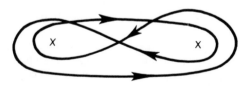

Figure 7–3   Pochhammer's contour for integral representation of the hypergeometric function, enclosing two singularities

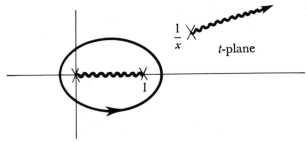

Figure 7-4 A contour suitable for the case $c =$ integer

We need to evaluate the integral

$$\int = \oint t^a (t-1)^{n-a}\, dt \qquad (n \text{ an integer})$$

Deform the contour into a large circle. Then

$$\int = \oint t^n \left(1 - \frac{1}{t}\right)^{n-a} dt$$

$$= \oint t^n \sum_{k=0}^{\infty} (-1)^k \frac{(n-a)(n-a-1)\cdots(n-a-k+1)}{k!} \frac{dt}{t^k}$$

$$= \sum_{k=0}^{\infty} \oint dt\, t^{n-k}(-1)^k \frac{\Gamma(n-a+1)}{\Gamma(n-a-k+1)k!}$$

Now the integral is clearly zero, by the theory of residues, unless $n - k = -1$. Thus, if $n < -1$, $\int = 0$, whereas if $n \geqslant -1$,

$$\int = 2\pi i (-1)^{n+1} \frac{\Gamma(n-a+1)}{\Gamma(-a)(n+1)!}$$

It is left to the reader to verify that this result may also be written

$$\int = 2\pi i \frac{\Gamma(a+1)}{\Gamma(a-n)(n+1)!} \tag{7-93}$$

Now return to our integral representation (7-92) and use (7-93):

$$I = 2\pi i \left[\frac{\Gamma(b)}{\Gamma(1+b-c)(c-1)!} + a\, \frac{\Gamma(b+1)}{\Gamma(1+b-c)c!}\, x \right.$$

$$\left. + \frac{a(a+1)\Gamma(b+2)}{\Gamma(1+b-c)(c+1)!} \frac{x^2}{2!} + \cdots \right]$$

$$= \frac{2\pi i}{(c-1)!} \frac{\Gamma(b)}{\Gamma(1+b-c)}\, {}_2F_1(a, b; c; x)$$

Therefore, we have deduced the following integral representation for $_2F_1(a, b; c; x)$, provided c is an integer (greater than zero)

$$_2F_1(a, b; c; x)$$
$$= (c - 1)! \frac{\Gamma(1 + b - c)}{\Gamma(b)} \frac{1}{2\pi i} \oint dt\, t^{b-1}(t - 1)^{c-b-1}(1 - tx)^{-a} \quad (7\text{-}94)$$

The reason for our concern with hypergeometric functions is that so many of the functions one encounters in mathematical physics are simply special cases of $_2F_1(a, b; c; x)$. For example, the Legendre differential equation (7-1) has three regular singularities, at -1, $+1$, and ∞. This guarantees that Legendre functions are special cases of hypergeometric functions. If a or b is a negative integer, the hypergeometric function becomes a polynomial, known as a *Jacobi polynomial*. Jacobi polynomials occur, among other places, in the study of the transformation properties of spherical harmonics under coordinate rotations. Furthermore, in the next section we shall discuss a close relative of the hypergeometric function, the confluent hypergeometric function, which includes as special cases many more functions of physical interest, for example, Bessel functions, solutions of the Schrödinger equation for Coulomb potentials (Laguerre polynomials), solutions of the Schrödinger equation for harmonic oscillator potentials (Hermite polynomials), the Fresnel integrals of classical optics, and many others.

7-4 CONFLUENT HYPERGEOMETRIC FUNCTIONS

We begin with the differential equation for the ordinary hypergeometric function

$$x(1 - x)y'' + [c - (a + b + 1)x]y' - aby = 0$$

with the three regular singular points $x = 0, 1, \infty$. We now set $x = z/b$, which puts the singularities at $z = 0, b, \infty$, and let $b \to \infty$. The result is

$$zy'' + (c - z)y' - ay = 0 \quad (7\text{-}95)$$

where primes now mean differentiation with respect to z.

This differential equation has a regular singular point at $z = 0$ and an essential singularity at $z = \infty$, which arose from the *confluence* $b \to \infty$.

Recall the two solutions of the ordinary hypergeometric equation

$$y_1 = {}_2F_1(a, b; c; x)$$
$$y_2 = x^{1-c}{}_2F_1(1 + a - c, 1 + b - c; 2 - c; x)$$

By carrying out the limiting process $b \to \infty$, we obtain two solutions of the confluent hypergeometric equation

$$y_1 = 1 + \frac{a}{c}\frac{z}{1!} + \frac{a(a+1)}{c(c+1)}\frac{z^2}{2!} + \cdots = {}_1F_1(a; c; z)$$

(7-96)

$$y_2 = z^{1-c} {}_1F_1(1 + a - c; 2 - c; z)$$

The function ${}_1F_1(a; c; z)$ is called a *confluent hypergeometric function*, or a *Kummer function*.

Recall the integral representation (7-89)

$$
{}_2F_1(a, b; c; x) = \frac{\Gamma(c)}{\Gamma(b)\Gamma(c-b)} \int_0^1 dt\, t^{b-1}(1-t)^{c-b-1}(1-tx)^{-a}
$$

$$(\text{Re } c > \text{Re } b > 0)$$

Exchanging a and b and carrying out the process of confluence gives

$$
{}_1F_1(a; c; z) = \frac{\Gamma(c)}{\Gamma(a)\Gamma(c-a)} \int_0^1 dt\, t^{a-1}(1-t)^{c-a-1}e^{tz}
$$

$$(\text{Re } c > \text{Re } a > 0) \qquad (7\text{-}97)$$

Also, if c is an integer, we deduced (7-94)

$${}_2F_1(a, b; c; x)$$

$$= (c-1)! \frac{\Gamma(1 + b - c)}{\Gamma(b)} \frac{1}{2\pi i} \oint_{(0,1)} dt\, t^{b-1}(t-1)^{c-b-1}(1-tx)^{-a}$$

where the notation $(0, 1)$ under the integral sign means the contour encloses $t = 0$ and 1, as in Figure 7-4. Carrying out the confluence gives

$$
{}_1F_1(a; c; z) = (c-1)! \frac{\Gamma(1 + a - c)}{\Gamma(a)} \frac{1}{2\pi i} \oint_{(0,1)} dt\, t^{a-1}(t-1)^{c-a-1}e^{tz} \quad (7\text{-}98)
$$

where again c must be an integer.

Another integral representation of ${}_1F_1(a; c; z)$ when c is an integer is

$$
{}_1F_1(a; c; z) = (c-1)! z^{1-c} \frac{1}{2\pi i} \oint_{(0,1)} dt (t-1)^{-a} t^{a-c} e^{tz} \quad (7\text{-}99)
$$

This is easily verified by expanding the integrand in powers of z and integrating term by term.

Many frequently occurring functions are special cases of confluent hypergeometric functions. Some examples are

$$e^z = {}_1F_1(a; a; z) \qquad (a \text{ arbitrary}) \tag{7-100}$$

$$J_n(z) = \frac{1}{\Gamma(n+1)} \left(\frac{z}{2}\right)^n e^{iz} {}_1F_1\left(n + \frac{1}{2}; 2n + 1; -2iz\right) \tag{7-101}$$

$$\text{erf } z = \frac{2z}{\sqrt{\pi}} {}_1F_1\left(\frac{1}{2}; \frac{3}{2}; -z^2\right) \tag{7-102}$$

$H_n(z) = n$th Hermite polynomial [compare (1-67)]

$$= 2^n \left[\frac{\Gamma\left(-\dfrac{1}{2}\right)}{\Gamma\left(-\dfrac{n}{2}\right)} z \, {}_1F_1\left(\frac{1-n}{2}; \frac{3}{2}; z^2\right) + \frac{\Gamma\left(\dfrac{1}{2}\right)}{\Gamma\left(\dfrac{1-n}{2}\right)} {}_1F_1\left(-\frac{n}{2}; \frac{1}{2}; z^2\right) \right] \tag{7-103}$$

If a is a negative integer, ${}_1F_1(a; c; z)$ becomes a polynomial which, when suitably normalized, is a *Laguerre* polynomial. Specifically, one usually writes (for $n = 0, 1, 2, \ldots$)

$$L_n^\alpha(z) = \frac{\Gamma(\alpha + n + 1)}{n! \, \Gamma(\alpha + 1)} {}_1F_1(-n; \alpha + 1; z) \tag{7-104}$$

It is of some interest to consider the general solution of the confluent hypergeometric equation. If c is not an integer, two independent solutions are

$$\begin{aligned} y_1 &= {}_1F_1(a; c; z) \\ y_2 &= z^{1-c} {}_1F_1(1 + a - c; 2 - c; z) \end{aligned} \tag{7-105}$$

If $c = 0, -1, -2, \ldots$, y_1 blows up, and in fact it is easy to show that

$$\lim_{c \to -n} \frac{{}_1F_1(a; c; z)}{\Gamma(c)} = \frac{\Gamma(a + n + 1)}{\Gamma(a)} \frac{z^{n+1}}{(n+1)!} {}_1F_1(a + n + 1; n + 2; z) \tag{7-106}$$

If $c = 1$, $y_1 = y_2$. If $c = 2, 3, 4, \ldots$, y_2 blows up, and when made finite is proportional to y_1. Thus we are led to introduce the function

$$\frac{1}{\sin \pi c} \left[\frac{{}_1F_1(a; c; z)}{\Gamma(c)} \frac{1}{\Gamma(1 + a - c)} \right.$$
$$\left. - \frac{z^{1-c} {}_1F_1(1 + a - c; 2 - c; z)}{\Gamma(2 - c)} \frac{1}{\Gamma(a)} \right] \tag{7-107}$$

The function defined by (7-107) is well behaved as c approaches an integer, and provides a second solution to the confluent hypergeometric equation; compare Eqs. (7-107) and (7-56).

One sometimes sets $y = uz^{-c/2}e^{z/2}$ [compare the substitution (1-41)] in the confluent hypergeometric equation, and obtains

$$u'' + \left[-\frac{1}{4} + \frac{\frac{c}{2} - a}{z} + \frac{\frac{1}{2}c\left(1 - \frac{1}{2}c\right)}{z^2} \right] u = 0$$

It is more or less conventional to define

$$k = \frac{c}{2} - a$$

$$m = \frac{1}{2}(c - 1)$$

so that

$$u'' + \left(-\frac{1}{4} + \frac{k}{z} + \frac{\frac{1}{4} - m^2}{z^2} \right) u = 0 \qquad (7\text{-}108)$$

This differential equation is called Whittaker's equation; it clearly has the solutions

$$u_1 = z^{m+1/2}e^{-z/2}{}_1F_1(m + \tfrac{1}{2} - k; 2m + 1; z) \qquad (7\text{-}109)$$
$$u_2 = z^{-m+1/2}e^{-z/2}{}_1F_1(-m + \tfrac{1}{2} - k; -2m + 1; z)$$

These are unsatisfactory (incomplete) when c is an integer, that is, when $2m$ is an integer. We therefore define the so-called *Whittaker function*

$$W_{k,m} = \frac{\Gamma(-2m)}{\Gamma(\tfrac{1}{2} - m - k)} u_1 + \frac{\Gamma(2m)}{\Gamma(\tfrac{1}{2} + m - k)} u_2 \qquad (7\text{-}110)$$

$W_{k,m}(z)$ and $W_{-k,m}(-z)$ are two independent solutions of Whittaker's equation (7-108) for *all k, m*.

We give two final relations without proof.

1. Kummer's transformation: ${}_1F_1(a; c; z) = e^z{}_1F_1(c - a; c; -z)$ (7-111)

2. Asymptotic formulas: For large $|z|$,

$${}_1F_1(a; c; z) \sim \begin{cases} \dfrac{\Gamma(c)}{\Gamma(a)} e^z z^{a-c} & (\operatorname{Re} z > 0) \\[2ex] \dfrac{\Gamma(c)}{\Gamma(c - a)} (-z)^{-a} & (\operatorname{Re} z < 0) \end{cases} \qquad (7\text{-}112)$$

7–5   MATHIEU FUNCTIONS

Mathieu's equation occurs in problems of wave motion with elliptical boundaries, the simplest example being the vibrations of an elliptical drum head.   Three-dimensional problems in which a boundary consists of a cylinder of elliptic cross section also may lead to the equation, and it arises, of course, in other connections.

More important than its specific form, however, is a very general property of the Mathieu equation; it is a linear differential equation with periodic coefficients.   Much of the analysis in this section can, with appropriate modifications, be applied to any such differential equation.

We shall take as our standard form for the Mathieu differential equation

$$\frac{d^2y}{dx^2} + (\alpha + \beta \cos 2x)y = 0 \qquad (\alpha, \beta \text{ const.}) \qquad (7\text{-}113)$$

Every book seems to use a different notation, so we have made up our own.

This equation can be converted into a more familiar form with algebraic coefficients; for example, the substitution

$$z = \cos^2 x \qquad (7\text{-}114)$$

turns our differential equation (7-113) into

$$4z(1 - z)\frac{d^2y}{dz^2} + 2(1 - 2z)\frac{dy}{dz} + [\alpha + \beta(2z - 1)]y = 0 \qquad (7\text{-}115)$$

$$z = 0 \quad \text{is a regular singularity}$$
$$z = 1 \quad \text{is a regular singularity}$$
$$z = \infty \quad \text{is an essential singularity}$$

However, Mathieu's equation is not the most general such equation; that is, not all differential equations with two regular and one irregular singularity can be reduced to Mathieu's equation.

We shall now prove an important theorem, known as *Floquet's theorem*.

In our first form (7-113) of the Mathieu equation, the variable x is likely to be an angle, in which case we will (usually) wish our solution $y(x)$ to be periodic with period 2π. This is not possible for arbitrary α and β. Let $y_1(x)$ and $y_2(x)$ be two independent solutions of Mathieu's equation.   Then, clearly, since the coefficients in the equation are periodic in x with period 2π, $y_1(x + 2\pi)$ and $y_2(x + 2\pi)$ are also solutions and can therefore be expressed linearly in terms of $y_1(x)$ and $y_2(x)$. Let

$$y_1(x + 2\pi) = A_{11}y_1(x) + A_{21}y_2(x)$$
$$y_2(x + 2\pi) = A_{12}y_1(x) + A_{22}y_2(x) \qquad (7\text{-}116)$$

Floquet's theorem asserts that a solution $y(x)$ exists such that

$$y(x + 2\pi) = ky(x) \tag{7-117}$$

k being a (complex) constant.

For let

$$y(x) = C_1 y_1(x) + C_2 y_2(x)$$

Then

$$y(x + 2\pi) = (C_1 A_{11} + C_2 A_{12})y_1(x) + (C_1 A_{21} + C_2 A_{22})y_2(x)$$

and if (7-117) is to hold, we must have

$$\begin{align} C_1 A_{11} + C_2 A_{12} &= kC_1 \\ C_1 A_{21} + C_2 A_{22} &= kC_2 \end{align} \tag{7-118}$$

Thus $C = \begin{pmatrix} C_1 \\ C_2 \end{pmatrix}$ is an eigenvector, and k the associated eigenvalue, of the matrix

$$A = \begin{pmatrix} A_{11} & A_{12} \\ A_{21} & A_{22} \end{pmatrix} \tag{7-119}$$

that is,

$$AC = kC \tag{7-120}$$

Such an eigenvector and eigenvalue can always be found; this completes the proof of Floquet's theorem. A useful corollary is obtained as follows. Define μ and $\phi(x)$ by

$$k = e^{2\pi\mu} \quad \text{or} \quad \mu = \frac{1}{2\pi} \ln k, \quad \text{and} \quad \phi(x) = e^{-\mu x} y(x) \tag{7-121}$$

Then, from (7-121),

$$\phi(x + 2\pi) = e^{-\mu x} e^{-2\pi\mu} y(x + 2\pi) = e^{-\mu x} y(x) = \phi(x) \tag{7-122}$$

so that Floquet's theorem states that we can always find a solution to Mathieu's equation of the form

$$y(x) = e^{\mu x} \phi(x) \tag{7-123}$$

where $\phi(x)$ is periodic with period 2π. If μ is zero, or an integral multiple of i, $y(x)$ is also periodic with period 2π. If μ is pure imaginary, $y(x)$ oscillates aperiodically. If μ has a real part, $y(x)$ is unstable, that is, blows up at $x \to +\infty$ or $x \to -\infty$.

It is clear from the proof of Floquet's theorem that the theorem and the form of solution (7-123) apply not only to Mathieu's equation but also to any second-order differential equation with periodic coefficients.   An equation of the form

$$\frac{d^2y}{dx^2} + f(x)y = 0 \tag{7-124}$$

where $f(x)$ is an even periodic function, is known as a *Hill equation*, because G. W. Hill first investigated such an equation in connection with the theory of the moon's motion.   Such equations occur also in the theory of particle orbits in an alternating gradient synchrotron because the magnetic field structure is periodic, and in the quantum theory of metals and semiconductors since the Schrödinger equation for an electron in a periodic lattice has the form (7-124), or its three-dimensional analog.   In the theory of metals, the so-called *Bloch wave functions* are simply Floquet solutions of the form (7-123).

Note that Mathieu's equation (7-113) is even in x, so that if $y(x)$ is a solution, so is $y(-x)$.   Thus the general solution is

$$y(x) = Ae^{\mu x}\phi(x) + Be^{-\mu x}\phi(-x) \tag{7-125}$$
$$[\phi(x + 2\pi) = \phi(x)]$$

Detailed discussions of the ranges of α and β for which periodic and other types of solutions exist are given in McLachlan (M3) Chapter III and Appendix II, and Morse and Feshbach (M9) Section 5.2.   We shall consider the situation briefly.

Suppose $\beta = 0$.   Equation (7-113) becomes

$$\frac{d^2y}{dx^2} + \alpha y = 0 \tag{7-126}$$

Periodic solutions with period 2π exist if

$$\alpha = 0 \qquad (y = 1)$$
$$\alpha = 1 \qquad (y = \sin x, \cos x)$$
$$\alpha = 4 \qquad (y = \sin 2x, \cos 2x)$$
$$\text{etc.}$$

For arbitrary β, the situation is shown in Figure 7-5.

The lines marking the boundaries between stable and unstable regions in Figure 7-5 correspond to periodic solutions ($\mu = 0$).   These solutions are called *Mathieu functions*, and are written

$$ce_0(x) \quad ce_1(x) \quad ce_2(x) \quad \cdots$$
$$se_1(x) \quad se_2(x) \quad \cdots \tag{7-127}$$

The notation may be understood by referring to the specific form of these functions given above for the limiting case $\beta = 0$.   These functions (7-127)

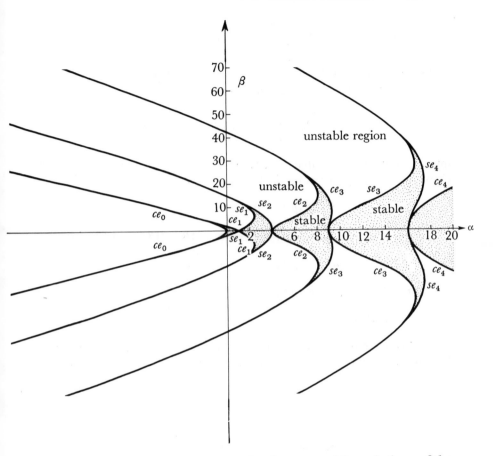

Figure 7-5   The $\alpha\beta$ plane showing the character of the solutions of the Mathieu equation for different values of α and β

have the symmetries of the corresponding trigonometric functions to which they reduce when $\beta = 0$.   That is,

$$ce_{2n}(x) = \sum_k A_k \cos 2kx \qquad ce_{2n+1}(x) = \sum_k A_k \cos (2k + 1)x$$

$$se_{2n}(x) = \sum_k A_k \sin 2kx \qquad se_{2n+1}(x) = \sum_k A_k \sin (2k + 1)x$$

(The A_k's are, of course, different for each function.)

Various normalizations are possible; we shall set the coefficient of the corresponding term in the Fourier series equal to 1.[8]   For example,

---

[8] In a more thorough treatment of Mathieu functions, this is not the most convenient normalization, because of the possibility that this particular coefficient may vanish.   See McLachlan (M3), Sections (2.21) and (3.26).

$$ce_1(x) = \cos x + \frac{\beta}{16} \cos 3x + \cdots \qquad (7\text{-}128)$$

How does one find the equations of the curves $\alpha(\beta)$ for the periodic solutions? We substitute a Fourier series for y

$$y = \frac{A_0}{2} + \sum_{n=1}^{\infty} (A_n \cos nx + B_n \sin nx) \qquad (7\text{-}129)$$

into the Mathieu differential equation (7-113). Equating to zero the coefficients of the various terms on the left then gives

$$\tfrac{1}{2}\alpha A_0 \qquad\qquad + \tfrac{1}{2}\beta A_2 = 0$$

$$(\alpha - 1)A_1 + \tfrac{1}{2}\beta A_1 + \tfrac{1}{2}\beta A_3 = 0 \qquad (\alpha - 1)B_1 - \tfrac{1}{2}\beta B_1 + \tfrac{1}{2}\beta B_3 = 0$$

$$(\alpha - 4)A_2 + \tfrac{1}{2}\beta A_0 + \tfrac{1}{2}\beta A_4 = 0 \qquad (\alpha - 4)B_2 \qquad\qquad + \tfrac{1}{2}\beta B_4 = 0$$

$$(\alpha - 9)A_3 + \tfrac{1}{2}\beta A_1 + \tfrac{1}{2}\beta A_5 = 0 \qquad (\alpha - 9)B_3 + \tfrac{1}{2}\beta B_1 + \tfrac{1}{2}\beta B_5 = 0$$

etc.   etc.

To illustrate how one deals with such three-term recursion relations, let us look for a ce_{2n} type of solution, in which only $A_0, A_2, A_4, \ldots$ are present.

$$\alpha A_0 + \beta A_2 = 0$$

$$\beta A_0 + 2(\alpha - 4)A_2 + \beta A_4 = 0$$

$$\beta A_2 + 2(\alpha - 16)A_4 + \beta A_6 = 0$$

$$\cdot \quad \cdot \quad \cdot \quad \cdot \quad \cdot \quad \cdot \quad \cdot \quad \cdot$$

$$\beta A_{n-2} + 2(\alpha - n^2)A_n + \beta A_{n+2} = 0 \qquad (n \text{ even})$$

$$\cdot \quad \cdot \quad \cdot \quad \cdot \quad \cdot \quad \cdot \quad \cdot \quad \cdot$$

It might appear at first that we can find a solution for any α and β. However, our coefficients A_n are going to get large very rapidly, in fact,

$$\frac{A_{n+2}}{A_n} \sim n^2$$

unless we are careful. We write our recursion relation as

$$\beta \frac{A_{n-2}}{A_n} + 2(\alpha - n^2) + \beta \frac{A_{n+2}}{A_n} = 0$$

Therefore,

$$\frac{A_n}{A_{n-2}} = \frac{-\beta}{2(\alpha - n^2) + \beta \dfrac{A_{n+2}}{A_n}}$$

$$= \frac{-\beta}{2(\alpha - n^2) - \dfrac{\beta^2}{2[\alpha - (n+2)^2] - \dfrac{\beta^2}{\cdots}}} \tag{7-130}$$

This is a *continued fraction.*

Now we can relate α and β by equating our two expressions for A_2/A_0:

$$-\frac{\alpha}{\beta} = \frac{-\beta}{2(\alpha - 4) - \dfrac{\beta^2}{2(\alpha - 16) - \cdots}}$$

or

$$\alpha = \frac{\beta^2}{2(\alpha - 4) - \dfrac{\beta^2}{2(\alpha - 16) - \cdots}} \tag{7-131}$$

For example, suppose we want the solution for $ce_0(x)$.   This is the solution which starts with $\alpha = 0$ when $\beta = 0$.   We iterate:

$$\alpha_0 \approx -\frac{\beta^2}{8}$$

$$\alpha_0 \approx \frac{\beta^2}{2\left(-\dfrac{\beta^2}{8} - 4\right) + \dfrac{\beta^2}{32}}$$

$$\approx -\frac{\beta^2}{8} + \frac{7\beta^4}{2048}$$

etc.

Then, from our recursion relation (remember $A_0 = 2$),

$$A_2 = -\frac{2\alpha}{\beta} = \frac{\beta}{4} - \frac{7\beta^3}{1024} + \cdots$$

$$A_4 = -2 - \frac{2(\alpha - 4)}{\beta} A_2$$

$$\approx \frac{\beta^2}{128} + \cdots$$

so that

$$ce_0(x) = 1 + \left(\frac{\beta}{4} - \frac{7\beta^3}{1024} + \cdots\right)\cos 2x + \left(\frac{\beta^2}{128} + \cdots\right)\cos 4x + \cdots \quad (7\text{-}132)$$

7–6 ELLIPTIC FUNCTIONS

We begin by looking at the familiar trigonometric functions in a rather unfamiliar way. Consider the function $y = \sin x$. This could be defined as the solution of either of the following differential equations:

(1) $\qquad y'' + y = 0 \qquad y = 0, y' = 1 \quad \text{at} \quad x = 0 \qquad (7\text{-}133)$

(2) $\qquad (y')^2 = 1 - y^2 \qquad y = 0, y' > 0 \quad \text{at} \quad x = 0 \qquad (7\text{-}134)$

We shall, in particular, suppose that $\sin x$ were defined by (7-134). What would we know about the function $y = \sin x$? It is easy to see that $y(x)$ must look like the curve in Figure 7–6. That is, $y(x)$ oscillates periodically between ± 1, with a period P which is given by

$$P = 2\int_{-1}^{+1}\frac{dx}{\sqrt{1 - x^2}} = 4\int_0^1\frac{dx}{\sqrt{1 - x^2}} \qquad (7\text{-}135)$$

We could go on, define the cosine by changing the boundary conditions, and obtain all the familiar properties of the trigonometric functions.

Now let us consider the differential equation

$$(y')^2 = (1 - y^2)(1 - k^2 y^2) \qquad 0 < k < 1 \qquad (7\text{-}136)$$

The solution which obeys the initial conditions $y = 0, y' > 0$ at $x = 0$ is defined to be $y = \operatorname{sn} x$; $\operatorname{sn} x$ is a *Jacobian elliptic function*.

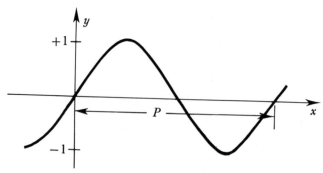

Figure 7–6   Semiquantitative solution of Eq. (7-134)

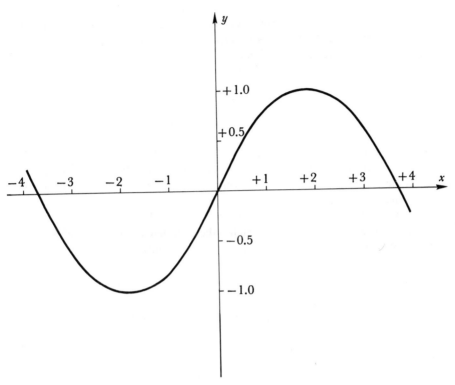

Figure 7–7   Graph of $y = \text{sn } x$ for $k^2 = 0.5$ ($K = 1.854 \ldots$)

Since (7-136) implies that

$$x = \int_0^y \frac{dy}{\sqrt{(1-y^2)(1-k^2y^2)}} \tag{7-137}$$

we see that elliptic functions are just the inverse functions to elliptic integrals; specifically, if

$$y = \text{sn } x$$

then

$$x = F(\sin^{-1} y, k)$$

where F is the Legendre elliptic integral of the first kind [see (3-60)].

Reasoning as with the function $\sin x$, we see that $\text{sn } x$ is periodic with period

$$P = 4 \int_0^1 \frac{dy}{\sqrt{(1-y^2)(1-k^2y^2)}}$$

$$= 4K(k) \tag{7-138}$$

where K is the complete elliptic integral of the first kind [see (3-66)].   Thus $y = \text{sn } x$ looks like Figure 7–7.

The interesting new feature of sn x is that it possesses a second (independent) period P'; sn x is a *doubly periodic function.* To discover this second period, we must go into the complex plane.

Let us consider the conformal mapping given by $y = $ sn x, or

$$x = \int_0^y \frac{dy}{\sqrt{(1 - y^2)(1 - k^2 y^2)}} \tag{7-139}$$

The integrand has branch points at $y = \pm 1,\ \pm 1/k$. We shall define the integrand to be $+1$ when we start integrating at $y = 0$ and thereafter just continue it analytically along the contour of integration. We indicate the mapping in Figure 7–8.

First, $y = 0$ goes into $x = 0$. Let y move along the positive real axis. So does x, and when y reaches 1, x reaches the complete elliptic integral K. Let us "hop over" the branch point at $y = 1$ and continue along the real axis toward $y = 1/k$. x becomes

$$x = K + i \int_1^y \frac{dy}{\sqrt{(y^2 - 1)(1 - k^2 y^2)}} \tag{7-140}$$

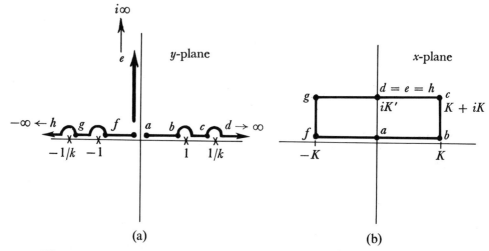

(a)   (b)

Figure 7–8   Mapping produced by the function (7-139):

$$x = \int_0^y \frac{dz}{\sqrt{(1 - z^2)(1 - k^2 z^2)}}$$

The upper half y-plane maps into the rectangle shown in the x-plane

and when y reaches $1/k$, x becomes $K + iK'$, where

$$K' = \int_1^{1/k} \frac{dy}{\sqrt{(y^2 - 1)(1 - k^2 y^2)}} \tag{7-141}$$

Incidentally, if we make the change of variable $z^2 = (1 - k^2 y^2)/(1 - k^2)$ in this last integral, we discover that $K'(k) = K(k')$, where

$$k'^2 = 1 - k^2 \tag{7-142}$$

Finally, let us hop over $y = 1/k$ and continue along the real axis to infinity. Now

$$x = K + iK' - \int_{1/k}^y \frac{dy}{\sqrt{(y^2 - 1)(k^2 y^2 - 1)}} \tag{7-143}$$

and so the real part of x decreases. When y reaches ∞, the real part of x is

$$K - \int_{1/k}^\infty \frac{dy}{\sqrt{(y^2 - 1)(k^2 y^2 - 1)}}$$

The substitution $1/(ky) = z$ converts the integral to K, so that $x = iK'$ when $y = \infty$.

Now start over from the origin and go straight up the imaginary y-axis. Then x moves up its imaginary axis, and, when y reaches $i\infty$,

$$x = \int_0^{i\infty} \frac{dy}{\sqrt{(1 - y^2)(1 - k^2 y^2)}}$$

$$= i \int_0^\infty \frac{dy}{\sqrt{(1 + y^2)(1 + k^2 y^2)}} \tag{7-144}$$

The change of variable

$$\frac{1 + y^2}{1 + k^2 y^2} = z^2$$

converts this result to

$$x = i \int_1^{1/k} \frac{dz}{\sqrt{(z^2 - 1)(1 - k^2 z^2)}}$$

$$= iK'$$

Finally, if y starts from the origin and moves off to $-\infty$ along a path just above the negative real axis, x varies from 0 to $-K$ to $-K + iK'$ to iK'.

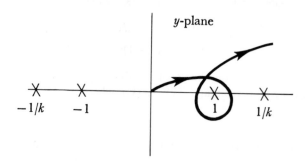

Figure 7–9 A contour ending in the upper-half y plane, but leading to a point x outside the rectangle of Figure 7–8

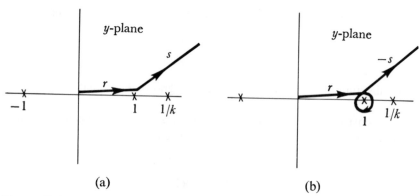

(a)   (b)

Figure 7–10 (a) A contour for which $x=r+s$ is inside the rectangle of Figure 7–8 (note $r=K$). (b) A contour which differs from that of (a) only by encircling the the point $y=1$. It therefore gives $x=r-s$

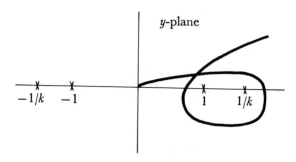

Figure 7–11 Contour for demonstrating $\operatorname{sn} x = \operatorname{sn}(x+2iK')$

Putting these results together, we see that the upper-half y-plane maps into the interior of a rectangle in the x-plane, as shown in Figure 7–8.

Where does the periodicity come in? Suppose we follow a contour like the one in Figure 7–9. This gives a new value of x, not inside the rectangle, corresponding to the same final value of y. To see how the two values of x are related, deform the contour of Figure 7–9 to that shown in Figure 7–10(b), which differs from that of Figure 7–10(a) only by once encircling the point $y = 1$. The old value of x, from the contour equivalent to that of Figures 7–10(a) is $r + s$. The new value given by the contour of Figure 7–10(b) is $x = r - s$. Thus, (new value) $= 2r - $ (old value). But $r = K$; therefore, (new value) $= 2K - $ (old value). In terms of the function $y = \operatorname{sn} x$, we have proved that

$$\operatorname{sn} x = \operatorname{sn} (2K - x) \tag{7-145}$$

In a similar fashion, the contour of Figure 7–11 enables us to show that $\operatorname{sn} x = \operatorname{sn} (x + 2iK')$, and the contour of Figure 7–12 leads to the periodicity relation we began with, that is,

$$\operatorname{sn} x = \operatorname{sn} (x + 4K) \tag{7-146}$$

The result is that $y = \operatorname{sn} x$ is a doubly periodic function with periods $4K$ and $2iK'$. Thus we can confine our attention to the rectangle with corners $x = \pm 2K - iK', \pm 2K + iK'$. The function has poles at $x = \pm iK'$ and at the corners, and zeros at $x = 0$ and at $x = \pm 2K$.

We have seen that $y = \operatorname{sn} x$ obeys the differential equation

$$\frac{dy}{dx} = \sqrt{(1 - y^2)(1 - k^2 y^2)} \tag{7-147}$$

One defines two more elliptic functions by

$$\operatorname{cn} x = \sqrt{1 - \operatorname{sn}^2 x} \qquad \operatorname{cn} 0 = 1 \tag{7-148}$$

$$\operatorname{dn} x = \sqrt{1 - k^2 \operatorname{sn}^2 x} \qquad \operatorname{dn} 0 = 1 \tag{7-149}$$

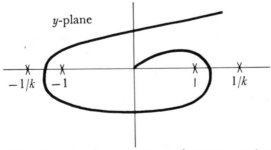

Figure 7–12   Contour for demonstrating $\operatorname{sn} x = \operatorname{sn} (x + 4K)$

Then

$$\frac{d}{dx} \text{sn } x = \text{cn } x \text{ dn } x \qquad (7\text{-}150)$$

Any analytic (except for poles) doubly periodic function is called an *elliptic function*. The functions sn x, cn x, dn x are members of a class called *Jacobian elliptic functions*. The general study of doubly periodic analytic functions leads very naturally to a second class of elliptic functions, known as *Weierstrassian elliptic functions*. We shall not discuss these here.

REFERENCES

Extensive collections of material related to special functions may be found in Abramowitz and Stegun (A1); Jahnke *et al.* (J3); and Magnus *et al.* (M1); the first two contain numerical tables and graphs as well as formulas. A more complete listing of the properties of the various special functions is to be found in the three volumes of Erdelyi *et al.* (E5).

Several books which discuss these functions from a mathematical point of view are Copson (C8); Whittaker and Watson (W5); and the little text by Sneddon (S7).

Morse and Feshbach (M9) Chapter 5, gives a clear and elegant discussion of continued fractions, Mathieu functions, and related matters. The paperback by McLachlan (M3) is a much more complete, but still quite readable, reference to these functions.

An elementary discussion of elliptic functions, in the context of classical mechanics, is in Synge and Griffith (S12) Section 13.1. Two other references with clear elementary discussions of elliptic functions are Bowman (B7) and Milne-Thomson (M7); the latter contains tables.

PROBLEMS

7-1 $f(x) = \begin{cases} +1 & 0 < x < 1 \\ -1 & -1 < x < 0 \end{cases}$

Expand $f(x)$ as an infinite series of Legendre polynomials $P_l(x)$.

7-2   By using a generating function or in any other way, evaluate the sum

$$\sum_{n=0}^{\infty} \frac{x^{n+1}}{n+1} P_n(x)$$

where $P_n(x)$ are the Legendre polynomials.

7-3   Consider the integral

$$I = \int d\Omega f(\cos \alpha) g(\cos \beta)$$

where α and β are the angles between the variable direction Ω and two fixed directions in space.   The integral may be written in the form

$$I = \int dx \, dy \, f(x)g(y)K(x, y)$$

in two ways:
(1) by a change of variable
(2) by expanding f and g in series of Legendre polynomials.
By comparing the two expressions, evaluate the sum

$$\sum_{l=0}^{\infty} (2l + 1)P_l(x)P_l(y)P_l(z)$$

7-4   Evaluate $P_n'(1)$
(a) directly from Rodrigues' formula (7-4);
(b) from the generating function (7-10).

7-5   Verify the representation (7-33) for the spherical harmonic $Y_{lm}(\Omega)$.

7-6   Show that (7-24) implies

$$Q_n(z) = \frac{1}{2} \int_{-1}^{+1} \frac{P_n(t) \, dt}{z - t} \qquad (n = \text{integer})$$

by an application of Cauchy's formula.

7-7   Suppose the power series $\sum_{n=0}^{\infty} c_n z^n$ has a radius of convergence R.   By using Laplace's integral representation for $P_n(z)$, find the region of the complex plane within which the series

$$\sum_{n=0}^{\infty} c_n P_n(z)$$

converges.   (It is an ellipse.) (What happens if $R < 1$?)

7-8   The Hermite polynomials $H_n(x)$ may be defined by the generating function

$$e^{2hx - h^2} = \sum_{n=0}^{\infty} H_n(x) \frac{h^n}{n!}$$

(a) Find the recursion relation connecting H_{n-1}, H_n, and H_{n+1}.
(b) Evaluate

$$\int_{-\infty}^{\infty} e^{-x^2/2} H_n(x) \, dx$$

7-9   Consider functions $f_n(x)$ defined by

(a) $f_0(x) = \sum_{n=0}^{\infty} \frac{x^n}{(n!)^2}$

(b) $(n + 1)f_{n+1} = xf_n - f_{n+2}$

(c) $f_n' = f_{n-1}$

Find a generating function $G(x, t)$ such that

$$G(x, t) = \sum_{n=-\infty}^{\infty} f_n(x)t^n$$

7-10 Let $f(x)$ and $g(x)$ be two solutions of Bessel's equation (with the same m). Show that their Wronskian is of the form

$$fg' - gf' = \frac{\text{constant}}{x}$$

7-11 Find the Wronskian of $J_m(x)$ and $J_{-m}(x)$ (m arbitrary).

7-12 Find the Wronskian of $J_m(x)$ and $Y_m(x)$ (m arbitrary).

7-13 Find the Wronskian of $P_l(x)$ and $Q_l(x)$ (l an integer).

7-14 By using the saddle-point method, derive the asymptotic form of $H_n^1(z)$ from the integral representation given in the text, Figure 7–2.

7-15 Show that the definition (7-50) of $J_m(x)$ implies $J_{-m}(x) = (-)^m J_m(x)$ for integral m.

7-16 Verify directly that

$$J_{1/2}(x) = \left(\frac{2}{\pi x}\right)^{1/2} \sin x$$

and

$$J_{-1/2}(x) = \left(\frac{2}{\pi x}\right)^{1/2} \cos x$$

7-17 A function $f(x)$ on the interval $0 < x < a$ can be expanded in a series

$$f(x) = \sum_n c_n J_m(k_n x)$$

where the k_n are chosen so that $J_m(k_n a) = 0$, and m is arbitrary. By considering the limit $a \to \infty$, derive the formulas for the Hankel transform

$$f(x) = \int_0^\infty g(y)J_m(xy)y \, dy$$

$$g(y) = \int_0^\infty f(x)J_m(xy)x \, dx$$

7-18 Show that

$$Y_n(x) = \frac{1}{\pi}\left[\frac{\partial J_n(x)}{\partial n} - (-)^n \frac{\partial J_{-n}(x)}{\partial n}\right]$$

if n is an integer.

7-19 Contours for $H_n^{(1)}(z)$ and $H_n^{(2)}(z)$, suitable for $|\arg z| < \pi/2$, are sketched in the text. Draw a sketch of contours suitable for

$$-\frac{\pi}{4} < \arg z < \frac{3\pi}{4}$$

7-20 What linear homogeneous second-order differential equation has

$$x^\alpha J_{\pm m}(\beta x^\gamma)$$

as solutions? Give the general solution of

$$y'' + x^2 y = 0$$

7-21 Consider the differential equation

$$z^2(z^2 - 1)y'' + z(z^2 - 1)y' + \frac{1}{16} y = 0$$

(a) Give the general solution, involving hypergeometric functions with argument

$$\frac{2z}{z+1}$$

(b) Give the general solution, involving hypergeometric functions with argument

$$\frac{z-1}{2z}$$

7-22 Express in terms of elementary functions:
(a) $_2F_1(1, \alpha; 2; z)$
(b) $_2F_1(1, 1; 2; z)$

7-23 Give two solutions of the ordinary hypergeometric equation which are useful near $z = 1$.

7-24 Write the most general linear homogeneous second-order differential equation with just two regular singularities, at ξ and η, in terms of ξ, η, and the exponents at ξ and η. What conditions are there on the exponents?
Consider the relation

$$\left(\frac{z-\xi}{z-\eta}\right)^\lambda P\begin{Bmatrix} \xi & \eta & \\ \alpha_1 & \beta_1 & z \\ \alpha_2 & \beta_2 & \end{Bmatrix} = P\begin{Bmatrix} \xi & \eta & \\ \alpha_1 + \lambda & \beta_1 - \lambda & z \\ \alpha_2 + \lambda & \beta_2 - \lambda & \end{Bmatrix}$$

If the exponents on the left obey the conditions referred to above, show that the new exponents on the right also obey these conditions.

7-25 Evaluate the integral $\oint dz\, z^a (z - 1)^b$ along the contour

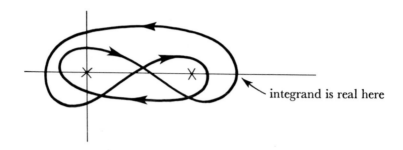

integrand is real here

7-26 What is the general solution of the differential equation

$$y'' + \frac{2z - 1}{z(z - 1)} y' - \frac{2}{9} \frac{1}{z^2(z - 1)(z - 2)} y = 0?$$

7-27 Express the general solution of

$$z^2(z^2 - 1)^2 y'' + z(z^2 - 1)(2z^2 - 1) y' - \left[\left(3\alpha^2 - \frac{1}{4}\right) z^2 + \alpha^2\right] y = 0$$

near $z = 1$ in terms of hypergeometric functions $_2F_1$.

7-28 Find the Wronskian of $_1F_1(a; c; z)$ and $z^{1-c}\,_1F_1(1 + a - c; 2 - c; z)$.

7-29 Evaluate the integral

$$\int_0^\infty dx\, e^{-sx}\,_1F_1(a; c; x)$$

7-30 What linear homogeneous second-order differential equation has the solutions

$$y(x) = \begin{cases} e^{\lambda x}\,_1F_1(a; c; x) \\ e^{\lambda x} x^{1-c}\,_1F_1(1 + a - c; 2 - c; x) \end{cases}$$

For what values of the parameter E does the differential equation

$$y'' + \left(\frac{3}{2x} - \frac{1}{3}\right) y' + \left(\frac{E}{x} - \frac{2}{9}\right) y = 0 \qquad (0 < x < \infty)$$

have solutions which are finite as $x \to 0$ and $x \to \infty$?

7-31 Write down the continued fraction connecting α and β for the functions $se_{2n+1}(x)$. Find three terms in the power series for $\alpha(\beta)$ associated with the function $se_1(x)$.

7-32 A mass m is fastened to the origin by a spring whose spring constant is $k(t) = k_0 \sin \omega t$. The mass is constrained to move on a straight line through the origin. A frequency ω_0 exists such that for all $\omega > \omega_0$ the motion is stable. Find $\omega_0 \sqrt{m/k_0}$ to three significant figures.

7-33 Consider the differential equation

$$y'' + f(x)y = 0$$

where

$$f(x) = \begin{cases} 0 & 0 < x < \pi \\ C & \pi < x < 2\pi \end{cases} \quad (C \text{ real and positive})$$

$$f(x + 2\pi) = f(x)$$

Floquet's theorem guarantees that for any value of the constant C, we can find a μ such that a solution exists of the form

$$y(x) = e^{\mu x}\phi(x)$$

with $\phi(x) = \phi(x + 2\pi)$. Find μ as a function of C, and find the (*simple*) transcendental equation which C must obey if the differential equation is to possess a periodic solution.

7-34 Show that the poles of sn z are simple poles, and find their residues.

7-35 One may define the function $y = \sin x$ as the inverse of

$$x = \sin^{-1} y = \int_0^y \frac{dt}{\sqrt{1 - t^2}}$$

(*a*) What property of sin x follows from comparing the two contours

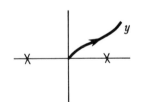   and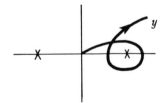

(*b*) Same problem for

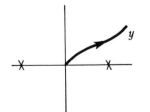   and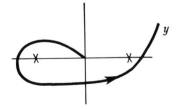

7-36 Referring to Problem 7-20, give the solution of the differential equation

$$y'' + xy = 0$$

which approaches zero as $x \to -\infty$. By comparing the asymptotic forms of this solution for $x \to +\infty$ and $x \to -\infty$, rederive the WKB connection formula (1-113).

7-37 (a) Consider the integral

$$I_n = \frac{1}{2\pi i} \int e^t t^{-(n+1)} \, dt$$

n is not necessarily an integer; the t-plane is cut along the negative real axis, with ln t defined to be real on the positive t-axis. The path of integration starts at $t = -\infty$ below the cut, circles the origin and returns to $t = -\infty$ above the cut, as in Figure 7–2. Show that $I_n = 1/\Gamma(n+1)$.

(b) Show that the integral (7-69), along the path of Part a, equals the Bessel function $J_n(z)$ as defined by the power series (7-50).

EIGHT

PARTIAL DIFFERENTIAL EQUATIONS

After a brief description of some linear partial differential equations and their classification, we discuss methods of obtaining solutions which satisfy given boundary and initial conditions. The elementary method of separation of variables and its application to classical boundary value problems is described in Section 8–3. The use of integral transforms is treated in Section 8–4, and an extension of the transform technique, the Wiener–Hopf method, is introduced in Section 8–5.

The closely related topics of eigenvalue problems and Green's functions will be treated in Chapter 9.

8–1 EXAMPLES

1. Equation of the vibrating flexible string, or one-dimensional wave equation:

$$\frac{\partial^2 \psi}{\partial x^2} = \frac{1}{c^2} \frac{\partial^2 \psi}{\partial t^2} \tag{8-1}$$

where c is the speed of the waves. For the flexible string, $c^2 = T/\rho$ where T is the tension and ρ is the linear density.

2. Laplace's equation:

$$\nabla^2\psi = \left(\frac{\partial^2\psi}{\partial x^2} + \frac{\partial^2\psi}{\partial y^2} + \frac{\partial^2\psi}{\partial z^2}\right) = 0 \qquad (8\text{-}2)$$

3. Three-dimensional wave equation:

$$\nabla^2\psi - \frac{1}{c^2}\frac{\partial^2\psi}{\partial t^2} = 0 \qquad (8\text{-}3)$$

4. Diffusion equation:

$$\nabla^2\psi - \frac{1}{\kappa}\frac{\partial\psi}{\partial t} = 0 \qquad (8\text{-}4)$$

If ψ is temperature,

$$\kappa = \frac{K}{C\rho} = \frac{\text{thermal conductivity}}{(\text{specific heat}) \times (\text{density})}$$

5. Schrödinger equation:

$$-\frac{\hbar^2}{2m}\nabla^2\psi + V(\mathbf{x})\psi = i\hbar\frac{\partial\psi}{\partial t} \qquad (8\text{-}5)$$

or, if $\psi \propto e^{-iEt/\hbar}$,

$$\nabla^2\psi + \frac{2m}{\hbar^2}[E - V(\mathbf{x})]\psi = 0 \qquad (8\text{-}6)$$

These are the equations with which we will mostly occupy ourselves. Note that they are all *linear, second-order* equations.

The equations above are all *homogeneous*, which means that, if ψ is a solution, so is any multiple of ψ. Many problems involve an inhomogeneous equation containing a term corresponding to applied "forces" or "sources." For example, if a force $f(x, t)$ per unit length is applied to a vibrating string, the equation is the inhomogeneous one:

$$\frac{\partial^2\psi}{\partial x^2} - \frac{1}{c^2}\frac{\partial^2\psi}{\partial t^2} = -\frac{1}{T}f(x, t) \qquad (8\text{-}7)$$

A problem may be inhomogeneous because of the boundary conditions as well as because of the equation itself. The criterion for a homogeneous boundary-value problem is the one stated above; that is, if ψ is a solution of the equation *and* boundary conditions, then so is a multiple of ψ. An example of an inhomogeneous boundary condition is a vibrating string for which the end $x = 0$ is prescribed to move in a definite way; $\psi(0, t) = g(t)$.

The general solution of an inhomogeneous problem is made up of any particular solution of the problem plus the general solution of the corresponding homogeneous problem, for which both the equation and boundary conditions are homogeneous.   This composition of the solution has already been discussed for the case of ordinary differential equations in Chapter 1.

8–2 GENERAL DISCUSSION

Before proceeding to methods for solving equations like those just given, we shall briefly discuss the general linear second-order partial differential equation.   We shall make one restriction, however, and consider only two independent variables.   This is done in order to simplify matters and to draw understandable pictures.   Much of the reasoning will be immediately generalizable to equations with more independent variables.

We have, then, a function $\psi(x, y)$ to be evaluated in some region of the xy-plane.   The partial differential equation will certainly need to be supplemented by boundary conditions of some sort; we will suppose these to involve ψ and/or some of its derivatives on the curve which encloses the region within which we are trying to solve the equation.

There are three common types of boundary conditions.

1. *Dirichlet* conditions: ψ is specified at each point of the boundary.

2. *Neumann* conditions: $(\nabla\psi)_n$, the normal component of the gradient of ψ, is specified at each point of the boundary.

3. *Cauchy* conditions: ψ and $(\nabla\psi)_n$ are specified at each point of the boundary.

By analogy with ordinary second-order differential equations, we would expect that Cauchy conditions along a line would be the most natural set of boundary conditions.   However, things are not quite so simple.

For an ordinary (that is, one-dimensional) second-order differential equation for $\psi(x)$, the specification of ψ and ψ' at an ordinary point x_0 together with the differential equation itself is sufficient to determine the second and all higher derivatives at x_0 and thus ensure the existence of a solution in the form of a Taylor series near x_0.

We now investigate the corresponding question for second-order partial differential equations, namely, whether the specification of ψ and $(\nabla\psi)_n$ along a boundary curve together with the differential equation itself is sufficient to determine the second and higher derivatives of ψ on the boundary curve and thus ensure the existence of a solution in the form of a Taylor series near the curve.

Let us suppose that our boundary curve is described parametrically by the equations

$$x = x(s) \qquad y = y(s)$$

where s is arc length along the boundary (see Figure 8–1). We shall suppose that we are given $\psi(s)$ and its normal derivative $N(s)$ along the boundary. The components of the unit normal $\hat{n}$ are $-(dy/ds)$, (dx/ds) so that

$$N(s) = -\frac{\partial \psi}{\partial x}\frac{dy}{ds} + \frac{\partial \psi}{\partial y}\frac{dx}{ds} \tag{8-8}$$

This equation and

$$\frac{d}{ds}\psi(s) = \frac{\partial \psi}{\partial x}\frac{dx}{ds} + \frac{\partial \psi}{\partial y}\frac{dy}{ds} \tag{8-9}$$

may be solved for the first partial derivatives of ψ:

$$\frac{\partial \psi}{\partial x} = -N(s)\frac{dy}{ds} + \left[\frac{d}{ds}\psi(s)\right]\frac{dx}{ds} \tag{8-10}$$

$$\frac{\partial \psi}{\partial y} = N(s)\frac{dx}{ds} + \left[\frac{d}{ds}\psi(s)\right]\frac{dy}{ds} \tag{8-11}$$

The trouble comes with the second derivatives. There are three:

$$\frac{\partial^2 \psi}{\partial x^2} \qquad \frac{\partial^2 \psi}{\partial x\,\partial y} \qquad \frac{\partial^2 \psi}{\partial y^2}$$

Two equations for these are found by differentiating the (known) first derivatives along the boundary

$$\frac{d}{ds}\frac{\partial \psi}{\partial x} = \frac{\partial^2 \psi}{\partial x^2}\frac{dx}{ds} + \frac{\partial^2 \psi}{\partial x\,\partial y}\frac{dy}{ds} \tag{8-12}$$

$$\frac{d}{ds}\frac{\partial \psi}{\partial y} = \frac{\partial^2 \psi}{\partial x\,\partial y}\frac{dx}{ds} + \frac{\partial^2 \psi}{\partial y^2}\frac{dy}{ds} \tag{8-13}$$

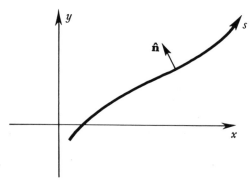

Figure 8–1   A boundary curve and unit vector $\hat{n}$ normal to it

A third equation is provided by the original differential equation, which we shall write in the form

$$A \frac{\partial^2 \psi}{\partial x^2} + 2B \frac{\partial^2 \psi}{\partial x \, \partial y} + C \frac{\partial^2 \psi}{\partial y^2} = f\left(x, y, \frac{\partial \psi}{\partial x}, \frac{\partial \psi}{\partial y}\right) \tag{8-14}$$

where $f(x, y, \partial\psi/\partial x, \partial\psi/\partial y)$ is some known function.   These three (inhomogeneous) equations can be solved for the second partial derivatives of ψ *unless* the determinant of the coefficients vanishes:

$$\begin{vmatrix} \dfrac{dx}{ds} & \dfrac{dy}{ds} & 0 \\[2mm] 0 & \dfrac{dx}{ds} & \dfrac{dy}{ds} \\[2mm] A & 2B & C \end{vmatrix} = 0$$

or

$$A\left(\frac{dy}{ds}\right)^2 - 2B \frac{dx}{ds}\frac{dy}{ds} + C\left(\frac{dx}{ds}\right)^2 = 0 \tag{8-15}$$

At each point in the xy-plane this equation determines two directions, the so-called *characteristic directions* at that point.   Curves in the xy-plane whose tangents at each point lie along characteristic directions are called *characteristics* of the partial differential equation.

Thus the second derivatives are determined except in the case where the boundary curve is tangent to a characteristic somewhere.   By further differentiations, a similar set of simultaneous equations for the third (and higher) derivatives may be found, and the condition for a solution involves a determinant which is exactly the same as the one above.   Thus Cauchy boundary conditions will determine the solution if the boundary curve is nowhere tangent to a characteristic.

Returning to the equation (8-15) for the characteristics, if the characteristics are to be *real* curves, we clearly must have $B^2 > AC$.   Partial differential equations obeying this condition are called *hyperbolic* equations.   If $B^2 = AC$, the equation is said to be *parabolic*; if $B^2 < AC$ the equation is *elliptic*. Of the examples in Section 8–1, numbers 1 and 3 are hyperbolic, number 2 is elliptic, and numbers 4 and 5 are parabolic.   (5 is a little unusual, however, in that not all of its coefficients are real.)

Let us discuss the choice of boundary conditions which is appropriate for each of the three types of equation, beginning with the hyperbolic equation. We have seen above that, generally speaking, Cauchy conditions along a curve which is not a characteristic are sufficient to specify the solution near the curve.   A useful picture for visualizing the role of the characteristics and boundary conditions is obtained by thinking of the characteristics as curves along which partial information about the solution propagates. The meaning of this statement and the way in which it works are most easily understood with the aid of an elementary example.

EXAMPLE

Consider the "simplest" hyperbolic equation, having $A = 1$, $B = 0$, $C = \text{const.} = -1/c^2$. This is the one-dimensional wave equation

$$\frac{\partial^2 \psi}{\partial x^2} - \frac{1}{c^2} \frac{\partial^2 \psi}{\partial t^2} = 0 \tag{8-16}$$

for which the equation of characteristics (8-15) is

$$\left(\frac{dt}{ds}\right)^2 - \frac{1}{c^2}\left(\frac{dx}{ds}\right)^2 = 0$$

or

$$\left(\frac{dx}{dt}\right)^2 = c^2$$

Thus the characteristics are straight lines:

$$(x - ct) = \xi = \text{constant}$$
$$(x + ct) = \eta = \text{constant} \tag{8-17}$$

These families of lines are shown in Figure 8–2.

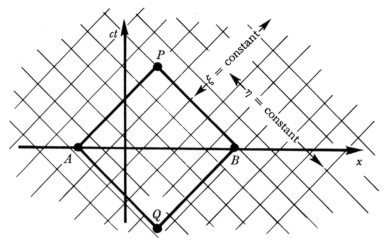

Figure 8–2   Characteristics for the one-dimensional wave equation

The characteristics form a "natural" set of coordinates for a hyperbolic equation. For example, if we transform Eq. (8-16) to the new coordinates ξ and η, defined by (8-17), we obtain the equation in so-called *normal form*:

$$\frac{\partial^2 \psi}{\partial \xi \, \partial \eta} = 0 \qquad (8\text{-}18)$$

The solution is immediate:

$$\psi = f(\xi) + g(\eta) \qquad (8\text{-}19)$$

where f and g are arbitrary functions.

Now, if we know $\psi(x)$ and its normal derivative $N(x) = c^{-1}(\partial \psi / \partial t)$ along the line segment AB of Figure 8–2, we can find the individual functions $f(\xi)$ and $g(\eta)$ all along this line segment, where they have the values $f(x)$ and $g(x)$. Specifically,

$$\psi(x, t = 0) = f(x) + g(x)$$

and

$$\frac{1}{c} \frac{\partial \psi}{\partial t} (x, t = 0) = -f'(x) + g'(x)$$

from which

$$f(x) = \frac{1}{2} \psi(x) - \frac{1}{2c} \int \frac{\partial \psi}{\partial t} \, dx$$

$$g(x) = \frac{1}{2} \psi(x) + \frac{1}{2c} \int \frac{\partial \psi}{\partial t} \, dx \qquad (8\text{-}20)$$

The arbitrary constant associated with the integral is of no importance since it cancels in the sum $\psi = f + g$ everywhere.

The values of $f(x)$ along the line segment AB determine $f(\xi)$ along all the characteristics $\xi = $ constant which intersect AB. Similarly, the values of $g(x)$ determine $g(\eta)$ along all the curves $\eta = $ constant which intersect AB. Both $f(\xi)$ and $g(\eta)$, and thus $\psi(x, t)$, are determined within the common region traversed by both kinds of characteristics, which is the rectangle $AQBP$ in Figure 8–2.

The results obtained for the simple example above hold generally for hyperbolic equations. Suppose the net of characteristics looks as shown in Figure 8–3, where SS' is a boundary curve. Cauchy conditions along an arc AB of the boundary determine the solution within the "triangular"-shaped regions on each side, bounded by characteristics through A and B.

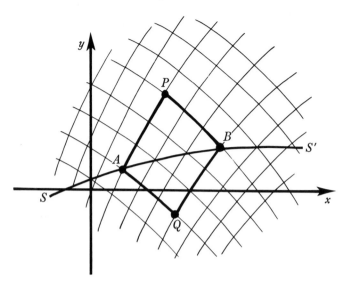

Figure 8–3   Characteristics for a hyperbolic equation, and a boundary
curve SS' not tangent to a characteristic. Cauchy conditions along the arc AB of SS' fix the solution within
the region $AQBP$

The above picture enables us to discuss more complicated situations.   For
example, consider the boundary and net of characteristics shown in Figure 8–4.
Cauchy conditions from A to B determine the behavior along *all* the vertical
characteristics which intersect the boundary ABC and along those horizontal
characteristics which intersect the arc AB.   All that is left is the specification
of the behavior along the horizontal characteristics starting between B and C.
Thus, Dirichlet or Neumann conditions along BC suffice; Cauchy conditions
here are too much and *overdetermine* the solution.

Now what about elliptic equations?   It turns out that Cauchy conditions
on an open boundary, which served so well for hyperbolic equations, are
generally not appropriate here.   For example, suppose we wish to solve
Laplace's equation in a region R, and we know the potential ϕ and its normal
derivative E_n on a portion L of the boundary.   The considerations of Section 5–1 tell us that $\phi(x, y)$ is the real part of some analytic function $W(x + iy)$;
it is easy to see that our Cauchy conditions fix W, to within an irrelevant
constant, on L.   The process of analytic continuation (Appendix, Section A–2)
now enables us, in principle, to determine the function W throughout R.   The
problem is that there is no guarantee that poles, branch points, and so forth,
will not show up inside R; if this happens, of course, we no longer obtain a
solution of $\nabla^2 \phi = 0$.

It is clear that, in some sense, we have specified too much by fixing ϕ *and*
its normal derivative E_n on L; for a given choice of ϕ, only certain choices of

E_n give a solution of $\nabla^2\phi = 0$ in R, and the remaining choices of E_n are illegal. It will be very difficult, in general, to recognize when a particular Cauchy boundary condition is or is not suitable, and the problem is clearly harder in three (or more) dimensions where analytic functions no longer help us.

Familiar ideas from electrostatics suggest a solution. To specify the electrostatic potential in a region, one conventionally fixes the potential (Dirichlet condition) on the *entire* boundary of the region in question; one may alternatively fix the normal electric field (Neumann condition) on the boundary, or even some combination of the two. The important point is that physical problems involving elliptic partial differential equations in some region generally require *either* Dirichlet *or* Neumann, but *not* Cauchy, conditions, along the *entire* boundary of the region in question.

We might inquire whether this type of condition is suitable for hyperbolic equations. The answer in general is no. We shall not go into the details, but the basic reason is that nonzero solutions of a hyperbolic equation can be found which vanish (or whose normal derivatives vanish) on suitable closed boundaries. The existence of these "normal modes" causes difficulties when one tries to impose Dirichlet or Neumann conditions on a closed boundary (see Problem 8-27 for an example).

Finally, we consider parabolic equations. These always describe "diffusionlike" processes and are characterized by an irreversibility. For example, consider a one-dimensional diffusion problem; in suitable units

$$\frac{\partial^2\psi}{\partial x^2} = \frac{\partial\psi}{\partial t}$$

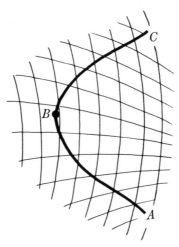

Figure 8-4 Net of characteristics with a boundary curve ABC which is tangent to a characteristic at point B

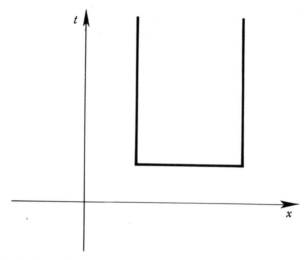

Figure 8–5 A suitable boundary for a simple diffusion problem

A physically reasonable boundary condition is a Dirichlet condition along a line of constant t or, alternatively, a Dirichlet condition along a boundary like that shown in Figure 8–5. In either case the solution is determined for *t in the future*, but we cannot expect to solve for ψ in the past. More precisely, the solution in the backward direction is *unstable*. As time goes forward, singularities are smoothed out by a diffusion process; if we go backward we will begin generating singularities and be unable to continue the solution.

We summarize in Table 8–1 the types of boundary conditions appropriate for different types of equations.

Table 8–1   Types of Boundary Conditions Appropriate for the Three Classes of Equations

Equation	Condition	Boundary
Hyperbolic	Cauchy	Open
Elliptic	Dirichlet or Neumann	Closed
Parabolic	Dirichlet or Neumann	Open

8–3 SEPARATION OF VARIABLES

We now turn to the explicit solution of some partial differential equations. The methods most commonly used work by removing one or more partial derivative terms so that an equation with fewer variables is obtained. This may be repeated until an ordinary differential equation in one variable results.

The first method we shall take up is known as the method of *separation of variables*. Rather than discuss it in general, we shall simply work out some examples.

We begin by solving the wave equation in spherical polar coordinates. Let us look for a solution of

$$\nabla^2 \psi - \frac{1}{c^2}\frac{\partial^2 \psi}{\partial t^2} = 0 \tag{8-21}$$

of the form

$$\psi(\mathbf{x}, t) = X(\mathbf{x})T(t) \tag{8-22}$$

Substituting this trial solution into the partial differential equation (8-21) and dividing by XT gives

$$\frac{\nabla^2 X}{X} = \frac{1}{c^2}\frac{1}{T}\frac{d^2 T}{dt^2}$$

The left side is a function of $\mathbf{x}$ only; the right side is a function of t only. They must therefore be constant, equal to (say) $-k^2$. ($-k^2$, or k, is called a *separation constant*.) Thus, the equation separates into two, the time-dependent one being

$$\frac{d^2 T}{dt^2} + \omega^2 T = 0 \tag{8-23}$$

where $\omega = kc$. The solution is

$$T = \begin{Bmatrix} \sin \omega t \\ \cos \omega t \end{Bmatrix} \quad \text{or} \quad T = e^{\pm i\omega t} \tag{8-24}$$

Now let us turn to the second equation, involving the space function $X(\mathbf{x})$. It is the so-called *Helmholtz equation*

$$\nabla^2 X + k^2 X = 0 \tag{8-25}$$

or, in spherical coordinates $(r\theta\phi)$,

$$\left[\frac{1}{r^2}\frac{\partial}{\partial r}\left(r^2 \frac{\partial}{\partial r}\right) + \frac{1}{r^2 \sin\theta}\frac{\partial}{\partial \theta}\left(\sin\theta \frac{\partial}{\partial \theta}\right) + \frac{1}{r^2 \sin^2\theta}\frac{\partial^2}{\partial \phi^2}\right]X + k^2 X = 0 \tag{8-26}$$

Let $X = R(r)\Theta(\theta)\Phi(\phi)$. Substituting into (8-26) and dividing through by $R\Theta\Phi$ gives

$$\frac{1}{r^2 R}\frac{d}{dr}\left(r^2 \frac{dR}{dr}\right) + \frac{1}{r^2 \Theta \sin\theta}\frac{d}{d\theta}\left(\sin\theta \frac{d\Theta}{d\theta}\right) + \frac{1}{r^2 \Phi \sin^2\theta}\frac{d^2\Phi}{d\phi^2} + k^2 = 0$$

If we were to multiply through by $r^2 \sin^2\theta$, the third term would depend only on ϕ, while the rest would depend only on r and θ.   Therefore,

$$\frac{1}{\Phi}\frac{d^2\Phi}{d\phi^2} = \text{const.} = -m^2 \qquad (8\text{-}27)$$

or

$$\frac{d^2\Phi}{d\phi^2} + m^2\Phi = 0$$

with solutions

$$\Phi = \begin{pmatrix} \sin m\phi \\ \cos m\phi \end{pmatrix} \quad \text{or} \quad \Phi = e^{\pm im\phi} \qquad (8\text{-}28)$$

The r, θ equation becomes

$$\frac{1}{r^2 R}\frac{d}{dr}\left(r^2\frac{dR}{dr}\right) + \frac{1}{r^2\Theta\sin\theta}\frac{d}{d\theta}\left(\sin\theta\frac{d\Theta}{d\theta}\right) - \frac{m^2}{r^2\sin^2\theta} + k^2 = 0$$

If we multiply through by r^2, the first and fourth terms depend only on r, while the second and third depend only on θ.   Thus

$$\frac{1}{\sin\theta}\frac{d}{d\theta}\left(\sin\theta\frac{d\Theta}{d\theta}\right) - \frac{m^2}{\sin^2\theta}\Theta = \text{const.} \times \Theta = -l(l+1)\Theta \quad (8\text{-}29)$$

and

$$\frac{1}{r^2}\frac{d}{dr}\left(r^2\frac{dR}{dr}\right) + \left[k^2 - \frac{l(l+1)}{r^2}\right]R = 0 \qquad (8\text{-}30)$$

If we set $\cos\theta = x$, the θ equation (8-29) becomes

$$(1-x^2)\frac{d^2\Theta}{dx^2} - 2x\frac{d\Theta}{dx} + \left[l(l+1) - \frac{m^2}{1-x^2}\right]\Theta = 0 \qquad (8\text{-}31)$$

This is just the associated Legendre equation (7-27).   Its solutions are

$$\Theta = P_l^m(x), \, Q_l^m(x) \quad \text{(associated Legendre functions)} \qquad (8\text{-}32)$$

The radial equation (8-30), with the change of dependent variable $R = u/\sqrt{r}$, becomes

$$\frac{d^2u}{dr^2} + \frac{1}{r}\frac{du}{dr} + \left[k^2 - \frac{(l+\frac{1}{2})^2}{r^2}\right]u = 0 \qquad (8\text{-}33)$$

which is just Bessel's equation (7-48) with $x = kr$ and $m = l + \frac{1}{2}$.   Therefore,

$$R = \frac{J_{l+1/2}(kr)}{\sqrt{r}} \qquad \frac{Y_{l+1/2}(kr)}{\sqrt{r}} \qquad (8\text{-}34)$$

It is conventional to define "*spherical Bessel functions*" by

$$j_l(x) = \sqrt{\frac{\pi}{2x}} J_{l+1/2}(x) \qquad n_l(x) = \sqrt{\frac{\pi}{2x}} Y_{l+1/2}(x) \qquad (8\text{-}35)$$

$$h_l^{(1,\,2)}(x) = j_l(x) \pm i n_l(x) \qquad (\text{"spherical Hankel functions"}) \qquad (8\text{-}36)$$

It can be shown that

$$j_l(x) = (-x)^l \left(\frac{1}{x}\frac{d}{dx}\right)^l \frac{\sin x}{x} \qquad (8\text{-}37)$$

$$n_l(x) = (-x)^l \left(\frac{1}{x}\frac{d}{dx}\right)^l \left(-\frac{\cos x}{x}\right) \qquad (8\text{-}38)$$

Abramowitz and Stegun (A1), in Chapter 10, give a convenient summary of the recursion relations, asymptotic behavior, and so on, of spherical Bessel functions.

If $k = 0$, so that $\partial\psi/\partial t = 0$ and we are really discussing Laplace's equation, the radial equation (8-30) becomes

$$R'' + \frac{2}{r}R' - \frac{l(l+1)}{r^2} R = 0 \qquad (8\text{-}39)$$

which has the solutions

$$R = \left\{ \begin{matrix} r^l \\ r^{-(l+1)} \end{matrix} \right\} \qquad (8\text{-}40)$$

Thus we have found the following solutions:

$$\nabla^2\psi - \frac{1}{c^2}\frac{\partial^2\psi}{\partial t^2} = 0 \Rightarrow \psi = \left\{ \begin{matrix} e^{+i\omega t} \\ e^{-i\omega t} \end{matrix} \right\} \cdot \left\{ \begin{matrix} e^{+im\phi} \\ e^{-im\phi} \end{matrix} \right\} \cdot \left\{ \begin{matrix} P_l^m(\cos\theta) \\ Q_l^m(\cos\theta) \end{matrix} \right\} \cdot \left\{ \begin{matrix} j_l(kr) \\ n_l(kr) \end{matrix} \right\} \qquad (8\text{-}41)$$

$$\nabla^2\psi = 0 \Rightarrow \psi = \left\{ \begin{matrix} e^{+im\phi} \\ e^{-im\phi} \end{matrix} \right\} \cdot \left\{ \begin{matrix} P_l^m(\cos\theta) \\ Q_l^m(\cos\theta) \end{matrix} \right\} \cdot \left\{ \begin{matrix} r^l \\ r^{-(l+1)} \end{matrix} \right\} \qquad (8\text{-}42)$$

where each bracket represents a linear combination of the two functions inside. Any linear combination of such solutions is again a solution, because of the linearity of the original differential equation.

EXAMPLE

To illustrate the usefulness of these solutions, we solve the following boundary-value problem. Consider the acoustic radiation from a "split sphere" antenna; that is, ψ obeys the wave equation (8-21), and at $r = a$,

$$\psi = \begin{cases} V_0\, e^{-i\omega_0 t} & 0 < \theta < \dfrac{\pi}{2} \\[2mm] -V_0\, e^{-i\omega_0 t} & \dfrac{\pi}{2} < \theta < \pi \end{cases} \qquad (8\text{-}43)$$

To begin, we will clearly only use solutions with $\omega = \omega_0$, so that

$$k = k_0 = \frac{\omega_0}{c}$$

Also, since the boundary conditions are axially symmetric, we need only consider solutions with $m = 0$. Since everything must be well behaved at $\cos \theta = \pm 1$, we use only P_l with integral l. Thus we have reduced our trial solution to

$$\psi = e^{-i\omega t} \sum_l P_l(\cos \theta)[A_l \, j_l(k_0 r) + B_l \, n_l(k_0 r)]$$

We now use the fact that at infinity only *outgoing* waves are present. For large x

$$j_l(x) \sim \frac{1}{x} \cos \left[x - \frac{\pi}{2}(l+1) \right]$$

$$n_l(x) \sim \frac{1}{x} \sin \left[x - \frac{\pi}{2}(l+1) \right]$$

$$h_l^{(1,\,2)}(x) \sim \frac{1}{x} e^{\pm i[x-(l+1)\pi/2]}$$

Thus the radial functions we want are the $h_l^{(1)}(k_0 r)$, since then as $r \to \infty$

$$\psi \sim \frac{1}{r} e^{i(k_0 r - \omega_0 t)}$$

which is an outgoing wave. We have reduced our solution to

$$\psi = e^{-i\omega_0 t} \sum_l A_l P_l(\cos \theta) h_l^{(1)}(k_0 r) \qquad (8\text{-}44)$$

We now impose the boundary condition at $r = a$:

$$\sum_l A_l P_l(\cos \theta) h_l^{(1)}(k_0 a) = \begin{cases} V_0 & 0 < \theta < \dfrac{\pi}{2} \\[2mm] -V_0 & \dfrac{\pi}{2} < \theta < \pi \end{cases}$$

To determine the A_l we multiply both sides of (8-44) by $P_m(\cos \theta) \, d\cos \theta$ and integrate from -1 to $+1$. Then, since

$$\int_0^\pi P_l(\cos \theta) P_m(\cos \theta) \sin \theta \, d\theta = \frac{2}{2l+1} \delta_{lm}$$

we obtain (compare Problem 7-1)

$$\frac{2}{2l+1} A_l h_l^{(1)}(k_0 a) = V_0 \int_0^{\pi/2} P_l(\cos \theta) \sin \theta \, d\theta - V_0 \int_{\pi/2}^\pi P_l(\cos \theta) \sin \theta \, d\theta$$

Clearly only odd l contribute.   If l is odd

$$A_l = \frac{(2l+1)V_0}{h_l^{(1)}(k_0 a)} \int_0^1 P_l(x)\, dx$$

But from the recursion relations on p. 171

$$\int P_l(x)\, dx = \frac{P_{l+1}(x) - P_{l-1}(x)}{2l+1}$$

Therefore,

$$A_l = \frac{V_0}{h_l^{(1)}(k_0 a)} [P_{l+1}(x) - P_{l-1}(x)]_0^1$$

We use the relations (7-15) and (7-16)

$$P_l(1) = 1 \qquad P_l(0) = \begin{cases} 0 & (l \text{ odd}) \\ (-1)^{1/2}\dfrac{l!}{2^l\left[\left(\dfrac{l}{2}\right)!\right]^2} & (l \text{ even}) \end{cases}$$

The result is

$$A_l = (-1)^{(l-1)/2}\frac{V_0}{h_l^{(1)}(k_0 a)}\frac{(l+1)(2l+1)(l-1)!}{2^{l+1}\left[\left(\dfrac{l+1}{2}\right)!\right]^2} \qquad (l \text{ odd})$$

so that our final solution is

$$\psi(r, \theta, t)$$

$$= V_0 e^{-i\omega t} \sum_{l \text{ odd}} (-1)^{(1-1)/2}\frac{(l+1)(2l+1)(l-1)!}{2^{l+1}\left[\left(\dfrac{l+1}{2}\right)!\right]^2}\frac{h_l^{(1)}(k_0 r)}{h_l^{(1)}(k_0 a)}P_l(\cos\theta)$$

$$(8\text{-}45)$$

As a second exercise in separating variables, let us treat the vibration of a round drum head.   The differential equation describing small oscillations is

$$\nabla^2 u = \frac{1}{c^2}\frac{\partial^2 u}{\partial t^2} \qquad (8\text{-}46)$$

Let us look for periodic solutions, which describe the *normal modes* of the drum.   These have the form $u(\mathbf{x}, t) = u(\mathbf{x})e^{-i\omega t}$, and substituting into (8-46) we find the equation for $u(\mathbf{x})$:

$$\nabla^2 u + k^2 u = 0 \qquad \text{where} \qquad k = \frac{\omega}{c} = \text{wave number} \qquad (8\text{-}47)$$

In two-dimensional polar coordinates

$$\nabla^2 = \frac{1}{r} \frac{\partial}{\partial r} \left(r \frac{\partial}{\partial r} \right) + \frac{1}{r^2} \frac{\partial^2}{\partial \theta^2} \tag{8-48}$$

so that (8-47) is

$$\frac{1}{r} \frac{\partial}{\partial r} \left(r \frac{\partial u}{\partial r} \right) + \frac{1}{r^2} \frac{\partial^2 u}{\partial \theta^2} + k^2 u = 0$$

Try a solution of the form

$$u(r, \theta) = R(r) \Theta(\theta)$$

Separating the variables gives

$$\frac{d^2\Theta}{d\theta^2} + n^2\Theta = 0 \Rightarrow \Theta = e^{\pm in\theta}$$

$$\frac{d^2R}{dr^2} + \frac{1}{r} \frac{dR}{dr} + \left(k^2 - \frac{n^2}{r^2} \right) R = 0 \tag{8-49}$$

The R equation is Bessel's equation, so that

$$R = \begin{Bmatrix} J_n(kr) \\ Y_n(kr) \end{Bmatrix} \tag{8-50}$$

Before proceeding with the solution of the drum-head problem, we shall give the generalization of this separation procedure for three-dimensional cylindrical coordinates. The Laplacian is

$$\nabla^2 = \frac{1}{\rho} \frac{\partial}{\partial \rho} \left(\rho \frac{\partial}{\partial \rho} \right) + \frac{1^2}{\rho^2} \frac{\partial^2}{\partial \phi^2} + \frac{\partial^2}{\partial z^2} \tag{8-51}$$

Solutions of Laplace's equation

$$\nabla^2 \psi = 0$$

are

$$\psi = \begin{Bmatrix} J_m(\alpha\rho) \\ Y_m(\alpha\rho) \end{Bmatrix} \begin{Bmatrix} e^{\alpha z} \\ e^{-\alpha z} \end{Bmatrix} \begin{Bmatrix} e^{im\phi} \\ e^{-im\phi} \end{Bmatrix} \tag{8-52}$$

while solutions of the Helmholtz equation

$$\nabla^2 \psi + k^2 \psi = 0$$

are given by

$$\psi = \begin{Bmatrix} J_m(\sqrt{k^2 - \alpha^2}\,\rho) \\ Y_m(\sqrt{k^2 - \alpha^2}\,\rho) \end{Bmatrix} \begin{Bmatrix} e^{i\alpha z} \\ e^{-i\alpha z} \end{Bmatrix} \begin{Bmatrix} e^{im\phi} \\ e^{-im\phi} \end{Bmatrix} \tag{8-53}$$

For $k = 0$, the solutions (8-53) reduce to solutions of the Laplace equation;

for $\alpha = 0$ they reduce to our two-dimensional drum-head solutions (8-49) and (8-50); for $k = 0$ *and* $\alpha = 0$, we recover the familiar electrostatic potentials $\rho^{\pm m} e^{\pm im\phi}$.

Returning to the drum problem, the amplitude of oscillation must have the form

$$u = J_m(kr)\begin{Bmatrix}\cos m\theta\\\sin m\theta\end{Bmatrix}e^{\pm i\omega t} \tag{8-54}$$

where the function $Y_m(kr)$ has been eliminated because it becomes infinite at $r \to 0$.   The requirement that our solution be single valued means that m must be an integer.   If the drum head is clamped at its outer edge $(r = R)$, we must have

$$u = 0 \qquad \text{at} \qquad r = R$$

Therefore,

$$J_m(kR) = 0$$

The zeros of Bessel functions are tabulated in various places, for example, Abramowitz and Stegun (A1).   Some of the zeros are (approximately) as follows:

$$J_0(x) = 0: x \approx 2.40,\ 5.52,\ 8.65,\ \ldots$$
$$J_1(x) = 0: x \approx 3.83,\ 7.02,\ 10.17,\ \ldots \tag{8-55}$$
$$J_2(x) = 0: x \approx 5.14,\ 8.42,\ 11.62,\ \ldots$$

Thus the lowest modes of our drum head are

$$k = \frac{2.40}{R} \qquad \omega = 2.40\,\frac{c}{R} \qquad u \propto J_0\!\left(2.40\,\frac{r}{R}\right)$$

$$k = \frac{3.83}{R} \qquad \omega = 3.83\,\frac{c}{R} \qquad u \propto J_1\!\left(3.83\,\frac{r}{R}\right)\begin{Bmatrix}\cos\theta\\\sin\theta\end{Bmatrix}$$

$$k = \frac{5.14}{R} \qquad \omega = 5.14\,\frac{c}{R} \qquad u \propto J_2\!\left(5.14\,\frac{r}{R}\right)\begin{Bmatrix}\cos 2\theta\\\sin 2\theta\end{Bmatrix}$$

$$k = \frac{5.52}{R} \qquad \omega = 5.52\,\frac{c}{R} \qquad u \propto J_0\!\left(5.52\,\frac{r}{R}\right)$$

To the right of each mode we have drawn the conventional type of picture, indicating the *nodes* or places where u is always zero.

Note that there are two independent modes belonging to the second and third frequencies.   This is an example of *degeneracy*, which we mentioned in Chapter 6 and which will continue to appear throughout the next several chapters.

As a final example of the use of separation of variables in solving boundary-value problems, let us find the temperature within a cube of side L, initially at temperature $T = 0$, which at time $t = 0$ is immersed in a heat bath at temperature $T = T_0$. We must solve the equation

$$\nabla^2 T = \frac{1}{\kappa} \frac{\partial T}{\partial t} \tag{8-56}$$

Suppose

$$T \propto e^{-\lambda t}$$

Then

$$\nabla^2 T + \frac{\lambda}{\kappa} T = 0$$

$$\frac{\partial^2 T}{\partial x^2} + \frac{\partial^2 T}{\partial y^2} + \frac{\partial^2 T}{\partial z^2} = -\frac{\lambda}{\kappa} T$$

Separation of variables gives

$$T \propto e^{i\alpha x} e^{i\beta y} e^{i\gamma z}$$

with

$$\alpha^2 + \beta^2 + \gamma^2 = \frac{\lambda}{\kappa}$$

Now one boundary condition is (choosing the origin at one corner of the cube)

$$T = T_0 \quad \text{for} \quad x = 0, L$$
$$T = T_0 \quad \text{for} \quad y = 0, L$$
$$T = T_0 \quad \text{for} \quad z = 0, L$$

Note that this is an inhomogeneous boundary condition, although a very simple one. A particular solution of (8-56) is $T = T_0$, and the complementary function must satisfy the corresponding homogeneous boundary conditions: $T = 0$ on the surface. Thus we write

$$T = T_0 + \sum_{lmn} C_{lmn} \sin \frac{l\pi x}{L} \sin \frac{m\pi y}{L} \sin \frac{n\pi z}{L} e^{-\lambda_{lmn} t} \tag{8-57}$$

where

$$\lambda_{lmn} = \kappa \frac{\pi^2}{L^2} (l^2 + m^2 + n^2)$$

To determine the constants C_{lmn}, we have the condition that $T = 0$ when $t = 0$. Therefore,

$$\sum_{lmn} C_{lmn} \sin \frac{l\pi x}{L} \sin \frac{m\pi y}{L} \sin \frac{n\pi z}{L} = -T_0$$

Multiply by $\sin(l'\pi x/L) \sin(m'\pi y/L) \sin(n'\pi z/L)$ and integrate over the whole cube. The result is

$$C_{lmn} = \begin{cases} -\dfrac{64T_0}{\pi^3 lmn} & \text{if } l, m, n \text{ all odd} \\ 0 & \text{otherwise} \end{cases}$$

Thus, for $t > 0$

$$T = T_0 \left\{ 1 - \frac{64}{\pi^3} \sum_{lmn \text{ odd}} \frac{1}{lmn} \sin \frac{l\pi x}{L} \sin \frac{m\pi y}{L} \sin \frac{n\pi z}{L} \right.$$

$$\left. \times \exp \left[-(l^2 + m^2 + n^2)\kappa \frac{\pi^2 t}{L^2} \right] \right\} \qquad (8\text{-}58)$$

For $t \gg L^2/\kappa$, only the first term in the sum matters, and

$$T \approx T_0 \left[1 - \frac{64}{\pi^3} \sin \frac{\pi x}{L} \sin \frac{\pi y}{L} \sin \frac{\pi z}{L} \exp \left(-\frac{3\kappa\pi^2}{L^2} t \right) \right] \qquad (8\text{-}59)$$

Our solution (8-58) is not very useful for small t; at the end of Section 8–4 we discuss a method which gives a solution which is useful for small t.

In all the examples above, the separation of variables has been achieved by looking for a solution in the form of a product of functions, each depending on fewer variables than are present in the original equation. For some problems, however, a separation of variables can be made by means of a solution having a different form; for example, a sum of functions of fewer variables.

EXAMPLE

Consider an infinite heat conducting slab of thickness D with one surface ($x = D$) insulated. If, initially, the temperature T is zero and then heat is supplied (for example, by radiation) at a constant rate, Q calories per sec per cm^2 at the surface $x = 0$, find the temperature as a function of position and time within the slab. This situation is illustrated in Figure 8–6.

We must solve the one-dimensional diffusion equation

$$\frac{\partial^2 T}{\partial x^2} - \frac{1}{\kappa} \frac{\partial T}{\partial t} = 0 \qquad \kappa = \frac{K}{C\rho} \qquad (8\text{-}60)$$

with the given inhomogeneous boundary conditions. Physical considerations tell us the qualitative behavior of the temperature with time. After an initial transient period, we expect the temperature to rise linearly with time at any

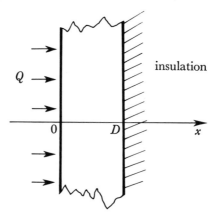

Figure 8–6 An infinite heat conducting slab of thickness D with the surface $x = D$ insulated and heat supplied at the surface $x = 0$ at a constant rate, Q calories per cm² per sec

position. Thus we try a particular solution of the form

$$T_p = u(x) + \alpha t \qquad (8\text{-}61)$$

which leads to a separation of variables, the x equation being

$$\frac{d^2 u}{dx^2} = \frac{\alpha}{\kappa} \qquad (8\text{-}62)$$

with the solution

$$u(x) = \frac{1}{2} \frac{\alpha}{\kappa} x^2 + ax + b$$

b may be arbitrarily chosen for the particular solution, while the constants α and a are determined by the boundary conditions:

$$-Ku'(0) = Q \qquad u'(D) = 0$$

A solution satisfying these conditions is

$$u(x) = \frac{1}{2} \frac{Q}{KD} (x - D)^2 \qquad \alpha = \frac{Q\kappa}{KD} = \frac{Q}{C\rho D}$$

giving the particular solution

$$T_p = \frac{1}{2} \frac{Q}{KD} (x - D)^2 + \frac{Q}{C\rho D} t \qquad (8\text{-}63)$$

We invite the student to finish the problem by finding the complementary function T_c and using it to satisfy the initial condition

$$T(x, 0) = T_p + T_c = 0 \quad \text{at} \quad t = 0$$

Separation of variables by the method just illustrated finds its main application in classical mechanics in connection with the Hamilton–Jacobi differential equation.   For these applications, we refer to Goldstein (G4).

An interesting example of an elementary problem whose solution requires some ingenuity is the solution of the two-dimensional Helmholtz equation inside an equilateral triangle.   To be specific, let us consider the Helmholtz equation

$$\nabla^2 u + k^2 u = 0 \qquad (8\text{-}64)$$

inside the equilateral triangle of side 1 shown in Figure 8–7a; the boundary condition will be taken to be $u = 0$ on the perimeter of the triangle.

We now reflect our triangle in its side AB, yielding the triangle ABC of Figure 8–7b.   At the same time, we "reflect" the function $u(\mathbf{x})$; that is, we define $u(\mathbf{x})$ at each point in ABC to be the negative of $u(\mathbf{x})$ at the "image

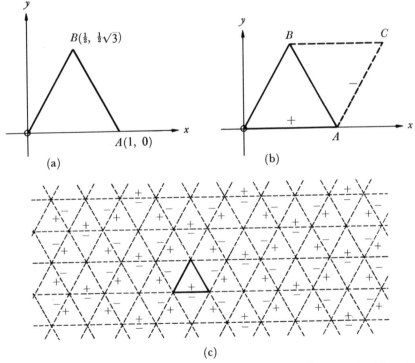

Figure 8–7   Steps   in   the   solution   of $\nabla^2 u + k^2 u = 0$ inside   an equilateral triangle. (a) Original triangle. (b) Original triangle OAB and its "image" ABC in side AB. (c) The xy-plane covered by repeated reflecting of triangles in their sides. The plus and minus signs denote the relative signs of $u(x, y)$ at various points

point" in OAB. We have symbolized this reflection process by the small + sign in OAB and the − sign at its image point in ABC.

The reason for this particular mapping prescription is that we wish the resulting function $u(\mathbf{x})$ to obey the Helmholtz equation throughout the two-triangle region $OACB$. If our original boundary condition had been that the normal derivative of u, rather than u itself, was to vanish on the perimeter of the triangle, then we would have had to define u so that it was *even* under reflection in AB, rather than odd.

This process of reflection can be repeated over and over, so that we fill up the xy-plane with equilateral triangles (Figure 8–7c). Equally important, we have extended our solution of $\nabla^2 u + k^2 u = 0$ inside the original triangle OAB into a solution of $\nabla^2 u + k^2 u = 0$ throughout the entire plane. Furthermore, this solution has a high degree of symmetry and periodicity, as symbolized by the plus and minus signs of Figure 8–7c.

To be precise about the periodicity, we can read off from Figure 8–7c the relations

$$u(x, y) = u(x + 3, y)$$

$$u(x, y) = u\left(x, y + \sqrt{3}\right)$$

Therefore we can expand $u(x, y)$ in the double Fourier series

$$u(x, y) = \sum_{l=-\infty}^{\infty} \sum_{m=-\infty}^{\infty} a_{lm} e^{\frac{2\pi i}{3} lx} e^{\frac{2\pi i}{\sqrt{3}} my} \tag{8-65}$$

which implies, with (8-64),

$$k^2 = 4\pi^2 \left(\frac{l^2}{9} + \frac{m^2}{3}\right) \tag{8-66}$$

If we are to have, then, a solution of the Helmholtz equation (8-64), with a definite value for k, there can appear in the expansion (8-65) only those Fourier coefficients a_{lm} for which l and m are related by (8-66).

There are several more symmetry relations. For example, it is obvious from Figure 8–7c that $u(x, y)$ is an odd function of y; $u(x, -y) = -u(x, y)$. From (8-65) we see that this implies

$$a_{lm} = -a_{l,-m} \tag{8-67}$$

Also, we can see from Figure 8–7c that $u(x, y)$ is left unchanged by a rotation through $120°$ about the origin; that is,

$$u\left(-\frac{1}{2} x + \frac{\sqrt{3}}{2} y, \ -\frac{\sqrt{3}}{2} x - \frac{1}{2} y\right) = u(x, y)$$

Table 8-2   Normal mode solutions of the Helmholtz equation which vanish on the surface of an equilateral triangle of side 1

(l, m)			k^2
$(3, 1)$ $(3, -1)^-$	$(-3, 1)$ $(-3, -1)^-$	$(0, -2)$ $(0, 2)^-$ }	$16\pi^2/3$
$(5, 1)$ $(5, -1)^-$	$(-4, 2)$ $(-4, -2)^-$	$(-1, -3)$ $(-1, 3)^-$ }	$112\pi^2/9$
$(6, 2)$ $(6, -2)^-$	$(-6, 2)$ $(-6, -2)^-$	$(0, -4)$ $(0, 4)^-$ }	$48\pi^2/3$
$\vdots$		$\vdots$	

Expressing this relation in terms of Fourier coefficients by means of (8-65) gives

$$a_{lm} = a_{l'm'} \qquad \text{where} \qquad \begin{cases} l' = -\tfrac{1}{2}l - \tfrac{3}{2}m \\ m' = \tfrac{1}{2}l - \tfrac{1}{2}m \end{cases} \tag{8-68}$$

A second rotation through 120° gives

$$a_{lm} = a_{l''m''} \qquad \text{where} \qquad \begin{cases} l'' = -\tfrac{1}{2}l + \tfrac{3}{2}m \\ m'' = -\tfrac{1}{2}l - \tfrac{1}{2}m \end{cases} \tag{8-69}$$

It is reassuring to verify that the l' and m' of (8-68) and the l'' and m'' of (8-69) give, according to (8-66), the same value of k^2 as do l and m.

It now remains only to enumerate the possible (l, m) pairs, satisfying the criteria (8-67), (8-68), and (8-69), and the associated values of k^2. The results are given in Table 8–2. The negative superscripts on (l, m) pairs in this table denote pairs whose Fourier coefficients a_{lm} must, according to (8-67), be of opposite sign to the others.

Note that the second and third solutions of Table 8–2 are *degenerate*; each yields another solution with the same value of k^2 simply by changing the sign of every l and m. The first solution, however, goes into (the negative of) itself under this change of sign, and so there is only one independent solution with this lowest value $16\pi^2/3$ for k^2.

8-4   INTEGRAL TRANSFORM METHODS

We shall next work a few problems to show how various integral transforms can be used to solve boundary-value problems. The essential feature is the transformation of the equation to one containing derivatives with respect to a smaller number of variables. Also, the boundary conditions may be inserted in a more automatic way.

EXAMPLE

Consider the temperature distribution in the semi-infinite region $x > 0$ when initially $T = 0$ and the plane $x = 0$ is maintained at $T = T_0$. The diffusion equation becomes

$$\frac{\partial^2 T}{\partial x^2} = \frac{1}{\kappa} \frac{\partial T}{\partial t}$$

The boundary conditions are

(1) $$T(x, 0) = 0$$

(2) $$T(0, t) = T_0$$

The fact that we are only interested in $t > 0$ suggests the use of the Laplace transform with respect to t. (We are also interested only in $x > 0$, but a Laplace transform with respect to x is not useful. Why?) Let

$$F(x, s) = \int_0^\infty e^{-st} T(x, t)\, dt$$

The Laplace transform of the differential equation is

$$\frac{\partial^2 F}{\partial x^2} = \frac{sF}{\kappa} \tag{8-70}$$

since $T = 0$ for $t = 0$.

The second boundary condition becomes

$$F(0, s) = \frac{T_0}{s} \tag{8-71}$$

Thus, solving (8-70) and making use of (8-71), we obtain

$$F(x, s) = \frac{T_0}{s} \exp\left(-\sqrt{s/\kappa}\, x\right)$$

as we certainly reject the solution $\sim \exp\left(+\sqrt{s/\kappa}\,x\right)$.

Now we must invert this transform to obtain our solution.

$$T(x, t) = \frac{1}{2\pi i} \int_{c-i\infty}^{c+i\infty} \frac{T_0}{s} \exp\left(-\sqrt{s/\kappa}\, x\right) e^{st}\, ds \tag{8-72}$$

The integrand is multivalued. We shall cut the s-plane along the negative real axis and consider the contour shown in Figure 8–8. Since the total integral on the closed contour of Figure 8–8 is zero, the contribution from ① is given by

$$① = -② - ③ - ④ - ⑤ - ⑥$$

② and ⑥ give no contribution.

$$④ = -2\pi i \frac{1}{2\pi i} T_0 = -T_0$$

$$③ = \frac{1}{2\pi i} \int_{-\infty}^{0} \frac{T_0}{s} \exp\left(-ix\sqrt{1/\kappa}\sqrt{-s}\right)e^{st}\,ds$$

$$= \frac{-1}{2\pi i} \int_{0}^{\infty} \frac{T_0}{s} \exp\left(-ix\sqrt{1/\kappa}\sqrt{s}\right)e^{-st}\,ds$$

Similarly,

$$⑤ = \frac{1}{2\pi i} \int_{0}^{\infty} \frac{T_0}{s} \exp\left(ix\sqrt{1/\kappa}\sqrt{s}\right)e^{-st}\,ds$$

Therefore,

$$③ + ⑤ = \frac{1}{\pi} \int_{0}^{\infty} \frac{T_0}{s} e^{-st} \sin x\sqrt{s/\kappa}\,ds$$

Let $s = z^2$.   Then

$$③ + ⑤ = \frac{2T_0}{\pi} \int_{0}^{\infty} \frac{dz}{z} e^{-tz^2} \sin x\sqrt{1/\kappa}\,z$$

and

$$T(x, t) = T_0\left(1 - \frac{2}{\pi} \int_{0}^{\infty} \frac{dz}{z} e^{-tz^2} \sin x\sqrt{1/\kappa}\,z\right)$$

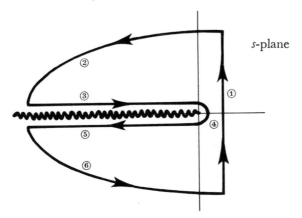

Figure 8–8   The s-plane with branch cut and contour used for evaluating the Laplace inversion integral (8-72)

It will be left to the student to do this integral and obtain[1]

$$T(x, t) = T_0\left[1 - \text{erf}\left(\frac{x}{2}\sqrt{\frac{1}{\kappa t}}\right)\right] \tag{8-73}$$

Note that we have obtained an answer in closed form, rather than an infinite series such as the separation of variables usually gives. This feature results, however, from the infinite range of the variable x rather than from the transform method. To solve this problem by the separation of variables technique, we would consider a slab of finite thickness D and in the end let $D \to \infty$. In this limiting process, the series solution would be transformed into an integral whose evaluation would give the result (8-73). It is instructive for the student to carry out this procedure.

As a second example of the use of integral transform methods in solving partial differential equations, consider another diffusion problem. Find the temperature distribution $T(x, t)$ in an infinite solid if we are given $T(x, 0) = f(x)$. Nothing depends on y or z, so that

$$\frac{\partial^2 T}{\partial x^2} = \frac{1}{\kappa}\frac{\partial T}{\partial t}$$

For variety, we shall Fourier transform the x variable.

$$T(x, t) = \int_{-\infty}^{\infty} \frac{dk}{2\pi} F(k, t)e^{ikx}$$

$$F(k, t) = \int_{-\infty}^{\infty} T(x, t)e^{-ikx}\,dx$$

Thus

$$-k^2 F(k, t) = \frac{1}{\kappa}\frac{\partial F(k, t)}{\partial t}$$

with the solution

$$F(k, t) = \phi(k)e^{-k^2\kappa t}$$

---

[1] Our evaluation of the integral (8-72) is straightforward and instructive, but a simpler method exists. Namely, if one differentiates both sides of (8-72) with respect to x, and changes the variable of integration to $u = \sqrt{s}$, one finds by a trivial contour deformation that

$$\frac{\partial T}{\partial x} = -\frac{T_0}{\sqrt{\pi\kappa t}}e^{-x^2/4\kappa t}$$

from which (8-73) follows directly.

The initial condition gives

$$F(k, 0) = \int_{-\infty}^{\infty} T(x, 0)e^{-ikx} \, dx$$

$$= \int_{-\infty}^{\infty} f(x)e^{-ikx} \, dx$$

Therefore,

$$\phi(k) = \int_{-\infty}^{\infty} f(x)e^{-ikx} \, dx$$

and

$$F(k, t) = \int_{-\infty}^{\infty} f(x) \, dx \, e^{-ikx} e^{-k^2 \kappa t}$$

Inverting the Fourier transform, we obtain

$$T(x, t) = \int_{-\infty}^{\infty} \frac{dk}{2\pi} e^{ikx} \int_{-\infty}^{\infty} dx' f(x') e^{-ikx'} e^{-k^2 \kappa t}$$

$$= \int_{-\infty}^{\infty} dx' f(x') \int_{-\infty}^{\infty} \frac{dk}{2\pi} e^{ik(x-x')} e^{-k^2 \kappa t} \qquad (8\text{-}74)$$

The k integral is easy and gives

$$\sqrt{\frac{1}{4\pi\kappa t}} \, e^{-(x-x')^2/4\kappa t}$$

Therefore,

$$T(x, t) = \int_{-\infty}^{\infty} dx' f(x') \sqrt{\frac{1}{4\pi\kappa t}} \, e^{-(x-x')^2/4\kappa t} \qquad (8\text{-}75)$$

The function

$$G(x, t; x') = \sqrt{\frac{1}{4\pi\kappa t}} \, e^{-(x-x')^2/4\kappa t}$$

is the *Green's function* for the above problem.   Green's functions will be discussed in more detail in the next chapter.

Suppose the initial heat " source " $f(x)$ (that is, the initial temperature distribution) consists of the *plane source* $\delta(x)$.   Then

$$T(x, t) = \sqrt{\frac{1}{4\pi\kappa t}} \, e^{-x^2/4\kappa t} = G(x, t; 0) \qquad (8\text{-}76)$$

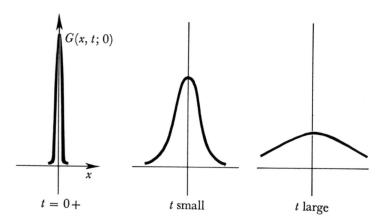

$t = 0+$ t small t large

Figure 8–9   The Green's function $G(x, t; 0)$ for the one-dimensional heat problem with given initial temperature distribution

This is a Gaussian dependence on x, with a width increasing as $\sqrt{t}$, as shown in Figure 8–9.

What is the distribution resulting from a *point source* $\delta(\mathbf{x})$ at time $t = 0$? We can get this from our preceding result as follows. Let $f(x, t)$ be the response to a plane source $\delta(x)$ at $t = 0$. Let $g(r, t)$ be the response to a point source $\delta(\mathbf{x})$ at $t = 0$. Then from Figure 8–10 we see

$$f(x, t) = \int_0^\infty 2\pi y \, dy \, g(r, t)$$

$$= \int_x^\infty 2\pi r \, dr \, g(r, t)$$

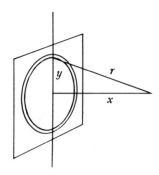

Figure 8–10   Relationships between variables x and r appropriate for a plane source and a point source, respectively

Therefore,

$$\frac{\partial f}{\partial x} = -2\pi x g(x, t)$$

$$g(r, t) = \frac{-1}{2\pi r} \left[\frac{\partial f(x, t)}{\partial x} \right]_{x=r} \tag{8-77}$$

Using our previously found response $f(x, t)$ for a plane source (8-76), we immediately obtain

$$g(r, t) = \left(\frac{1}{4\pi\kappa t} \right)^{3/2} e^{-r^2/4\kappa t} \tag{8-78}$$

as the response to a point initial distribution $T(\mathbf{x}, 0) = \delta(\mathbf{x})$.

This point-source distribution is another example of a Green's function about which we will say more later.   However, we shall illustrate the usefulness of Green's functions in two diffusion problems.

As a first example, consider the response of the semi-infinite solid $x > 0$ to a point initial temperature distribution at $x = a$, $y = z = 0$, if the entire solid was initially $(t = 0)$ at temperature $T = 0$ (except at the above point) and the boundary $x = 0$ is maintained at $T = 0$.   The answer is obtained by superimposing a source function about $x = a$, $y = z = 0$:

$$\left(\frac{1}{4\pi\kappa t} \right)^{3/2} \exp \left[-\frac{(x - a)^2 + y^2 + z^2}{4\kappa t} \right] \tag{8-79}$$

and a negative source function about $x = -a$, $y = z = 0$:

$$-\left(\frac{1}{4\pi\kappa t} \right)^{3/2} \exp \left[-\frac{(x + a)^2 + y^2 + z^2}{4\kappa t} \right] \tag{8-80}$$

This second "fictitious" source is called an *image*.   This *method of images* should be familiar from electrostatics where it is often used.

As a second example, consider the problem of a cube initially at zero temperature, whose sides are maintained at zero temperature, with $T(\mathbf{x}, t = 0) = \delta(\mathbf{x})$, $\mathbf{x} = 0$ being the center of the cube.   We can solve for the temperature distribution at later times either by separating variables or by superimposing an infinite set of images.   The first solution is useful for large t; the second is useful for small t.

8–5   WIENER–HOPF METHOD

Historically, this method was developed in order to solve a particular type of integral equation, but it also lends itself to the solution of certain boundary-value problems which are very difficult to solve by more conventional techniques.

As used here, the method is an extension of the integral transform method, for which we need to consider the analytic continuation of Fourier transforms off the real axis of the transform variable. Thus we consider the Fourier transform of $f(x)$:

$$F(k) = \int_{-\infty}^{\infty} f(x)e^{-ikx} \, dx \tag{8-81}$$

for complex $k = \omega + i\gamma$. Then

$$F(k) = \int_{-\infty}^{\infty} f(x)e^{\gamma x}e^{-i\omega x} \, dx \tag{8-82}$$

The factor $e^{\gamma x}$ may cause the integral to diverge at either $x \to +\infty$ or $x \to -\infty$ unless the function $f(x)$ prevents the divergence, which it may do for a limited range of γ. This range of $\gamma = \text{Im } k$ depends on the asymptotic behavior of $f(x)$ in the following ways:

1. If

$$f(x)e^{\alpha x} \to 0 \qquad \text{as} \qquad x \to +\infty$$

and

$$f(x)e^{\beta x} \to 0 \qquad \text{as} \qquad x \to -\infty \qquad \text{with } \alpha > \beta$$

$$\tag{8-83}$$

then $F(k)$ is an analytic function of k in the strip $\beta < \text{Im } k < \alpha$.

2. If $f(x) = 0$ for all $x < 0$, then clearly β can be an arbitrary negative number. In this case, not only is $F(k)$ analytic for $-\infty < \text{Im } k < \alpha$, but also $F(k) \to 0$ uniformly as $\text{Im } k \to -\infty$, since now $F(k) = \int_0^{\infty} f(x)e^{\gamma x}e^{-i\omega x} \, dx$.

To see how much stronger this second statement is, consider the function e^{-x^2}. The β of part 1 may be taken to be $-\infty$, and indeed the Fourier transform $F(k) \sim e^{-k^2/4}$ is analytic in the entire lower half plane. But $f(x)$ does not *vanish* for $x < 0$, and, sure enough, $F(k)$ does not vanish uniformly as $\text{Im } k \to -\infty$.

In passing, we may point out that everything we have said about the connection of the behavior of $f(x)$ at $\pm \infty$ to the analyticity of $F(k)$ can be applied in the reverse direction; the behavior of $F(k)$ at $\pm \infty$ (actually at $\text{Re } k \to \pm \infty$) determines the region of analyticity of the function $f(x)$ given by the inversion integral.

In addition to these properties of Fourier transforms, we must discuss a simple decomposition of a function of a complex variable into the sum of two functions. This decomposition is the characteristic feature of any Wiener–Hopf solution, whether applied to a boundary-value problem, an integral equation, or anything else.

Suppose a certain function $F(k)$ is analytic in the strip

$$\beta < \text{Im } k < \alpha$$

and goes to zero at the ends of the strip (Re $k \to \pm \infty$).   Then we may write $F(k)$ as the sum of two functions

$$F(k) = F_+(k) + F_-(k) \tag{8-84}$$

where $F_+(k)$ is analytic for Im $k > \beta$ and goes to zero as Im $k \to +\infty$, while $F_-(k)$ is analytic for Im $k < \alpha$ and goes to zero as Im $k \to -\infty$.

For, by Cauchy's theorem, with k inside the strip,

$$F(k) = \frac{1}{2\pi i} \int_{-\infty+i\beta}^{\infty+i\beta} \frac{F(k')}{k'-k}\, dk' - \frac{1}{2\pi i} \int_{-\infty+i\alpha}^{\infty+i\alpha} \frac{F(k')}{k'-k}\, dk' \tag{8-85}$$

The first integral is analytic for Im $k > \beta$ (also for Im $k < \beta$, but this does not concern us) and may be identified with $F_+(k)$, while the second term may be similarly identified as $F_-(k)$.   It is easily shown that the decomposition is unique.

The decomposition can sometimes be achieved by inspection.   For example, suppose

$$F(k) = \frac{1}{(k+i)\sqrt{k-i}} \tag{8-86}$$

This function has a pole at $k = -i$ and a branch point at $k = +i$, but is analytic in the strip $|\text{Im } k| < 1$.   Let us fix the square-root function by agreeing that

$$\sqrt{0-i} = e^{-i\pi/4} = \frac{1-i}{\sqrt{2}}$$

Then we may write

$$F(k) = \frac{1}{k+i}\left(\frac{1}{\sqrt{k-i}} - \frac{1}{\sqrt{-i-i}}\right) + \frac{1}{\sqrt{-i-i}}\frac{1}{k+i}$$

$$= \underbrace{\frac{1}{k+i}\left(\frac{1}{\sqrt{k-i}} - \frac{1+i}{2}\right)}_{F_-(k)} + \underbrace{\frac{1+i}{2}\frac{1}{k+i}}_{F_+(k)} \tag{8-87}$$

where $F_-(k)$ has been found by removing the pole at $k = -i$.

We shall now illustrate the Wiener–Hopf method by working a problem.[2] Let us solve the two-dimensional Laplace equation

$$\nabla^2 \phi = 0 \tag{8-88}$$

[2] We state our problem in mathematical terms.   For a physical application, see Problem 8-26.

in the upper-half plane ($y > 0$), subject to the following boundary conditions:

(1) $\phi \to 0$ as $y \to \infty$

(2) $\phi = e^{-ax}\ (a > 0)$ for $y = 0, x > 0$

(3) $\dfrac{\partial \phi}{\partial y} = Ce^{bx}\ (b > 0)$ for $y = 0, x < 0$

Let

$$\phi(x, 0) = f(x) \qquad \frac{\partial \phi}{\partial y}(x, 0) = g(x) \tag{8-89}$$

Then (2) and (3) tell us that

$$f(x) = \begin{cases} e^{-ax} & (x > 0) \\ ? & (x < 0) \end{cases} \qquad g(x) = \begin{cases} ? & (x > 0) \\ Ce^{bx} & (x < 0) \end{cases} \tag{8-90}$$

Solving the equation (8-88) and boundary condition (1) by a Fourier transform with respect to x, we write

$$\phi(x, y) = \int_{-\infty}^{\infty} \frac{dk}{2\pi} e^{ikx} F(k, y)$$

where the transform function $F(k, y)$ satisfies

$$\frac{\partial^2 F}{\partial y^2} - k^2 F = 0$$

Thus

$$F(k, y) = \Phi(k)e^{-|k|y} \qquad \text{for } k \text{ real}$$

In order to extend the function $F(k, y)$ into the complex k-plane, we replace the function $|k|$ for k real by

$$\sqrt{k^2 + \lambda^2} \tag{8-91}$$

with cuts running to $\pm i\infty$ as shown in Figure 8–11, and the root defined by $\sqrt{k^2 + \lambda^2} \to +k$ as Re $k \to +\infty$. When we are through, we will let λ go to zero.

Thus

$$\phi(x, y) = \int_{-\infty}^{\infty} \frac{dk}{2\pi} e^{ikx} \exp\left(-\sqrt{k^2 + \lambda^2}\,y\right)\Phi(k) \tag{8-92}$$

We wish to find $\Phi(k)$ to satisfy the boundary conditions (8-90).

$$f(x) = \int_{-\infty}^{\infty} \frac{dk}{2\pi} e^{ikx}\Phi(k) \tag{8-93}$$

$$g(x) = -\int_{-\infty}^{\infty} \frac{dk}{2\pi} e^{ikx}\sqrt{k^2 + \lambda^2}\,\Phi(k) \tag{8-94}$$

Inverting these last two Fourier transforms gives

$$\Phi(k) = \int_{-\infty}^{\infty} dx\, e^{-ikx} f(x)$$

$$= \left(\int_{-\infty}^{0} + \int_{0}^{\infty} \right) dx\, e^{-ikx} f(x)$$

$$= \Phi_{+}(k) + \int_{0}^{\infty} dx\, e^{-ikx} e^{-ax}$$

$$= \Phi_{+}(k) + \frac{1}{i} \frac{1}{k - ia} \tag{8-95}$$

and similarly,

$$\Phi(k)\sqrt{k^2 + \lambda^2} = \psi_{-}(k) + \frac{C}{i} \frac{1}{k + ib} \tag{8-96}$$

where $\Phi_{+}(k)$ is the Fourier transform of a function which vanishes for $x > 0$, while $\psi_{-}(k)$ is the Fourier transform of a function which vanishes for $x < 0$.

Equating these two expressions for $\Phi(k)$ gives

$$\sqrt{k^2 + \lambda^2}\, \Phi_{+}(k) + \frac{1}{i} \frac{\sqrt{k^2 + \lambda^2}}{k - ia} = \psi_{-}(k) + \frac{C}{i} \frac{1}{k + ib}$$

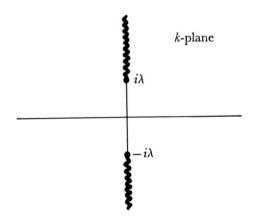

Figure 8–11 **The k-plane with cuts used to specify the function** $\sqrt{k^2 + \lambda^2}$

We rewrite this equation in the form

$$\underbrace{\sqrt{k+i\lambda}\,\Phi_+(k)}_{F_+(k)} - \underbrace{\frac{\psi_-(k)}{\sqrt{k-i\lambda}}}_{F_-(k)} = \underbrace{\frac{C}{i}\frac{1}{k+ib}\frac{1}{\sqrt{k-i\lambda}} - \frac{1}{i}\frac{\sqrt{k+i\lambda}}{k-ia}}_{F(k)} \quad (8\text{-}97)$$

In this form, the two terms on the left satisfy the conditions appropriate for the notation $F_+(k)$ and $F_-(k)$ [compare (8-84)], while the right side is a specific function. We must decompose the right side $F(k) = F_+(k) + F_-(k)$ in order to find $\Phi_+(k)$ and $\psi_-(k)$, from either of which $\Phi(k)$ may be obtained. It is this procedure which is the essential point of the Wiener–Hopf method. In this particular example, each piece of $F(k)$ can be decomposed by inspection.

First we have

$$\frac{1}{k+ib}\frac{1}{\sqrt{k-i\lambda}} = \frac{1}{k+ib}\left(\frac{1}{\sqrt{k-i\lambda}} - \frac{1}{\sqrt{-ib-i\lambda}}\right) + \frac{1}{\sqrt{-ib-i\lambda}}\frac{1}{k+ib}$$

$$= \frac{1}{k+ib}\left(\frac{1}{\sqrt{k-i\lambda}} - \frac{1+i}{\sqrt{2}}\frac{1}{\sqrt{b+\lambda}}\right) + \frac{1+i}{\sqrt{2}}\frac{1}{\sqrt{b+\lambda}}\frac{1}{k+ib}$$

Similarly,

$$\frac{\sqrt{k+i\lambda}}{k-ia} = \frac{1}{k-ia}\left(\sqrt{k+i\lambda} - \sqrt{ia+i\lambda}\right) + \sqrt{ia+i\lambda}\,\frac{1}{k-ia}$$

$$= \frac{1}{k-ia}\left(\sqrt{k+i\lambda} - \frac{1+i}{\sqrt{2}}\sqrt{a+\lambda}\right) + \frac{1+i}{\sqrt{2}}\sqrt{a+\lambda}\,\frac{1}{k-ia}$$

Therefore,

$$F_+(k) = \sqrt{k+i\lambda}\,\Phi_+(k)$$

$$= \frac{C}{i}\frac{1+i}{\sqrt{2}}\frac{1}{\sqrt{b+\lambda}}\frac{1}{k+ib} - \frac{1}{i}\frac{1}{k-ia}\left(\sqrt{k+i\lambda} - \frac{1+i}{\sqrt{2}}\sqrt{a+\lambda}\right)$$

$$(8\text{-}98)$$

Substituting this result into our expression (8-95) for $\Phi(k)$ in terms of $\Phi_+(k)$ gives

$$\Phi(k) = \frac{1-i}{\sqrt{2}}\frac{1}{\sqrt{k+i\lambda}}\left(\frac{\sqrt{a+\lambda}}{k-ia} + \frac{C}{\sqrt{b+\lambda}}\frac{1}{k+ib}\right)$$

We can now pass to the limit $\lambda \to 0$. Note that

$$\sqrt{k+i\lambda} \to \begin{cases} \sqrt{k} & (k>0) \\ i\sqrt{-k} & (k<0) \end{cases}$$

Thus

$$\Phi(k) = \begin{cases} \dfrac{1-i}{\sqrt{2}}\dfrac{1}{\sqrt{k}}\left(\dfrac{\sqrt{a}}{k-ia} + \dfrac{C}{\sqrt{b}}\dfrac{1}{k+ib}\right) & (k>0) \\[4mm] \dfrac{-1-i}{\sqrt{2}}\dfrac{1}{\sqrt{-k}}\left(\dfrac{\sqrt{a}}{k-ia} + \dfrac{C}{\sqrt{b}}\dfrac{1}{k+ib}\right) & (k<0) \end{cases} \qquad (8\text{-}99)$$

The solution to our problem is then given by

$$\phi(x, y) = \int_{-\infty}^{\infty} \frac{dk}{2\pi} e^{ikx} e^{-|k|y} \Phi(k)$$

$$= \int_{-\infty}^{0} \frac{-1-i}{2\pi\sqrt{2}} \frac{1}{\sqrt{-k}}\left(\frac{\sqrt{a}}{k-ia} + \frac{C}{\sqrt{b}}\frac{1}{k+ib}\right) e^{ikx} e^{ky}\, dk$$

$$+ \int_{0}^{\infty} \frac{1-i}{2\pi\sqrt{2}} \frac{1}{\sqrt{k}}\left(\frac{\sqrt{a}}{k-ia} + \frac{C}{\sqrt{b}}\frac{1}{k+ib}\right) e^{ikx} e^{-ky}\, dk$$

$$= 2\,\mathrm{Re} \int_{0}^{\infty} \frac{1-i}{2\pi\sqrt{2}} \frac{1}{\sqrt{k}}\left(\frac{\sqrt{a}}{k-ia} + \frac{C}{\sqrt{b}}\frac{1}{k+ib}\right) e^{ikx} e^{-ky}\, dk \qquad (8\text{-}100)$$

We must evaluate the integral

$$I(u, a) = \int_{0}^{\infty} \frac{dx}{\sqrt{x}}\frac{e^{-ux}}{x+a}$$

$$= e^{ua} \int_{a}^{\infty} \frac{dy}{\sqrt{y-a}}\frac{e^{-uy}}{y}$$

$$= e^{ua} J(u, a) \qquad (8\text{-}101)$$

where

$$J(u, a) = \int_{a}^{\infty} \frac{dy}{\sqrt{y-a}}\frac{e^{-uy}}{y}$$

$$\frac{\partial J}{\partial u} = -\int_{a}^{\infty} \frac{dy}{\sqrt{y-a}} e^{-uy}$$

$$= -2\int_{0}^{\infty} dz\, e^{-u(z^2+a)} \qquad (y=z^2+a)$$

$$= -\sqrt{\pi}\,\frac{e^{-ua}}{\sqrt{u}}$$

$$J(u, a) = -\sqrt{\pi} \int_{0}^{u} \frac{dt}{\sqrt{t}} e^{-ta} + C(a)$$

When $u = 0$,

$$J(u, a) = \int_a^\infty \frac{dy}{y\sqrt{y - a}}$$

$$= 2 \int_0^\infty \frac{dz}{z^2 + a} = \frac{\pi}{\sqrt{a}}$$

Thus $C(a) = \pi/\sqrt{a}$, and

$$J(u, a) = \frac{\pi}{\sqrt{a}} - \sqrt{\pi} \int_0^u \frac{dt}{\sqrt{t}} e^{-ta} \qquad \left(\text{let } t = \frac{z^2}{a} \right)$$

$$= \frac{\pi}{\sqrt{a}} - \frac{2\sqrt{\pi}}{\sqrt{a}} \int_0^{\sqrt{ua}} dz\, e^{-z^2}$$

$$= \frac{\pi}{\sqrt{a}} (1 - \text{erf } \sqrt{ua}) \tag{8-102}$$

Therefore,

$$I(u, a) = \frac{\pi}{\sqrt{a}} e^{au}(1 - \text{erf } \sqrt{ua}) \tag{8-103}$$

If we use this result to evaluate the integrals in our expression (8-100) for $\phi(x, y)$, we obtain

$$\phi(x, y) = \text{Re} \left[e^{-az}(1 - \text{erf } \sqrt{-az}) - i\frac{C}{b} e^{bz}(1 - \text{erf } \sqrt{bz}) \right] \tag{8-104}$$

where $z = x + iy$. The student should figure out which sign is to be taken for each square root, either from first principles or by checking against the boundary conditions. The error function for complex argument has been tabulated by Fried and Conte (F7) and by Faddeyeva and Terent'ev (F1).

REFERENCES

Morse and Feshbach (M9) give many examples of partial differential equations in physics as well as discussions of their classification, methods of solution, and a detailed account of the role of boundary conditions. Other useful general references are Courant and Hilbert (C10) Volume II; Sommerfeld (S9); and Goertzel and Tralli (G3) Part Two.

The separation of variables technique is discussed in Margenau and Murphy (M2) Chapter 7; and Sokolnikoff and Redheffer (S8) Chapter 6.

A collection of the results of separating the wave equation in various coordinate systems is in Magnus *et al.* (M1) Section 12.2.

Some examples of the use of integral transforms in solving partial differential equations may be found in the little book by Tranter (T8).

Special examples of Wiener–Hopf methods are in Morse and Fesbach (M9) Section 8.5; a more general and yet fairly readable account is in the monograph by Noble (N3).

PROBLEMS

8-1 Find the lowest frequency of oscillation of acoustic waves in a hollow sphere of radius R.   The boundary condition is $\partial\psi/\partial r = 0$ at $r = R$ and ψ obeys the differential equation

$$\nabla^2\psi = \frac{1}{c^2}\frac{\partial^2\psi}{\partial t^2}$$

8-2 Assume that the neutron density n inside U_{235} obeys the differential equation

$$\nabla^2 n + \lambda n = \frac{1}{\kappa}\frac{\partial n}{\partial t} \qquad (n = 0 \text{ on surface})$$

(*a*) Find the critical radius R_0 such that the neutron density inside a U_{235} sphere of radius R_0 or greater is unstable and increases exponentially with time.

(*b*) Suppose two hemispheres, each just barely stable, are brought together to form a sphere.   This sphere is unstable, with

$$n \sim e^{t/\tau}$$

Find the "time-constant" τ of the resulting explosion.

8-3 A sphere of radius R is at temperature $T = 0$ throughout.   At time $t = 0$, it is immersed in a liquid bath at temperature T_0.   Find the subsequent temperature distribution $T(r, t)$ inside the sphere.   [Let κ = thermal conductivity/(density $\times$ specific heat).]

8-4 Find the three lowest eigenvalues of the Schrödinger equation for a particle confined in a cylindrical box of radius a and height h.

$$-\frac{\hbar^2}{2m}\nabla^2\psi = E\psi \qquad (\psi = 0 \text{ on walls}) \qquad a \approx h$$

8-5 A long hollow conductor has a rectangular cross section, with sides a and $2a$.   One side of length $2a$ is charged to a potential V_0, and the other three sides are grounded ($V = 0$).

(a) By using a conformal transformation, find σ, the charge density at the point P, the midpoint of its side, and also find the total charge per unit length on that side.

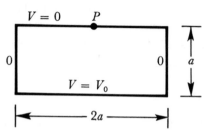

(b) Use separation of variable techniques on Laplace's equation to solve the same problem.

8-6 A quantity u satisfies the wave equation

$$\nabla^2 u - \frac{1}{c^2}\frac{\partial^2 u}{\partial t^2} = 0$$

inside a hollow cylindrical pipe of radius a, and $u = 0$ on the walls of the pipe.   If at the end $z = 0$, $u = u_0 e^{-i\omega_0 t}$, waves will be sent down the pipe with various spatial distributions (modes).   Find the phase velocity of the fundamental mode as a function of the frequency ω_0 and interpret the result for small ω_0.

8-7 Give a complete set of solutions of the equation

$$\nabla^4 \phi = \nabla^2(\nabla^2 \phi) = 0$$

in (a) spherical coordinates (θ, ϕ functions orthogonal)
   (b) rectangular coordinates (x, y functions orthogonal)

8-8 Find the lowest frequency of a drum head in the shape of an isosceles right triangle with sides a, a, $a\sqrt{2}$.
   Hint: Consider the modes of a square drum.

8-9 A half cube of material (cut on a diagonal as shown below) is heated to temperature T_0 and then plunged into an oil bath which keeps the surface at zero temperature. Find the approximate temperature $T(\mathbf{x}, t)$ at late times when a single term is adequate. What happens if the two sides adjacent to the right angle are not of equal length?

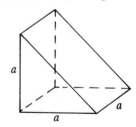

8-10 A half cylinder of metal of length a and radius a is initially at temperature T_0. At time $t = 0$ the metal is immersed in a bath which maintains the surface at temperature $T = 0$. Find an approximate expression for $T(\rho, \theta, z, t)$ for large t.

8-11 The equation describing elastic waves in an isotropic medium is

$$(\lambda + 2\mu)\nabla\nabla \cdot \mathbf{a} - \mu\nabla \times (\nabla \times \mathbf{a}) = \rho \, \frac{\partial^2 \mathbf{a}}{\partial t^2}$$

where $\mathbf{a}$ is the displacement from equilibrium, ρ is the density, and λ and μ the elastic constants of the medium. Find the lowest frequency of oscillation of an isotropic elastic sphere of radius R, given that
(a) $a(\mathbf{x})$ is of the form $f(r)\hat{\mathbf{e}}_r$ and
(b) the boundary condition at $r = R$ is

$$\lambda\nabla \cdot \mathbf{a} + 2\mu\frac{\partial a_r}{\partial r} = 0$$

In order to get a definite number at the end, you may set $\lambda = \mu$.

8-12 Consider a drum head in the shape of a sector of a circle of radius R and angle β.
(a) Which mode (that is, first, third, ninety-eighth, etc.) is the one illustrated? Give the answer for $\beta = \pi/2, \pi, 3\pi/2$.

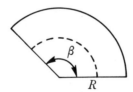

(b) Sketch on one graph the frequencies of the first half-dozen modes as functions of β.

8-13 The cube $|x| < L/2$, $|y| < L/2$, $|z| < L/2$ is immersed in a heat bath at temperature $T = 0$ and allowed to come to equilibrium. At time $t = 0$, a pulse of energy is released at the origin, so that $T = \delta(\mathbf{x})$. Find two expressions for the temperature as a function of time at the point $(L/4, 0, 0)$, one useful for short times and one useful for long times. In each case, you need keep only the two largest exponentials. At what time is the temperature at that point a maximum?

8-14 A cylinder of length l has a semicircular cross section of radius a (see figure below). The cylinder has density ρ, specific heat C, and thermal conductivity K. It is initially at temperature T_0, and its surface is maintained at that temperature by immersion in an oil bath. At time

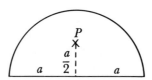

$t = 0$, a pulse of energy is released at a point halfway between the ends of the cylinder, marked P in the figure above. If the total energy released is E, find the temperature distribution in the cylinder for large t.

8-15 The temperature in a homogeneous sphere of radius a obeys the differential equation

$$\nabla^2 T = \frac{1}{\kappa} \frac{\partial T}{\partial t}$$

By external means, the surface temperature of the sphere is forced to behave as shown in the figure below.

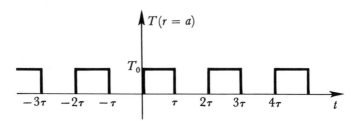

The alternations extend to $t = \pm\infty$. Find the temperature $T(t)$ at the center of the sphere.

8-16 Outside an infinitely long cylinder of radius a, a potential function $u(\mathbf{x}, t)$ satisfies the wave equation

$$\nabla^2 u = \frac{1}{c^2} \frac{\partial^2 u}{\partial t^2}$$

The cylinder is split along its length, and on its surface

$$u = \begin{cases} u_0\, e^{-i\omega_0 t} & (0 < \phi < \pi) \\ -u_0\, e^{-i\omega_0 t} & (\pi < \phi < 2\pi) \end{cases}$$

Find u outside the cylinder if only outgoing waves are present at large distances.

8-17 Use a cosine transform with respect to y to find the steady-state temperature distribution in a semi-infinite solid $x > 0$ when the temperature on the surface $x = 0$ is unity for $-a < y < a$, and zero outside this strip.

8-18   A semi-infinite body $x < 0$ has thermal conductivity K_1, density ρ_1, and specific heat C_1.   It is initially at temperature T_0.   At time $t = 0$, it is placed in thermal contact with the semi-infinite body $x > 0$, which has parameters K_2, ρ_2, C_2, and is initially at temperature $T = 0$.   Find $T_2(x, t)$, the temperature in the second body

$$\frac{\partial^2 T}{\partial x^2} = \frac{C\rho}{K} \frac{\partial T}{\partial t}$$

8-19   A straight wire of radius a is immersed in an infinite volume of liquid. Initially the wire and the liquid have temperature $T = 0$.   At time $t = 0$, the wire is suddenly raised to temperature T_0 and maintained at that temperature.   Find $F(r, s)$, the Laplace transform of the resulting temperature distribution $T(r, t)$ in the liquid.

8-20   Use Fourier transforms to find the motion of an infinitely long stretched string with initial displacement $\phi(x)$ and initial velocity zero.

$$\frac{\partial^2 y}{\partial x^2} = \frac{1}{c^2} \frac{\partial^2 y}{\partial t^2}$$

8-21   An infinite stretched membrane has areal density ρ and tension T. Initially it has a given displacement $f(\mathbf{r})$ and zero initial velocity everywhere.   Find the subsequent motion.

8-22   In a long rod of square cross section ($0 \leqslant x \leqslant s$; $0 \leqslant y \leqslant s$) the temperature obeys the differential equation

$$\nabla^2 T = \frac{1}{\kappa} \frac{\partial T}{\partial t}$$

At the end $z = 0$, the temperature is $T = T_0 \cos \omega t$ (independent of x and y).
(a) Find the temperature along the rod, if the sides are insulated.
(b) Find the temperature along the rod, if the sides are maintained at zero temperature.
In both cases, neglect transients.

8-23   In an absorbing medium, the neutron density obeys the differential equation

$$\kappa \nabla^2 n - \frac{n}{T} = \frac{\partial n}{\partial t}$$

where T is a constant.   At $t = 0$, a burst of neutrons is produced on the yz-plane of an infinite medium, so that

$$n(\mathbf{x}, 0) = \delta(x) \qquad [\text{not } \delta(\mathbf{x})]$$

Find the neutron density, and the total number of neutrons (per unit area), for later times.

8-24 Assume that in a block of uranium 235 the neutron density $n(\mathbf{x}, t)$ obeys the differential equation

$$\nabla^2 n + n = \frac{\partial n}{\partial t}$$

and that $n = 0$ at the surface. Find the minimum volume of material which can be made into a cylindrical piece which is supercritical.

8-25 Decompose the function $F(k) = (k^2 + 1)^{-1/2}$ into (the sum of) two functions $F_\pm(k)$, with $F_+(k)$ analytic for $\text{Im } k > -1$ and $F_-(k)$ analytic for $\text{Im } k < 1$.

8-26 Consider the semi-infinite region $y > 0$. For $x > 0$, the surface $y = 0$ is maintained at a temperature $T_0 e^{-x/l}$. For $x < 0$, the surface $y = 0$ is insulated, so that no heat flows in or out. Find the equilibrium temperature at the point $(-l, 0)$.

8-27 Consider the motion of a stretched string, whose displacement $y(x, t)$ obeys the differential equation

$$\frac{1}{c^2} \frac{\partial^2 y}{\partial t^2} = \frac{\partial^2 y}{\partial x^2}$$

The boundary conditions are
(a) the ends of the string are fixed; $y(0, t) = y(L, t) = 0$
(b) the shape of the string is given at $t = 0$; $y(x, 0) = f(x)$
(c) the shape of the string is given at another time T; $y(x, T) = g(x)$.
This is an example of the application of a Dirichlet boundary condition on a closed boundary (in the $x - t$ plane) to a hyperbolic differential equation.

Does the problem have a solution? Discuss separately the cases:

(a) $\dfrac{cT}{L} = $ integer

(b) $\dfrac{cT}{L} = $ rational number

(c) $\dfrac{cT}{L} = $ irrational number

8-28 The temperature in an infinite cylindrical rod of radius a satisfies the conditions

(a) $\nabla^2 T = \alpha^{-1} \dfrac{\partial T}{\partial t}$ $\left(\alpha = \dfrac{K}{C\rho} \right)$

(b) $T = 0$ at $t = 0$

(c) $T = T_0 \cos \phi$ at $\rho = a$

Find $T(\rho, \phi, t)$ for $t > 0$.

8-29 Find the lowest three values of k^2 for which the two-dimensional Helmholtz equation

$$\nabla^2 \phi + k^2 \phi = 0$$

has a nontrivial solution inside a $30° - 60° - 90°$ right triangle whose sides are a, $a\sqrt{3}$, and $2a$. The boundary condition is that $\phi = 0$ on the perimeter of the triangle.

8-30 Solve the problem which led to Table 8–2, but with the boundary condition $(\text{grad } u)_n = 0$ on the perimeter of the triangle.

NINE

EIGENFUNCTIONS, EIGENVALUES, AND GREEN'S FUNCTIONS

Eigenvalue problems and some general properties of the eigenfunctions of Hermitian differential operators are discussed in the first two sections of this chapter. The solution of inhomogeneous problems is treated in Section 9–4, with emphasis on the use of Green's functions for describing particular solutions. Various methods for finding Green's functions are presented, and in Section 9–5 some examples from electrodynamics are given.

9–1 SIMPLE EXAMPLES OF EIGENVALUE PROBLEMS

We have already discussed (in Chapter 6) eigenvalues of matrices. These were values of the parameter λ for which nontrivial solutions v of the equation

$$Mv = \lambda v \qquad (9\text{-}1)$$

exist, where M is the matrix under consideration. We shall now generalize this idea and consider the eigenvalue problem for *any* linear operator M, whether it be a matrix, differential operator, or integral operator. The basic equation is just the one written above, with v being whatever kind of object M can operate on.   If M is a matrix, v is a column vector; if M is a differential operator, v is a function, and so forth.   Solutions v of the above equation are called *eigenfunctions, eigenvectors*, and so on.

Eigenvalue problems have a great deal in common whether we are dealing with matrices or more general linear operators.   The most important such problems in physics involve three-dimensional second-order partial differential operators.   The examples and notation of this chapter will reflect the importance of such problems, but the student should not forget that most of our techniques may be generalized to any other type of eigenvalue problem.

As a very simple example, consider the problem of finding the eigenvalues and eigenvectors of the matrix

$$M = \begin{pmatrix} 1 & 1 \\ 1 & 1 \end{pmatrix} \tag{9-2}$$

The equation

$$Mv = \lambda v$$

is equivalent to the two equations

$$(1 - \lambda)v_1 + v_2 = 0$$
$$v_1 + (1 - \lambda)v_2 = 0$$

if we set

$$v = \begin{pmatrix} v_1 \\ v_2 \end{pmatrix}$$

The eigenvalues and associated eigenvectors are

$$\lambda_1 = 0 \quad v^{(1)} = \begin{pmatrix} 1 \\ -1 \end{pmatrix} \quad \lambda_2 = 2 \quad v^{(2)} = \begin{pmatrix} 1 \\ 1 \end{pmatrix} \tag{9-3}$$

We may note several facts.

1. Because the original problem is homogeneous, eigensolutions are only determined up to an arbitrary constant multiplicative factor.

2. If λ equals an eigenvalue, the corresponding inhomogeneous problem

$$(M - \lambda)v = u \quad (u \neq 0) \tag{9-4}$$

generally has no solution.   In fact, a solution will exist only if u satisfies certain conditions; with the matrix M of (9-2), for example, and $\lambda = 0$, u_1 must equal u_2 in order that (9-4) have a solution.   In other words, u must be "orthogonal" to the eigenvector $v^{(1)}$ belonging to the eigenvalue $\lambda = 0$; this trivial observation may be generalized [compare Eq. (9-29)].

3. The "dot product" of our two eigenvectors is zero.

4. An arbitrary 2-component vector can be written as a linear combination of the eigenvectors.

These four properties are quite general and will occur over and over as we encounter other eigenvalue problems.

EXAMPLE

Consider the equation for a string, fastened at $x = 0$ and $x = L$, vibrating with angular frequency ω (one-dimensional Helmholtz equation).

$$\frac{d^2u}{dx^2} + k^2u = 0 \qquad \left(k = \frac{\omega}{c}\right)$$

Since $u = 0$ when $x = 0$, $u = a \sin kx$.   Since $u = 0$ when $x = L$, $k = n\pi/L$, where $n = 1, 2, 3, \ldots$.   These are the eigenvalues k.   Technically if we, consider d^2/dx^2 to be the linear operator, then we should call $-k^2$ the eigenvalue in this problem.   This distinction is not very important, however, and we shall consider $-k^2$, or k^2, or k to be entitled to the name eigenvalue.

Note that a differential equation needs both a region of interest and boundary conditions in order to define an eigenvalue problem, and that the problem must be homogeneous.   The student is invited to discover the analogies, in this second example, of the four properties mentioned after the first example.

Finally, consider a three-dimensional example.

EXAMPLE

$$\nabla^2u + k^2u = 0 \text{ inside the sphere } r = R \qquad (9\text{-}5)$$

with the boundary condition: $u = 0$ on $r = R$.

Solutions are

$$u = j_l(kr)P_l^m(\cos \theta)e^{\pm im\phi} \qquad m, l \text{ integers}, l \geqslant m$$

Eigenvalues are fixed by the condition

$$j_l(kR) = 0$$

Thus the eigenvalues are

$$k \approx \frac{3.14}{R}, \frac{4.49}{R}, \ldots$$

9–2   GENERAL DISCUSSION

Consider a linear differential operator L and the eigenvalue problem

$$Lu(\mathbf{x}) = \lambda u(\mathbf{x})$$

We suppose that a region Ω has been specified, and suitable boundary conditions imposed.

L is said to be *Hermitian* if

$$\int_\Omega u^*(\mathbf{x})Lv(\mathbf{x})\,d^3x = \left[\int_\Omega v^*(\mathbf{x})Lu(\mathbf{x})\,d^3x\right]^*$$

where the asterisk denotes complex conjugation and u and v are arbitrary functions *obeying the boundary conditions*.   The quantity on the left side of the above equation is often called the "u, v matrix element of L," or the "matrix element of L between u and v," or just L_{uv}.   The connection with Hermitian matrices, as defined in Chapter 6, should be obvious.

Suppose L is Hermitian.   Consider a particular eigenvalue λ_i and an eigenfunction u_i belonging to λ_i.   Then

$$Lu_i(\mathbf{x}) = \lambda_i u_i(\mathbf{x})$$

The corresponding equation for the pair λ_j, $u_j(\mathbf{x})$ is

$$Lu_j(\mathbf{x}) = \lambda_j u_j(\mathbf{x})$$

Then

$$\int_\Omega u_j^*(\mathbf{x})Lu_i(\mathbf{x})\,d^3x = \lambda_i \int_\Omega u_j^*(\mathbf{x})u_i(\mathbf{x})\,d^3x$$

$$\int_\Omega u_i^*(\mathbf{x})Lu_j(\mathbf{x})\,d^3x = \lambda_j \int_\Omega u_i^*(\mathbf{x})u_j(\mathbf{x})\,d^3x$$

Since L is Hermitian, the left sides are complex conjugates of each other. Therefore,

$$(\lambda_i - \lambda_j^*) \int_\Omega u_j^*(\mathbf{x})u_i(\mathbf{x})\,d^3x = 0 \tag{9-6}$$

At this point it should have become clear that we are just repeating the steps of the proof in Section 6–5 of certain properties of the eigenvalues and eigenvectors of a Hermitian matrix.   We draw the same conclusions from the relation (9-6) by considering the two cases $i = j$ and $\lambda_i \neq \lambda_j$. Namely:

1. The eigenvalues of a Hermitian differential operator are real.

2. Eigenfunctions of a Hermitian differential operator, belonging to different eigenvalues, are orthogonal.

By $u(\mathbf{x})$ and $v(\mathbf{x})$ being orthogonal, we mean [compare Eq. (6-94)]

$$u \cdot v \equiv \int_{\Omega} u^*(\mathbf{x})v(\mathbf{x}) \, d^3x = 0 \qquad (9\text{-}7)$$

The most familiar set of orthogonal functions is the set of trigonometric functions. They are the eigenfunctions associated with the eigenvalue problem

$$\frac{d^2u}{dx^2} + \lambda u = 0 \qquad 0 \leqslant x \leqslant 2\pi$$

$$u(0) = u(2\pi) \qquad u'(0) = u'(2\pi) \qquad (9\text{-}8)$$

Is the operator d^2/dx^2 on the interval $0 \leqslant x \leqslant 2\pi$ with the periodic boundary conditions of (9-8) Hermitian? Let us see.

$$\int_0^{2\pi} u^* \frac{d^2}{dx^2} v \, dx = u^* \frac{dv}{dx} \Big|_0^{2\pi} - \int_0^{2\pi} \frac{du^*}{dx} \frac{dv}{dx} \, dx$$

The integrated part vanishes because of the periodicity of u and v. Integrating once more by parts,

$$\int_0^{2\pi} u^* \frac{d^2}{dx^2} v \, dx = -\frac{du^*}{dx} v \Big|_0^{2\pi} + \int_0^{2\pi} \frac{d^2u^*}{dx^2} v \, dx$$

Again the integrated part vanishes, so that

$$\int_0^{2\pi} u^* \frac{d^2}{dx^2} v \, dx = \left(\int_0^{2\pi} v^* \frac{d^2}{dx^2} u \, dx \right)^*$$

and d^2/dx^2 is indeed Hermitian, when the periodic boundary conditions of (9-8) are specified.

As another illustration of an eigenvalue problem involving a Hermitian operator, consider the so-called *Sturm–Liouville differential equation*

$$\frac{d}{dx} \left[p(x) \frac{du(x)}{dx} \right] - q(x)u(x) + \lambda \rho(x)u(x) = 0 \qquad (9\text{-}9)$$

$p(x)$, $q(x)$, and $\rho(x)$ are real functions, and in addition $\rho(x)$ is assumed to be nonnegative on the interval in question. The function $u(x)$ will be required to vanish at both ends of the interval.[1]   We must consider two points.

1. Is the operator $L = p(d^2/dx^2) + p'(d/dx) - q$ Hermitian? The verification proceeds just as for the simpler case of d^2/dx^2; we shall leave it for the student.

---

[1] Other boundary conditions could be specified; for example, $u(x)$ could vanish at one end and its derivative at the other. The student is invited to construct the most general boundary condition for which the succeeding arguments remain valid.

2. What is the effect of $\rho(x)$ not being 1 ?   It is straightforward to repeat all the preceding arguments with a so-called *density function* $\rho(x)$ present in the term containing the eigenvalue.   Orthogonality now means

$$u_i \cdot u_j \equiv \int_a^b u_i^*(x)u_j(x)\rho(x)\, dx = 0 \qquad \text{if} \qquad \lambda_i \neq \lambda_j \qquad (9\text{-}10)$$

Often two or more eigenfunctions belong to the same eigenvalue.   This situation is referred to as *degeneracy*, and the eigenvalue in question is said to be degenerate.   (Compare the corresponding discussion for matrices on p. 152.)   Note that an arbitrary linear combination of eigenfunctions belonging to a degenerate set is again an eigenfunction with the same eigenvalue.   By using the Gram–Schmidt procedure (see p. 152) we can always construct an orthogonal set of eigenfunctions.

Thus it is possible to arrange things so that the eigenfunctions of a Hermitian operator form an *orthonormal* set, being *orthogonal*

$$u_i \cdot u_j = 0 \qquad (i \neq j)$$

and *normalized*

$$u_i \cdot u_i = 1$$

These two conditions may be written

$$u_i \cdot u_j = \left[\int d^3x\, u_i^*(\mathbf{x})u_j(\mathbf{x})\rho(\mathbf{x})\right] = \delta_{ij} \qquad (9\text{-}11)$$

Property 4 noted at the beginning of this chapter, together with the example of Fourier series, suggest that it is possible to expand any function, obeying the appropriate conditions, in a series of eigenfunctions.   That is, the eigenfunctions of a Hermitian operator form a *complete set* under very general conditions.   This is just the infinite dimensional generalization of the theorem we have already proved in Chapter 6, p. 158.   We shall not prove this property here but it is in fact true for all the commonly encountered differential equations in physics.   The student interested in a mathematical treatment of the subject should consult a reference such as Courant and Hilbert (C10) Volume I, Chapters II and IV, or Titchmarsh (T2).

Let us therefore consider an expansion of the form

$$f(\mathbf{x}) = \sum_n c_n u_n(\mathbf{x}) \qquad (9\text{-}12)$$

If we have chosen the u_n to be orthonormal, we can determine the c_n very easily:

$$u_m \cdot f = \sum_n c_n u_m \cdot u_n = \sum_n c_n \delta_{mn} = c_m \qquad (9\text{-}13)$$

A useful formal relation follows by substituting this result back into the original series:

$$f(\mathbf{x}) = \sum_n u_n(\mathbf{x}) u_n \cdot f$$

$$= \sum_n u_n(\mathbf{x}) \int_\Omega d^3x' u_n^*(\mathbf{x}') f(\mathbf{x}') \rho(\mathbf{x}')$$

$$= \int_\Omega d^3x' f(\mathbf{x}') \rho(\mathbf{x}') \sum_n u_n(\mathbf{x}) u_n^*(\mathbf{x}')$$

Therefore,

$$\rho(\mathbf{x}') \sum_n u_n(\mathbf{x}) u_n^*(\mathbf{x}') = \delta(\mathbf{x} - \mathbf{x}') \qquad (9\text{-}14)$$

This is sometimes called the *completeness relation* and sometimes the *closure* property of the eigenfunctions. Note the interesting comparison with the orthogonality relation (9-11).

9–3 SOLUTIONS OF BOUNDARY-VALUE PROBLEMS AS EIGENFUNCTION EXPANSIONS

There is a close relationship between the expansion of a function in terms of eigenfunctions of a differential operator and the solution of a partial differential equation obtained by either the method of separation of variables or the method of integral transforms.

Suppose a partial differential equation contains a differential operator L with respect to one of the variables. If the solution is expressed as a sum over the eigenfunctions of L, then the combination of derivatives L will be eliminated from the equation and replaced by the eigenvalues, thus reducing the equation to one with fewer variables.

EXAMPLE

Consider the drum-head problem treated in Chapter 8, beginning with Eq. (8-46).

$$\frac{1}{r} \frac{\partial}{\partial r} \left(r \frac{\partial u}{\partial r} \right) + \frac{1}{r^2} \frac{\partial^2 u}{\partial \theta^2} - \frac{1}{c^2} \frac{\partial^2 u}{\partial t^2} = 0 \qquad (9\text{-}15)$$

We could eliminate the derivatives with respect to r, θ, or t first; choose θ. The operator $L_\theta = d^2/d\theta^2$ has eigenfunctions $e^{\pm in\theta}$ belonging to eigenvalues $-n^2$, where n is an integer for the boundary conditions $u(2\pi) = u(0)$,

$u'(2\pi) = u'(0)$.   Thus we express u as the sum

$$u = \sum_{n=0}^{\infty} f_n(r, t)(e^{in\theta} + a_n e^{-in\theta}) \tag{9-16}$$

where the coefficients f_n are functions of the remaining variables r and t.   The differential equation is reduced to

$$\frac{1}{r}\frac{\partial}{\partial r}\left(r\frac{\partial f_n}{\partial r}\right) - \frac{n^2}{r^2}f_n - \frac{1}{c^2}\frac{\partial^2 f_n}{\partial t^2} = 0 \tag{9-17}$$

Next, eliminate the r derivatives by expanding f_n in eigenfunctions of the operator

$$L_r = \frac{\partial^2}{\partial r^2} + \frac{1}{r}\frac{\partial}{\partial r} - \frac{n^2}{r^2} \tag{9-18}$$

with the boundary conditions: $u(r = 0)$ is finite and $u(r = R) = 0$.   These are the Bessel functions $J_n(k_{jn} r)$ with eigenvalues $-k_{jn}^2$ such that $J_n(k_{jn} R) = 0$. Thus

$$f_n(r, t) = \sum_{j=1}^{\infty} g_{jn}(t)J_n(k_{jn} r) \tag{9-19}$$

where the coefficients g_{jn} depend on the remaining variable t.   The differential equation is reduced to

$$k_{jn}^2 g_{jn}(t) + \frac{1}{c^2}\frac{d^2 g_{jn}}{dt^2} = 0 \tag{9-20}$$

with solutions

$$g_{jn}(t) = e^{\pm i\omega_{jn}t} \qquad \omega_{jn} = ck_{jn} \tag{9-21}$$

Collecting results, we have the solution (8-54) obtained by separation of variables

$$u(r, \theta, t) = \sum_{j,n=1}^{\infty} A_{jn} J_n(k_{jn} r)(\sin n\theta + B_n \cos n\theta)(\sin \omega_{jn} t + C_{jn} \cos \omega_{jn} t) \tag{9-22}$$

9-4   INHOMOGENEOUS PROBLEMS. GREEN'S FUNCTIONS

As pointed out in Chapter 8, a problem may be inhomogeneous because of the differential equation or because of the boundary conditions.   Usually a transformation from one form of inhomogeneity to the other is possible, as we shall see near the end of this section.

First, let us consider the inhomogeneous equation

$$Lu(\mathbf{x}) - \lambda u(\mathbf{x}) = f(\mathbf{x}) \tag{9-23}$$

over a domain Ω, with L a Hermitian differential operator, with $u(\mathbf{x})$ subject to the usual type of (homogeneous) boundary conditions, and λ a given constant. To solve the equation, expand $u(\mathbf{x})$ and $f(\mathbf{x})$ in eigenfunctions of the operator L.

$$u(\mathbf{x}) = \sum_n c_n u_n(\mathbf{x}) \qquad f(\mathbf{x}) = \sum_n d_n u_n(\mathbf{x}) \tag{9-24}$$

Equation (9-23) becomes

$$\sum_n c_n(\lambda_n - \lambda)u_n(\mathbf{x}) = \sum_n d_n u_n(\mathbf{x})$$

Therefore, since the eigenfunctions $u_n(\mathbf{x})$ are linearly independent,

$$c_n = \frac{d_n}{\lambda_n - \lambda}$$

or, since

$$d_n = u_n \cdot f$$

we may write this as

$$c_n = \frac{u_n \cdot f}{\lambda_n - \lambda}$$

Therefore,

$$u(\mathbf{x}) = \sum_n \frac{u_n u_n \cdot f}{\lambda_n - \lambda}$$

$$= \sum_n \frac{u_n(\mathbf{x})}{\lambda_n - \lambda} \int_\Omega u_n^*(\mathbf{x}')f(\mathbf{x}')\, d^3x' \tag{9-25}$$

This expression may be written in the form

$$u(\mathbf{x}) = \int_\Omega G(\mathbf{x}, \mathbf{x}')f(\mathbf{x}')\, d^3x' \tag{9-26}$$

where the so-called *Green's function* is given by

$$G(\mathbf{x}, \mathbf{x}') = \sum_n \frac{u_n(\mathbf{x})u_n^*(\mathbf{x}')}{\lambda_n - \lambda} \tag{9-27}$$

Note that a Green's function is determined by a differential operator (nearly always Hermitian), a particular region, and suitable boundary conditions. Sometimes one writes $G(\mathbf{x}, \mathbf{x}'; \lambda)$ to emphasize the dependence of G on λ as well as on $\mathbf{x}$ and $\mathbf{x}'$.

Let us find a differential equation obeyed by $G(\mathbf{x}, \mathbf{x}')$. Suppose $f(\mathbf{x})$ in the above derivation is taken to be $\delta(\mathbf{x} - \mathbf{x}_0)$. Then we obtain for the solution

$$u(\mathbf{x}) = \int_\Omega G(\mathbf{x}, \mathbf{x}') \, \delta(\mathbf{x}' - \mathbf{x}_0) \, d^3x = G(\mathbf{x}, \mathbf{x}_0)$$

Therefore, $G(\mathbf{x}, \mathbf{x}')$ is the solution of

$$LG(\mathbf{x}, \mathbf{x}') - \lambda G(\mathbf{x}, \mathbf{x}') = \delta(\mathbf{x} - \mathbf{x}') \tag{9-28}$$

subject to the appropriate boundary conditions. This fact could also be recognized by applying the operator $L - \lambda$ to the infinite series representation (9-27) for $G(\mathbf{x}, \mathbf{x}')$ and using the completeness relation (9-14) (with $\rho = 1$).

The Green's function thus has a simple physical significance. It is the solution of the problem for a unit point "source" $f(\mathbf{x}) = \delta(\mathbf{x} - \mathbf{x}')$.

We notice in (9-27) a feature which first appeared at the very beginning of this chapter. If λ equals an eigenvalue (so that a nontrivial solution of the *homogeneous* equation exists) there are serious difficulties involved in solving the inhomogeneous equation.   When λ equals an eigenvalue λ_n, the Green's function (9-27) is infinite, and there is no solution $u(\mathbf{x})$ unless the right-hand side of the inhomogeneous equation obeys the condition

$$\int u_n^*(\mathbf{x}) f(\mathbf{x}) \, d^3x = 0 \tag{9-29}$$

Let us consider a specific example, a string of length l vibrating with (angular) frequency ω.   The equation and boundary conditions are

$$\frac{d^2u}{dx^2} + k^2u = 0 \qquad u(0) = u(l) = 0 \tag{9-30}$$

$$k = \frac{\omega}{c}$$

What is the Green's function?
First method: Let $k^2 = -\lambda$

$$\frac{d^2u}{dx^2} = \lambda u$$

Eigenvalues are

$$\lambda_n = -\left(\frac{n\pi}{l}\right)^2 \qquad \text{with } n = 1, 2, 3, \dots$$

Normalized eigenfunctions are

$$u_n = \sqrt{\frac{2}{l}} \sin \frac{n\pi x}{l}$$

Therefore, the general formula

$$G(x, x') = \sum_n \frac{u_n(x)u_n^*(x')}{\lambda_n - \lambda}$$

becomes

$$G(x, x') = \frac{2}{l} \sum_n \frac{\sin(n\pi x/l)\sin(n\pi x'/l)}{k^2 - (n\pi/l)^2} \tag{9-31}$$

The above result illustrates an important general symmetry relation for Green's functions, which follows from Eq. (9-27):

$$G(x', x) = [G(x, x')]^* \tag{9-32}$$

The physical significance of this result, for real Green's functions, is the *reciprocity* relation, that the response at x to a unit point disturbance at x' is the same as the response at x' to a unit point disturbance at x.

Second method: Try solving the differential equation

$$\frac{d^2u}{dx^2} + k^2u = \delta(x - x') \tag{9-33}$$

which $G(x, x')$ obeys. In interpreting the physical significance of the result, it is useful to keep in mind that (9-33) is the time-independent equation for a vibrating string subject to a force $\propto -\delta(x - x')e^{-i\omega t}$. For x equal to anything but x',

$$\frac{d^2u}{dx^2} + k^2u = 0$$

Therefore,

$$G(x, x') = \begin{cases} a \sin kx & (x < x') \\ b \sin k(x - l) & (x > x') \end{cases} \tag{9-34}$$

How do we determine the constants a and b?

Integrate the differential equation (9-33), which $G(x, x')$ obeys, from $x' - \varepsilon$ to $x' + \varepsilon$, where ε is infinitesimal. The result is

$$\frac{dG}{dx}\Big|_{x'-\varepsilon}^{x'+\varepsilon} = 1 \tag{9-35}$$

Integrating again gives

$$G\Big|_{x'-\varepsilon}^{x'+\varepsilon} = 0 \tag{9-36}$$

That is, $G(x, x')$, as a function of x, is continuous at $x = x'$ but its first derivative jumps by $+1$ at that point. Using the expressions (9-34) for G, these conditions give

$$a \sin kx' = b \sin k(x' - l)$$

$$ka \cos kx' + 1 = kb \cos k(x' - l)$$

The solution of these simultaneous equations is

$$a = \frac{\sin k(x' - l)}{k \sin kl} \qquad b = \frac{\sin kx'}{k \sin kl}$$

and the Green's function is

$$G(x, x') = \frac{1}{k \sin kl} \begin{cases} \sin kx \sin k(x' - l) & 0 < x < x' \\ \sin kx' \sin k(x - l) & x' < x < l \end{cases}$$

or, in a more concise notation,

$$G(x, x') = \frac{-1}{k \sin kl} \sin kx_< \sin k(l - x)_< \qquad (9\text{-}37)$$

where $\qquad x_< = \min(x, x') \qquad (l - x)_< = \min[(l - x), (l - x')]$

The same result is obtained if we solve

$$\frac{d^2u}{dx^2} + k^2u = f(x)$$

by variation of parameters (see Problem 1-26). Note the same symmetry between x and x' that we obtained in (9-31).

Now let's work out a two-dimensional Green's function. We could just write down the formal sum involving normalized eigenfunctions, analogous to (9-31), but we shall use the second approach and solve the differential equation.

The problem is that of a circular drum, for which the equation and boundary conditions are

$$\nabla^2u + k^2u = 0, \qquad u = 0 \quad \text{when} \quad r = R \qquad (9\text{-}38)$$

It is clear from physical considerations that $G(\mathbf{x}, \mathbf{x}')$ can depend only on r, r', and θ, the angle between $\mathbf{x}$ and $\mathbf{x}'$. Choose $\mathbf{x}'$ on the axis of polar coordinates, as shown in Figure 9–1.

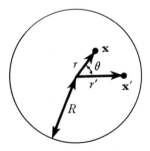

Figure 9–1   Coordinates for the circular-drum Green's function

Now G is the solution of

$$\nabla^2 G + k^2 G = \delta(\mathbf{x} - \mathbf{x}') \qquad (9\text{-}39)$$

where $\delta(\mathbf{x} - \mathbf{x}')$ is the two-dimensional delta function; $\int \delta(\mathbf{x} - \mathbf{x}')\, d^2x = 1$ for any area of integration which includes $\mathbf{x}'$.

For $\mathbf{x} \neq \mathbf{x}'$,

$$\nabla^2 G + k^2 G = 0 \qquad (9\text{-}40)$$

The solution of this equation satisfying the boundary conditions may be written in the form

$$G = \begin{cases} \sum_m A_m J_m(kr) \cos m\theta & (r < r') \\[2mm] \sum_m B_m [J_m(kr) Y_m(kR) - Y_m(kr) J_m(kR)] \cos m\theta & (r > r') \end{cases} \qquad (9\text{-}41)$$

Note that we choose the solution for $r > r'$ so that it vanishes automatically at $r = R$; we have also used the physically obvious fact that G is an *even* function of θ.

We determine the constants A_m and B_m by fitting our solutions (9-41) together along the circle $r = r'$. As in our one-dimensional example, G is continuous but its derivative (that is, its *gradient*) has a discontinuity at the point $\mathbf{x} = \mathbf{x}'$. To find the singularity, integrate the differential equation (9-39) for G over an infinitesimal area which includes the point $\mathbf{x} = \mathbf{x}'$. We obtain

$$\int \nabla^2 G \, d^2x = \int (\nabla G)_n \, dl = 1 \qquad (9\text{-}42)$$

where we have used the two-dimensional analog of Gauss' theorem:

$$\int_V \nabla \cdot \mathbf{u} \, d^3x = \int_\Sigma \mathbf{u} \cdot d\mathbf{S}$$

Σ being the surface which encloses the volume V.

For an area enclosing the point $\mathbf{x}'$, we shall choose one shaped as shown in Figure 9–2. Neglecting the ends, (9-42) gives us

$$\int_{r'+\varepsilon} \frac{\partial G}{\partial r} \, dl - \int_{r'-\varepsilon} \frac{\partial G}{\partial r} \, dl = 1$$

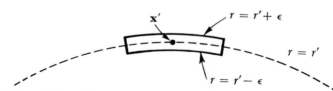

Figure 9–2   The area for application of Gauss' theorem

where l measures arc length. Since $dl = r'\, d\theta$, this gives

$$\int d\theta \left(\frac{\partial G}{\partial r} \bigg|_{r'+\varepsilon} - \frac{\partial G}{\partial r} \bigg|_{r'-\varepsilon} \right) = \frac{1}{r'}$$

(provided the range of integration includes the point $\mathbf{x}'$; otherwise we get zero). Therefore,

$$\left(\frac{\partial G}{\partial r} \bigg|_{r'+\varepsilon} - \frac{\partial G}{\partial r} \bigg|_{r'-\varepsilon} \right) = \frac{1}{r'} \delta(\theta)$$

Let

$$\frac{\partial G}{\partial r} \bigg|_{r'+\varepsilon} - \frac{\partial G}{\partial r} \bigg|_{r'-\varepsilon} = \sum_m C_m \cos m\theta$$

We multiply both sides by $\cos m'\theta$, in the usual way, and integrate over θ from $-\pi$ to π. The result is

$$\frac{1}{r'} = \pi C_{m'}\, \varepsilon_{m'} \qquad \text{where} \qquad \varepsilon_{m'} = \begin{cases} 2 & \text{if} & m' = 0 \\ 1 & \text{if} & m' > 0 \end{cases} \tag{9-43}$$

Thus

$$\frac{\partial G}{\partial r} \bigg|_{r'+\varepsilon} - \frac{\partial G}{\partial r} \bigg|_{r'-\varepsilon} = \frac{1}{\pi r'} \sum_m \frac{1}{\varepsilon_m} \cos m\theta \tag{9-44}$$

This is the condition on the discontinuity in the gradient of G that we were after.

Now we can write down two simultaneous equations for the A_m and B_m of (9-41). From the continuity of G at $r = r'$, we obtain

$$A_m J_m(kr') = B_m[J_m(kr')Y_m(kR) - Y_m(kr')J_m(kR)]$$

From the condition (9-44) on the discontinuity in $\partial G/\partial r$, we obtain

$$B_m[J'_m(kr')Y_m(kR) - Y'_m(kr')J_m(kR)] - A_m J'_m(kr') = \frac{1}{\pi \varepsilon_m\, kr'}$$

The solution is

$$A_m = \frac{J_m(kR)Y_m(kr') - J_m(kr')Y_m(kR)}{2\varepsilon_m J_m(kR)} \tag{9-45}$$

$$B_m = \frac{-J_m(kr')}{2\varepsilon_m J_m(kR)} \tag{9-46}$$

where we have used the relation (found in Problem 7-12)

$$J_m(x)Y'_m(x) - J'_m(x)Y_m(x) = \frac{2}{\pi x} \tag{9-47}$$

Note that $G(\mathbf{x}, \mathbf{x}') = G(\mathbf{x}', \mathbf{x})$, as before.

Let us now return to the original Green's function equation (9-39) and solve it by another method.

$$\nabla^2 G + k^2 G = \delta(\mathbf{x} - \mathbf{x}') \qquad (9\text{-}48)$$

where $\delta(\mathbf{x} - \mathbf{x}')$ is the two-dimensional delta-function located at the "source" point $\mathbf{x}'$.

There are two difficulties in finding the solution: one is to insure the proper singularity at the source point $\mathbf{x}'$ and the other is to satisfy the boundary conditions. It is sometimes convenient to separate the two difficulties by finding a solution of the form

$$G = u(\mathbf{x}, \mathbf{x}') + v(\mathbf{x}, \mathbf{x}') \qquad (9\text{-}49)$$

where u has the singularity at $\mathbf{x}'$ [that is, obeys Eq. (9-48)] without necessarily satisfying the boundary conditions, while v is "smooth" at $\mathbf{x}'$ [that is, obeys the homogeneous equation (9-40)] but is adjusted to make $G(\mathbf{x}, \mathbf{x}')$ satisfy the boundary conditions. The function $u(\mathbf{x}, \mathbf{x}')$ is sometimes called a *fundamental solution* of (9-48).

The singularity required at $\mathbf{x}'$ may be found by integrating Eq. (9-48) over a small area element centered at $\mathbf{x}'$ and finding u in the form $u(\rho)$ where $\rho = |\mathbf{x} - \mathbf{x}'|$. From Gauss' theorem [compare (9-42)]

$$\int_0^\rho \nabla^2 G \cdot 2\pi\rho \, d\rho = 2\pi\rho \frac{\partial G}{\partial \rho}$$

so that

$$2\pi\rho \frac{\partial G}{\partial \rho} + k^2 \int_0^\rho G \cdot 2\pi\rho \, d\rho = 1$$

and

$$G(\rho) \to \frac{1}{2\pi} \ln \rho + \text{constant} \qquad \text{as } \rho \to 0 \qquad (9\text{-}50)$$

$u(\mathbf{x}, \mathbf{x}')$ should then have this behavior near $\rho = 0$. The singular solution of Eq. (9-40), $Y_0(k\rho)$, has the behavior for small ρ:

$$Y_0(k\rho) \to \frac{2}{\pi} \ln \rho + \text{constant}$$

Thus

$$G = \tfrac{1}{4} Y_0(k\rho) + v(\mathbf{x}, \mathbf{x}') \qquad (9\text{-}51)$$

We now write

$$v = \sum_n A_n J_n(kr) \cos n\theta$$

with A_n chosen so that G satisfies the boundary conditions. That is,

$$G(r = R) = 0 = \tfrac{1}{4}Y_0(k\rho_R) + \sum_n A_n J_n(kR) \cos n\theta$$

$$A_n = -\frac{1}{4\pi J_n(kR)\varepsilon_n} \int_0^{2\pi} Y_0(k\rho_R) \cos n\theta \, d\theta$$

where

$$\rho_R^2 = R^2 + r'^2 - 2Rr' \cos \theta$$

(see Figure 9–3) and ε_n is defined in (9-43). The result is

$$G(\mathbf{x}, \mathbf{x}') = \tfrac{1}{4}Y_0(k\rho) - \sum_{n=0}^{\infty} \frac{J_n(kr) \cos n\theta}{2\pi J_n(kR)\varepsilon_n} \int_0^{\pi} Y_0(k\rho_R) \cos n\theta \, d\theta \qquad (9\text{-}52)$$

This form of the Green's function is convenient for some purposes; it is easy to visualize for low frequencies ω and source positions $\mathbf{x}'$ not too near the edge.

In summary, we recall the three forms found above for a Green's function:

1. The formal sum over eigenfunctions (9-27)

2. A solution of the homogeneous equation and boundary conditions on either side of a "surface" containing the source point; the two solutions being matched on this surface in such a way as to produce the source-point singularity

3. The form $G(\mathbf{x}, \mathbf{x}') = u(\mathbf{x}, \mathbf{x}') + v(\mathbf{x}, \mathbf{x}')$, where u is the fundamental solution, and v takes care of the boundary conditions.

If a problem is inhomogeneous because of the boundary conditions rather than the differential equation, the solution may still be written in terms of a Green's function. In fact, the examples in Chapter 8 of heat problems with initial temperature distributions were of this type, since initial conditions are boundary conditions in time (see also Problem 9-5).

Alternatively, a homogeneous equation with inhomogeneous boundary

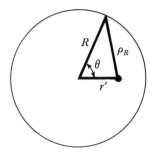

Figure 9–3   The coordinate ρ_R for the drum head

conditions may be transformed into an inhomogeneous equation with homogeneous boundary conditions. (The reverse is also true.) The transformation to accomplish this is not unique, and one should be careful to choose one which leads to a simple differential equation.

EXAMPLE

Consider the heat problem on p. 235. An infinite slab of thickness D and initial temperature zero has the surface $x = D$ insulated, and heat is supplied at the surface $x = 0$ at a constant rate, Q calories per sec per cm^2. The mathematical problem is

$$\frac{\partial^2 u(x, t)}{\partial x^2} - \frac{1}{\kappa} \frac{\partial u(x, t)}{\partial t} = 0 \qquad (9\text{-}53)$$

$$u(x, 0) = 0 \qquad \frac{\partial u}{\partial x}\bigg|_{x=D} = 0 \qquad \frac{\partial u}{\partial x}\bigg|_{x=0} = -\frac{Q}{K}$$

where $\kappa = K/C\rho$, $K =$ thermal conductivity, $C =$ specific heat, and $\rho =$ density.

The problem may be transformed to one with homogeneous boundary conditions (in x) by a change of variable:

$$v(x, t) = u(x, t) - w(x)$$

where $w(x)$ satisfies the conditions

$$\frac{dw}{dx}\bigg|_{x=D} = 0 \qquad \frac{dw}{dx}\bigg|_{x=0} = -\frac{Q}{K}$$

and also $w(x)$ should be chosen so that the operation d^2w/dx^2 in the differential equation is easy to perform and gives a simple result. The simplest choice is probably the parabola

$$w(x) = \frac{Q}{2KD}(x - D)^2 \qquad (9\text{-}54)$$

Then the differential equation (9-53) becomes

$$\frac{\partial^2 v}{\partial x^2} - \frac{1}{\kappa} \frac{\partial v}{\partial t} = -\frac{Q}{KD} \qquad (9\text{-}55)$$

A particular solution is

$$v_p = \frac{Q\kappa}{KD}t \qquad u_p(x, t) = \frac{Q}{C\rho D}t + \frac{Q}{2KD}(x - D)^2 \qquad (9\text{-}56)$$

which is the same solution we found before in (8-63).

As a final application of Green's functions, we observe that the use of a Green's function enables us to convert a partial differential equation into an integral equation. For example, consider the equation

$$\nabla^2 u(\mathbf{x}) = f(\mathbf{x})u(\mathbf{x}) \tag{9-57}$$

in some region Ω, with suitable boundary conditions. Let $G(\mathbf{x}, \mathbf{x}')$ be the Green's function of Laplace's equation[2] for the particular region and boundary conditions. Then

$$\nabla^2 G(\mathbf{x}, \mathbf{x}') = \delta(\mathbf{x} - \mathbf{x}') \tag{9-58}$$

and the solution of

$$\nabla^2 u(\mathbf{x}) = g(\mathbf{x})$$

is

$$u(\mathbf{x}) = \int d^3x' \, G(\mathbf{x}, \mathbf{x}')g(\mathbf{x}') \tag{9-59}$$

Thus our original differential equation (9-57) is equivalent to the integral equation

$$u(\mathbf{x}) = \int d^3x' \, G(\mathbf{x}, \mathbf{x}')f(\mathbf{x}')u(\mathbf{x}') \tag{9-60}$$

Note that the integral equation has the boundary conditions "built-in." A specific example of this application will be given in Chapter 11 [see Eq. (11-18)].

9–5   GREEN'S FUNCTIONS IN ELECTRODYNAMICS

We shall discuss briefly two Green's functions which are important in electrodynamics.

First, consider Laplace's equation

$$\nabla^2 \phi = 0 \tag{9-61}$$

in an infinite region. The boundary condition is that $\phi \to 0$ as $r \to \infty$. To find the Green's function, we must solve

$$\nabla^2 \phi(\mathbf{x}) = \delta(\mathbf{x} - \mathbf{x}') \tag{9-62}$$

Clearly, ϕ can depend only on the scalar quantity $r = |\mathbf{x} - \mathbf{x}'|$. Thus we take our origin of (spherical) coordinates to be at the point $\mathbf{x}'$.

---

[2] We might equally well call $G(\mathbf{x}, \mathbf{x}')$ the Green's function for Poisson's equation. In general we may label a Green's function by either its homogeneous or inhomogeneous equation.

For all $\mathbf{x} \neq \mathbf{x}'$, $\nabla^2 \phi(\mathbf{x}) = 0$. Therefore,

$$\phi \sim \left\{ \begin{matrix} r^l \\ r^{-(l+1)} \end{matrix} \right\} P_l^m(\cos \theta) e^{\pm im\phi}$$

Because of the spherical symmetry and the boundary condition at infinity, the only possibility is

$$\phi = \frac{A}{r}$$

To find A, integrate the differential equation (9-62) over the volume of a sphere of radius R surrounding the origin. This gives

$$1 = \int d\mathbf{S} \cdot \nabla \phi$$

$$= 4\pi r^2 \left(\frac{-A}{r^2} \right) \Rightarrow A = -\frac{1}{4\pi}$$

The Green's function (that is, potential of a point charge or Coulomb potential) is therefore

$$G(\mathbf{x}, \mathbf{x}') = \frac{-1}{4\pi |\mathbf{x} - \mathbf{x}'|} \tag{9-63}$$

This is a "fundamental solution" of Eq. (9-62).

As a second example, consider the wave equation in an infinite region.

$$\nabla^2 \phi(\mathbf{x}, t) - \frac{1}{c^2} \frac{\partial^2 \phi(\mathbf{x}, t)}{\partial t^2} = 0 \tag{9-64}$$

We wish to solve the equation

$$\nabla^2 \phi - \frac{1}{c^2} \frac{\partial^2 \phi}{\partial t^2} = \delta(\mathbf{x} - \mathbf{x}') \, \delta(t - t') \tag{9-65}$$

First, it is clear that the solution depends only on $\mathbf{x} - \mathbf{x}'$ and $t - t'$; in other words, the wave equation (9-64) possesses *translational invariance* in $\mathbf{x}$ and t. Therefore we can, without loss of generality, set $\mathbf{x}' = 0$, $t' = 0$. Now introduce the Fourier transform Φ of ϕ:

$$\phi(\mathbf{x}, t) = \int \frac{d^3 k}{(2\pi)^3} \int \frac{d\omega}{2\pi} \, \Phi(\mathbf{k}, \omega) e^{i(\mathbf{k} \cdot \mathbf{x} - \omega t)} \tag{9-66}$$

$$\Phi(\mathbf{k}, \omega) = \int d^3 x \int dt \, \phi(\mathbf{x}, t) e^{-i(\mathbf{k} \cdot \mathbf{x} - \omega t)} \tag{9-67}$$

The Fourier transform of our differential equation (9-65) is

$$\left(-k^2 + \frac{\omega^2}{c^2} \right) \Phi = 1$$

so that

$$\Phi(\mathbf{k}, \omega) = \frac{c^2}{\omega^2 - k^2 c^2} \tag{9-68}$$

and

$$\phi(\mathbf{x}, t) = \int \frac{d^3 k}{(2\pi)^3} \int \frac{d\omega}{2\pi} c^2 \frac{e^{i(\mathbf{k} \cdot \mathbf{x} - \omega t)}}{\omega^2 - k^2 c^2} \tag{9-69}$$

If we choose the axis of polar coordinates in $\mathbf{k}$-space along the vector $\mathbf{x}$, the angular integrations are straightforward, and we obtain

$$\phi(\mathbf{x}, t) = \frac{1}{(2\pi)^3} \frac{c^2}{ir} \int_{-\infty}^{\infty} k \, dk \int_{-\infty}^{\infty} d\omega \frac{e^{i(kr - \omega t)}}{\omega^2 - k^2 c^2} \qquad (r = |\mathbf{x}|) \quad (9-70)$$

We shall now do the ω integral. A problem arises, in that there are two poles on the path of integration, as shown in Figure 9–4. What path of integration do we follow? The uncertainty is directly related to our failure to specify the boundary conditions for our original problem. To see this, note that the only difference between going over, under, or through each pole is that ϕ picks up a different amount of

$$\frac{e^{ik(r \pm ct)}}{r}$$

But these functions are the (spherically symmetric) solutions of the homogeneous equation (9-64).

Now our differential equation (9-65) involves a disturbance localized at $t = 0$, $\mathbf{x} = 0$. A very reasonable boundary condition is that $\phi = 0$ for $t < 0$; that is, nothing happens *before* the disturbance. In order to satisfy this

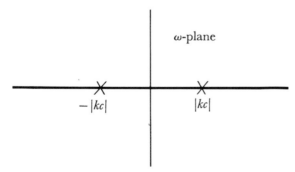

Figure 9–4   The ω-plane showing the location of the poles of the integrand in (9-70)

condition, we make our contour go *over* both poles in Figure 9–4. Then, when $t < 0$ and we complete the contour by an *upper* semicircle, we enclose no poles.

When $t > 0$, we must complete the contour by a *lower* semicircle, and

$$\phi(\mathbf{x}, t) = \frac{1}{(2\pi)^3} \frac{c^2}{ir} \int_{-\infty}^{\infty} k \, dk \, e^{ikr} \int_{-\infty}^{\infty} d\omega \, \frac{e^{-i\omega t}}{\omega^2 - k^2 c^2}$$

$$= \frac{1}{(2\pi)^3} \frac{c}{2} \frac{1}{ir} \int_{-\infty}^{\infty} dk \, e^{ikr} \int_{-\infty}^{\infty} d\omega e^{-i\omega t} \left(\frac{1}{\omega - kc} - \frac{1}{\omega + kc} \right)$$

$$= -\frac{c}{8\pi^2 r} \int_{-\infty}^{\infty} dk \, e^{ikr} (e^{-ikct} - e^{ikct})$$

$$= -\frac{c}{4\pi r} [\delta(r - ct) - \delta(r + ct)] \tag{9-71}$$

The second delta function doesn't contribute, because r and t are both positive. Therefore the desired Green's function is

$$G(\mathbf{x} - \mathbf{x}', t - t') = \begin{cases} 0 & \text{if } t < t' \\ -\dfrac{c}{4\pi |\mathbf{x} - \mathbf{x}'|} \delta[|\mathbf{x} - \mathbf{x}'| - c(t - t')] & \text{if } t > t' \end{cases} \tag{9-72}$$

Let us briefly illustrate the use of this Green's function. The solution of

$$\nabla^2 \phi - \frac{1}{c^2} \frac{\partial^2 \phi}{\partial t^2} = f(\mathbf{x}, t) \tag{9-73}$$

is

$$\phi(\mathbf{x}, t) = -\frac{c}{4\pi} \int d^3 x' \, dt' f(\mathbf{x}', t') \frac{\delta[|\mathbf{x} - \mathbf{x}'| - c(t - t')]}{|\mathbf{x} - \mathbf{x}'|}$$

The t' integration may be done because of the δ-function, with the result

$$\phi(\mathbf{x}, t) = -\frac{1}{4\pi} \int d^3 x' \frac{f\left(\mathbf{x}', t - \dfrac{1}{c} |\mathbf{x} - \mathbf{x}'|\right)}{|\mathbf{x} - \mathbf{x}'|} \tag{9-74}$$

This is called a *retarded potential* because the source term in the integrand is evaluated at the previous time $t - (1/c)|\mathbf{x} - \mathbf{x}'|$, $(1/c)|\mathbf{x} - \mathbf{x}'|$ being the time for its influence to travel a distance $|\mathbf{x} - \mathbf{x}'|$ at the speed c. For a specific example, let $f(\mathbf{x}', t') = \delta[\mathbf{x}' - \xi(t')]$. This describes a point source moving along the path $\xi(t)$.   Then

$$\phi(\mathbf{x}, t) = -\frac{1}{4\pi} \int d^3x' \int dt' \frac{\delta[\mathbf{x}' - \xi(t')] \delta\left(t - t' - \frac{1}{c}|\mathbf{x} - \mathbf{x}'|\right)}{|\mathbf{x} - \mathbf{x}'|} \qquad (9\text{-}75)$$

Clearly, the only contribution to the integral comes when $\mathbf{x}'$ and t' obey the equations

$$\mathbf{x}' = \xi(t')$$

$$t - t' = \frac{1}{c}|\mathbf{x} - \mathbf{x}'|$$

The question is: What contribution arises from this point?
    Let us consider more generally the evaluation of

$$\int dx\, dy\, dz\, dt\, \delta[f_1(xyzt)]\, \delta[f_2(xyzt)]\, \delta[f_3(xyzt)]\, \delta[f_4(xyzt)]$$

This integral is done by changing the variables of integration from $xyzt$ to $f_1 f_2 f_3 f_4$.   The integral becomes

$$\int df_1\, df_2\, df_3\, df_4 \left|\frac{\partial(xyzt)}{\partial(f_1 f_2 f_3 f_4)}\right| \delta(f_1)\, \delta(f_2)\, \delta(f_3)\, \delta(f_4) = \left|\frac{\partial(xyzt)}{\partial(f_1 f_2 f_3 f_4)}\right| \qquad (9\text{-}76)$$

evaluated at $f_1 = f_2 = f_3 = f_4 = 0$, where this last expression involves the Jacobian

$$\begin{vmatrix} \dfrac{\partial x}{\partial f_1} & \dfrac{\partial x}{\partial f_2} & \dfrac{\partial x}{\partial f_3} & \dfrac{\partial x}{\partial f_4} \\[2mm] \dfrac{\partial y}{\partial f_1} & \dfrac{\partial y}{\partial f_2} & \dfrac{\partial y}{\partial f_3} & \dfrac{\partial y}{\partial f_4} \\[2mm] \dfrac{\partial z}{\partial f_1} & \dfrac{\partial z}{\partial f_2} & \dfrac{\partial z}{\partial f_3} & \dfrac{\partial z}{\partial f_4} \\[2mm] \dfrac{\partial t}{\partial f_1} & \dfrac{\partial t}{\partial f_2} & \dfrac{\partial t}{\partial f_3} & \dfrac{\partial t}{\partial f_4} \end{vmatrix} = \frac{\partial(xyzt)}{\partial(f_1 f_2 f_3 f_4)} \qquad (9\text{-}77)$$

A useful theorem from calculus tells us that

$$\frac{\partial(xyzt)}{\partial(f_1 f_2 f_3 f_4)} = \left[\frac{\partial(f_1 f_2 f_3 f_4)}{\partial(xyzt)}\right]^{-1} \qquad (9\text{-}78)$$

Now what is this Jacobian in our particular case?

$$f_1 = x' - \xi_x(t') \qquad f_3 = z' - \xi_z(t')$$

$$f_2 = y' - \xi_y(t') \qquad f_4 = t - t' - \frac{1}{c}|\mathbf{x} - \mathbf{x}'|$$

The evaluation of the determinant is straightforward and gives

$$\frac{\partial(f_1 f_2 f_3 f_4)}{\partial(x'y'z't')} = -1 - \frac{1}{c}\frac{\dot{\boldsymbol{\xi}} \cdot (\mathbf{x}' - \mathbf{x})}{|\mathbf{x}' - \mathbf{x}|}$$

Therefore,

$$\phi(\mathbf{x}, t) = -\frac{1}{4\pi}\frac{1}{|\mathbf{x} - \mathbf{x}'| + \frac{1}{c}\dot{\boldsymbol{\xi}}(t') \cdot (\mathbf{x}' - \mathbf{x})} \tag{9-79}$$

where $\mathbf{x}' = \boldsymbol{\xi}(t')$ and $t' = t - (1/c)|\mathbf{x} - \mathbf{x}'|$. In electrodynamics, this is referred to as the *Lienard–Wiechert potential.*

Actually, the integral (9-75) may be done more simply by doing first the d^3x' integral, and then the dt' integral. The only result we need is that given in Problem 4-6; the details are left as an exercise (Problem 9-13).

REFERENCES

Clear and reasonably elementary discussions of eigenvalues and eigenfunctions of differential equations may be found in Chapter 8 of Margenau and Murphy (M2) as well as throughout the book by Sagan (S1).

Chapters 6 and 7 of Morse and Feshbach (M9) contain a large amount of material on these and related topics, but the abstract notation and wide variety of problems considered do not make for easy reading.

Green's functions arise most frequently (in classical physics at least) in the study of electromagnetic theory. Some examples of texts which discuss Green's functions in this context are those by Jackson (J1) and Smythe (S6).

The subject of radiation from a moving charge is discussed in any work on classical electromagnetic theory such as the previous two or the one by Landau and Lifshitz (L1).

PROBLEMS

9-1   What is the solution of

$$\frac{d^2y}{dx^2} + \frac{2}{x}\frac{dy}{dx} - \frac{l(l+1)}{x^2}y = \delta(x-a) \qquad (a > 0)$$

on the interval $0 < x < \infty$, subject to the boundary conditions $y(0) = y(\infty) = 0$? l is a positive integer.

9-2 $f_n(x)$ is a polynomial of order n ($n = 0, 1, 2, \ldots$) and these polynomials are mutually orthogonal on the range 0 to ∞, with weight function e^{-x}; that is, $\int_0^\infty e^{-x}f_n(x)f_m(x)\,dx = 0$ if $m \neq n$. Find a differential equation satisfied by $f_n(x)$ of the form

$$x \frac{d^2 f_n}{dx^2} + g(x) \frac{df_n}{dx} + \lambda_n f_n(x) = 0$$

9-3 Find the Green's function $G(\mathbf{x}, \mathbf{x}')$ for the Helmholtz equation

$$\nabla^2 u + k^2 u = 0$$

inside the sphere $r = a$, with the boundary condition $u(r = a) = 0$. Find G by solving the equation, not just as a formal sum over eigenfunctions like (9-27). Note that G can depend only on r, r', and θ, the angle between $\mathbf{x}$ and $\mathbf{x}'$.

9-4 Find the Green's function of the Helmholtz equation in the cube $0 \leqslant x, y, z \leqslant L$ by solving the equation

$$\nabla^2 u + k^2 u = \delta(\mathbf{x} - \mathbf{x}') \qquad (u = 0 \text{ on surface of cube})$$

9-5 Green's functions can be written for *homogeneous* equations with *inhomogeneous* boundary conditions. To illustrate this, consider the Helmholtz equation

$$\nabla^2 u + k^2 u = 0$$

inside the circle $r = R$, with the boundary condition $u = f(\theta)$ at $r = R$, $f(\theta)$ being some given function. The solution can be written

$$u(\mathbf{x}) = \int_0^{2\pi} G(\mathbf{x}, \theta') f(\theta') \, d\theta'$$

Find $G(\mathbf{x}, \theta)$.

9-6 A stretched string of length L has the end $x = 0$ fastened down, and the end $x = L$ moves in a prescribed manner:

$$y(L, t) = \cos \omega_0 t \qquad t > 0$$

Initially, the string is stretched in a straight line with unit displacement at the end $x = L$ and zero initial velocities everywhere. Find the displacement $y(x, t)$ as a function of x and t after $t = 0$.

9-7 The solution of

$$y'' + \omega^2 y = g(x) \qquad 0 \leqslant x \leqslant 2\pi$$

subject to the boundary conditions

$$y(0) = y(2\pi) \qquad y'(0) = y'(2\pi)$$

can be written in the form

$$y(x) = \int_0^{2\pi} G(x, x', \omega) g(x') \, dx'$$

Find the Green's function $G(x, x', \omega)$ in closed form. This Green's function is of practical importance in treating the effects of magnet errors on the periodic orbits in a synchrotron.

9-8   A stretched string with length L, linear density ρ, and tension T vibrates with resistance in response to a driving force $F(x)e^{i\omega t}$. It thus satisfies the equation

$$\frac{\partial^2 y}{\partial t^2} = c^2 \frac{\partial^2 y}{\partial x^2} - 2\alpha \frac{\partial y}{\partial t} + \frac{1}{\rho} F(x)e^{i\omega t}$$

Write the solution in terms of a Green's function and find the Green's function in closed form (that is, not a series).

9-9   A point charge e moves along the x-axis with uniform velocity v, so that its position x at time t is given by $x = vt$. Derive an explicit expression for the potential $\phi(t)$ at the point $(0, b, 0)$. The differential equation obeyed by $\phi(\mathbf{x}, t)$ is

$$\nabla^2 \phi - \frac{1}{c^2} \frac{\partial^2 \phi}{\partial t^2} = -4\pi\rho$$

where ρ is the charge density.

9-10   Show that the (retarded) Green's function for the Helmholtz equation

$$\nabla^2 u(\mathbf{x}) + k^2 u(\mathbf{x}) = 0$$

with the boundary condition that $u(\mathbf{x})e^{-i\omega t}$ represents outgoing waves only at infinity is

$$G(\mathbf{x}, \mathbf{x}') = -\frac{e^{ikr}}{4\pi r}$$

where $r = |\mathbf{x} - \mathbf{x}'|$. What is the connection between this result and the Green's function (9-72)?

9-11   We have shown [Eq. (9-32)] that the Green's function for a Hermitian differential operator L satisfies the symmetry relation

$$G(\mathbf{x}, \mathbf{x}') = [G(\mathbf{x}', \mathbf{x})]^*$$

Show this by a more direct method than used in the text, beginning with the differential equations for $G(\mathbf{x}, \mathbf{x}')$ and $G(\mathbf{x}, \mathbf{x}'')$, that is,

$$LG(\mathbf{x}, \mathbf{x}') = \delta(\mathbf{x} - \mathbf{x}')$$
$$LG(\mathbf{x}, \mathbf{x}'') = \delta(\mathbf{x} - \mathbf{x}'')$$

9-12   Consider the solution $u(\mathbf{x})$ of the two-dimensional inhomogeneous

Helmholtz equation

$$\nabla^2 u(\mathbf{x}) + k^2 u(\mathbf{x}) = f(\mathbf{x})$$

inside the circle $r = R$, the boundary condition being $u(\mathbf{x}) = g(\theta)$ at $r = R$. $f(\mathbf{x})$ and $g(\theta)$ are given functions. By manipulating the differential equations obeyed by $u(\mathbf{x})$ and $G(\mathbf{x}, \mathbf{x}')$, where the latter is the Green's function (9-41), obtain an explicit expression for $u(\mathbf{x})$ in terms of $f(\mathbf{x})$, $g(\theta)$, and $G(\mathbf{x}, \mathbf{x}')$.

Discuss and verify the relationship between $G(\mathbf{x}, \mathbf{x}')$ and the solution of Problem 9-5.

9-13 Integrate (9-75) over d^3x' and dt', in that order, and obtain (9-79) directly.

TEN

PERTURBATION THEORY

This short chapter presents an elementary account of perturbation theory as applied to eigenvalue problems. Its use in finding approximate eigenvalues and eigenfunctions is illustrated by examples from classical physics and quantum mechanics.

10–1 CONVENTIONAL NONDEGENERATE THEORY

Consider an eigenvalue problem

$$Lu = \lambda u \tag{10-1}$$

where L can be a matrix or a linear differential operator. We want to find the eigenvectors u_n and eigenvalues λ_n:

$$Lu_n = \lambda_n u_n \tag{10-2}$$

Suppose L is nearly equal to an L^0 whose eigenvectors u_n^0 and eigenvalues λ_n^0 we know:

$$L^0 u_n^0 = \lambda_n^0 u_n^0 \tag{10-3}$$

Let

$$L = L^0 + Q \tag{10-4}$$

where Q is small, in some sense. Then we assume

$$\lambda_n = \lambda_n^0 + \lambda_n^{(1)} + \lambda_n^{(2)} + \cdots \tag{10-5}$$

$$u_n = u_n^0 + \sum_m a_{mn}^{(1)} u_m^0 + \sum_m a_{mn}^{(2)} u_m^0 + \cdots \tag{10-6}$$

where $\lambda_n^{(1)}$ and $a_{mn}^{(1)}$ are small like Q, $\lambda_n^{(2)}$ and $a_{mn}^{(2)}$ are of order Q^2, and so on. We have also assumed that the unperturbed eigenvectors u_m^0 form a complete set; for convenience, we assume they form an orthonormal set:

$$u_m^0 \cdot u_n^0 = \delta_{mn} \tag{10-7}$$

We now substitute our trial forms (10-5) and (10-6) for λ_n and u_n into the original eigenvalue problem (10-2) and obtain

$$(L^0 + Q)\left(u_n^0 + \sum_m a_{mn}^{(1)} u_m^0 + \cdots\right) = (\lambda_n^0 + \lambda_n^{(1)} + \cdots)\left(u_n^0 + \sum_m a_{mn}^{(1)} u_m^0 + \cdots\right) \tag{10-8}$$

The zero-order part of this equation is just

$$L^0 u_n^0 = \lambda_n^0 u_n^0 \tag{10-9}$$

that is, nothing new. The first-order terms give

$$Q u_n^0 + \sum_m a_{mn}^{(1)} \lambda_m^0 u_m^0 = \lambda_n^{(1)} u_n^0 + \lambda_n^0 \sum_m a_{mn}^{(1)} u_m^0 \tag{10-10}$$

Dot u_n^0 into both sides of (10-10). We obtain

$$u_n^0 \cdot Q u_n^0 = \lambda_n^{(1)} \tag{10-11}$$

a very important result. The corrected eigenvalue λ_n is then[1]

$$\lambda_n = \lambda_n^0 + u_n^0 \cdot Q u_n^0 + O(Q^2)$$

What about the $a_{mn}^{(1)}$? Dot $u_p^0 (p \neq n)$ into our first-order equation (10-10). The result is

$$u_p^0 \cdot Q u_n^0 = (\lambda_n^0 - \lambda_p^0) a_{pn}^{(1)}$$

For the moment, assume $\lambda_p^0 \neq \lambda_n^0$; this will certainly be the case if the unperturbed eigenvalue λ_n^0 is nondegenerate.[2] Then

$$a_{pn}^{(1)} = \frac{u_p^0 \cdot Q u_n^0}{\lambda_n^0 - \lambda_p^0} \tag{10-12}$$

We see from our results (10-11) and (10-12) that the perturbation Q may be considered "small" if the numbers $u_p^0 \cdot Q u_n^0$ are small compared to the differences $\lambda_n^0 - \lambda_p^0$.

---

[1] The symbol $O(x)$ means "a quantity of order x."
[2] The degenerate case is considered in Section 10–3.

Digression on notation: We shall abbreviate $u_p^0 \cdot Qu_n^0$ by Q_{pn}. This representation of a general *linear operator* Q by the *matrix* Q_{pn}, with the aid of the "coordinate system" u_n^0, is the sort of thing we discussed in Chapter 6, in rather abstract language; see also Section 9–2 of Chapter 9.

Note that a Hermitian differential operator will be represented by a Hermitian matrix.

In this notation, our results (10-11) and (10-12) may be written

$$\lambda_n^{(1)} = Q_{nn} \tag{10-13}$$

and

$$a_{mn}^{(1)} = \frac{Q_{mn}}{\lambda_n^0 - \lambda_m^0} \qquad (m \neq n) \tag{10-14}$$

What about $a_{nn}^{(1)}$? This is not fixed by our first-order equation (10-10). Suppose, however, that we want our new eigenvectors to form an orthonormal set.

$$u_m \cdot u_n = \left(u_m^0 + \sum_{m'} a_{m'm}^{(1)} u_{m'}^0 + \cdots \right) \cdot \left(u_n^0 + \sum_{n'} a_{n'n}^{(1)} u_{n'}^0 + \cdots \right)$$

$$= \delta_{mn} + a_{nm}^{(1)*} + a_{mn}^{(1)} + \cdots \tag{10-15}$$

If Q is not Hermitian, the first-order terms here will not vanish in general. If Q is Hermitian, as will usually be the case, the first-order terms in (10-15) will cancel for $m \neq n$. For $m = n$, the requirement that

$$u_n \cdot u_n = 1$$

imposes the condition

$$\mathrm{Re}\, a_{nn}^{(1)} = 0 \tag{10-16}$$

The imaginary part of $a_{nn}^{(1)}$ is not determined at all. The reason for this is that one can always multiply an eigenvector by a complex number of absolute value 1

$$u_n \rightarrow e^{i\delta} u_n$$

$$= (1 + i\delta + \cdots) u_n \tag{10-17}$$

without affecting the orthonormality of the eigenvectors.

We shall now consider some examples.

EXAMPLE

$$L = \begin{pmatrix} 2 + \varepsilon & 1 - \varepsilon \\ 1 - \varepsilon & 2 + 2\varepsilon \end{pmatrix} \qquad (\varepsilon \ll 1) \tag{10-18}$$

$$L^0 = \begin{pmatrix} 2 & 1 \\ 1 & 2 \end{pmatrix} \qquad Q = \varepsilon \begin{pmatrix} 1 & -1 \\ -1 & 2 \end{pmatrix}$$

The unperturbed (normalized) eigenvectors and eigenvalues are

$$u_1^0 = \sqrt{\frac{1}{2}} \begin{pmatrix} 1 \\ 1 \end{pmatrix} \qquad u_2^0 = \sqrt{\frac{1}{2}} \begin{pmatrix} 1 \\ -1 \end{pmatrix}$$

$$\lambda_1^0 = 3 \qquad\qquad \lambda_2^0 = 1$$

Then, by our general formalism

$$\lambda_1^{(1)} = Q_{11} = \frac{1}{2} (1 \quad 1) Q \begin{pmatrix} 1 \\ 1 \end{pmatrix} = \frac{\varepsilon}{2}$$

$$\lambda_2^{(1)} = Q_{22} = \frac{5\varepsilon}{2}$$

so that

$$\lambda_1 = 3 + \frac{\varepsilon}{2} + \cdots$$

$$\lambda_2 = 1 + \frac{5\varepsilon}{2} + \cdots$$

Also

$$a_{21}^{(1)} = \frac{Q_{21}}{3 - 1} = -\frac{\varepsilon}{4}$$

$$a_{12}^{(1)} = \frac{Q_{12}}{1 - 3} = \frac{\varepsilon}{4}$$

so that

$$u_1 = \sqrt{\frac{1}{2}} \begin{pmatrix} 1 \\ 1 \end{pmatrix} - \frac{\varepsilon}{4} \sqrt{\frac{1}{2}} \begin{pmatrix} 1 \\ -1 \end{pmatrix} + \cdots$$

$$\approx \sqrt{\frac{1}{2}} \begin{pmatrix} 1 - \dfrac{\varepsilon}{4} \\ 1 + \dfrac{\varepsilon}{4} \end{pmatrix}$$

$$u_2 = \sqrt{\frac{1}{2}} \begin{pmatrix} 1 \\ -1 \end{pmatrix} + \frac{\varepsilon}{4} \sqrt{\frac{1}{2}} \begin{pmatrix} 1 \\ 1 \end{pmatrix} + \cdots$$

$$\approx \sqrt{\frac{1}{2}} \begin{pmatrix} 1 + \dfrac{\varepsilon}{4} \\ -1 + \dfrac{\varepsilon}{4} \end{pmatrix}$$

These results may be checked by solving the perturbed problem exactly.

EXAMPLE

Consider a circular drum head of radius a with a point mass m attached a distance b from the center.   Find the shift in the lowest frequency.
The basic equation is

$$T\nabla^2 u = \rho \frac{\partial^2 u}{\partial t^2} \tag{10-19}$$

If $u \sim e^{-i\omega t}$, then

$$\frac{T}{\rho} \nabla^2 u = -\omega^2 u$$

This is a standard eigenvalue problem with operator $(T/\rho)\nabla^2$ and eigenvalue $-\omega^2$.   Let

$$\rho = \rho_0 + \Delta\rho$$

ρ_0 being a uniform density.   Then the unperturbed operator is

$$\frac{T}{\rho_0} \nabla^2$$

and the perturbation Q is[3]

$$Q = -\frac{T\Delta\rho}{\rho_0^2} \nabla^2 \tag{10-20}$$

The lowest unperturbed eigenvalue and eigenfunction are (see p. 233)

$$\lambda_0^0 = -\left(\frac{2.40}{a}\right)^2 \frac{T}{\rho_0} \qquad u_0^0 = \frac{J_0\left(2.40 \frac{r}{a}\right)}{\sqrt{\pi a^2} J_1(2.40)}$$

Note that we have normalized u_0^0 to agree with our initial assumption (10-7)

$$\int_0^{2\pi} d\theta \int_0^a r \, dr (u_0^0)^2 = 1$$

[3] One may object that the expansion $(\rho_0 + \Delta\rho)^{-1} \approx \rho_0^{-1} - \rho_0^{-2}\Delta\rho$, valid for small $\Delta\rho$, is not obviously appropriate when $\Delta\rho$ is a delta function.   The procedure is nevertheless correct; the reader whom physical intuition does not convince may construct a rigorous proof.   (See also Problem 10-9.)

Then

$$\lambda_0^{(1)} = \int_0^{2\pi} d\theta \int_0^a r \, dr \, u_0^0 (Q u_0^0)$$

But

$$Q u_0^0 = - \frac{T \Delta \rho}{\rho_0^2} \nabla^2 u_0^0$$

$$= - \frac{\Delta \rho}{\rho_0} \lambda_0^0 u_0^0 \qquad (10\text{-}21)$$

Also, in the problem at hand, $\Delta \rho$ is m times a delta function at $r = b$. Thus

$$\lambda_0^{(1)} = - m \frac{\lambda_0^0}{\rho_0} [u_0^0 (r = b)]^2$$

or

$$\frac{\lambda_0^{(1)}}{\lambda_0^0} = - \frac{m}{\rho_0} [u_0^0 (r = b)]^2$$

$$= - \frac{m}{M} \frac{\left[J_0 \left(2.40 \frac{b}{a} \right) \right]^2}{[J_1 (2.40)]^2}$$

where $M = \rho_0 \pi a^2$ is the total mass of the drum head.   The frequency shift is given by

$$\frac{\delta \omega}{\omega} = \frac{1}{2} \frac{\lambda_0^{(1)}}{\lambda_0^0} = - \frac{m}{2M} \frac{\left[J_0 \left(2.40 \frac{b}{a} \right) \right]^2}{[J_1 (2.40)]^2} \qquad (10\text{-}22)$$

EXAMPLE

The Schrödinger equation for the hydrogen atom is

$$\left[- \frac{\hbar^2}{2m} \nabla^2 + V(\mathbf{x}) \right] \psi(\mathbf{x}) = E \psi(\mathbf{x})$$

where m is the (reduced) mass of the electron and the energy E is an eigenvalue. $V(\mathbf{x})$ is the Coulomb potential $-e^2/r$.   Let us consider the $2p$ state with magnetic quantum number $+1$

$$\psi_{2p}(\mathbf{x}) = C(x + iy)e^{-r/2a_0} \qquad E_{2p} = - \frac{e^2}{8a_0}$$

a_0 is the Bohr radius, equal to $\hbar^2/(me^2)$; the constant C must equal $1/\sqrt{64\pi a_0^5}$ if $\psi_{2p}(\mathbf{x})$ is to be normalized.   What is the shift in energy (ΔE) if the small perturbation $\Delta V(\mathbf{x}) = \lambda z^2/a_0^2$ is applied?
   Our formula (10-13) yields immediately

$$\Delta E = \int d^3 x \, \psi_{2p}^*(\mathbf{x}) \frac{\lambda z^2}{a_0^2} \psi_{2p}(\mathbf{x})$$

$$= \frac{\lambda C^2}{a_0^2} \int_0^\infty r^2 \, dr \int d\Omega (x^2 + y^2) z^2 e^{-r/a_0}$$

$$= 6\lambda$$

10-2   A REARRANGED SERIES

   The preceding formalism, while straightforward enough, is rather cumbersome in higher orders.   We shall briefly consider a different approach, which leads to a rather elegant "rearrangement" of the previous series.
   As before, we wish to solve

$$Lu = \lambda u$$

$$(L^0 + Q)u = \lambda u$$

We shall write this in the form

$$L^0 u - \lambda u = -Qu \tag{10-23}$$

   Now we have already seen in Chapter 9 that the solution of

$$(L_0 - \lambda)u = f \tag{10-24}$$

is

$$u(\mathbf{x}) = \int G(\mathbf{x}, \mathbf{x}') f(\mathbf{x}') \, d^3 x'$$

$$= \sum_n \frac{u_n^0 (u_n^0 \cdot f)}{\lambda_n^0 - \lambda} \tag{10-25}$$

Therefore Eq. (10-23) has the formal solution

$$u = \sum_n \frac{u_n^0 (u_n^0 \cdot Qu)}{\lambda - \lambda_n^0} \tag{10-26}$$

   The trouble, of course, is that u appears on the right side as well as on the left.
   Suppose $u \to u_m^0$, $\lambda \to \lambda_m^0$ as $Q \to 0$.   Then we denote this solution by u_m and

remove the mth term from the sum in (10-26).   This gives

$$u_m = cu_m^0 + \sum_{n \neq m} \frac{u_n^0(u_n^0 \cdot Qu_m)}{\lambda_m - \lambda_n^0} \tag{10-27}$$

where

$$c = \frac{u_m^0 \cdot Qu_m}{\lambda_m - \lambda_m^0} \tag{10-28}$$

We now solve the implicit equation (10-27) for u_m by iteration:

$$
\begin{aligned}
u_m &= cu_m^0 + \sum_{n \neq m} \frac{u_n^0(u_n^0 \cdot Qcu_m^0)}{\lambda_m - \lambda_n^0} + \sum_{n \neq m} \sum_{p \neq m} \frac{u_n^0(u_n^0 \cdot Qu_p^0)(u_p^0 \cdot Qcu_m^0)}{(\lambda_m - \lambda_n^0)(\lambda_m - \lambda_p^0)} + \cdots \\
&= c\left[u_m^0 + \sum_{n \neq m} \frac{u_n^0 Q_{nm}}{\lambda_m - \lambda_n^0} + \sum_{n \neq m} \sum_{p \neq m} \frac{u_n^0 Q_{np} Q_{pm}}{(\lambda_m - \lambda_n^0)(\lambda_m - \lambda_p^0)} + \cdots \right] \tag{10-29}
\end{aligned}
$$

Thus, from (10-28) and (10-29),

$$
\begin{aligned}
c(\lambda_m - \lambda_m^0) &= u_m^0 \cdot Qu_m \\
&= c\left[Q_{mm} + \sum_{n \neq m} \frac{Q_{mn} Q_{nm}}{\lambda_m - \lambda_n^0} + \sum_{n \neq m} \sum_{p \neq m} \frac{Q_{mn} Q_{np} Q_{pm}}{(\lambda_m - \lambda_n^0)(\lambda_m - \lambda_p^0)} + \cdots \right]
\end{aligned}
$$

from which we obtain an elegant implicit series for λ_m:

$$\lambda_m = \lambda_m^0 + Q_{mm} + \sum_{n \neq m} \frac{Q_{mn} Q_{nm}}{\lambda_m - \lambda_n^0} + \sum_{n \neq m} \sum_{p \neq m} \frac{Q_{mn} Q_{np} Q_{pm}}{(\lambda_m - \lambda_n^0)(\lambda_m - \lambda_p^0)} + \cdots \tag{10-30}$$

This series is an improvement over the one deduced previously in that we can write down successive terms by inspection; it suffers, of course, from the disadvantage that it contains λ_m on the right as well as the left and must be solved by iteration or some other type of approximation.

10–3   DEGENERATE PERTURBATION THEORY

The methods discussed so far need some modifications when the eigenvalue under consideration is degenerate.   For example, formula (10-12),

$$a_{pn}^{(1)} = \frac{u_p^0 \cdot Qu_n^0}{\lambda_n^0 - \lambda_p^0}$$

yields an infinite $a_{pn}^{(1)}$ if u_p^0 is degenerate with u_n^0, that is, if $\lambda_n^0 = \lambda_p^0$.   (Unless, of course, $u_p^0 \cdot Qu_n^0$ happens to vanish; we shall see below that in fact it *does* vanish.)

The difference between the degenerate and nondegenerate situations begins back with the zero-order eigenfunction u_n^0. If λ_n^0 is nondegenerate, u_n^0 is uniquely defined by the requirement that it be normalized, except for its phase. If, however, there is a set of eigenfunctions u_α^0 all belonging to the eigenvalue λ_n^0, what linear combination of these should be taken as u_n^0?

Another way to put the problem is to consider the perturbation being gradually turned off. In general, the perturbation will have removed the degeneracy, and, as the perturbation goes away and $u_n \to u_n^0$ (by definition of u_n^0), what linear combination of the u_α^0 does u_n^0 turn out to be?

A simple physical example of this problem is provided by a circular drum head. We saw in Chapter 8 that the second unperturbed frequency is degenerate, and involves the two modes shown in Figure 10–1. Now suppose we inquire as to the effect of a small mass point fastened to the drum head. It is physically clear that the two zero-order solutions which are appropriate to this question have their nodes along, and perpendicular to, the diameter through the mass point, respectively.

As in this example, the correct zero-order eigenfunctions may often be written down from symmetry or other physical considerations. In any case, they are determined by the perturbation theory as we shall now see.

We express u_n^0 as the linear combination

$$u_n^0 = \sum_{\alpha=1}^{N} a_{\alpha n}^0 u_\alpha^0 \qquad (10\text{-}31)$$

where the u_α^0 form an arbitrarily chosen (orthonormal) set. We also recall the first-order equation (10-10)

$$Qu_n^0 + \sum_m a_{mn}^{(1)}\lambda_m^0 u_m^0 = \lambda_n^{(1)}u_n^0 + \lambda_n^0 \sum_m a_{mn}^{(1)}u_m^0 \qquad (10\text{-}32)$$

If we insert (10-31) into (10-32) and dot into it u_β^0, a member of the degenerate set (that is, $\lambda_\beta^0 = \lambda_\alpha^0 = \lambda_n^0$), we obtain

$$u_\beta^0 \cdot \sum_{\alpha=1}^{N} a_{\alpha n}^0 Qu_\alpha^0 + a_{\beta n}^{(1)}\lambda_\beta^0 = \lambda_n^{(1)} \sum_{\alpha=1}^{N} a_{\alpha n}^0 u_\beta^0 \cdot u_\alpha^0 + \lambda_n^0 a_{\beta n}^{(1)}$$

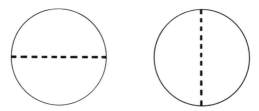

Figure 10–1   Two modes belonging to the second normal frequency of a circular drum

or

$$\sum_{\alpha=1}^{N} Q_{\beta\alpha} a_{\alpha n}^0 = \lambda_n^{(1)} a_{\beta n}^0 \tag{10-33}$$

This is the β-component of the matrix equation

$$Q a_n^0 = \lambda_n^{(1)} a_n^0 \tag{10-34}$$

in a coordinate system with base vectors u_α^0. Thus the a_n^0, which represent the u_n^0 in this system, are the eigenvectors of the $N \times N$ matrix Q, and the first-order corrections $\lambda_n^{(1)}$ are the corresponding eigenvalues.

If we now change to a new system with basis u_n^0, Q will be diagonal, and its diagonal elements will be the first-order changes $\lambda_n^{(1)}$. That is,

$$Q_{n'n} = \lambda_n^{(1)} \delta_{n'n} \tag{10-35}$$

Also, by dotting u_m^0 (with $\lambda_m^0 \neq \lambda_n^0$) into (10-32), we find for the first-order corrections to u_n^0 the same result as in the nondegenerate theory:

$$a_{mn}^{(1)} = \frac{Q_{mn}}{\lambda_n^0 - \lambda_m^0}$$

We now illustrate the application of perturbation theory in a degenerate problem.

EXAMPLE

Consider the circular drum-head problem mentioned above. The second unperturbed eigenvalue is

$$\lambda_2^0 = -\left(\frac{3.83}{a}\right)^2 \frac{T}{\rho_0} \tag{10-36}$$

and two normalized orthogonal eigenfunctions belonging to λ_2^0 are

$$u_\alpha^0 = \sqrt{\frac{2}{\pi a^2}} \frac{J_1\left(3.83 \dfrac{r}{a}\right) \cos\theta}{J_0(3.83)}$$

$$u_\beta^0 = \sqrt{\frac{2}{\pi a^2}} \frac{J_1\left(3.83 \dfrac{r}{a}\right) \sin\theta}{J_0(3.83)}$$

The perturbation Q is given by

$$Q = -\frac{T\Delta\rho}{\rho_0^2} \nabla^2$$

where $\Delta\rho$ is the density due to a point mass m, having polar coordinates (b, ψ). It is straightforward to verify that Q is represented by the matrix

$$\frac{2m}{\pi a^2 \rho_0} \frac{\left[J_1\left(3.83\,\dfrac{b}{a}\right)\right]^2}{[J_0(3.83)]^2} \frac{T}{\rho_0}\left(\frac{3.83}{a}\right)^2 \begin{pmatrix} \cos^2\psi & \sin\psi\cos\psi \\ \sin\psi\cos\psi & \sin^2\psi \end{pmatrix} \quad (10\text{-}37)$$

The 2×2 matrix in brackets has eigenvalues and eigenvectors

$$\lambda_1 = 1 \qquad\qquad \lambda_2 = 0$$

$$a_1^0 = \begin{pmatrix} \cos\psi \\ \sin\psi \end{pmatrix} \qquad a_2^0 = \begin{pmatrix} -\sin\psi \\ \cos\psi \end{pmatrix}$$

Therefore, the correct zero-order eigenfunctions are

$$\begin{aligned} u_1^0 &= (\cos\psi)u_\alpha^0 + (\sin\psi)u_\beta^0 \\ u_2^0 &= (-\sin\psi)u_\alpha^0 + (\cos\psi)u_\beta^0 \end{aligned} \quad (10\text{-}38)$$

and, in the system with these base vectors, Q becomes

$$\frac{2m}{\pi a^2 \rho_0} \frac{\left[J_1\left(3.83\,\dfrac{b}{a}\right)\right]^2}{[J_0(3.83)]^2} \frac{T}{\rho_0}\left(\frac{3.83}{a}\right)^2 \begin{pmatrix} 1 & 0 \\ 0 & 0 \end{pmatrix}$$

The diagonal elements are $\lambda_1^{(1)}$ and $\lambda_2^{(1)}$, and u_1^0 and u_2^0 are just the zero-order modes deduced earlier in this section by physical intuition.

One can extend such calculations to higher-order effects of perturbations on degenerate eigenvalues and eigenfunctions, but things get complicated pretty fast and we shall not go into these matters here.

REFERENCES

The most readable discussions of elementary perturbation theory are to be found in modern books on quantum mechanics such as those by Merzbacher (M4) Chapter 16; Schiff (S2) Section 25; and Landau and Lifshitz (L2) Chapter VI.

Another clear discussion, again within the context of quantum mechanics, is in Margenau and Murphy (M2) Chapter 11.

As usual, an exhaustive discussion, including numerous variations of the simple techniques discussed in this chapter, may be found in Morse and Feshbach (M9) Chapter 9.

PROBLEMS

10-1 Find the three lowest frequencies of oscillation of a string of length L, tension T, and mass per unit length ρ, if a mass m is fastened to the string a distance $L/4$ from one end.

$$\frac{\partial^2 y}{\partial x^2} - \frac{\rho}{T}\frac{\partial^2 y}{\partial t^2} = 0 \qquad (m \ll \rho L)$$

10-2   A square drum head of side a has surface density ρ_0 except in the corner $0 < x < a/2$, $0 < y < a/2$, where the density is 10% greater. Find first-order approximate values for the three lowest frequencies of oscillation.

$$T\nabla^2 u = \rho \frac{\partial^2 u}{\partial t^2}$$

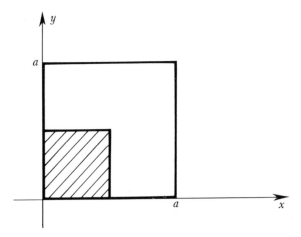

10-3   A square drum head with tension T, surface density ρ, and side a is perturbed by fastening two small masses, each of mass m, to the drum head as shown below.   Find the three lowest frequencies of the drum, neglecting terms of order $(m/\rho a^2)^2$.

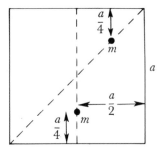

10-4   A particle of mass m obeying the two-dimensional Schrödinger equation is confined to the interior of an equilateral triangle of side s.

Find the first-order shift in the ground-state energy due to the perturbing potential $V(\mathbf{x}) = \lambda \, \delta(\mathbf{x} - \mathbf{x}_0)$, where $\mathbf{x}_0$ is the center of the triangle.

10-5   Find $\lambda(\alpha)$, the lowest eigenvalue of

$$\Delta^2 \phi + \lambda(1 + \alpha r^2)\phi = 0 \qquad (0 \leqslant r \leqslant R)$$

$$(\phi = 0 \text{ on surface of sphere } r = R)$$

for small α.

10-6   (a) Use perturbation theory to obtain an approximate value for the lowest eigenvalue of the Schrödinger equation

$$\left[-\frac{\hbar^2}{2m} \nabla^2 + V(\mathbf{x}) \right] \psi(\mathbf{x}) = E\psi(\mathbf{x})$$

if $V(\mathbf{x}) = \alpha r^2 + \beta z^2$ (β small).

(b) Verify your answer by separating the differential equation in rectangular coordinates and solving exactly.

10-7   Using the perturbation theory of Section 10–2, calculate $\alpha(\beta)$ for the periodic solution of the Mathieu equation which reduces to $\sin x$ as $\beta \to 0$.   Retain terms up to and including β^4.

10-8   Using perturbation theory, calculate $\alpha(\beta)$ for the periodic solution of the Mathieu equation which reduces to a constant as $\beta \to 0$.   Retain terms through β^4.

10-9   (a) Sometimes an eigenvalue problem is perturbed by an operator Q in the eigenvalue term, that is

$$L^0 u = \lambda(1 + Q)u$$

so that (10-23) becomes

$$L^0 u - \lambda u = \lambda Q u$$

What is the expansion corresponding to (10-30) for this problem?

(b) Consider the eigenvalue problem

$$\frac{d^2 u}{dx^2} + \omega^2 \left[1 + \varepsilon \frac{d}{dx} \right] u = 0 \qquad u(0) = u(\pi) = 0 \qquad \varepsilon \ll 1$$

The lowest eigenvalue is $\omega^2 = 1 + \alpha \varepsilon^2$; calculate the constant α by perturbation theory.   (The result in Problem 3-31 may be of assistance here.)

10-10   Assume that L^0 and Q of (10-4) are Hermitian, and show that the second-order correction $\lambda^{(2)}$ to the *lowest* eigenvalue is always negative.

ELEVEN

INTEGRAL EQUATIONS

Some integral equations are easy to reduce to algebraic or differential equations, and may thus be solved by elementary means. Examples of these are given in Sections 11–2 and 11–5. The general series solutions of Neumann and Fredholm are given in Section 11–3 and the Schmidt–Hilbert theory, with a discussion of eigenvalue problems involving integral operators, is in Section 11–4. We conclude the chapter, in Section 11–6, with a brief account of integral equations encountered in dispersion theory.

11–1   CLASSIFICATION

An integral equation is an equation in which an unknown function appears under an integral sign. We shall consider only a few simple types of such equations.

The general *linear* integral equation involving a single unknown function $f(x)$ may be written

$$\lambda \int_a^b K(x, y) f(y) \, dy + g(x) = h(x) f(x) \tag{11-1}$$

$h(x)$ and $g(x)$ are known functions of x, λ is a constant parameter, often playing the role of an eigenvalue, and $K(x, y)$ is called the *kernel* of the

299

integral equation. If $h(x) = 0$, we have a *Fredholm equation of the first kind*; if $h(x) = 1$, we have a *Fredholm equation of the second kind.* In either case, if $g(x) = 0$, the equation is *homogeneous.*

Sometimes $K(x, y)$ is zero for $y > x$. In that case, the upper limit on the integral is x, and the equation is called a *Volterra equation.*

It is often convenient to write Eq. (11-1) in symbolic form:

$$\lambda Kf + g = hf \tag{11-2}$$

where K is the *operator* which means "multiply by the kernel $K(x, y)$ and integrate over y from a to b." Equations in this form may be easily compared with operator equations involving matrix or differential operators.

Recall that we have met integral equations before, in Section 6–7 as a direct generalization of matrix operator equations, and in Section 9–4 in connection with Green's functions.

Although most of the integral equations we consider in this chapter are linear, we shall consider some nonlinear equations near the end of the chapter. Many integral equations encountered in practice, both linear and nonlinear, may be solved by rather trivial methods, such as those discussed in Sections 11–2 and 11–5.

11–2 DEGENERATE KERNELS

If the kernel $K(x, y)$ is of the form

$$K(x, y) = \sum_{i=1}^{n} \phi_i(x)\psi_i(y) \tag{11-3}$$

it is said to be *degenerate.* Integral equations with degenerate kernels may be solved by elementary techniques. Rather than giving a general discussion, we shall simply give an example.

EXAMPLE

$$f(x) = x + \lambda \int_0^1 (xy^2 + x^2y)f(y)\, dy \tag{11-4}$$

Define

$$A = \int_0^1 y^2 f(y)\, dy \qquad B = \int_0^1 y f(y)\, dy \tag{11-5}$$

Then (11-4) becomes

$$f(x) = x + \lambda Ax + \lambda Bx^2 \tag{11-6}$$

Now substitute (11-6) back into the defining equations (11-5) for A and B:

$$A = \tfrac{1}{4} + \tfrac{1}{4}\lambda A + \tfrac{1}{5}\lambda B$$
$$B = \tfrac{1}{3} + \tfrac{1}{3}\lambda A + \tfrac{1}{4}\lambda B \qquad (11\text{-}7)$$

The solution of Eqs. (11-7) is

$$A = \frac{60 + \lambda}{240 - 120\lambda - \lambda^2} \qquad B = \frac{80}{240 - 120\lambda - \lambda^2}$$

so that the solution of our original integral equation is

$$f(x) = \frac{(240 - 60\lambda)x + 80\lambda x^2}{240 - 120\lambda - \lambda^2} \qquad (11\text{-}8)$$

Note that there are two values of λ for which our solution (11-8) becomes infinite. This sounds familiar; we call these the *eigenvalues* of the integral equation. The homogeneous equation has nontrivial solutions only if λ is one of these eigenvalues; these solutions are called eigenfunctions of the operator K.

Thus, if our kernel is degenerate, the problem of solving an integral equation is reduced to that of solving a system of algebraic equations, a much more familiar subject. If the degenerate kernel (11-3) contains N terms, we see that there will be N eigenvalues, not necessarily all different.

By observing that any reasonably behaved kernel can be written as an infinite series of degenerate kernels, Fredholm deduced a set of theorems which we shall state without proof. They should seem quite reasonable, however, from the student's previous experience with eigenvalues and algebraic equations. The theorems, as stated, refer to real kernels.

1. *Either* the inhomogeneous equation

$$f(x) = g(x) + \lambda \int_a^b K(x, y)f(y)\, dy \qquad (11\text{-}9)$$

has a (unique) solution for *any* function $g(x)$ (that is, λ *is not* an eigenvalue) *or* the homogeneous equation

$$f(x) = \lambda \int_a^b K(x, y)f(y)\, dy \qquad (11\text{-}10)$$

has at least one nontrivial solution (λ *is* an eigenvalue and the solution is an eigenfunction).

2. If λ is not an eigenvalue (first alternative), then λ is also not an eigenvalue of the "transposed" equation

$$f(x) = g(x) + \lambda \int_a^b K(y, x)f(y)\, dy \qquad (11\text{-}11)$$

while if λ is an eigenvalue (second alternative), λ is also an eigenvalue of the transposed equation, that is,

$$f(x) = \lambda \int_a^b K(y, x)f(y)\, dy \tag{11-12}$$

has at least one nontrivial solution.

3. If λ is an eigenvalue, the inhomogeneous equation (11-9) has a solution if, and only, if

$$\int_a^b \phi(x)g(x)\, dx = 0 \tag{11-13}$$

for every function $\phi(x)$ which obeys the transposed homogeneous equation (11-12).

We shall not prove these theorems here.[1] They are proved quite nicely in Courant and Hilbert (C10) Vol. 1. Statement 2, for example, is the analog of the fact that a matrix and its transpose have the same eigenvalues. The necessity of the condition (11-13) follows immediately if we multiply (11-9) by $\phi(x)$ and integrate over x.

11-3 NEUMANN AND FREDHOLM SERIES

A straightforward approach to solving the integral equation

$$f(x) = g(x) + \lambda \int_a^b K(x, y)f(y)\, dy \tag{11-14}$$

is iteration; we begin with the approximation

$$f(x) \approx g(x)$$

This is substituted into the original equation under the integral sign to obtain a second approximation, and the process is then iterated. The resulting series,

$$f(x) = g(x) + \lambda \int_a^b K(x, y)g(y)\, dy + \lambda^2 \int_a^b dy \int_a^b dy'\, K(x, y)K(y, y')g(y') + \cdots \tag{11-15}$$

is known as the *Neumann series*, or *Neumann solution*, of the integral equation (11-14). It will converge for sufficiently small λ, provided the kernel $K(x, y)$ is bounded.

---

[1] The theorems strictly apply only to integral equations with *bounded* kernels and *finite* limits of integration, that is, to *nonsingular* integral equations. The theory of *singular* integral equations is a different matter.

EXAMPLE

In the quantum mechanical theory of scattering by a potential $V(\mathbf{r})$ we seek a solution of the Schrödinger equation

$$\nabla^2 \psi(\mathbf{r}) - \frac{2m}{\hbar^2} V(\mathbf{r})\psi(\mathbf{r}) + k^2\psi(\mathbf{r}) = 0 \tag{11-16}$$

with boundary conditions that $\psi(\mathbf{r})e^{-iEt/\hbar}$ represent an incident plane wave, with wave vector $\mathbf{k}_0$, plus outgoing waves at $r \to \infty$. $k^2 = k_0^2 = 2mE/\hbar^2$.
    The equation

$$\nabla^2\psi + k^2\psi = f(\mathbf{r})$$

with outgoing wave boundary conditions for the function $\psi(\mathbf{r})e^{-i\omega t}$ has the Green's function of Problem 9-10:

$$G(\mathbf{r}, \mathbf{r}') = -\frac{1}{4\pi} \frac{e^{ik|\mathbf{r}-\mathbf{r}'|}}{|\mathbf{r} - \mathbf{r}'|} \tag{11-17}$$

Thus, as in Section 9-4, we may transform the scattering problem into the integral equation

$$\psi(\mathbf{r}) = e^{i\mathbf{k}_0 \cdot \mathbf{r}} - \frac{2m}{4\pi\hbar^2} \int d^3\mathbf{r}' \frac{e^{ik|\mathbf{r}-\mathbf{r}'|}}{|\mathbf{r} - \mathbf{r}'|} V(\mathbf{r}')\psi(\mathbf{r}') \tag{11-18}$$

where the term $e^{i\mathbf{k}_0 \cdot \mathbf{r}}$ is the complementary function, adjusted to fit the boundary conditions. The solution may be written as the Neumann series (11-15). The first iteration gives already a very important result, known as the *Born approximation,*

$$\psi(\mathbf{r}) \approx e^{i\mathbf{k}_0 \cdot \mathbf{r}} - \frac{m}{2\pi\hbar^2} \int d^3\mathbf{r}' \frac{e^{ik|\mathbf{r}-\mathbf{r}'|}}{|\mathbf{r} - \mathbf{r}'|} V(\mathbf{r}')e^{i\mathbf{k}_0 \cdot \mathbf{r}'} \tag{11-19}$$

Returning to the integral equation (11-14), a more elegant, if more complicated, series solution was found by Fredholm, by subdividing the interval $a < x < b$, replacing the integral by a sum, solving the resulting algebraic equations, and then passing to the limit of infinitely many subdivisions. The result is that the solution of the integral equation (11-14) is

$$f(x) = g(x) + \lambda \int_a^b R(x, y, \lambda)g(y)\,dy \tag{11-20}$$

where $R(x, y, \lambda)$, the so-called *resolvent kernel*, is the ratio of two infinite series.

$$R(x, y, \lambda) = \frac{D(x, y, \lambda)}{D(\lambda)} \tag{11-21}$$

$$D(x, y, \lambda) = K(x, y) - \lambda \int dz \begin{vmatrix} K(x, y) & K(x, z) \\ K(z, y) & K(z, z) \end{vmatrix}$$

$$+ \frac{\lambda^2}{2!} \int dz\, dz' \begin{vmatrix} K(x, y) & K(x, z) & K(x, z') \\ K(z, y) & K(z, z) & K(z, z') \\ K(z', y) & K(z', z) & K(z', z') \end{vmatrix} - + \cdots \qquad (11\text{-}22)$$

$$D(\lambda) = 1 - \lambda \int dz K(z, z) + \frac{\lambda^2}{2!} \int dz\, dz' \begin{vmatrix} K(z, z) & K(z, z') \\ K(z', z) & K(z', z') \end{vmatrix} - + \cdots \qquad (11\text{-}23)$$

These formulas, which are rather complicated, have another form which is quite elegant and similar to perturbation formalisms used in various areas of modern physics.   Let us represent each kernel K by a line. $K(x, y)$ will be a vertical line, $|$.   If we have two K's, with a common variable integrated, we join the two lines and put a cross to indicate the junction.   For example,

$$\int dz K(x, z) K(z, y) \to \quad \text{⽸}$$

$$\int dz K(z, z) \qquad \to \quad \text{⨂}$$

$$\int dz K(x, y) K(z, z) \to \quad | \;\; \text{⨂}$$

The Fredholm solution can be written, by means of these pictures,

$$R = \frac{\left| - \left(\text{⨂} \text{⽸} \right) + \frac{1}{2} \left(\text{⨂⨂} + 2\, \text{⽸}^{\!\!\!*} - \text{⽸⨂} - 2\, \text{⽸⨂} \right) + \cdots \right.}{1 - \text{⨂} + \frac{1}{2} \left(\text{⨂⨂} - \text{⽸⽸} \right) + \cdots}$$

$$\qquad\qquad (11\text{-}24)$$

We have left out the powers of λ in the terms; this can be taken care of by the rule that a factor of λ is to be associated with each cross (that is, with each integration).

Inspection suggests the rule, which is also obvious from the original formulation in terms of determinants, that each term in the denominator except the 1 is obtained by "tying together the ends" in the previous term in the numerator, dividing by the number of K's present, and changing the sign. Also note that we in fact obtain the Neumann series (11-15) if we carry out the division in (11-24)

$$R = \left| + \text{⽸} + \text{⽸}^{\!\!\!*} + \text{⽸⽸} + \cdots \right.$$

$$\qquad\qquad (11\text{-}25)$$

[In other words, the product of (11-25) and the denominator of (11-24) is the numerator of the latter expression.] The two properties mentioned in this paragraph completely define the Fredholm solution.

For example, let us construct the next terms. The third-order term in the denominator is obtained by tying together the ends of the second-order term in the numerator; the result is

$$-\frac{1}{6}\left(\text{⊗⊗⊗} + 2\ \text{⊗} - 3\ \text{⊗ ⊗} \right)$$

and to get the third-order term in the numerator, we collect third-order terms in the product of the denominator with the series (11-25). The result is

$$\text{⊗}\ -\ \text{⊗}\ +\ \frac{1}{2}\text{⊗⊗}\ -\ \frac{1}{2}\ \text{⊗⊗}$$

$$-\frac{1}{6}\Big|\text{⊗⊗⊗}\ -\ \frac{1}{3}\Big|\text{⊗}\ +\ \frac{1}{2}\Big|\text{⊗⊗}$$

The denominator has the remarkable form

$$D(\lambda) = \exp\left(-\text{⊗} - \frac{1}{2}\text{⊗⊗} - \frac{1}{3}\text{⊗} - \cdots \right) \tag{11–26}$$

as may be seen by expanding this expression; see Problem 11-10.

The importance of the Fredholm solution is that both power series (11-22) and (11-23) are guaranteed to converge (unlike the Neumann series, which often diverges). The eigenvalues can be found by looking for the zeros of the denominator function $D(\lambda)$.

11–4 SCHMIDT–HILBERT THEORY

We shall now consider an approach quite different from the Neumann and Fredholm series. This approach is based on considering the eigenfunctions and eigenvalues of the homogeneous integral equation.

A kernel $K(x, y)$ is said to be *symmetric* if

$$K(x, y) = K(y, x)$$

and *Hermitian* if

$$K(x, y) = K^*(y, x)$$

The eigenvalues of a Hermitian kernel are real, and eigenfunctions belonging to different eigenvalues are orthogonal; two functions $f(x)$ and $g(x)$ are said to be orthogonal if

$$\int f^*(x)g(x)\, dx = 0$$

The proofs of these statements are practically identical with the corresponding proofs in Chapter 6, and will be left to the student to fill in. The identity is obvious if we write the integral equation in the symbolic notation (11-2). The eigenvalue equation is then

$$\lambda K f = f \tag{11-27}$$

We shall restrict ourselves to Hermitian kernels, and begin by considering a degenerate one:

$$K(x, x') = \sum_{\alpha=1}^{N} f_\alpha(x)f_\alpha^*(x') \tag{11-28}$$

The corresponding homogeneous integral equation is

$$u(x) = \lambda \int_a^b K(x, x')u(x')\, dx'$$

$$= \lambda \sum_\alpha f_\alpha(x) \int f_\alpha^*(x')u(x')\, dx' \tag{11-29}$$

Let us define

$$\int f_\alpha^*(x')u(x')\, dx\,' = c_\alpha \tag{11-30}$$

so that

$$u(x) = \lambda \sum_\alpha c_\alpha f_\alpha(x) \tag{11-31}$$

We see that any eigenfunction $u(x)$ may be expressed linearly in terms of the N functions $f_\alpha(x)$, and it is clear, for example, that there can be no more than N linearly independent eigenfunctions.[2]

---

[2] There may be fewer than N independent eigenfunctions for several reasons. The functions $f_\alpha(x)$ need not be linearly independent; Problem 11-11 is relevant here. Also it is conventional to exclude infinity as an eigenvalue [compare the sentence immediately following Eq. (11-33)], and with it, of course, any eigenfunctions belonging to that eigenvalue.

Substituting expression (11-31) into (11-30), and defining

$$A_{\alpha\beta} = \int f_\alpha^*(x) f_\beta(x)\, dx = A_{\beta\alpha}^* \tag{11-32}$$

we obtain

$$c_\alpha = \lambda \sum_\beta A_{\alpha\beta}\, c_\beta \tag{11-33}$$

Thus the eigenvalues λ are the reciprocals of the (nonzero) eigenvalues of the Hermitian matrix $A_{\alpha\beta}$. We know from Chapter 6 that the numbers $c_\alpha^{(i)}$, where the superscript i distinguishes the different eigenvectors of the matrix A, obey the orthogonality condition

$$c^{(i)} \cdot c^{(j)} = \sum_\alpha c_\alpha^{(i)*} c_\alpha^{(j)} = 0 \qquad \text{if} \qquad i \neq j \tag{11-34}$$

To see what normalization is convenient, let $u^{(i)}(x)$ denote the ith eigenfunction, and similarly for $u^{(j)}(x)$. Then

$$\int u^{(i)}(x)^* u^{(j)}(x)\, dx = \lambda_i \lambda_j \sum_{\alpha\beta} c_\alpha^{(i)*} c_\beta^{(j)} \int f_\alpha(x)^* f_\beta(x)\, dx$$

$$= \lambda_i \lambda_j \sum_{\alpha\beta} c_\alpha^{(i)*} c_\beta^{(j)} A_{\alpha\beta}$$

$$= \lambda_i \sum_\alpha c_\alpha^{(i)*} c_\alpha^{(j)}$$

Therefore, we choose

$$\sum_\alpha c_\alpha^{(i)*} c_\alpha^{(j)} = \frac{\delta_{ij}}{\lambda_i} \tag{11-35}$$

so that

$$\int u^{(i)}(x)^* u^{(j)}(x)\, dx = \delta_{ij} \tag{11-36}$$

that is, our eigenfunctions are orthonormal.

Note that according to (11-35) the numbers $\sqrt{\lambda_i}\, c_\alpha^{(i)}$ form a unitary matrix. Therefore,

$$\sum_i \lambda_i c_\alpha^{(i)*} c_\beta^{(i)} = \delta_{\alpha\beta} \tag{11-37}$$

This is a useful orthogonality property. For example, it enables us to "solve" the relation

$$u^{(i)}(x) = \lambda_i \sum_\alpha c_\alpha^{(i)} f_\alpha(x)$$

for the $f_\alpha(x)$:

$$f_\alpha(x) = \sum_i c_\alpha^{(i)*} u^{(i)}(x) \tag{11-38}$$

This in turn enables us to express the kernel $K(x, x')$ given by (11-28) elegantly in terms of its eigenfunctions:

$$\begin{aligned}
K(x, x') &= \sum_\alpha f_\alpha(x) f_\alpha^*(x') \\
&= \sum_\alpha \sum_{ij} c_\alpha^{(i)*} u^{(i)}(x) c_\alpha^{(j)} u^{(j)}(x')^* \\
&= \sum_i \frac{u^{(i)}(x) u^{(i)}(x')^*}{\lambda_i}
\end{aligned} \tag{11-39}$$

This result, although shown here only for a degenerate kernel, is true in general for Hermitian kernels for which the numbers of positive and negative eigenvalues are both infinite. See mathematics books for proofs and conditions; Courant and Hilbert (C10) Vol. 1, Chapter III is particularly nice.

This approach can be extended to the inhomogeneous equation. We make use of a theorem which follows immediately from (11-39) but which also holds under more general conditions, as shown, for example, in Courant and Hilbert (C10): Any function which can be represented "sourcewise" in terms of the kernel $K(x, y)$, that is, any function $\phi(x)$ of the form

$$\phi(x) = \int K(x, y) \psi(y) \, dy \qquad (\psi \text{ arbitrary and } K \text{ Hermitian}) \tag{11-40}$$

can be expanded in a series of the eigenfunctions of $K(x, y)$:

$$\phi(x) = \sum_i c_i u^{(i)}(x) \tag{11-41}$$

where

$$c_i = \int u^{(i)}(x)^* \phi(x) \, dx = \frac{1}{\lambda_i} \int u^{(i)}(x)^* \psi(x) \, dx \tag{11-42}$$

In general, the eigenfunctions $u^{(i)}(x)$ do not form a complete set of functions. Not *any* function, but only sourcewise representable functions, can be expanded in terms of them. This is not an unfamiliar situation. For example, consider ordinary vectors in three dimensions, and an operator P which projects a vector $\mathbf{r}$ into the xy-plane. P has two independent eigenvectors, say, $\mathbf{e}_x$ and $\mathbf{e}_y$, and any vector $\mathbf{r}'$ of the form $\mathbf{r}' = P\mathbf{r}$ can be expressed in terms of them, although an arbitrary vector $\mathbf{r}$ cannot.

Now we can solve the inhomogeneous equation

$$\phi(x) = g(x) + \lambda \int K(x, y) \phi(y) \, dy \tag{11-43}$$

We may expand

$$\phi(x) - g(x) = \sum_i c_i u^{(i)}(x) \tag{11-44}$$

where, by (11-42),

$$c_i = \int dx\, u^{(i)}(x)^* [\phi(x) - g(x)] = \lambda \frac{1}{\lambda_i} \int u^{(i)}(x)^* \phi(x)\, dx \tag{11-45}$$

Thus, if we define

$$d_i = \int dx\, u^{(i)}(x)^* \phi(x) \qquad e_i = \int dx\, u^{(i)}(x)^* g(x) \tag{11-46}$$

(11-45) reads

$$c_i = d_i - e_i = \frac{\lambda}{\lambda_i} d_i$$

Therefore,

$$d_i = \frac{\lambda_i}{\lambda_i - \lambda} e_i$$

and

$$c_i = \frac{\lambda}{\lambda_i - \lambda} e_i$$

Substituting back into (11-44) gives for $\phi(x)$

$$\phi(x) = g(x) + \lambda \sum_i \frac{e_i u^{(i)}(x)}{\lambda_i - \lambda}$$

or

$$\phi(x) = g(x) + \lambda \sum_i \frac{u^{(i)}(x)}{\lambda_i - \lambda} \int u^{(i)}(y)^* g(y)\, dy \tag{11-47}$$

If we compare this last result with the solution (11-20), we see that we have obtained a series representation for the resolvent kernel:

$$R(x, y, \lambda) = \sum_i \frac{u^{(i)}(x) u^{(i)}(y)^*}{\lambda_i - \lambda} \tag{11-48}$$

Symbolically, the original equation (11-43) was

$$\phi = g + \lambda K\phi \tag{11-49}$$

which has the formal solution

$$\phi = (1 - \lambda K)^{-1} g \tag{11-50}$$

Note that the power series expansion of the operator in this expression gives just the Neumann series.

$$\phi = g + \lambda Kg + \lambda^2 K^2 g + \lambda^3 K^3 g + \cdots \tag{11-51}$$

Since, on the other hand, according to (11-20)

$$\phi = (1 + \lambda R)g \tag{11-52}$$

we have the formal identity

$$(1 - \lambda K)^{-1} = 1 + \lambda R \tag{11-53}$$

It is interesting to compare our present results with those obtained for partial differential equations in Section 9–4. Suppose a differential operator L has eigenvalues κ_i and eigenfunctions $u^{(i)}$ which satisfy given boundary conditions. The inhomogeneous differential equation has the form

$$L\phi + f = \kappa\phi \tag{11-54}$$

where the inhomogeneous term f is a known function. The solution in terms of the Green's function is given, according to (9-26) and (9-27), by

$$\phi(\mathbf{x}) = \int G(\mathbf{x}, \mathbf{x}') f(\mathbf{x}') \, d^3(\mathbf{x}') \tag{11-55}$$

where[3]

$$G(\mathbf{x}, \mathbf{x}') = -\sum_i \frac{u^{(i)}(\mathbf{x}) u^{(i)}(\mathbf{x}')^*}{\kappa_i - \kappa} \tag{11-56}$$

Note that $G(\mathbf{x}, \mathbf{x}')$ is an integral kernel operator, and Eq. (11-55) is symbolically

$$\phi = Gf \tag{11-57}$$

The differential equation (11-54) and the integral equation (11-49) have the same symbolic form if we set $\kappa = 1/\lambda$ and $\lambda f = g$ and identify L with K. The solutions of these equations, according to (11-57) and (11-52), are

$$\phi = (1 + \lambda R)g$$

and

$$\phi = Gf = \frac{1}{\lambda} Gg$$

We may therefore expect that the operators $(1 + \lambda R)$ and $(1/\lambda)G$ can be written in the same form in terms of the corresponding eigenfunctions and

---

[3] The difference in sign between (9-27) and (11-56) results from our placing the inhomogeneous term f in (11-54) on the left side of the equation.

eigenvalues.   To verify this, write (11-56) with $\kappa_i = 1/\lambda_i$ and $\kappa = 1/\lambda$.   Then

$$-\frac{1}{\kappa_i - \kappa} = \frac{\lambda\lambda_i}{\lambda_i - \lambda} = \lambda\left(1 + \frac{\lambda}{\lambda_i - \lambda}\right)$$

and (11-56) becomes

$$G(\mathbf{x}, \mathbf{x}') = \lambda \sum_i u^{(i)}(\mathbf{x})u^{(i)}(\mathbf{x}')^* + \lambda^2 \sum_i \frac{u^{(i)}(\mathbf{x})u^{(i)}(\mathbf{x}')^*}{\lambda_i - \lambda}$$

$$= \lambda\delta(\mathbf{x} - \mathbf{x}') + \lambda^2 \sum_i \frac{u^{(i)}(\mathbf{x})u^{(i)}(\mathbf{x}')^*}{\lambda_i - \lambda}$$

where we have used the completeness relation (9-14).   The integral kernel $\delta(\mathbf{x} - \mathbf{x}')$ is equivalent to the identity operator, so we have

$$\frac{1}{\lambda} G = 1 + \lambda \sum_i \frac{u^{(i)}(\mathbf{x})u^{(i)}(\mathbf{x}')^*}{\lambda_i - \lambda} \tag{11-58}$$

which has the same form as $1 + \lambda R$ with R given by the expansion (11-48).

11–5 MISCELLANEOUS DEVICES

Volterra equations can often be turned into differential equations by differentiating.

EXAMPLE

$$u(x) = x + \int_0^x xy\, u(y)\, dy = x + x f(x) \tag{11-59}$$

where

$$f(x) = \int_0^x y\, u(y)\, dy$$

Then

$$f'(x) = x\, u(x) = x[x + x f(x)]$$

Solving this differential equation gives

$$f(x) = -1 + Ce^{x^3/3}$$

so that

$$u(x) = Cxe^{x^3/3} \tag{11-60}$$

To find C, we substitute back into the original integral equation.   The result is $C = 1$.

If the kernel is a function of $(x - y)$ only, a so-called *displacement kernel*, and if the limits are $-\infty$ to $+\infty$, we can use Fourier transforms. Consider the equation

$$f(x) = \phi(x) + \lambda \int_{-\infty}^{\infty} K(x - y)f(y)\, dy \tag{11-61}$$

Take Fourier transforms (indicated by bars)

$$\int_{-\infty}^{\infty} dx f(x)e^{-ikx} = \bar{f}(k), \text{ etc.} \tag{11-62}$$

The transform of our integral equation (11-61) is

$$\bar{f}(k) = \bar{\phi}(k) + \lambda \bar{K}(k)\bar{f}(k) \tag{11-63}$$

Therefore,

$$\bar{f}(k) = \frac{\bar{\phi}(k)}{1 - \lambda \bar{K}(k)} \tag{11-64}$$

If we can invert this transform, we can solve the problem.

If the limits are 0 to x with a displacement kernel, and our functions vanish for $x < 0$, we use a Laplace transform, since it has the appropriate form of convolution integral for this case [see Eq. (4-58)].

The *Wiener–Hopf* technique, already discussed (Section 8–5) in connection with partial differential equations, can also be used on certain integral equations, said to be of the Wiener–Hopf type:

$$f(x) = \phi(x) + \int_{0}^{\infty} K(x - y)f(y)\, dy \qquad (-\infty < x < \infty) \tag{11-65}$$

The displacement kernel and half-infinite range of integration are the significant features. Let $f(x) = f_+(x) + f_-(x)$, where $f_+ = 0$ for $x > 0$ and $f_- = 0$ for $x < 0$. Then

$$f_+(x) + f_-(x) = \phi(x) + \int_{-\infty}^{\infty} K(x - y)f_-(y)\, dy \tag{11-66}$$

Now take Fourier transforms;

$$\bar{f}_+(k) + \bar{f}_-(k) = \bar{\phi}(k) + \bar{K}(k)\bar{f}_-(k)$$

which we shall write in the form

$$\bar{f}_-(k)[1 - \bar{K}(k)] + \bar{f}_+(k) = \bar{\phi}(k) \tag{11-67}$$

Now assume that

$$|K(x)| \lesssim \begin{cases} e^{-\alpha x} & \text{as } x \to \infty \\ e^{\beta x} & \text{as } x \to -\infty \end{cases} \tag{11-68}$$

Therefore, $\overline{K}(k)$ is analytic in the strip $-\beta < \text{Im } k < \alpha$. We shall also assume that the function $\phi(x)$ is at least this well behaved at infinity, so $\overline{\phi}(k)$ is also analytic in this strip.

By considering the original integral equation (11-65), it seems reasonable to assume that $\overline{f}(k)$ is at least analytic in some strip within $-\beta < \text{Im } k < \alpha$, if not in the entire strip. Thus, we suppose that everything in (11-67) is analytic in some strip

$$\gamma < \text{Im } k < \delta$$

Now we "factor" $1 - \overline{K}(k)$ into two functions

$$1 - \overline{K}(k) = \frac{A(k)}{B(k)}$$

where $A(k)$ is analytic for $\text{Im } k < \delta$, and $B(k)$ is analytic for $\text{Im } k > \gamma$. Note that such a factorization is essentially equivalent to decomposing

$$\ln [1 - \overline{K}(k)] = \ln A(k) - \ln B(k) \tag{11-69}$$

into the difference of two functions, each regular in the appropriate region; we already know, in principle anyway, how to do this [see Eq. (8-85)].

Our Fourier-transformed equation (11-67) becomes

$$\overline{f}_-(k)A(k) + \overline{f}_+(k)B(k) = B(k)\overline{\phi}(k) \tag{11-70}$$

We decompose $B(k)\overline{\phi}(k)$ in the now familiar way:

$$B(k)\overline{\phi}(k) = C(k) + D(k) \tag{11-71}$$

where $C(k)$ is analytic for $\text{Im } k > \gamma$, and $D(k)$ for $\text{Im } k < \delta$. Thus

$$\overline{f}_+(k)B(k) - C(k) = -\overline{f}_-(k)A(k) + D(k) \tag{11-72}$$

where the function on the left is analytic for $\text{Im } k > \gamma$, and the function on the right is analytic for $\text{Im } k < \delta$. Thus they must both equal an entire function. Furthermore, $\overline{f}_+(k)$ and $\overline{f}_-(k)$ actually go to zero at $\text{Im } k = \pm\infty$, and the functions $A, \ldots, D$ can usually be shown to behave at worst like polynomials. Thus, both sides of the above equation are equal to some polynomial. This fixes f_+ and f_-, except possibly for some unknown constants which may be determined by physical reasoning or substitution in the original integral equation.

This method is far from being a mechanical prescription; considerable thought must be exerted at each stage to insure that the calculation is making sense. We shall illustrate some of the difficulties by solving a rather simple example.

EXAMPLE

$$f(x) = e^{-|x|} + \lambda \int_0^\infty e^{-|x-y|} f(y)\, dy \qquad (11\text{-}73)$$

The Fourier transform of the function $e^{-|x|}$, which appears both as kernel and inhomogeneous term, is

$$\int_{-\infty}^\infty dx\, e^{-|x|} e^{-ikx} = \frac{2}{1+k^2}$$

so that the Fourier transform of our integral equation (11-73) is

$$\bar{f}_+(k) + \bar{f}_-(k) = \frac{2}{1+k^2} + \frac{2\lambda}{1+k^2} \bar{f}_-(k)$$

or

$$\frac{k^2 - \xi^2}{k^2 + 1} \bar{f}_-(k) + \bar{f}_+(k) = \frac{2}{1+k^2} \qquad (11\text{-}74)$$

where $\xi^2 = 2\lambda - 1$. For definiteness, we shall assume ξ real ($\lambda > \frac{1}{2}$).

Our next step is to represent the coefficient $(k^2 - \xi^2)/(k^2 + 1)$ as a quotient $A(k)/B(k)$. The singularities and zeros of this coefficient are at $\pm i$ and $\pm \xi$; thus two natural possibilities arise for the common strip within which no singularities are to occur. We shall concentrate on the strip $-1 < \mathrm{Im}\, k < 0$ for now, and leave consideration of the other possibility, $0 < \mathrm{Im}\, k < 1$, for later.

How then can we write

$$\frac{k^2 - \xi^2}{k^2 + 1} = \frac{A(k)}{B(k)}$$

where $A(k)$ is to have no singularities or zeros in $\mathrm{Im}\, k < 0$, while $B(k)$ is to have no zeros or singularities in $\mathrm{Im}\, k > -1$? The answer is trivial;

$$A(k) = \frac{k^2 - \xi^2}{k - i} \qquad\qquad B(k) = k + i$$

and multiplying our basic equation (11-74) by $B(k)$ yields [compare Eq. (11-70)]

$$\frac{k^2 - \xi^2}{k - i} \bar{f}_-(k) + (k + i)\bar{f}_+(k) = \frac{2}{k - i} \qquad (11\text{-}75)$$

Next, we are instructed to decompose the right-hand side of (11-75) into two pieces [compare (11-71)]

$$\frac{2}{k - i} = C(k) + D(k)$$

with $C(k)$ regular for $\operatorname{Im} k > -1$ and $D(k)$ regular for $\operatorname{Im} k < 0$. The decomposition is trivial; we can take

$$C(k) = 0 \qquad D(k) = \frac{2}{k-i}$$

More generally, as remarked above, we can add a constant to $C(k)$ provided we subtract the same constant from $D(k)$; a constant, after all, has no singularities *anywhere*.

If we then take

$$C(k) = iA \qquad \text{and} \qquad D(k) = \frac{2}{k-i} - iA \qquad (11\text{-}76)$$

with A an arbitrary constant, our problem is solved.   For, from (11-75), we must have

$$\frac{k^2 - \xi^2}{k - i} \bar{f}_-(k) = \frac{2}{k-i} - iA$$

$$\bar{f}_-(k) = \frac{2}{k^2 - \xi^2} - \frac{iA(k - i)}{k^2 - \xi^2}$$

$$f_-(x) = -\frac{2}{\xi} \sin \xi x + A\left(\cos \xi x + \frac{1}{\xi} \sin \xi x\right)$$

and

$$(k + i)\bar{f}_+(k) = iA$$

$$\bar{f}_+(k) = \frac{iA}{k + i}$$

$$f_+(x) = Ae^x$$

so that the solution to our integral equation (11-73) is

$$f(x) = \begin{cases} Ae^x & (x < 0) \\ -\dfrac{2}{\xi} \sin \xi x + A\left(\cos \xi x + \dfrac{1}{\xi} \sin \xi x\right) & (x > 0) \end{cases} \qquad (11\text{-}77)$$

with A an arbitrary constant.

Suppose we had guessed the wrong strip, namely $0 < \operatorname{Im} k < 1$; what would have happened then?   The solution would have collapsed, for reasons we leave to the reader to discover.   Also, why were we able to introduce one, and only one, arbitrary constant into (11-76)?   The reason is that the next function, after a constant, having no singularities in the finite complex plane is a linear function, say Bk.   If, however, we had added Bk to $C(k)$ or $D(k)$, and subtracted it from the other, our $\bar{f}_\pm(k)$ would not have approached zero as Im $k \to \pm\infty$, and this behavior would violate what we know must hold for $\bar{f}_\pm(k)$.

Practical applications of this method tend to be quite long and complicated. Morse and Feshbach (M9) in Chapter 8 and Noble (N3) give some detailed applications of Wiener–Hopf techniques to problems of physical interest.

11–6   INTEGRAL EQUATIONS IN DISPERSION THEORY

In Chapter 5, we encountered relations of the form

$$F(x) = \frac{1}{\pi} \int \frac{g(x')\, dx'}{x' - x - i\varepsilon} \tag{11-78}$$

so-called *dispersion relations*. Mathematicians call them *Hilbert transforms*. The integral runs along some part of the real axis (usually), and the function

$$f(z) = \frac{1}{\pi} \int \frac{g(x')\, dx'}{x' - z}$$

is analytic everywhere except for this part of the real axis, which forms a branch line. Note that

$$f(x + i\varepsilon) = F(x)$$

In order to find $f(z)$ or $F(x)$, we need some information about $g(x)$. If $g(x)$ is real, it is the imaginary part of $F(x)$. More generally, $g(x)$ is often called the *absorptive part* of $F(x)$, for physical reasons we cannot go into here.

As our first example of the solution of such integral equations, suppose we are told that

$$g(x) = |F(x)|^2 h(x)$$

where $h(x)$ is some given real function.[4]   We must therefore solve the non-linear integral equation

$$F(x) = \frac{1}{\pi} \int \frac{|F(x')|^2 h(x')\, dx'}{x' - x - i\varepsilon} \tag{11-79}$$

---

[4] This is a typical condition imposed on an elastic scattering amplitude by *unitarity*. For example, $F(x)$ is often required by unitarity to have the form

$$F(x) \sim \frac{e^{i\delta(x)} \sin \delta(x)}{h(x)}$$

where $\delta(x)$ and $h(x)$ are real. It follows that Im $F(x) = |F(x)|^2 h(x)$.

This looks hard, but is simplified markedly by considering, instead of the analytic function $f(z)$, its reciprocal

$$\phi(z) = \frac{1}{f(z)}$$

Provided $f(z)$ never vanishes (a crucial assumption!), its reciprocal is analytic everywhere except for the branch line.   Furthermore, the discontinuity of ϕ across the branch line is

$$\phi(x + i\varepsilon) - \phi(x - i\varepsilon) = 2i \text{ Im } \phi(x + i\varepsilon)$$

$$= -2i\frac{\text{Im} f(x + i\varepsilon)}{|f(x+i\varepsilon)|^2} = -2i\frac{\text{Im} F(x)}{|F(x)|^2}$$

$$= -2ih(x)$$

Thus, by a simple application of Cauchy's formula, just as in the derivation of the original dispersion relation,

$$\phi(z) = -\frac{1}{\pi}\int\frac{h(x')\,dx'}{x' - z}$$

and the solution to our original problem is

$$F(x) = \frac{1}{\phi(x + i\varepsilon)} = -\left[\frac{1}{\pi}\int\frac{h(x')\,dx'}{x' - x - i\varepsilon}\right]^{-1} \tag{11-80}$$

provided $\phi(z)$ does not vanish anywhere.
   As a second example, suppose that the function $g(x)$ is given by

$$g(x) = h^*(x)F(x) \tag{11-81}$$

where $h(x) = e^{i\delta} \sin \delta$, $\delta(x)$ being a given real function.[5]   Again we consider

$$f(z) = \frac{1}{\pi}\int\frac{h^*(x')F(x')\,dx'}{x' - z}$$

and observe that

$$f(x + i\varepsilon) - f(x - i\varepsilon) = 2ih^*(x)F(x)$$

Our dispersion relation may therefore be written

$$\frac{f(x + i\varepsilon) - f(x - i\varepsilon)}{2ih^*(x)} = f(x + i\varepsilon)$$

which, because of the identity

$$1 - 2ih^*(x) = e^{-2i\delta(x)}$$

---

[5] This again is a typical condition which may be required of a scattering amplitude by unitarity.

may be written

$$\frac{f(x + i\varepsilon)}{f(x - i\varepsilon)} = e^{2i\delta(x)}$$

We now consider, instead of the reciprocal of f, its logarithm

$$L(z) = \ln f(z)$$

Then

$$L(x + i\varepsilon) - L(x - i\varepsilon) = 2i\delta(x)$$

Now we are back on familiar ground. L is an analytic function of z (provided f does not vanish) except along the cut, and so

$$L(z) = \frac{1}{\pi} \int \frac{\delta(x') \, dx'}{x' - z}$$

and

$$F(x) = e^{L(x + i\varepsilon)} = \exp\left[\frac{1}{\pi} \int \frac{\delta(x') \, dx'}{x' - x - i\varepsilon}\right] \tag{11-82}$$

REFERENCES

Numerous useful discussions of linear integral equations exist; several that have proved especially useful to the authors are: Margenau and Murphy (M2); Lovitt (L9); Courant and Hilbert (C10); Irving and Mullineux (I3); Morse and Feshbach (M9). Of these references, Margenau and Murphy is particularly simple, concise, and understandable; Courant and Hilbert makes typically elegant reading, while Irving and Mullineux has a large number of approximate methods and interesting examples.

Physical applications of the solution of integral equations resulting from dispersion relations may be found, for example, in Castillejo et al. (C3) and Omnes (O1).

PROBLEMS

11-1   Solve the integral equations

(a) $u(x) = e^x + \lambda \int_0^1 xt \, u(t) \, dt$

(b) $u(x) = \lambda \int_0^\pi \sin(x - t) \, u(t) \, dt$ (two eigenvalues; two solutions)

11-2   Solve the integral equation

$$f(x) = x + \lambda \int_0^1 y(x + y) f(y) \, dy$$

keeping terms through λ^2,

(a) by Fredholm's method

(b) by Neumann's method

11-3   Solve the integral equation

$$f(x) = e^{-|x|} + \lambda \int_{-\infty}^{\infty} e^{-|x-y|} f(y) \, dy$$

where $f(x)$ is to remain finite for $x \to \pm\infty$.

11-4   (a) Find the eigenfunction(s) and eigenvalue(s) of the integral equation

$$f(x) = \lambda \int_0^1 e^{x-y} f(y) \, dy$$

(b) Solve the integral equation

$$f(x) = e^x + \lambda \int_0^1 e^{x-y} f(y) \, dy$$

11-5   Solve integral equation (11-73)

$$f(x) = e^{-|x|} + \lambda \int_0^{\infty} e^{-|x-y|} f(y) \, dy \qquad (-\infty < x < \infty; \lambda > \tfrac{1}{2})$$

directly, without using Fourier transforms, and compare your solution with (11-77).

11-6   Write the diagrams (with coefficients) for the sixth-order numerator and denominator of the Fredholm expansion.

11-7   Solve the integral equations

(a) $f(x) = x^2 + \int_0^1 xy f(y) \, dy$

(b) $f(x) = x + \int_0^x f(y) \, dy$

11-8   Solve the integral equation

$$f(x) = e^{-|x|} + \lambda \int_0^{\infty} f(y) \cos xy \, dy$$

Hint: Take the cosine transform of the integral equation.

11-9   Find the (Fredholm) integral equation whose solutions give the normal frequencies and normal modes of a vibrating stretched string of length L, with ends fixed.

11-10 (*a*) By considering its eigenvalues, show that an arbitrary matrix M obeys the equation

$$\det [e^M] = \exp [\operatorname{Tr} M]$$

(*b*) Without inquiring into the rigor required to pass from matrices to integral kernel operators, show that (11-26) may be written formally

$$D(\lambda) = \det (1 - \lambda K)$$

and that the numerator $D(x, y, \lambda)$ of the Fredholm solution (11-21) has the (formal) form

$$D(x, y, \lambda) = \frac{K \det (1 - \lambda K)}{1 - \lambda K}$$

11-11 Let $\phi_\alpha(x)$ be a complete set of orthogonal functions, and expand the kernel $K(x, x')$ in the series

$$K(x, x') = \sum_{\alpha\beta} c_{\alpha\beta} \, \phi_\alpha(x)\phi_\beta^*(x')$$

(*a*) What is the condition that the coefficients $c_{\alpha\beta}$ must satisfy if $K(x, x')$ is to be a *Hermitian* kernel?
(*b*) Show that such a Hermitian kernel can always be written in the form (11-28), provided only a finite number of coefficients $c_{\alpha\beta}$ are nonzero.

11-12 In a one-dimensional problem (the so-called *Milne problem*) involving the transfer of radiation up through a scattering atmosphere, $I(x, \theta)$, the intensity of radiation in a direction making an angle θ with the vertical, at height x, obeys the integral equation

$$\cos \theta \, \frac{\partial I(x, \theta)}{\partial x} = -\kappa \, I(x, \theta) + \frac{\kappa}{4\pi} B(x) \tag{1}$$

where κ is the reciprocal of the scattering mean free path and $B(x) = 2\pi \int I(x, \theta) \, d(\cos \theta)$ is the *total* radiation intensity at height x. The first term on the right of (1) represents the decrease in $I(x, \theta)$ due to scattering, while the second term represents the increase in intensity due to radiation scattered from other directions.

(*a*) Let the atmosphere extend downward from its surface at $x = 0$ to $x = -\infty$, and assume the boundary condition

$$I(0, \theta) = 0 \quad \text{for} \quad -1 < \cos \theta < 0$$

(that is, no radiation is incident on the atmosphere from above).

Derive

$$I(x, \theta) = \begin{cases} \dfrac{\kappa}{4\pi} \displaystyle\int_x^0 e^{\frac{\kappa(x'-x)}{\cos\theta}} B(x') \dfrac{dx'}{-\cos\theta} & (-1 < \cos\theta < 0) \\[20pt] \dfrac{\kappa}{4\pi} \displaystyle\int_{-\infty}^x e^{\frac{\kappa(x'-x)}{\cos\theta}} B(x') \dfrac{dx'}{\cos\theta} & (0 < \cos\theta < 1) \end{cases}$$

and give the physical significance of these relations.

(b) Show that $B(x)$ obeys a linear homogeneous integral equation

$$B(x) = \int_{-\infty}^0 dx' K(x - x') B(x')$$

and evaluate the kernel $K(x - x')$ in terms of the exponential integral function $E_1(z) = \int_1^\infty (dt/t)e^{-tz}$. [Note that this integral equation is essentially of the Wiener–Hopf form (11-65); its solution may be found in Morse and Feshbach (M9) Section 8.5.]

11-13 The integral equation

$$\int_{-T/2}^{T/2} dt' \phi(t') e^{-\Gamma|t-t'|} = \lambda \phi(t) \qquad (-\tfrac{1}{2}T < t < \tfrac{1}{2}T)$$

is of interest in the theory of photon counting statistics [see Bédard (B1)]. Show that the eigenvalues λ_k of this equation are given by

$$\Gamma\lambda_k = \frac{2}{1 + u_k^2}$$

where the u_k are the roots of the transcendental equation

$$\tan(\Gamma T u_k) = \frac{2u_k}{u_k^2 - 1}$$

TWELVE

CALCULUS OF VARIATIONS

The fundamental problem of the calculus of variations is to find a function $y(x)$ such that a functional, or function of this function, has a stationary value for small variations in $y(x)$. We begin this chapter with a standard demonstration of how a simple variational problem leads to the Euler–Lagrange differential equation. Extensions to different types of variational problems are treated in Section 12–2 including problems involving auxiliary constraints on the allowed variations of the function $y(x)$.

Finally, in Section 12–3, we discuss some relations between the calculus of variations and eigenvalue problems involving differential equations, matrix equations, and integral equations.

12–1 EULER–LAGRANGE EQUATION

The basic problem of the calculus of variations will be illustrated by some examples. Consider a function

$$F\left(y, \frac{dy}{dx}, x\right) \tag{12-1}$$

and the integral

$$I = \int_a^b F\left(y, \frac{dy}{dx}, x\right) dx = I[y(x)] \qquad (12\text{-}2)$$

Such an integral is often called a *functional*. It is a generalization of a function in that it is a number which depends on a function rather than on another number.

We now wish to choose the function $y(x)$ so that the functional

$$I[y(x)]$$

is a maximum, or minimum, or (more generally) is stationary. That is, we want to find a $y(x)$ such that if we replace $y(x)$ by $y(x) + \xi(x)$, I is unchanged to order ξ, provided ξ is sufficiently small.

In order to reduce this problem to the familiar one of making an ordinary *function* stationary, consider the replacement

$$y(x) \to y(x) + \alpha\eta(x) \qquad (12\text{-}3)$$

where α is small and $\eta(x)$ arbitrary. If $I[y(x)]$ is to be stationary, then we must have

$$\left.\frac{dI}{d\alpha}\right|_{\alpha=0} = 0 \qquad (12\text{-}4)$$

for *all* $\eta(x)$.

Now

$$I(\alpha) = \int_a^b F(y + \alpha\eta, \, y' + \alpha\eta', \, x) \, dx$$

$$= I(0) + \alpha \int_a^b \left(\frac{\partial F}{\partial y}\eta + \frac{\partial F}{\partial y'}\eta'\right) dx + O\,(\alpha^2) \qquad (12\text{-}5)$$

Thus we require (for *all* η)

$$\int_a^b \left(\frac{\partial F}{\partial y}\eta + \frac{\partial F}{\partial y'}\eta'\right) dx = 0 \qquad (12\text{-}6)$$

In order to simplify the problem, we shall at first impose a further condition. We shall only allow variations which vanish at the end points a and b. That is, of all functions joining the two *fixed* points P and Q, we want to select that one which makes $I[y(x)]$ stationary (see Figure 12–1). In the notation (12-3) which we have been using, we require $\eta(a) = \eta(b) = 0$.

Now integrate the second term in (12-6) by parts, whereupon the equation becomes

$$0 = \eta \frac{\partial F}{\partial y'}\bigg|_a^b + \int_a^b \left(\frac{\partial F}{\partial y} - \frac{d}{dx}\frac{\partial F}{\partial y'}\right)\eta \, dx \qquad (12\text{-}7)$$

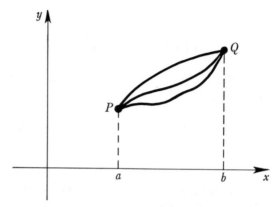

Figure 12–1   Various curves $y(x)$ joining the fixed end points P and Q

The integrated part vanishes because η vanishes at both ends.   Therefore, if the integral is to vanish for *arbitrary* $\eta(x)$, we must require

$$\frac{\partial F}{\partial y} - \frac{d}{dx}\frac{\partial F}{\partial y'} = 0 \tag{12-8}$$

The left-hand side of this equation is often written $\delta F/\delta y$, and called the *variational*, or *functional*, derivative of F with respect to y.[1]

This differential equation (12-8) is called the *Euler–Lagrange* equation. When combined with the appropriate boundary conditions, it is equivalent to the original variational problem.

Many laws of physics may thus be stated *either* in terms of a differential equation *or* as an equivalent variational principle.   An important example is well known in classical mechanics; Hamilton's (variational) principle,

$$\delta \int_{t_1}^{t_2} L\, dt = 0 \tag{12-9}$$

is equivalent to Lagrange's equations of motion,

$$\frac{\partial L}{\partial q_j} - \frac{d}{dt}\frac{\partial L}{\partial \dot{q}_j} = 0 \tag{12-10}$$

where the Lagrangian L is a function of the (generalized) coordinates q_i and velocities $\dot{q}_i$ and the time.   Lagrange's equations are, in turn, equivalent to Newton's.

In certain cases, we can integrate the Euler–Lagrange equation once. For example, suppose F does not depend on y.   Then the Euler–Lagrange

---

[1] Alternatively, one sometimes defines the expression on the left side of Eq. (12-8) as the variational derivative $\delta I/\delta y$ of the corresponding *functional I*.

equation (12-8) is

$$\frac{d}{dx}\frac{\partial F}{\partial y'} = 0 \qquad \text{so that} \qquad \frac{\partial F}{\partial y'} = \text{constant} \qquad (12\text{-}11)$$

As a second example, suppose F does not depend on x. Again, we can reduce the Euler–Lagrange equation to a first-order equation, as follows:

$$\frac{dF}{dx} = \underset{0}{\underbrace{\frac{\partial F}{\partial x}}} + y'\frac{\partial F}{\partial y} + y''\frac{\partial F}{\partial y'} \qquad (12\text{-}12)$$

The Euler–Lagrange equation, when multiplied by y', is

$$y'\frac{\partial F}{\partial y} - y'\frac{d}{dx}\frac{\partial F}{\partial y'} = 0$$

Adding and subtracting $y''(\partial F/\partial y')$, we obtain

$$\frac{d}{dx}\left(F - y'\frac{\partial F}{\partial y'}\right) = 0$$

which gives the first-order equation

$$F - y'\frac{\partial F}{\partial y'} = \text{constant} \qquad (12\text{-}13)$$

We consider an example.   Let A and B be two points in a vertical plane, A being higher than B. Along what curve connecting A and B will a particle slide (without friction) from A to B in the shortest time? This is a famous problem proposed by John Bernoulli. The curve is known as the *brachisto-chrone* (Greek for "shortest time").

An obvious simplification is to put A at the origin and measure y downward, as shown in Figure 12-2. We wish to minimize the time of descent.

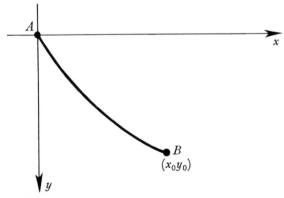

Figure 12–2   The brachistochrone

Now the particle speed is

$$v = \frac{ds}{dt}$$

where s denotes arc length along the path. Thus

$$dt = \frac{ds}{v}$$

and the total time is

$$t = \int_A^B \frac{ds}{v}$$

The element of path length is $ds = \sqrt{1 + y'^2}\, dx$, and, if the particle starts from rest at A, conservation of energy tells us that

$$v = \sqrt{2gy}$$

Therefore,

$$t = \int_0^{x_0} \sqrt{\frac{1 + y'^2}{2gy}}\, dx \tag{12-14}$$

This is a case in which the integrand does not contain x. Omitting the irrelevant factor $\sqrt{2g}$,

$$F = \sqrt{\frac{1 + y'^2}{y}}$$

and the equation [see (12-13)]

$$F - y' \frac{\partial F}{\partial y'} = \text{constant}$$

gives

$$y(1 + y'^2) = C$$

The expression $1 + y'^2$ may be simplified by the substitution

$$y' = \text{ctn}\, \theta$$

Then

$$y = \frac{C}{1 + y'^2} = C \sin^2 \theta = \frac{C}{2}(1 - \cos 2\theta)$$

and

$$\frac{dx}{d\theta} = \frac{1}{y'} \frac{dy}{d\theta} = C \tan \theta \sin 2\theta = C(1 - \cos 2\theta)$$

Integrating this equation,

$$x = \frac{C}{2}(2\theta - \sin 2\theta)$$

where $\theta = 0$ at the origin, $x = y = 0$. Letting $C = 2A$ and $2\theta = \phi$, we have

$$\begin{aligned} x &= A(\phi - \sin \phi) \\ y &= A(1 - \cos \phi) \end{aligned} \tag{12-15}$$

These are the parametric equations for a cycloid, the curve traced by a point on the rim of a wheel rolling on the x-axis.

12–2 GENERALIZATIONS OF THE BASIC PROBLEM

The following generalization of our basic problem is quite trivial. Suppose we wish to maximize the functional

$$I[y(x), z(x)] = \int_a^b F(y, y', z, z', x)\, dx \tag{12-16}$$

subject to fixed end-point conditions:

$$\begin{aligned} y(a) &= y_1 & y(b) &= y_2 \\ z(a) &= z_1 & z(b) &= z_2 \end{aligned}$$

We just write down an Euler–Lagrange equation for each dependent variable separately:

$$\begin{aligned} \frac{\delta F}{\delta y} &= \frac{\partial F}{\partial y} - \frac{d}{dx}\frac{\partial F}{\partial y'} = 0 \\ \frac{\delta F}{\delta z} &= \frac{\partial F}{\partial z} - \frac{d}{dx}\frac{\partial F}{\partial z'} = 0 \end{aligned} \tag{12-17}$$

A second simple generalization is the inclusion of higher derivatives. Suppose we want to maximize

$$I[y(x)] = \int_a^b F(y, y', y'', x)\, dx \tag{12-18}$$

with y *and* y' held fixed at a and b. The condition is easily shown to be

$$\frac{\delta F}{\delta y} = \frac{\partial F}{\partial y} - \frac{d}{dx}\frac{\partial F}{\partial y'} + \frac{d^2}{dx^2}\frac{\partial F}{\partial y''} = 0 \tag{12-19}$$

Note that we have generalized our definition of a variational derivative.

Next, we consider the case of more than one independent variable. Consider the functional

$$I[f(x, y)] = \iint_G F(f, f_x, f_y, x, y)\, dx\, dy \tag{12-20}$$

where G is some region of the xy-plane, $f_x = \partial f/\partial x$, and $f_y = \partial f/\partial y$. The function f is to be varied in such a way that its values on the boundary of G remain constant. Then the condition for $I[f]$ to be stationary is

$$\frac{\delta F}{\delta f} = \frac{\partial F}{\partial f} - \frac{d}{dx}\frac{\partial F}{\partial f_x} - \frac{d}{dy}\frac{\partial F}{\partial f_y} = 0 \tag{12-21}$$

Again we have extended the definition of variational derivative.

Some care is necessary in interpreting the various derivatives in Eq. (12-21). In calculating $\partial F/\partial f$, $\partial F/\partial f_x$, and so forth, in Eq. (12-21) all the explicit variables in F are considered independent, that is, f, f_x, f_y, x, and y. For the derivative d/dx in $(d/dx)(\partial F/\partial f_x)$, x and y are the independent variables, and y is held constant in the d/dx differentiation. Thus $[df(x, y)]/dx$ means $\partial f/\partial x$ and the Euler–Lagrange equation (12-21) is a conventional partial differential equation for $f(x, y)$.

We saw in the brachistochrone example that it was convenient to express the solution parametrically, that is, in the form

$$x = x(t) \qquad y = y(t)$$

where t is the parameter. We may thus think of the problem as the determination of these two functions, rather than the single function $y(x)$. If we think of x and y as being functions of t, the functional

$$I = \int_a^b F(y, y', x)\, dx \tag{12-22}$$

becomes

$$I = \int_{t_0}^{t_1} F\left(y, \frac{\dot{y}}{\dot{x}}, x\right)\dot{x}\, dt$$

$$= \int_{t_0}^{t_1} \mathscr{F}(x, y, \dot{x}, \dot{y})\, dt \tag{12-23}$$

where $\dot{x} = dx/dt$, $\dot{y} = dy/dt$, $\mathscr{F}(x, y, \dot{x}, \dot{y}) = \dot{x}F(y, \dot{y}/\dot{x}, x)$. The Euler equations are then

$$\frac{\partial \mathscr{F}}{\partial x} - \frac{d}{dt}\frac{\partial \mathscr{F}}{\partial \dot{x}} = 0$$

$$\frac{\partial \mathscr{F}}{\partial y} - \frac{d}{dt}\frac{\partial \mathscr{F}}{\partial \dot{y}} = 0 \tag{12-24}$$

Equations (12-24) turn out not to be independent, that is, one of them implies the other.   In fact, the student can show that

$$\dot{x}\left(\frac{\partial \mathscr{F}}{\partial x} - \frac{d}{dt}\frac{\partial \mathscr{F}}{\partial \dot{x}}\right) + \dot{y}\left(\frac{\partial \mathscr{F}}{\partial y} - \frac{d}{dt}\frac{\partial \mathscr{F}}{\partial \dot{y}}\right) = 0 \tag{12-25}$$

We now consider a very important generalization, to *variable end points*. To begin with, suppose we want to maximize the functional

$$I = \int_a^b F(y, y', x)\, dx$$

but we allow $y(b)$ to be arbitrary.   As before, if y is given an increment $\eta(x)$, the change in I is

$$\delta I = \int_a^b \left(\frac{\partial F}{\partial y}\eta + \frac{\partial F}{\partial y'}\eta'\right) dx$$

$$= \eta\frac{\partial F}{\partial y'}\bigg|_a^b + \int_a^b \left(\frac{\partial F}{\partial y} - \frac{d}{dx}\frac{\partial F}{\partial y'}\right)\eta\, dx \tag{12-26}$$

Clearly, the Euler–Lagrange equation must still hold; otherwise, we could find a variation η with $\eta(b) = 0$ which changed I.   But in addition, if δI is to vanish for arbitrary $\eta(b)$, we must have

$$\frac{\partial F}{\partial y'}\bigg|_{x=b} = 0 \tag{12-27}$$

If both end points are free, clearly $\partial F/\partial y'$ must vanish at both ends.

As another possibility, suppose y is fixed at $x = a$, but the other end point is free to lie anywhere on the curve

$$g(x, y) = 0$$

Now

$$\delta I = \frac{\partial F}{\partial y'}\eta\bigg|_a^b + \int_a^b \left(\frac{\partial F}{\partial y} - \frac{d}{dx}\frac{\partial F}{\partial y'}\right)\eta\, dx + F(b)\,\Delta x \tag{12-28}$$

where Δx is the displacement of the upper end point, as indicated in Figure 12–3.   Note that $\Delta y = \eta(b) + y'(b)\Delta x$.   The requirement $\Delta g = 0$ at the end of the curve gives

$$\frac{\partial g}{\partial x}\Delta x + \frac{\partial g}{\partial y}\Delta y = 0$$

$$\frac{\partial g}{\partial x}\Delta x + \frac{\partial g}{\partial y}(\eta + y'\,\Delta x) = 0$$

or

$$\left(\frac{\partial g}{\partial x} + y'\frac{\partial g}{\partial y}\right)\Delta x + \frac{\partial g}{\partial y}\eta = 0$$

Also, according to (12-28) the condition $\delta I = 0$ gives, besides the Euler–Lagrange equation, the end condition

$$F\Delta x + \frac{\partial F}{\partial y'}\eta = 0 \qquad \text{at} \qquad x = b \qquad (12\text{-}29)$$

Eliminating Δx and η from Eqs. (12-28) and (12-29) leads to the condition we are after, namely,

$$\left(F - y'\frac{\partial F}{\partial y'}\right)\frac{\partial g}{\partial y} - \frac{\partial F}{\partial y'}\frac{\partial g}{\partial x} = 0 \qquad (12\text{-}30)$$

where everything is evaluated at the end point in question.

As an example of the use of this condition, consider the problem of finding the curve along which a particle will descend most quickly from a given point *to a given curve*, rather than to a given point. As in (12-14),

$$F = \sqrt{\frac{1 + y'^2}{y}}$$

so that the condition (12-30) at the variable end point is

$$\frac{1}{\sqrt{y(1 + y'^2)}}\frac{\partial g}{\partial y} - \frac{y'}{\sqrt{y(1 + y'^2)}}\frac{\partial g}{\partial x} = 0$$

or

$$y' = \left(\frac{\partial g}{\partial y}\right)\left(\frac{\partial g}{\partial x}\right)^{-1} \qquad (12\text{-}31)$$

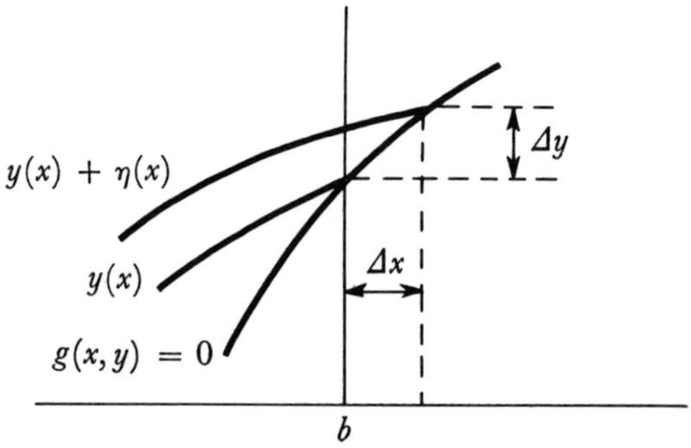

Figure 12–3   Variation of the end point b along the curve $g(x, y) = 0$

This just tells us the intuitively obvious fact that the curve of quickest descent must intersect the "destination" curve at right angles. The student may show that this transversality of $y(x)$ and $g(x)$ results whenever the integrand $F(x, y, y')$ is of the form $f(x, y)\sqrt{1 + y'^2}$.

We shall consider one final generalization of our basic problem, that of maximizing *one* functional

$$I[y(x)] = \int_a^b F(y, y', x)\, dx$$

subject to the condition, or constraint, that *another* functional

$$J[y(x)] = \int_a^b G(y, y', x)\, dx$$

be held constant. These are sometimes called *isoperimetric problems* because the classic example is the problem of finding a curve of fixed length which encloses a maximum area. Another example is to find the curve $y(x)$ of given length L which maximizes the volume obtained by rotating $y(x)$ about the x-axis.

The conventional technique for handling problems like this involves the use of *Lagrange multipliers*. Let us review the elementary use of Lagrange multipliers. Suppose we want to maximize a function of two variables $f(x, y)$. We must satisfy the conditions

$$f_x = 0 \qquad f_y = 0$$

Now suppose we want to maximize $f(x, y)$, *subject to the condition*

$$g(x, y) = \text{constant}$$

To begin with, $df = f_x\, dx + f_y\, dy = 0$. If dx and dy were independent, we could conclude $f_x = f_y = 0$. However, they are not independent, but are constrained by the condition

$$dg = g_x\, dx + g_y\, dy = 0$$

Therefore,

$$f_x/g_x = f_y/g_y$$

If we call the common ratio λ, we have

$$f_x - \lambda g_x = 0 \qquad f_y - \lambda g_y = 0 \qquad\qquad (12\text{-}32)$$

These are just the equations we would get if we tried to maximize the function $f - \lambda g$ ($\lambda = $ constant) without the constraint. λ is called a Lagrange multiplier. The solutions will clearly depend on λ; λ is adjusted so that $g(x, y)$ takes on the correct value.

The result has a simple geometrical meaning.   Consider a contour map of $f(x, y)$, as shown in Figure 12–4.   Also shown is the line $g(x, y) = $ constant. If there were no constraint, the solution for the maximum of f would clearly be the point A (the "top of the mountain").   With the constraint that we must stay on the curve $g(x, y) = $ constant, the solution clearly is the point B, where the curves $g \doteq$ constant and $f = $ constant are parallel.   This is just the result (12-32) deduced above.

The same technique works for an arbitrary number of constraints (of course, there must be fewer constraints than independent variables!).   If we want to maximize $f(x, y, z, \ldots)$ subject to

$$g_1(x, y, z, \ldots) = \text{constant}$$

$$g_2(x, y, z, \ldots) = \text{constant} \qquad (12\text{-}33)$$

$$\cdot \quad \cdot \quad \cdot \quad \cdot \quad \cdot \quad \cdot \quad \cdot$$

just maximize the function $f - \lambda_1 g_1 - \lambda_2 g_2 - \cdots$ where the λ's are undetermined (constant) Lagrange multipliers.

This method also works for *nonholonomic*, that is, nonintegrable constraints. Suppose we have the constraint on our variations

$$A(x, y, \ldots) \, dx + B(x, y, \ldots) \, dy + C(x, y, \ldots) \, dz + \cdots = 0 \quad (12\text{-}34)$$

which cannot be integrated (or we just do not know how).   Then $f(x, y, z, \ldots)$ is maximized by solving

$$f_x = \lambda A$$
$$f_y = \lambda B$$
$$\text{etc.}$$

Now we return to our calculus of variations problem.   It is straightforward to show that the way to maximize $\int F \, dx$, subject to the constraint

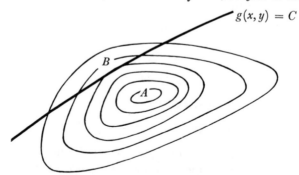

Figure 12–4   **Level contours for a function** $f(x, y)$ **and a curve** $g(x, y) = C$

$\int G\,dx$ = constant, is to solve the Euler–Lagrange equation for $F + \lambda G$, where λ is an undetermined (constant) Lagrange multiplier.

There are several ways to show this. One argument proceeds by considering the variations in the two functionals resulting from a change δy in $y(x)$.

$$\delta I = \int \frac{\delta F}{\delta y}\, \delta y\, dx$$

$$\delta J = \int \frac{\delta G}{\delta y}\, \delta y\, dx \qquad (12\text{-}35)$$

Now we require that for all δy such that $\delta J = 0$, δI also vanish. It is clear that this is only possible if the ratio of $\delta F/\delta y$ and $\delta G/\delta y$ is a constant, independent of x. This is equivalent to making

$$\int (F + \lambda G)\, dx \qquad (12\text{-}36)$$

stationary.

Another way of obtaining the above result is by using the following arguments.

1. We wish to find $y(x)$ which makes $I[y(x)] = \int_{x_0}^{x_1} F(x, y, y')\, dx$ a maximum under variations $\delta y(x)$ restricted by the condition

$$J[y(x)] = \int_{x_0}^{x_1} G(x, y, y')\, dx = J_0 = \text{constant}$$

2. An exactly equivalent problem is to find $y(x)$ which makes $I[y(x)] + \lambda J[y(x)]$ a maximum under these restricted variations, where λ is any constant.

3. The function $y(x, \lambda)$ which makes $I + \lambda J$ a maximum for *arbitrary* variations δy is found from the usual Euler–Lagrange equation for the integrand $F + \lambda G$.

4. If a value λ_1 is found, for which this solution $y(x, \lambda_1)$ satisfies the condition $J[y(x, \lambda_1)] = J_0$, this $y(x)$ also maximizes $I + \lambda J$ under the restricted variations and is the solution to our problem.

12-3 CONNECTIONS BETWEEN EIGENVALUE PROBLEMS AND THE CALCULUS OF VARIATIONS

Suppose we want to solve a particular differential equation. It often turns out that we can invent a functional whose Euler–Lagrange equation is just the differential equation in question. For example, consider Laplace's equation

$$\nabla^2 \phi = 0$$

in a volume V, with ϕ given on the boundary of V. Solving this equation is equivalent to minimizing the functional

$$I[\phi(x)] = \tfrac{1}{2} \int_V (\nabla\phi)^2 \, d^3x \qquad (12\text{-}37)$$

as we may easily verify. The integrand of (12-37) is

$$F = \frac{1}{2}(\nabla\phi)^2 = \frac{1}{2}\left[\left(\frac{\partial\phi}{\partial x}\right)^2 + \left(\frac{\partial\phi}{\partial y}\right)^2 + \left(\frac{\partial\phi}{\partial z}\right)^2 \right]$$

so that

$$\frac{\partial F}{\partial\phi} = 0 \qquad \frac{\partial F}{\partial\left(\dfrac{\partial\phi}{\partial x}\right)} = \frac{\partial\phi}{\partial x} \qquad \text{etc.}$$

and the Euler–Lagrange equation is

$$\sum_{xyz} \frac{d}{dx} \frac{\partial F}{\partial\left(\dfrac{\partial\phi}{\partial x}\right)} = \nabla^2\phi = 0$$

As another example, the Sturm–Liouville equation (9-9)

$$\frac{d}{dx}\left[p(x)\frac{du}{dx} \right] - q(x)u(x) + \lambda\rho(x)u(x) = 0 \qquad (a < x < b) \qquad (12\text{-}38)$$

is the Euler–Lagrange equation for the functional

$$I[u(x)] = \int_a^b [pu'^2 + (q - \lambda\rho)u^2] \, dx \qquad (12\text{-}39)$$

This is not the best form, however, since λ is generally unknown. We may alternatively minimize

$$I[u(x)] = \int_a^b (pu'^2 + qu^2) \, dx \qquad (12\text{-}40)$$

subject to the condition

$$J[u(x)] = \int_a^b \rho u^2 \, dx = \text{constant} \qquad (12\text{-}41)$$

The eigenvalue λ then enters as a Lagrange multiplier. Since (12-41) is just a normalization condition for $u(x)$, it is clear that this last variational procedure is equivalent to minimizing

$$K[u(x)] = \frac{I[u(x)]}{J[u(x)]} \qquad (12\text{-}42)$$

Furthermore, if we multiply Eq. (12-38) by $u(x)$, integrate from a to b, and perform an integration by parts, we see that the stationary value of $K[u]$ is in fact just equal to the eigenvalue λ.

It is also possible to show (see, for example, Morse and Feshbach (M9) Section 6.3) that the eigenvalues λ_n of a Sturm–Liouville differential equation have the following properties:

1. There is a smallest eigenvalue λ_0.
2. As $n \to \infty$, $\lambda_n \to +\infty$.
3. More precisely, as $n \to \infty$, $\lambda_n \sim$ const. $\times\ n^2$.

Now, according to (12-42) and the subsequent remark, the stationary values of the functional

$$K[u(x)] = \frac{\int_a^b (pu'^2 + qu^2)\, dx}{\int_a^b \rho u^2\, dx} \tag{12-43}$$

are the eigenvalues λ_n. Thus the *absolute minimum* of K is the lowest eigenvalue λ_0. This is a very powerful method for the calculation of the lowest eigenvalue. The method is readily generalized to problems in several dimensions.

Note that a trial function $u(x)$ which is good "to first order," when inserted in (12-43), yields an approximate eigenvalue which is good "to second order." For let

$$u = u_0 + c_1 u_1 + c_2 u_2 + \cdots \tag{12-44}$$

where the u_i are the true eigenfunctions and the c_i are small. If the u_i are normalized, so is u except for second-order terms. Then our estimate for λ_0 is

$$\int_a^b [p(u_0' + c_1 u_1' + \cdots)^2 + q(u_0 + c_1 u_1 + \cdots)^2]\, dx$$

But

$$\int_a^b (pu_i'^2 + qu_i^2)\, dx = \lambda_i$$

and

$$\int_a^b (pu_i' u_j' + qu_i u_j)\, dx = 0$$

for $i \neq j$ (prove it!). Thus our estimate for λ_0 is

$$(\lambda_0 + c_1^2 \lambda_1 + c_2^2 \lambda_2 + \cdots)[1 + O\,(c^2)] \tag{12-45}$$

which differs from λ_0 by second-order terms, as stated. It is interesting to evaluate the second-order terms, for which we need also

$$\int_a^b \rho u^2 \, dx = 1 + c_1^2 + c_2^2 + \cdots$$

The estimate (12-43) is then

$$K = \frac{\lambda_0 + c_1^2 \lambda_1 + c_2^2 \lambda_2 + \cdots}{1 + c_1^2 + c_2^2 + \cdots}$$

$$K = \lambda_0 + c_1^2(\lambda_1 - \lambda_0) + c_2^2(\lambda_2 - \lambda_0) + \cdots$$

This result displays the typical features of a variational calculation, that K cannot be less than λ_0 and is equal to λ_0 only if all the $c_i = 0$; that is, $u = u_0$.

EXAMPLE

Let us estimate the lowest eigenvalue λ of

$$\frac{d^2 u}{dx^2} + \lambda u = 0 \qquad u = 0 \text{ at } x = 0 \text{ and } x = 1 \qquad (12\text{-}46)$$

The correct answer, of course, is $\lambda = \pi^2$.
    Consider the trial function

$$u = x(1 - x)$$

Then, by our variational principle,

$$\lambda \leqslant \frac{\int_0^1 u'^2 \, dx}{\int_0^1 u^2 \, dx} = \frac{\frac{1}{3}}{\frac{1}{30}} = 10$$

which is pretty close to (but larger than) π^2. We could improve our estimate by setting

$$u = x(1 - x)(1 + c_1 x + \cdots)$$

and varying the parameters c_i to minimize K.

    The crux of the method is the choice of a trial function. A good trial function should satisfy three conditions:
    1. It should (in fact, *must*) satisfy the boundary conditions.
    2. It should look qualitatively like the expected solution.
    3. It should lead to reasonably simple calculations in the evaluation of the functional $K[u(x)]$.

EXAMPLE

As a second example, let us estimate the lowest frequency of a circular drum head of radius a.

$$\nabla^2 u + k^2 u = 0 \qquad k^2 = \frac{\omega^2}{c^2} \tag{12-47}$$

$$k^2 \leqslant \frac{\int (\nabla u)^2 dx\, dy}{\int u^2\, dx\, dy}$$

Try

$$u = 1 - \frac{r}{a}$$

Then our variational estimate is

$$k^2 \leqslant \frac{\pi}{\pi a^2/6} = \frac{6}{a^2}$$

The correct lowest eigenvalue is $k^2 = (2.405\ldots/a)^2 \approx 5.78/a^2$.

EXAMPLE

Finally, we consider as a third example the so-called *Rayleigh–Ritz variational method* for determining the lowest energy eigenvalue E of the Schrödinger equation

$$H\psi = E\psi \tag{12-48}$$

where H is the (self-adjoint) Hamiltonian, or energy operator. The associated variational problem is the minimizing of the functional[2]

$$\frac{\psi \cdot H\psi}{\psi \cdot \psi} \tag{12-49}$$

For example, let us try to find the ground state (lowest energy) of a quantum-mechanical harmonic oscillator; the Schrödinger equation is

$$H\psi = -\frac{d^2\psi}{dx^2} + x^2\psi = E\psi \qquad (-\infty < x < \infty) \tag{12-50}$$

As a trial function we shall take

$$\psi = (1 + \alpha x^2)e^{-x^2}$$

---

[2] It is left to the student to verify the equivalence of minimizing the functionals (12-42) and (12-49).

Then

$$E \leqslant \frac{\psi \cdot H\psi}{\psi \cdot \psi} = \frac{\dfrac{5}{4} - \dfrac{\alpha}{8} + \dfrac{43\alpha^2}{64}}{1 + \dfrac{\alpha}{2} + \dfrac{3\alpha^2}{16}}$$

The minimum is found by solving the quadratic equation

$$23\alpha^2 + 56\alpha - 48 = 0$$

$$\alpha = 0.6718$$

This gives $E < 1.034$ (correct answer: $E = 1$).   Note that this method assumes H is a Hermitian operator (see Section 9–2).

All we have discussed so far concerns the lowest eigenvalue only.   How can we get a variational principle, for example, for the next-to-the-lowest eigenvalue?   That's easy; just carry out the same procedure, but restrict the trial functions by requiring them to be orthogonal to the lowest eigenfunction. The generalization to higher eigenfunctions is straightforward.

Matrices also have eigenvalues.   What is the variational approach to these?   The analogy is immediate.   Suppose we maximize the function

$$I(x) = \frac{x \cdot Mx}{x \cdot x} = \frac{\sum_{ij} M_{ij} x_i x_j}{\sum_i x_i^2} \tag{12-51}$$

or, equivalently, maximize

$$\sum_{ij} M_{ij} x_i x_j$$

subject to

$$\sum_i x_i^2 = \text{constant}$$

The resulting equations may be written

$$Mx = \lambda x \tag{12-52}$$

where again the eigenvalue λ enters as a Lagrange multiplier.

The variational approach enables one to demonstrate the completeness of the eigenfunctions of a Sturm–Liouville equation.   Suppose we have a function $f(x)$ which we wish to represent as a linear combination of the eigenfunctions $u_i(x)$ of the Sturm–Liouville equation

$$(pu')' - qu + \lambda\rho u = 0 \qquad (a < x < b)$$

The functions u_i obey the orthonormality condition

$$\int_a^b u_i u_j \rho \, dx = \delta_{ij}$$

Then one derives heuristically

$$f(x) = \sum_i c_i u_i(x) \qquad c_i = \int_a^b u_i f \rho \, dx \qquad (12\text{-}53)$$

and we wish to consider the validity of this expansion.

Let

$$\Delta_n(x) = f(x) - \sum_{i=1}^n c_i u_i(x) \qquad (12\text{-}54)$$

and

$$\delta_n^2 = \int_a^b \Delta_n^2 \rho \, dx \qquad (12\text{-}55)$$

The function $\psi_n(x) = \Delta_n(x)/\delta_n$ is clearly a normalized function which is orthogonal to $u_1, u_2, \ldots, u_n$. Now the minimum of the functional

$$I[\psi(x)] = \int_a^b (\rho\psi'^2 + q\psi^2) \, dx$$

subject to the condition that ψ is normalized and orthogonal to the first n eigenfunctions $u_i(x)$ is λ_{n+1}. Thus

$$I[\psi_n(x)] \geqslant \lambda_{n+1}$$

or

$$\frac{I[\Delta_n(x)]}{\delta_n^2} \geqslant \lambda_{n+1} \qquad (12\text{-}56)$$

But it is easily seen that $I[\Delta_n(x)]$ is bounded as n goes to infinity. First we observe that, for any function $g(x)$ obeying the boundary conditions, a simple integration by parts gives

$$I[g(x)] = \int_a^b g(x)[Lg(x)] \, dx$$

where L is the Sturm–Liouville differential operator

$$L = -\frac{d}{dx} p \frac{d}{dx} + q$$

Then, using the definition (12-54) of $\Delta_n(x)$

$$I[\Delta_n] = \int_a^b f[Lf] \, dx - 2\sum_{i=1}^n c_i \int_a^b f[Lu_i] \, dx + \sum_{i=1}^n \sum_{j=1}^n c_i c_j \int_a^b u_i[Lu_j] \, dx$$

$$= I[f] - \sum_{i=1}^n \lambda_i c_i^2$$

Since $I[\Delta_n]$ is clearly positive, as are the λ_i for i sufficiently large, the boundedness of $I[\Delta_n]$ is apparent.

It now follows from (12-56) and the fact (stated on p. 335) that $\lambda_n \to \infty$ as $n \to \infty$, that

$$\lim_{n \to \infty} \delta_n^2 = 0$$

that is,

$$\lim_{n \to \infty} \int_a^b \left[f(x) - \sum_i c_i u_i(x) \right]^2 \rho \, dx = 0 \tag{12-57}$$

One usually describes this result by saying that the series

$$\sum_{n=1}^{\infty} c_n u_n(x) \tag{12-58}$$

converges "in the mean" to $f(x)$. For all physical purposes, this is just as good as ordinary ("point-by-point") convergence.

We shall conclude this chapter by indicating how variational principles can also be applied to *integral* equations. The problem of finding the maximum (or minimum) of the functional

$$I[\phi(x)] = \int_a^b \int_a^b K(x, y)\phi(x)\phi(y) \, dx \, dy \qquad [K(x, y) = K(y, x)] \tag{12-59}$$

subject to the normalization condition

$$\int_a^b \phi^2(x) \, dx = \text{constant} \tag{12-60}$$

gives immediately

$$\lambda \int_a^b K(x, y)\phi(y) \, dy = \phi(x) \tag{12-61}$$
$$(\text{proof will be left to the student})$$

Equation (12-61) is just a linear integral equation in standard form. Notice that for a function $\phi(x)$ obeying this integral equation, the functional

$$\frac{I[\phi(x)]}{\int_a^b \phi^2(x) \, dx} \tag{12-62}$$

equals $1/\lambda$. Thus, the eigenvalues of a symmetric kernel may be characterized variationally, just like the eigenvalues of a Sturm–Liouville differential equation.

REFERENCES

A long, clear, and useful discussion of the calculus of variations is in Courant and Hilbert (C10) Vol. 1, Chapter IV. An exhaustive treatment is also given by Forsyth (F4). A more recent text by Akhiezer (A3) should also be quite useful.

Briefer treatments may be found in Jeffreys and Jeffreys (J4) Chapter 10; Wilf (W7) Chapter 7; Margenau and Murphy (M2) Chapter 6; and Irving and Mullineux (I3) Chapter VII.

The usefulness of variational methods in quantum mechanics is discussed in any of the standard texts, such as Schiff (S2) Section 32 and Landau and Lifshitz (L2) Section 20.

Finally, Morse and Feshbach (M9) (particularly Chapter 3 and Sections 9.4 and 11.4) contains discussions of numerous applications of variational techniques to physical problems.

PROBLEMS

12-1 A uniform string of length 2 meters hangs from two supports at the same height, 1 meter apart. By minimizing the potential energy of the string, find the equation describing the curve it forms and, in particular, find the vertical distance between the supports and the lowest point on the string.

12-2 An object of mass m is in the two-dimensional force field

$$\mathbf{F} = -\frac{GMm}{r^2}\,\hat{\mathbf{r}}$$

Curves along which the mass falls between given points in the shortest time are solutions of a differential equation of the form

$$\frac{dr}{d\theta} = f(r)$$

Find the function $f(r)$.

12-3 Fermat's principle of least time states that a ray of light between two given points travels along the path which it can traverse in the shortest time. Consider a region in which the index of refraction varies linearly with height:

$$n = n_0(1 + \alpha z)$$

By what angle θ is a point P apparently lowered when viewed from a point P' at the same height ($z = 0$) as P at a horizontal distance d?

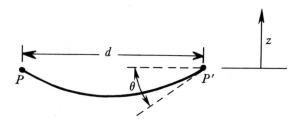

In order to get an explicit answer, you may assume $\alpha d \ll 1$.

12-4 Verify (12-25), and state clearly why it could have been written down immediately without calculation, that is, why it is obviously true.

12-5 Verify the second sentence after (12-31).

12-6 A real field $\phi(\mathbf{x}, t)$ obeys the variational principle

$$\delta \int L\, d^3x\, dt = 0$$

where

$$L(\mathbf{x}, t) = \tfrac{1}{2}(\dot{\phi}^2 - |\nabla\phi|^2 - \mu^2\phi^2)$$

Give the differential equation of motion obeyed by $\phi(\mathbf{x}, t)$.

12-7 By varying a suitable trial function, obtain an estimate for the lowest-energy eigenvalue of the one-dimensional Schrödinger equation for a particle of mass m in the potential

$$V(x) = kx^4 \qquad (-\infty < x < \infty)$$

12-8 Obtain an improvement on the estimate obtained in the text for the lowest natural frequency of a circular drum head of radius a by trying a solution of the form

$$u = 1 - \left(\frac{r}{a}\right)^n$$

and varying n. Express your result as an approximation for the lowest zero of $J_0(x)$.

12-9 Find the minimum value of the quadratic form

$$x^2 + 2y^2 + 3z^2 + 2xy + 2xz$$

subject to the condition

$$x^2 + y^2 + z^2 = 1$$

12-10 By using the simplest possible continuous polynomial trial function, obtain an upper bound for the lowest eigenvalue k_0^2 of

$$\nabla^2\phi + k^2\phi = 0$$

for the two-dimensional region below. $\phi = 0$ on all boundaries. Use a separate polynomial expression for the three numbered subregions; no polynomial should be higher than third degree.

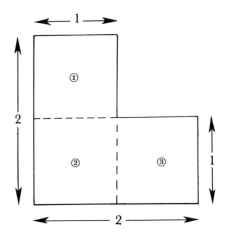

12-11 Obtain an upper limit for the lowest eigenvalue λ_0 of

$$\nabla^2 u + \lambda r^2 u = 0 \qquad (\text{3 dimensions, } 0 < r < a, u = 0 \text{ on } r = a)$$

12-12 Consider the eigenvalue problem

$$\nabla^2 u + k^2 u = 0 \qquad (\text{two dimensions})$$

inside an equilateral triangle of side a.   The boundary condition is that $u = 0$ on the triangle boundary.

(a) Construct a third-degree polynomial that satisfies the boundary condition, and thereby obtain an *upper* limit for the lowest eigenvalue k_0^2.

(b) It can be shown that for a given area, a circle gives the lowest eigenvalue for this equation.   Use this result to obtain a *lower* limit for k_0^2 in our triangle problem.

(c) Improve the estimate of (b) by considering a square instead of a circle.

12-13 A curve $y(x)$ of length $2a$ is drawn between the points $(0, 0)$ and $(a, 0)$ in such a way that the solid obtained by rotating the curve about the x-axis has the largest possible volume.   Find $y(\tfrac{1}{2}a)$.

12-14 Two circles of unit radius, each normal to the line through their centers, are a distance d apart.   A soap film is formed between them, as shown below; energetic considerations require the film to assume a shape of minimum surface area.

(a) Show that for any value of d less than a critical value d_c, there are *two* surfaces satisfying the appropriate Euler–Lagrange equation, while for $d > d_c$ there is no such surface. Evaluate d_c.
For a typical $d < d_c$, sketch the two surfaces. Which has the lesser area?

(c) Show that in a certain range $d_0 < d < d_c$, the minimum surface as given by the Euler–Lagrange equation has an area larger than 2π, so that the surface is "metastable"; that is, the surface is stable against small perturbations but not against *arbitrary* perturbation. Evaluate d_0.

(d) What happens when d is increased beyond d_c?

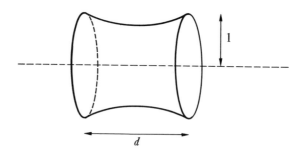

12-15 A string has length l, tension T, and mass $\rho(x)$ per unit length. Give a variational estimate of ω_0^2, the square of the lowest frequency. Give a general result; do not plug in a specific trial function.

12-16 By using as a trial function a polynomial of degree two, find an upper bound for the lowest eigenvalue λ_0 of the differential equation

$$\frac{d^2y}{dx^2} + \lambda xy = 0 \qquad y(0) = 0 \qquad y(1) = 0$$

THIRTEEN

NUMERICAL METHODS

In this chapter we present some methods for numerically carrying out the processes of interpolation, integration, solution of (ordinary) differential equations, finding roots of functions, and summing series.

13–1 INTERPOLATION

Suppose we have a table of values of some function $f(x)$ (Table 13–1).

Table 13–1

x	$\sin x$
0	0.00000
0.1	0.09983
0.2	0.19867
0.3	0.29552

How do we go about finding $\sin x$ for values of x not in the table, like $x = 0.1432$? We may think of the problem in graphical terms as that of drawing a smooth curve through the four given points and then reading off values for intermediate values of x.

For analytical convenience, the "smooth curve" is usually assumed to be a

345

polynomial.   That is, we arbitrarily set

$$f(x) = A + Bx + Cx^2 + Dx^3 + \cdots \tag{13-1}$$

and use the given points to evaluate the coefficients A, B, C, ... In our case we are given four points, so we could fix the coefficients through D, if we assume all the rest are zero.

Straightforward evaluation of coefficients in the manner outlined above would be rather tedious; we would have to solve four simultaneous linear equations just to do the interpolation suggested above.   Many shortcuts have been devised, and we shall sketch a couple of approaches to the problem.

We shall suppose we are given a table of values for $y = f(x)$ (Table 13–2), in which the values of x are equally spaced (as they actually are in most tables).   That is,

$$x_2 = x_1 + h \qquad x_3 = x_2 + h = x_1 + 2h \qquad \text{etc.} \tag{13-2}$$

Now define an operator E by $Ey_n = y_{n+1}$, and an operator Δ by $\Delta = E - 1$. Thus

$$\Delta y_n = (E - 1)y_n = y_{n+1} - y_n \tag{13-3}$$

$$\Delta^2 y_n = \Delta y_{n+1} - \Delta y_n = (y_{n+2} - y_{n+1}) - (y_{n+1} - y_n) \tag{13-4}$$

Note that

$$y_2 = y(x_1 + h) = Ey_1$$

$$y_3 = y(x_1 + 2h) = E^2 y_1$$

and, in general,

$$y_{n+1} = y(x_1 + nh) = E^n y_1 \tag{13-5}$$

We obtain an interpolation formula if we formally let n be an arbitrary number, not necessarily an integer, in the last equation above.   That is, set

$$y(x_1 + \alpha h) = E^\alpha y_1$$

$$= (1 + \Delta)^\alpha y_1$$

$$= \left[1 + \alpha\Delta + \frac{\alpha(\alpha - 1)}{2!} \Delta^2 + \cdots \right] y_1 \tag{13-6}$$

Table 13–2

x_1	y_1
x_2	y_2
x_3	y_3
$\vdots$	$\vdots$

One generally keeps only the first few terms in this infinite series; this then gives an approximation to $y(x)$ which is just the same as the polynomial approximation mentioned at the beginning of this section. If, for example, we neglect third and higher differences, we obtain

$$y(x_1 + \alpha h) \approx \left[1 + \alpha \Delta + \frac{\alpha(\alpha - 1)}{2!} \Delta^2\right] y_1$$

$$\approx \left[1 + \alpha(E - 1) + \frac{\alpha(\alpha - 1)}{2!} (E - 1)^2\right] y_1$$

$$\approx \left[\frac{(\alpha - 1)(\alpha - 2)}{2} + \alpha(2 - \alpha)E + \frac{\alpha(\alpha - 1)}{2} E^2\right] y_1$$

$$\approx \frac{(\alpha - 1)(\alpha - 2)}{2} y_1 + \alpha(2 - \alpha)y_2 + \frac{\alpha(\alpha - 1)}{2} y_3 \qquad (13\text{-}7)$$

If one is using a table with printed first and second differences, the first form is probably most convenient; otherwise the last form is more useful.

The last form, when examined closely, is seen to have a structure which, if one were clever, one ought to have written down immediately, by inspection. Since we assumed third and higher differences vanished, y must be quadratic (in α). Furthermore, when $x = x_2$ ($\alpha = 1$), $y = y_2$, and when $x = x_3$ ($\alpha = 2$), $y = y_3$. In each case there is no y_1 in the answer, and so the coefficient of y_1 must be

$$\text{constant} \times (\alpha - 1)(\alpha - 2)$$

The constant is $\frac{1}{2}$ in order that $y = y_1$ when $x = x_1$ ($\alpha = 0$). The other coefficients in the last line of (13-7) may be found by similar arguments.

As another example of this reasoning, we may write down immediately the corresponding formula keeping third differences,

$$y = \frac{(\alpha - 1)(\alpha - 2)(\alpha - 3)}{-6} y_1 + \frac{\alpha(\alpha - 2)(\alpha - 3)}{2} y_2$$

$$+ \frac{\alpha(\alpha - 1)(\alpha - 3)}{-2} y_3 + \frac{\alpha(\alpha - 1)(\alpha - 2)}{6} y_4 \qquad (13\text{-}8)$$

The coefficients in the formulas (13-7) and (13-8) are known as Lagrangian interpolation coefficients; large tables[1] of them exist as functions of α.

All the analysis so far has assumed equally spaced values of x, the independent variable. However, Lagrangian coefficients can also be written

---

[1] See, for example, Abramowitz and Stegun (A1) Chapter 25 or the NBS volume (N1).

down by inspection for the general case.   For example, from

$$
\begin{array}{cc}
x_1 & y_1 \\
x_2 & y_2 \\
x_3 & y_3
\end{array}
$$

we can form the reasonable "smooth-curve" approximation for y:

$$y(x) = \frac{(x - x_2)(x - x_3)}{(x_1 - x_2)(x_1 - x_3)}\, y_1 + \frac{(x - x_1)(x - x_3)}{(x_2 - x_1)(x_2 - x_3)}\, y_2$$

$$+ \frac{(x - x_1)(x - x_2)}{(x_3 - x_1)(x_3 - x_2)}\, y_3 \quad (13\text{-}9)$$

Sometimes one wishes to make an estimate of the error involved in using a particular interpolation formula.   We shall illustrate the method by estimating the error in formula (13-7) derived above:

$$y = y_1 + \alpha \Delta y_1 + \frac{\alpha(\alpha - 1)}{2!} \Delta^2 y_1$$

By Taylor's theorem

$$y_2 = y_1 + hy_1' + \tfrac{1}{2}h^2 y_1'' + \tfrac{1}{6}h^3 y_1''' + \cdots$$

$$y_3 = y_1 + 2hy_1' + 2h^2 y_1'' + \tfrac{4}{3}h^3 y_1''' + \cdots$$

Then

$$\Delta y_1 = y_2 - y_1 = hy_1' + \tfrac{1}{2}h^2 y_1'' + \tfrac{1}{6}h^3 y_1''' + \cdots$$

$$\Delta^2 y_1 = y_3 - 2y_2 + y_1 = h^2 y_1'' + h^3 y_1''' + \cdots \quad (13\text{-}10)$$

Thus our estimate (13-7) for $y(x_1 + \alpha h)$ is the series

$$y_1 + \alpha\left(hy_1' + \frac{1}{2}h^2 y_1'' + \frac{1}{6}h^3 y_1''' + \cdots\right) + \frac{1}{2}\alpha(\alpha - 1)(h^2 y_1'' + h^3 y_1''' + \cdots)$$

$$= y_1 + \alpha hy_1' + \frac{1}{2}\alpha^2 h^2 y_1'' + \frac{\alpha}{6}(3\alpha - 2)h^3 y_1''' + \cdots \quad (13\text{-}11)$$

On the other hand, the (exact) Taylor series for $y(x_1 + \alpha h)$ is

$$y_1 + \alpha hy_1' + \tfrac{1}{2}\alpha^2 h^1 y_1'' + \tfrac{1}{6}\alpha^3 h^3 y_1''' + \cdots \quad (13\text{-}12)$$

and the error is given by the difference between (13-11) and (13-12), namely,

$$\frac{\alpha}{6}(\alpha - 1)(\alpha - 2)h^3 y_1''' + \text{small terms} \quad (13\text{-}13)$$

More precisely, the error is $(\alpha/6)(\alpha - 1)(\alpha - 2)h^3 y'''$, where the third derivative is evaluated somewhere on the interval $x_1 \leqslant x \leqslant x_3$.

How can we estimate y'''?   This leads naturally to the question of *numerical differentiation*, which is closely related to interpolation.   The easiest way to produce formulas for derivatives is to introduce the operator D:

$$Dy_n = y'_n \qquad (13\text{-}14)$$

(we are once more considering equally spaced arguments)

What is the connection between D and our other operators E and Δ?   The connecting link is Taylor's series:

$$
\begin{aligned}
y_{n+1} = Ey_n &= y_n + hy'_n + \tfrac{1}{2}h^2 y''_n + \tfrac{1}{6}h^3 y'''_n + \cdots \\
&= (1 + hD + \tfrac{1}{2}h^2 D^2 + \tfrac{1}{6}h^3 D^3 + \cdots)y_n \\
&= e^{hD} y_n
\end{aligned}
$$

Thus

$$E = e^{hD} \qquad (13\text{-}15)$$

or

$$D = \frac{1}{h}\ln E = \frac{1}{h}\ln(1 + \Delta)$$

$$= \frac{1}{h}\left(\Delta - \frac{\Delta^2}{2} + \frac{\Delta^3}{3} - + \cdots\right) \qquad (13\text{-}16)$$

To answer the question above about y''', we observe from (13-16) that

$$D^n \approx \left(\frac{\Delta}{h}\right)^n \qquad (13\text{-}17)$$

so that

$$y''' \approx \frac{1}{h^3}\Delta^3 y \qquad (13\text{-}18)$$

13-2   NUMERICAL INTEGRATION

The principle of numerical integration is essentially the same as that of interpolation and numerical differentiation; the function is approximated by a polynomial and the polynomial is integrated exactly.   One generally does not try to fit the entire function by a single polynomial.   Instead, the interval over which we are integrating is split up into many smaller subintervals, and a separate approximating polynomial is used in each subinterval.

The simplest method of this type is the *trapezoidal rule.*   If the integral in question is

$$\int_a^b f(x)\, dx \tag{13-19}$$

we divide the interval $a < x < b$ into n subintervals of length $h = (b - a)/n$. In each subinterval, the function is replaced by a straight line connecting the values at each end.   The result is

$$\int_a^b f(x)\, dx \approx h(\tfrac{1}{2}y_0 + y_1 + y_2 + \cdots + y_{n-1} + \tfrac{1}{2}y_n) \tag{13-20}$$

where $y_0 = f(a)$, $y_1 = f(a + h)$, $\ldots$, $y_{n-1} = f(b - h)$, $y_n = f(b)$.

A more accurate and useful formula is *Simpson's rule.*   The interval $a < x < b$ is subdivided into an *even* number of subintervals. We fit a parabola to $y(x)$ at $x = a$, $a + h$, $a + 2h$; another at $x = a + 2h$, $a + 3h$, $a + 4h$; and so on.

What is the area under a typical parabola, such as that of Figure 13–1? Let the area be

$$A = h(\alpha y_0 + \beta y_1 + \gamma y_2)$$

We must find the coefficients α, β, γ.   By symmetry, $\alpha = \gamma$.   Consider the special parabola $y = x^2$ from -1 to $+1$.   Then $y_0 = y_2 = 1$, $y_1 = 0$, and

$$A = \tfrac{2}{3} = \alpha + \gamma$$

Therefore,

$$\alpha = \gamma = \tfrac{1}{3}$$

To find β, consider the case when $y_0 = y_1 = y_2 = 1$.   The approximating curve is the horizontal line $y = 1$, and

$$A = 2h = h(\tfrac{1}{3} + \beta + \tfrac{1}{3})$$

Therefore,

$$\beta = \tfrac{4}{3}$$

and we have derived the formula

$$A = \frac{h}{3}(y_0 + 4y_1 + y_2) \tag{13-21}$$

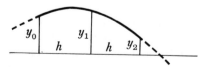

Figure 13–1   An element of area involved in Simpson's rule

When applied to the more general case of an integral from a to b, subdivided into n pieces, the result is

$$\int_a^b f(x)\,dx \approx \frac{h}{3}(y_0 + 4y_1 + 2y_2 + 4y_3 + 2y_4 + \cdots + 4y_{n-1} + y_n) \quad (13\text{-}22)$$

One could go on and consider fitting by cubics, or quartics, in an effort to improve accuracy, but in practice one generally sticks to Simpson's rule, subdividing more finely if necessary to improve the accuracy.

Another way to improve the accuracy is to give up the requirement of equally spaced abscissas and adjust the positions where we evaluate the function, as well as the coefficients, in order to fit a polynomial of as high degree as possible. For example, in Simpson's rule, taking only three points, we obtain a formula which is guaranteed exact for polynomials up to and including *second* order. By allowing the points x_i, where the function is evaluated, to vary, we have altogether *six* parameters and should be able to fit any polynomial through *fifth* order. This approach is known as *Gaussian integration*; we shall consider it briefly.

Let us construct an integration formula

$$\int_a^b f(x)\,dx \approx c_1 f(x_1) + c_2 f(x_2) + c_3 f(x_3) \quad (13\text{-}23)$$

which is to be exact for any polynomial of fifth order or less. The c_i and x_i could be determined directly from this condition, but the resulting six equations in six unknowns are very messy (for example, nonlinear in the x_i). Instead we shall use a trick to get the x_i directly, leaving the c_i to be evaluated later.

How do we choose the x_i? Consider a polynomial whose roots are the x_i:

$$g(x) = (x - x_1)(x - x_2)(x - x_3) \quad (13\text{-}24)$$

Then, Gaussian integration by formula (13-23), when applied to the integral

$$\int_a^b g(x)\,dx \quad (13\text{-}25)$$

clearly gives zero. In fact, Gaussian integration applied to the integral

$$\int_a^b g(x)\phi(x)\,dx \quad (13\text{-}26)$$

gives zero for any function $\phi(x)$.

Now our method is supposed to be exact for polynomials up through fifth order. Since $g(x)$ is just third order, it follows that

$$\int_a^b g(x)\,dx = 0 \quad (13\text{-}27)$$

must be *exactly* true.   Similarly,

$$\int_a^b xg(x)\, dx = 0 \tag{13-28}$$

and

$$\int_a^b x^2 g(x)\, dx = 0 \tag{13-29}$$

On the other hand, $\int_a^b x^3 g(x)\, dx$ need not vanish, although its Gaussian estimate does, since $x^3 g(x)$ is a sixth-order polynomial and our formula is only guaranteed through fifth-order polynomials.

The three "orthogonality" conditions (13-27), (13-28), and (13-29) fix $g(x)$, except for an arbitrary multiplicative constant, and so fix the x_i.   If we choose, as we may without loss of generality, $a = -1$ and $b = +1$, $g(x)$ is just the Legendre polynomial $P_3(x)$.   (Why?)   The generalization to more than three points is straightforward.

The weights, or coefficients c_i in (13-23), are determined by similar reasoning; that is, we require that the method be exact for $f(x) = 1, x, x^2$.   This gives the three linear equations

$$\begin{aligned}
b - a &= c_1 + c_2 + c_3 \\
\tfrac{1}{2}(b^2 - a^2) &= c_1 x_1 + c_2 x_2 + c_3 x_3 \\
\tfrac{1}{3}(b^3 - a^3) &= c_1 x_1^2 + c_2 x_2^2 + c_3 x_3^2
\end{aligned} \tag{13-30}$$

These three equations can be solved for c_1, c_2, c_3, since the x_i are now known to be the zeros of $P_3(x)$.

Gaussian integration can be generalized to include a weight function $\rho(x)$. That is, we can construct a formula

$$\int_a^b f(x)\rho(x)\, dx \approx c_1 f(x_1) + c_2 f(x_2) + \cdots + c_n f(x_n) \tag{13-31}$$

where the $2n$ constants c_i and x_i are chosen so that the formula is exact for $f(x) = 1, x, x^2, \ldots, x^{2n-1}$.   Everything goes as before; for example, the x_i are the roots of a polynomial $g_n(x)$ such that

$$\int_a^b g_n(x)\rho(x)\, dx = 0$$

$$\int_a^b g_n(x)\rho(x)x\, dx = 0$$

$$\vdots$$

$$\int_a^b g_n(x)\rho(x)x^{n-1}\, dx = 0 \tag{13-32}$$

If $\rho \neq 1$, g_n will of course no longer be the Legendre polynomial P_n.
Occasionally one encounters an *infinite integral*, of the form

$$I = \int_a^\infty f(x)\, dx \tag{13-33}$$

There are several ways to do such an integral numerically.

In the first place, we can use Simpson's rule

$$I = \tfrac{1}{3}h[f(a) + 4f(a + h) + 2f(a + 2h) + \cdots] \tag{13-34}$$

and just add up contributions out to a point such that further contributions
are negligible. This method works best if $f(x)$ approaches zero rapidly and
smoothly.

Another approach is Gaussian integration, with a suitable weight function.
For example, if our integrand behaves roughly like e^{-x}, a formula

$$\int_a^\infty f(x)e^{-x}\, dx \approx c_1 f(x_1) + c_2 f(x_2) + \cdots + c_n f(x_n) \tag{13-35}$$

will be useful.

Finally, a change of variable often helps. For example,

$$\int_0^\infty f(x)\, dx = \int_0^1 f\left(\frac{t}{1-t}\right) \frac{dt}{(1-t)^2} \tag{13-36}$$

If $f(x)$ goes to zero at least as fast as $1/x^2$ as $x \to \infty$, our new integrand is
finite at $t = 1$ and may be integrated by standard methods.

13-3 NUMERICAL SOLUTION OF DIFFERENTIAL EQUATIONS

There are many techniques available for integrating differential equations
numerically. We shall consider only a single first-order equation

$$\frac{dy}{dx} = f(x, y) \tag{13-37}$$

and give two typical and widely used methods which can be applied to it.

The first is the *Runge–Kutta method*. Suppose $y = y_0$ at $x = x_0$. We
choose an interval δx and calculate successively

$$\begin{aligned}
k_1 &= \delta x f(x_0, y_0) \\
k_2 &= \delta x f(x_0 + \tfrac{1}{2}\delta x, y_0 + \tfrac{1}{2}k_1) \\
k_3 &= \delta x f(x_0 + \tfrac{1}{2}\delta x, y_0 + \tfrac{1}{2}k_2) \\
k_4 &= \delta x f(x_0 + \delta x, y_0 + k_3)
\end{aligned} \tag{13-38}$$

Table 13–3

x_0	y_0	y_0'
$x_1 (= x_0 + \delta x)$	y_1	y_1'
$x_2 (= x_1 + \delta x)$	y_2	y_2'
$x_3 (= x_2 + \delta x)$	y_3	y_3'

Then at $x = x_0 + \delta x$, $y \approx y_0 + \frac{1}{6}(k_1 + 2k_2 + 2k_3 + k_4)$. (This is approximately the Simpson's rule integral of y' from x_0 to $x_0 + \delta x$.) We can now proceed to find y at $x_0 + 2\delta x$ in exactly the same way, and so on.

A second method, the *Bashforth–Adams–Milne method*, assumes that the solution is started somehow and that we have a table such as Table 13–3. We proceed as follows.

1. "Predict" y_4 by $y_4 = y_0 + \frac{4}{3}\delta x(2y_1' - y_2' + 2y_3')$.

[This formula may be obtained by expressing y' in terms of α ($x = x_0 + \alpha \delta x$), using our previous interpolation formula (13-8), and then integrating y' to obtain $y_4 - y_0$.]

2. Calculate $y_4' = f(x_4, y_4)$.

3. "Correct" y_4 by $y_4 = y_2 + \frac{1}{3}\delta x(y_2' + 4y_3' + y_4')$ (Simpson's rule).

A comparison of the predicted and corrected values of y_4 gives an indication of how accurately we are doing. More precisely, the error in our prediction formula is

$$\tfrac{28}{90}(\delta x)^5 y^{(v)} \tag{13-39}$$

while the error in the correction formula is

$$\tfrac{1}{90}(\delta x)^5 y^{(v)} \tag{13-40}$$

Thus a good rule of thumb is the following. If the difference between the predicted and corrected values of y_4 is less than 14 in the last decimal place, we are probably all right. Otherwise, we should either correct again, or (better) make the interval δx smaller.

A property which an integration method should have, besides smallness of error, is *stability*. We shall not go into this deeply, but discuss a simple example to show its importance. Consider the possibility of solving the differential equation

$$y' = f(x, y) \tag{13-41}$$

by using Simpson's rule only. That is, from

x_0	y_0	y_0'
x_1	y_1	y_1'

we extrapolate to x_2 by the equation

$$y_2 = y_0 + \tfrac{1}{3}\delta x(y_0' + 4y_1' + y_2')$$
$$= y_0 + \tfrac{1}{3}\delta x[y_0' + 4y_1' + f(x_2, y_2)] \qquad (13\text{-}42)$$

We solve this equation for y_2 by trial and error (or otherwise) and then go on to y_3 in the same way.

To show that a serious defect is lurking unseen in this rather reasonable looking, if simple-minded, method, imagine applying it to the trivial differential equation

$$y' = \alpha y \qquad (\alpha = \text{constant}) \qquad (13\text{-}43)$$

The extrapolation formula

$$y_2 = y_0 + \tfrac{1}{3}\delta x(y_0' + 4y_1' + y_2')$$

becomes

$$y_2 = y_0 + \frac{\alpha\delta x}{3}(y_0 + 4y_1 + y_2) \qquad (13\text{-}44)$$

This is a linear *difference equation*, with constant coefficients.   Such equations have many analogies with linear differential equations with constant coefficients (discussed in Section 1–1).   The standard approach is to try a solution of the form

$$y_n = k^n \qquad (13\text{-}45)$$

where k is a constant to be determined.   Substituting (13-45) into the difference equation (13-44) yields a quadratic equation for k

$$(1 - \tfrac{1}{3}\alpha\,\delta x)k^2 - \tfrac{4}{3}\alpha\,\delta x\,k - (1 + \tfrac{1}{3}\alpha\,\delta x) = 0$$

For small $\alpha\delta x$, the solutions are

$$k \approx \begin{cases} 1 + \alpha\delta x \\ -1 + \tfrac{1}{3}\alpha\delta x \end{cases}$$

Thus, the general solution to our difference equation is an arbitrary linear combination of

$$y_n \approx (1 + \alpha\delta x)^n \approx e^{\alpha x}$$

and

$$y_n \approx (-1 + \tfrac{1}{3}\alpha\delta x)^n \approx (-1)^n e^{-\alpha x/3} \qquad (13\text{-}46)$$

The first solution is fine and is in fact the correct solution of the original differential equation (13-43).   If α is negative, however, the second solution (13-46) is a wildly oscillating one which will sooner or later swamp the correct solution.   Thus our method is *unstable*.   By solving a first-order differential equation with a second-order difference equation, we have introduced an extraneous unstable solution.

An investigation of the stability of the other two methods described in this section is more involved; we refer to a mathematical book on numerical analysis, such as Todd (T5) Chapter 9.

We shall digress briefly, however, to touch upon an amusing application of stability considerations in numerical computation. A classic example is provided by the recursion relation (7-52) obeyed by Bessel functions

$$J_{m+1}(x) = \frac{2m}{x} J_m(x) - J_{m-1}(x) \tag{13-47}$$

At first sight this recursion relation might appear to provide a useful means for calculating, say, $J_{10}(1)$ on a digital computer. One need only have starting values of $J_0(1)$ and $J_1(1)$; repeated application of (13-47) should then yield $J_2(1), J_3(1), \ldots, J_{10}(1), \ldots$.

The defect with this method is its instability. Let us try to analyze (13-47) in the form

$$y_{m+1} = \frac{2m}{x} y_m - y_{m-1} \tag{13-48}$$

for $m \to \infty$, as we analyzed the difference equation (13-44). There, however, we could solve the equation exactly because it was linear with constant coefficients; Eq. (13-48) although linear has a nonconstant (that is, depending on m) coefficient, and our simple method no longer works.

The trial-and-error tactics of Section 1-3 can be employed here to good advantage. For example, let us try neglecting the second term on the right of (13-48). The resulting equation

$$y_{m+1} = \frac{2m}{x} y_m$$

has the solution

$$y_m = \text{constant} \cdot \left(\frac{2}{x}\right)^m \Gamma(m) \tag{13-49}$$

and the neglect of y_{m-1} in (13-48) is now easily justified a posteriori, assuming $m \to \infty$.

Similarly, if we neglect y_{m+1} in (13-48), we obtain

$$y_m = \frac{x}{2m} y_{m-1}$$

from which

$$y_m = \text{constant} \cdot \frac{1}{\Gamma(m+1)} \left(\frac{x}{2}\right)^m \tag{13-50}$$

and again we can justify the neglect of y_{m+1}.

The difference equation (13-48) thus has one solution whose behavior for large m is described by (13-49) and another whose behavior is given by (13-50). Now, according to (7-50),

$$J_m(x) \approx \frac{1}{\Gamma(m+1)}\left(\frac{x}{2}\right)^m \qquad \text{for} \qquad x \ll m$$

so that the second solution found above, namely (13-50), is in fact the desired solution $y_m = J_m(x)$. Our analysis has, however, revealed the presence of another solution, (13-49), which becomes very large as m increases. Thus the inevitable round-off errors will avalanche catastrophically, and our conclusion, easily verified by a numerical example (Table 13–4), is that the recursion relation (13-47) is not very useful for calculating Bessel functions of large order from known values of Bessel functions of low order.

A most striking conclusion emerges, however, if we consider using recursion relation (13-48) in the *opposite* direction, namely for *decreasing m*. Again one solution is growing and the other shrinking, but now it is the *growing* solution which is the desired result $J_m(x)$, while the unwanted extraneous solution dies out rapidly. In other words, our recursive calculation is now *stable*.

We give a numerical illustration of this second procedure in Table 13–5. Note that we begin, quite arbitrarily, with $y_9 = 1$ and $y_{10} = 0$. The fact that these are abysmally poor estimates of $J_9(1)$ and $J_{10}(1)$ is irrelevant; we are assured that the errors will die off (exponentially!).

Note that the final entry, y_0, in Table 13–5 is 146,181,170, which is surely a poor approximation to $J_0(1)$! The discrepancy, however, is simply the arbitrary constant in (13-50). We therefore divide y_0 by the known value of $J_0(1)$ to obtain a "scale factor"; the arithmetic is shown in Table 13–5. To verify the usefulness of the calculation, we have multiplied our y_m by this scale

Table 13–4   Recursion Relation (13-48) for m Increasing $(x = 1)$

m	y_m	$J_m(1)$
0	.76519 76866	.76519 76866
1	.44005 05857	.44005 05857
2	.11490 34848	.11490 34849
3	.01956 33535	.01956 33540
4	.00247 66362	.00247 66390
5	.00024 97361	.00024 97577
6	.00002 07248	.00002 09383
7	−.00000 10385	.00000 15023
8	−.00003 52638	.00000 00942
9	−.00056 52593	.00000 00052
10	−.01073 81267	.00000 00003

Table 13–5   Recursion Relation (13-48) for Decreasing m ($x = 1$)

m	y_m	$y_m \cdot \dfrac{J_0(1)}{y_0}$	$J_m(1)$
10	0	.00000 000	.00000 000
9	1	.00000 001	.00000 001
8	18	.00000 009	.00000 009
7	287	.00000 150	.00000 150
6	4,000	00002 094	.00002 094
5	47,713	.00024 976	.00024 976
4	473,130	.00247 664	.00247 664
3	3,737,327	.01956 335	.01956 335
2	21,950,832	.11490 348	.11490 348
1	84,066,001	.44005 059	.44005 059
0	146,181,170	.76519 769	.76519 769

$$\text{Scale factor} = \frac{J_0(1)}{y_0} = \frac{.76519\ 7687}{146,181,170} = 5.2345\ 8450 \times 10^{-9}$$

factor; the results (in the third column) agree with $J_m(1)$ (in the fourth column) to eight decimals. If we want more significant figures, or $J_m(x)$ for larger values of m, we should simply begin at $m = 15$ or $m = 20$ instead of $m = 10$.[2]

We have thus learned that use of recursion relation (13-47) for descending m, beginning with two arbitrary starting values, plus a table of $J_0(x)$, enables us to calculate by simple arithmetic any $J_m(x)$, for integer m. This procedure and analogous procedures are actually used in digital computer subroutines to evaluate Bessel functions and other special functions.

13–4   ROOTS OF EQUATIONS

One is often faced with the problem of finding more or less accurately an x such that

$$f(x) = 0$$

where $f(x)$ is sufficiently complicated that a direct solution is not possible.

A simple way to do this is to make a table which straddles the desired root (Table 13–6). Then, from the standpoint of the inverse function, the problem is simply interpolation with unequal spacings of the argument, and the Lagrange method may be used. The root for this example would be,

---

[2] For example, if we begin with $y_{20} = 0$, $y_{19} = 1$, and continue down to y_0, in order to get a scale factor by comparing with $J_0(1)$, we will have evaluated $J_{10}(1)$ [and also $J_9(1), \ldots, J_1(1)$] to ten-figure accuracy.

Table 13–6

x	$f(x)$
1	1.76
2	0.41
3	−0.16
4	−0.32

by the equation analogous to (13-9) for four points,

$$\frac{(0 - 0.41)(0 + 0.16)(0 + 0.32)}{(1.76 - 0.41)(1.76 + 0.16)(1.76 + 0.32)} \times (1)$$

$$+ \frac{(0 - 1.76)(0 + 0.16)(0 + 0.32)}{(0.41 - 1.76)(0.41 + 0.16)(0.41 + 0.32)} \times (2)$$

$$+ \frac{(0 - 1.76)(0 - 0.41)(0 + 0.32)}{(-0.16 - 1.76)(-0.16 - 0.41)(-0.16 + 0.32)} \times (3)$$

$$+ \frac{(0 - 1.76)(0 - 0.41)(0 + 0.16)}{(-0.32 - 1.76)(-0.32 - 0.41)(-0.32 + 0.16)} \times (4) = 2.37$$

A better method, if $f(x)$ is calculable for arbitrary x, is to iterate, so that we can get any accuracy we want.   Linear interpolation is generally sufficient. For example, let us find the positive root of

$$f(x) = (5 - x)e^x - 5 = 0$$

First, try

$$x_1 = 5 \qquad f(5) = -5$$
$$x_2 = 4.5 \qquad f(4.5) \approx 40$$

A linear interpolation between these values is shown in Figure 13–2.   Next,

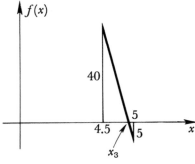

Figure 13–2   A linear interpolation between points (x_1, y_1) and (x_2, y_2) to find x_3

try

$$x_3 = \frac{40 \times 5 + 5 \times (4.5)}{45} = 4.944 \qquad f(4.944) = 2.859$$

$$x_4 = \frac{(2.859) \times 5 + 5 \times (4.944)}{7.859} = 4.964 \qquad f(4.964) = 0.154$$

etc.

(This method is sometimes called *regula falsi*.)

Alternatively, we may obtain an improved estimate of x by *Newton's method*, involving $f(x)$ and $f'(x)$ (see Figure 13–3):

$$x_{n+1} = x_n - \frac{f(x_n)}{f'(x_n)}$$

This may fail if the function has a point of inflection, or other bad behavior, near the zero.

A very simple-minded approach that is often worth trying on an equation of the form

$$x = f(x)$$

is straightforward iteration:

$$x_1 = f(x_0)$$
$$x_2 = f(x_1)$$
$$x_3 = f(x_2)$$
etc.

If the scheme converges, it is often the fastest way to estimate x. Convergence requires

$$|f'(x)| < 1$$

near the root.

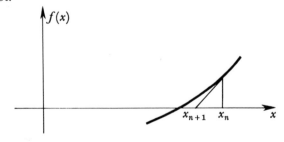

Figure 13–3 Illustration of Newton's method of obtaining an improved estimate, x_{n+1}, for the root of $f(x)$

Finally, we shall describe *Horner's method* for evaluating (real) roots of polynomials. It is based on two algorithms, one for reducing the roots of a polynomial by a constant and one for multiplying each root by ten.[3] We shall give a numerical example of this method, which is most useful in pencil-and-paper calculations; a digital computer generally uses other methods.[4]

In order to reduce each root of a function $f(x)$ by the constant a, we have simply to construct the new function $g(x) = f(x + a)$; clearly every root of $g(x)$ is less by a than a root of $f(x)$. If $f(x)$ is a polynomial, we may write

$$f(x) = a_n x^n + a_{n-1}x^{n-1} + \cdots + a_0$$
$$g(x) = b_n x^n + b_{n-1}x^{n-1} + \cdots + b_0$$

and a simple prescription is available for calculating the b_n if we know the a_n. For

$$f(x) = g(x - a) = b_n(x - a)^n + b_{n-1}(x - a)^{n-1} + \cdots + b_0$$

Thus b_0 is the remainder when $f(x)$ is divided by $x - a$. If the resulting quotient is again divided by $(x - a)$, the remainder is b_1, and so on.

The division of a polynomial by $x - a$ is most easily accomplished by the process of *synthetic division*. For example, a conventional representation of the division of $2x^3 - 7x^2 + x + 3$ by $x - 3$ is

$$
\begin{array}{r|l}
2x^3 - 7x^2 + x + 3 & x - 3 \\
2x^3 - 6x^2 & 2x^2 - x - 2 \\
\cline{1-1}
\quad - x^2 + x & \\
\quad - x^2 + 3x & \\
\cline{1-1}
\qquad - 2x + 3 & \\
\qquad - 2x + 6 & \\
\cline{1-1}
\qquad\qquad -3 &
\end{array}
$$

Synthetic division merely means the abbreviation of this computation to

$$
\begin{array}{r|l}
2 - 7 + 1 + 3 & 3 \\
+ 6 - 3 - 6 & \\
\cline{1-1}
2 - 1 - 2 & - 3
\end{array}
$$

---

[3] More precisely, one algorithm constructs a new polynomial whose roots are the roots of the original polynomial, decreased by a specified constant, while the other constructs a new polynomial whose roots are the roots of the original polynomial multiplied by 10.

[4] For a more complete discussion of root-finding techniques, the reader should consult a specialized reference such as Todd (T5) Chapter 7; Ralston (R2) Chapter 8; or Ostrowski (O3).

Note that detached coefficients are used throughout; the first three numbers on the last line represent the quotient $(2x^2 - x - 2)$, while the last number (-3) is the remainder. Incidentally, this remainder is of course also the result of substituting $x = 3$ into the polynomial $2x^3 - 7x^2 + x + 3$.

Now we have seen that the polynomial whose roots are three less than the roots of $2x^3 - 7x^2 + x + 3$ may be found by repeated division by $x - 3$; the repeated synthetic division looks like this:

$$
\begin{array}{rrrr|l}
2 & -7 & 1 & 3 & \underline{\,3} \\
 & +6 & -3 & -6 & \\ \hline
2 & -1 & -2 & -3 & \\
 & +6 & +15 & & \\ \hline
2 & +5 & +13 & & \\
 & +6 & & & \\ \hline
2 & +11 & & &
\end{array}
$$

(13-51)

The desired polynomial is $2x^3 + 11x^2 + 13x - 3$.

By noting that our original polynomial $2x^3 - 7x^2 + x + 3$ changes sign between $x = 3$ and $x = 4$, we decide that it has a root in this interval.[5] The first step in Horner's method is the reduction of its roots by 3, as in (13-51). The resulting polynomial, $2x^3 + 11x^2 + 13x - 3$, then has a root between 0 and 1; we therefore multiply all roots by 10 in preparation for the calculation of the next digit in the root. It is trivial to verify that the desired polynomial, whose roots are 10 times the roots of $2x^3 + 11x^2 + 13x - 3$, is $2x^3 + 110x^2 + 1300x - 3000$.

We now proceed as before. Our new polynomial changes sign between 1 and 2; the roots are reduced by 1, again multiplied by 10, and so on. Several steps of the complete calculation are exhibited below.

$$
\begin{array}{rrrr|l}
2 & -7 & 1 & 3 & \underline{\,3} \\
 & +6 & -3 & -6 & \\ \hline
2 & -1 & -2 & -3 & \\
 & +6 & +15 & & \\ \hline
2 & +5 & +13 & & \\
 & +6 & & & \\ \hline
2 & +11 & & &
\end{array}
$$

[5] For more sophisticated methods of locating roots of polynomials, such as Sturm's method, the references mentioned at the end of this chapter should be consulted.

$$2 + 110 + 1300 - 3000 \quad \underline{|1}$$
$$\underline{+ \quad 2 + 112 \; + 1412}$$
$$2 + 112 + 1412 \underline{| - 1588}$$
$$\underline{+ \quad 2 + \quad 114}$$
$$2 + 114 \underline{| + 1526}$$
$$\underline{+ \quad 2}$$
$$2 \underline{| + 116}$$

$$2 + 1160 + 152600 - 1588000 \quad \underline{|9}$$
$$\underline{+ \quad 18 + \quad 10602 + 1468818}$$
$$2 + 1178 + 163202 \underline{| - 119182}$$
$$\underline{+ \quad 18 + \quad 10764}$$
$$2 + 1196 \underline{| + 173966}$$
$$\underline{+ \quad 18}$$
$$2 \underline{| + 1214}$$
$$2 + 12140 + 17396600 - 119182000$$
$$\cdots$$

So far the root is 3.19 ...; we can clearly extract several more digits simply by dividing the last coefficient of our polynomial by the next-to-the-last:

$$\frac{119{,}182{,}000}{17{,}396{,}000} = 6.85 \ldots$$

Thus $x \approx 3.19685$.   The actual root is $3.19682 \ldots$

13–5   SUMMING SERIES

To begin with, one can just add up terms until the remaining ones are so small they can be neglected.   This works fine if the series is converging rapidly, but many frequently occurring series converge too slowly for this simple method to be useful.

For example, consider the sum

$$\sum_{n=1}^{\infty} \frac{1}{n^2} \tag{13-52}$$

Table 13–7

n	$\dfrac{1}{n^2}$	Σ
1	1.0000	1.0000
2	0.2500	1.2500
3	0.1111	1.3611
4	0.0625	1.4236
5	0.0400	1.4636

The first few terms are shown in Table 13–7. Clearly, we are going to have to write down about 100 terms in order to get four decimals.

Infinite integrals can be used to estimate infinite series. For example,

$$\int_6^\infty \frac{dx}{x^2} < \sum_{n=6}^\infty \frac{1}{n^2} < \int_5^\infty \frac{dx}{x^2} \tag{13-53}$$

as is clear from Figure 13–4. Thus

$$0.1667 < \sum_{n=6}^\infty \frac{1}{n^2} < 0.2000$$

and, combining this result with the first five terms which we did by hand above, we get

$$1.6303 < \sum_{n=1}^\infty \frac{1}{n^2} < 1.6636 \tag{13-54}$$

Things still are not satisfactory; the average of the two limits, 1.6470, is not very accurate. The correct answer, of course, is [see (2-37)]

$$\frac{\pi^2}{6} = 1.6449 \ldots$$

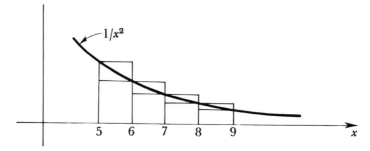

Figure 13–4   Relations between a series and an integral showing the inequalities (13-53)

The *Euler–Maclaurin formula* provides a more accurate comparison of a series and the corresponding integral.   It is most easily derived by using the E and D operators discussed in Section 13–1 of this chapter.   First, one writes down the trapezoidal approximation

$$\int_a^b f(x)\, dx \approx h\left[\frac{1}{2} f(a) + f(a + h) + f(a + 2h) + \cdots \right.$$

$$\left. + f(a + (n - 1)h) + \frac{1}{2} f(b)\right]$$

$$\approx h\left[\frac{1}{2} + E + E^2 + \cdots + E^{n-1} + \frac{1}{2} E^n\right] f(a)$$

$$\approx h(E^n - 1)\left(\frac{1}{2} + \frac{1}{E - 1}\right) f(a) \tag{13-55}$$

The exact value of the integral, on the other hand, is

$$\int_a^b f(x)\, dx = g(b) - g(a)$$

where

$$g(x) = \int f(x)\, dx$$

Then

$$Dg(a) = f(a)$$

and

$$\int_a^b f(x)\, dx = g(b) - g(a) = (E^n - 1)g(a) = \frac{(E^n - 1)}{D} f(a) \tag{13-56}$$

Thus

(integral) − (approximation)

$$= \frac{(E^n - 1)}{D}\left[1 - hD\left(\frac{1}{2} + \frac{1}{E - 1}\right)\right] f(a)$$

$$= \left(\frac{E^n - 1}{D}\right)\left[1 - hD\left(\frac{1}{2} + \frac{1}{e^{hD} - 1}\right)\right] f(a)$$

$$= \frac{E^n - 1}{D}\left[-\frac{B_2}{2!}(hD)^2 - \frac{B_4}{4!}(hD)^4 - \cdots\right] f(a)$$

$$= -\frac{B_2}{2!} h^2 f'(x)\Big|_a^b - \frac{B_4}{4!} h^4 f'''(x)\Big|_a^b - \cdots$$

where the B_n are the Bernoulli numbers, discussed in Chapter 2 [see (2-17) and (2-20)]. Our final result may be written

$$\int_a^b f(x)\,dx = h\left[\frac{1}{2}f(a) + f(a+h) + \cdots + f(b-h) + \frac{1}{2}f(b)\right]$$
$$- \frac{B_2}{2!}h^2 f'(x)\Big|_a^b - \frac{B_4}{4!}h^4 f'''(x)\Big|_a^b - \cdots \tag{13-57}$$

This formula can be used to estimate integrals by improving the trapezoidal approximation, or to estimate series. We shall illustrate the latter application only. With $a = 6$, $b = \infty$, $h = 1$, and $f(x) = 1/x^2$, (13-57) gives

$$\int_6^\infty \frac{dx}{x^2} = \left[\frac{1}{2}\frac{1}{6^2} + \frac{1}{7^2} + \frac{1}{8^2} + \cdots\right] + \frac{1}{6x^3}\Big|_6^\infty - \frac{1}{30x^5}\Big|_6^\infty + - \cdots$$

Therefore,

$$\sum_{n=6}^\infty \frac{1}{n^2} = \int_6^\infty \frac{dx}{x^2} + \frac{1}{2}\frac{1}{6^2} - \frac{1}{6x^3}\Big|_6^\infty + \frac{1}{30x^5}\Big|_6^\infty - + \cdots$$
$$= \frac{1}{6} + \frac{1}{72} + \frac{1}{1296} - \cdots$$
$$= 0.1813$$

Adding this to $\sum_{n=1}^5 1/n^2$, we get

$$\sum_{n=1}^\infty \frac{1}{n^2} \approx 1.6449 \tag{13-58}$$

which is correct to four decimals.

Suppose we try to avoid doing *any* of the sum by hand!

$$\int_1^\infty \frac{dx}{x^2} = \left[\frac{1}{2}\frac{1}{1} + \frac{1}{4} + \frac{1}{9} + \cdots\right] + \frac{1}{6x^3}\Big|_1^\infty - \frac{1}{30x^5}\Big|_1^\infty$$
$$+ \frac{1}{42x^7}\Big|_1^\infty - \frac{1}{30x^9}\Big|_1^\infty + \frac{5}{66x^{11}}\Big|_1^\infty - + \cdots$$

$$\sum_{n=1}^\infty \frac{1}{n^2} = 1 + \frac{1}{2} + \frac{1}{6} - \frac{1}{30} + \frac{1}{42} - \frac{1}{30} + \frac{5}{66} - + \cdots \tag{13-59}$$

This series is not converging! The reason is that the Euler–Maclaurin formula usually yields an *asymptotic* series (as defined in Section 3–5), not a *convergent* series. This generally does not matter for numerical purposes, provided the terms fall off rapidly to begin with.

We shall close this chapter by illustrating how several tricks can be used to improve convergence very drastically. Suppose we want to evaluate π.

A useful series is

$$\tan^{-1} x = x - \frac{x^3}{3} + \frac{x^5}{5} - + \cdots \tag{13-60}$$

Setting $x = 1$ gives

$$\frac{\pi}{4} = 1 - \frac{1}{3} + \frac{1}{5} - + \cdots \qquad \text{(Gregory's series)}$$

This converges pretty slowly. Try $x = \sqrt{\frac{1}{3}}$

$$\frac{\pi}{6} = \sqrt{\frac{1}{3}} \left(1 - \frac{1}{3 \times 3} + \frac{1}{5 \times 9} - \frac{1}{7 \times 27} + - \cdots \right)$$

This is clearly better. To do even better, set $x = \frac{1}{5}$

$$\alpha = \tan^{-1} \frac{1}{5} = \frac{1}{5} - \frac{1}{3 \times 5^3} + \frac{1}{5 \times 5^5} - + \cdots$$

$$2\alpha = \tan^{-1} \frac{5}{12}$$

$$4\alpha = \tan^{-1} \frac{120}{119} \left(\approx \frac{\pi}{4} \, ! \right)$$

In fact, $4\alpha = \pi/4 + \beta$, where (taking tangents)

$$\frac{120}{119} = \frac{1 + \tan \beta}{1 - \tan \beta} \Rightarrow \beta = \tan^{-1} \frac{1}{239}$$

Thus

$$\pi = 16\alpha - 4\beta$$

$$= 16 \tan^{-1} \frac{1}{5} - 4 \tan^{-1} \frac{1}{239}$$

$$= 16 \left(\frac{1}{5} - \frac{1}{3 \times 5^3} + \frac{1}{5 \times 5^5} - + \cdots \right)$$

$$- 4 \left(\frac{1}{239} - \frac{1}{3 \times 239^3} + - \cdots \right) \tag{13-61}$$

This expresses π in terms of two rapidly converging series.

As a second example, suppose we want to evaluate ln 2. We could begin with

$$\ln (1 + x) = x - \frac{x^2}{2} + \frac{x^3}{3} - + \cdots \tag{13-62}$$

which gives

$$\ln 2 = 1 - \tfrac{1}{2} + \tfrac{1}{3} - + \cdots$$

This series converges very slowly. For one improvement, we observe that

$$\ln\left(\frac{1+x}{1-x}\right) = 2\left(x + \frac{x^3}{3} + \frac{x^5}{5} + \cdots\right)$$

Setting $x = \frac{1}{3}$ gives

$$\ln 2 = 2\left(\frac{1}{3} + \frac{1}{3 \times 3^3} + \frac{1}{5 \times 3^5} + \cdots\right)$$

This converges much better.

If we want series that really converge fast, we need numbers near 1. For example,

$$\ln 3 - \ln 2 = \ln \tfrac{3}{2} \qquad (x = \tfrac{1}{5})$$
$$2 \ln 3 - 3 \ln 2 = \ln \tfrac{9}{8} \qquad (x = \tfrac{1}{17})$$

These give ln 2 and ln 3 in terms of very rapidly converging series.

The Euler transformation (see Section 2–3) is another device which can sometimes be used to improve convergence of a series.

REFERENCES

The classic work in this field is Boole (B5). Two successors worthy of mention are Milne-Thomson (M6) and Whittaker and Robinson (W4). Although written before the era of electronic digital computers, these books contain much interesting material.

Among the numerous recent works on the subject, Booth (B6) is a convenient and fairly up-to-date collection of useful numerical formulas. For more complete treatments of numerical methods, see Todd (T5); Hildebrand (H8); Hamming (H4); Ralston (R2); or Henrici (H7). The work of Ostrowski (O3) is excellent but rather specialized. The two small volumes of Noble (N4) give an elementary presentation of these topics.

The problem of numerical evaluation of roots, and particularly roots of polynomials, is discussed in most of the above references, and also in texts on the theory of equations. Conkwright (C7) and Weisner (W3) are typical books in which fuller presentations of Horner's method and more powerful and general approaches are found.

PROBLEMS

13-1   Using the table on the opposite page, estimate:

(a) $f(1.5)$
(b) $f'(1.5)$

(c) the root of $f(x) = 0$ between 2 and 3

x	$f(x)$
1	4.721
2	2.160
3	−0.357
4	−2.099

13-2   Suppose a function takes on the values y_{-1}, y_0, y_1 at three equally spaced values of its argument x.
(a) Estimate y_{max}, the maximum (or minimum) value of y, in terms of y_{-1}, y_0, and y_1.
(b) If $y(x)$ possesses a small third derivative, show that the error Δy_{max} in the estimate of part (a) may be written

$$\Delta y_{max} \approx C \frac{y'y'''}{y''}\left[\left(\frac{y'}{y''}\right)^2 - h^2\right]$$

where h is the spacing between x values.   What is the value of the constant C?

13-3   Construct the two-dimensional analog of the linear interpolation method illustrated in Figure 13–2.   That is, it is desired to solve simultaneously the equations

$$f(x, y) = 0 \quad \text{and} \quad g(x, y) = 0$$

Three (x, y) pairs are guessed, with the results

$$f(x_1, y_1) = A_1 \quad f(x_2, y_2) = A_2 \quad f(x_3, y_3) = A_3$$
$$g(x_1, y_1) = B_1 \quad g(x_2, y_2) = B_2 \quad g(x_3, y_3) = B_3$$

Evaluate x and y in terms of x_i, y_i, A_i, and B_i, $(i = 1, 2, 3)$.

13-4   Find $y(0.4)$ from the table below, using the Runge–Kutta method discussed in the text.

$$y' = x - y^2$$

x	y	y'
0	1.00000	−1.00000
0.1	0.91379	−0.73501
0.2	0.85119	−0.52452
0.3	0.80762	−0.35225

13-5   Find $y(0.4)$ from the table in Problem 13-4, using Milne's method. (One prediction, one correction.)

13-6 Suppose we want to construct a "predictor" formula, so that from

$$x_0 \quad y_0 \quad y_0'$$
$$x_1 \quad y_1 \quad y_1'$$

with equally spaced arguments, we can predict y_2.

(a) Give a formula involving y_0, y_0', and y_1'. Discuss its error and stability.

(b) Give a formula involving y_0, y_1, y_0', and y_1'. Discuss its error and stability.

13-7 (a) What two asymptotic behaviors (for large n) are possible for exact solutions of the recursion relation (b) in Problem 7-9?

(b) Which of these behaviors do the functions $f_n(x)$ of that problem exhibit?

(c) Beginning with the (nonsense) values $f_{10}(1) = 0$, $f_9(1) = 1$, iterate backwards recursively to obtain $f_1(1)$, given that $f_0(1) = 2.2796$. Check your answer by summing the power series for $f_1(1)$.

13-8 Estimate the error per subdivision when using Simpson's rule for numerical integration.

13-9 Give a Gaussian estimate for the integral

$$I = \int_0^\pi f(x) \sin x \, dx$$

of the form

$$I \approx c_1 f(x_1) + c_2 f(x_2)$$

treating $\sin x$ as a weight function and approximating $f(x)$ by a polynomial. What is the error?

13-10 Evaluate by a two-point Gaussian formula:

(a) $\displaystyle \int_1^2 \frac{dx}{x}$

(b) $\displaystyle \int_0^\infty \frac{e^{-x}\, dx}{1 + x^2}$ (treat e^{-x} as a weight function)

In each case, give also the correct answer to three significant figures.

13-11 For many integrals of the form $\int_a^b f(x)\, dx$ it is reasonable to require of the integration method that it be exact for polynomials, but for integrals of the form $\int_0^{2\pi} f(\phi)\, d\phi$, where ϕ is a geometrical angle, the "natural expansion" of $f(\phi)$ is not a power series but a Fourier series, and it would appear more suitable to require that the integration method be exact for as many terms of the Fourier series as possible. For a given number of integration points, what does this imply for their locations and coefficients?

13-12 Consider a rather simple version of the Runge–Kutta method for the differential equation $dy/dx = f(x, y)$, starting at x_0, y_0:

$$k_1 = \delta x f(x_0, y_0)$$

$$k_2 = \delta x f(x_0 + \lambda \delta x, y_0 + \mu k_1)$$

$$y(x_0 + \delta x) \approx y_0 + \rho k_1 + \sigma k_2$$

$$\text{error} \approx C(\delta x)^3 y'' \frac{\partial f}{\partial y}$$

Determine the constants λ, μ, ρ, σ, C.

13-13 Evaluate

$$\sum_{n=1}^{\infty} \frac{1}{1 + n^2}$$

to four decimal places, using the Euler–Maclaurin formula.

13-14 Find the two-point Gaussian formula for

$$\int_0^1 x^2 f(x) \, dx$$

treating x^2 as a weight function.

13-15 The Euler–Maclaurin formula was derived as a correction to the trapezoidal rule for numerical integration. Derive the analogous formula which begins with Simpson's rule. Only one correction term involving derivatives at the end points need be retained.

FOURTEEN

PROBABILITY AND STATISTICS

We begin this chapter with a brief discussion of the fundamental laws of probability and a review of some facts about permutations and combinations. Probability distributions are treated in Sections 14–4, 14–5, and 14–6, including the use of characteristic functions. The interpretation of experimental results is discussed in Section 14–7, where the maximum likelihood method of fitting data is described.

14–1 INTRODUCTION

It is an experimental fact that if an ordinary coin is flipped N times, coming up heads N_h times, then the ratio

$$\frac{N_h}{N}$$

is nearly $\frac{1}{2}$, and the larger we make N the closer the ratio approaches $\frac{1}{2}$. We express this result by the following statement. The *probability* of a coin coming up heads is $\frac{1}{2}$. More generally, if an experiment is performed N times, with N_s "successes," and if the ratio

$$\frac{N_s}{N}$$

appears to approach a limit as $N \to \infty$, we say the probability of success is N_s/N.

This sort of probability is called a posteriori probability. We make no attempt to predict the result; we just measure it.

If one attempts to *predict* probabilities, one is led to the definition of a priori probability. If an experiment has N possible outcomes, *all equally likely*, and N_s of these lead to "success," then the a priori probability of success is N_s/N. The key phrase, of course, is "equally likely." We shall not attempt to *define* when two events are equally likely; we simply assume that one can recognize equally likely phenomena.

Of course, the two types of probability should always give the same answer. If the a posteriori probability comes out different from the a priori probability, we conclude that certain events were mistakenly assumed to be equally likely. In what follows, we shall not distinguish between the two types of probability.

14–2   FUNDAMENTAL PROBABILITY LAWS

Let $P(A)$ be the probability of something (called A) occurring when an experiment is performed. $P(A)$ certainly lies between 0 and 1. If A is *certain* to happen, $P(A) = 1$. If A will certainly *not* happen, $P(A) = 0$.

To illustrate some more complicated possibilities, consider an experiment with n equally likely outcomes, involving two events A and B. Let

$n_1 =$ number of outcomes in which A occurs, but not B

$n_2 =$ number of outcomes in which B occurs, but not A

$n_3 =$ number of outcomes in which both A and B occur

$n_4 =$ number of outcomes in which neither A nor B occurs

Since we have exhausted all possibilities,

$$n_1 + n_2 + n_3 + n_4 = n$$

Now the probabilities of A and B are obviously given by

$$P(A) = \frac{n_1 + n_3}{n} \qquad P(B) = \frac{n_2 + n_3}{n} \tag{14-1}$$

We can define more complicated probabilities. The probability of *either A or B* (or both) occurring is

$$P(A + B) = \frac{n_1 + n_2 + n_3}{n} \tag{14-2}$$

The probability of *both A and B* occurring is called the *joint* probability of A and B:

$$P(AB) = \frac{n_3}{n} \qquad (14\text{-}3)$$

Finally, we may define *conditional* probabilities; the probability that A occurs, *given that B occurs*, is

$$P(A\,|\,B) = \frac{n_3}{n_2 + n_3} \qquad (14\text{-}4)$$

Similarly,

$$P(B\,|\,A) = \frac{n_3}{n_1 + n_3} \qquad (14\text{-}5)$$

Two important rules may be extracted from this simple example, and in fact are easily seen to be true in general:

$$P(A + B) = P(A) + P(B) - P(AB) \qquad (14\text{-}6)$$
$$P(AB) = P(B)P(A\,|\,B) = P(A)P(B\,|\,A) \qquad (14\text{-}7)$$

An example of (14-6) is the probability that when one card is drawn from each of two decks, at least one will be an ace.   The answer is

$$P = \tfrac{1}{13} + \tfrac{1}{13} - \tfrac{1}{169} = \tfrac{25}{169}$$

(Note that each card in a deck is assumed to be "equally likely" to be drawn.)

An example of (14-7) is the probability of drawing two hearts, when two cards are drawn successively from the same deck of cards.

$$P(2 \text{ hearts}) = P(1 \text{ heart})P(1 \text{ heart}\,|\,1 \text{ heart})$$

$$= \tfrac{1}{4} \times \tfrac{12}{51} = \tfrac{1}{17}$$

If two events are *mutually exclusive*, which means $P(AB) = 0$, then (14-6) gives

$$P(A + B) = P(A) + P(B) \qquad (14\text{-}8)$$

If $P(AB) = P(A)P(B)$, we say the events A and B are *statistically independent*. Note that this implies [by (14-7)]

$$P(A\,|\,B) = P(A) \qquad P(B\,|\,A) = P(B) \qquad (14\text{-}9)$$

We can deduce some interesting relations among conditional probabilities. For example, from

$$P(AB) = P(B)P(A\,|\,B) \qquad (14\text{-}10)$$

and

$$P(AB) = P(A)P(B \mid A)$$

we deduce

$$P(B \mid A) = \frac{P(B)}{P(A)} P(A \mid B) \tag{14-11}$$

Writing a similar relation for A and C, and dividing, we obtain a result known as *Bayes' theorem*:

$$\frac{P(B \mid A)}{P(C \mid A)} = \frac{P(B)}{P(C)} \frac{P(A \mid B)}{P(A \mid C)} \tag{14-12}$$

As a rather simple example of this theorem, consider the old problem of the three drawers A, B, C containing two gold coins, one gold and one silver coin, and two silver coins, respectively.   A drawer is picked at random, and a coin is picked from it at random.   The coin is gold.   What is the probability that the other coin is also gold?   If D denotes the event of drawing the first gold coin, we wish to calculate $P(A \mid D)$.   Now

$$P(C \mid D) = 0$$

and

$$\frac{P(A \mid D)}{P(B \mid D)} = \frac{P(A)}{P(B)} \frac{P(D \mid A)}{P(D \mid B)} = \frac{1/3}{1/3} \times \frac{1}{1/2} = 2$$

Since $P(A \mid D) + P(B \mid D) + P(C \mid D) = 1$, we have

$$P(A \mid D) = 2/3 \qquad P(B \mid D) = 1/3 \qquad P(C \mid D) = 0 \tag{14-13}$$

14–3   COMBINATIONS AND PERMUTATIONS

We shall now review some results which are probably familiar to the student.

The number of arrangements, or *permutations*, of n objects is $n!$, since the first position can be occupied by any one of the n objects, the second by any one of the $(n - 1)$ remaining objects, and so on.   The number of ordered subsets containing m objects out of n is, by similar reasoning,

$$n(n - 1) \cdots (n - m + 1) = \frac{n!}{(n - m)!} \tag{14-14}$$

If we simply ask for the number of subsets containing m objects o ut of n without regard to the order in which they appear ("number of *combinations* of n objects taken m at a time"), we must divide the above result by $m!$ since each combination may be arranged in $m!$ ways and so appears that many times in the above enumeration (14-14). Thus the number of combinations of n things taken m at a time is the *binomial coefficient*

$$\binom{n}{m} = \frac{n!}{m!(n-m)!} = \binom{n}{n-m} \tag{14-15}$$

Trickier questions can be asked. For example, how many *distinguishable* arrangements are there of n objects, if n_1 of the objects are identical, another n_2 are identical (but different from the n_1 objects), and so on. The total number of permutations is $n!$, but each distinguishable permutation appears $n_1!, n_2!, \dots$ times, so that the number of distinguishable permutations is

$$\frac{n!}{n_1! n_2! \cdots} \tag{14-16}$$

As another example, how many combinations of n objects taken m at a time are there *if repetitions are allowed?* For example, the number of combinations of three objects taken two at a time, with repetitions allowed, is six:

$$11 \quad\quad 12 \quad\quad 13 \quad\quad 22 \quad\quad 23 \quad\quad 33$$

To derive a general formula, we first note that the number of combinations of n kinds of objects, taken m at a time, with repetitions allowed, is equal to the number of ways m identical balls can be distributed among n boxes. In the above example, there would be three boxes

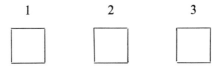

and denoting the two "balls" by crosses, we have the six distributions shown in Table 14–1. These just correspond to the six combinations enumerated above, in the same order.

Now to enumerate the number of ways m identical objects can be distributed among n boxes, we just write down m crosses and $(n-1)$ vertical lines, in any order. For example, if $m = 6$ and $n = 4$, the arrangement

$$xx| \quad |xxxx|$$

corresponds to two balls in the first box, none in the second, four in the

Table 14–1

1	2	3
xx		
x	x	
x		x
	xx	
	x	x
		xx

third, and none in the fourth. The total number of arrangements of m (identical) crosses and $(n - 1)$ (identical) lines is given by (14-16):

$$\frac{(n + m - 1)!}{(n - 1)!\,m!} = \binom{n + m - 1}{m} \tag{14-17}$$

In the example above, $n = 3$ and $m = 2$ so that the number should be

$$\binom{4}{2} = 6$$

which checks.

Many more complicated combinatorial questions can be asked; we have just tried to illustrate some techniques that are useful. Actually, the relatively simple examples we have discussed are useful in deriving the so-called distribution functions (Bose–Einstein, Fermi–Dirac, and Maxwell–Boltzmann) for systems of noninteracting particles in statistical mechanics.

For a discussion of these matters, we refer the reader to a book on statistical mechanics such as those listed at the end of this chapter.

14–4   THE BINOMIAL, POISSON, AND GAUSSIAN DISTRIBUTIONS

We now return to probability theory and ask for the probability that exactly m heads will come up if a (fair) coin is flipped n times. The results of n successive coin flips can be represented by a sequence of n letters, each either h or t, for example,

$$hhth \cdots t \tag{14-18}$$

The probability of any such outcome is $(\tfrac{1}{2})^n$; the number of such arrangements with exactly m heads is

$$\frac{n!}{m!(n - m)!} = \binom{n}{m}$$

so that the probability we are after is

$$P(m) = \binom{n}{m}\left(\frac{1}{2}\right)^n \tag{14-19}$$

Generalizing slightly, we may ask for the probability of exactly m "successes" and $(n - m)$ "failures" in n repetitions of an experiment, if the probability of a success is p and the probability of a failure is $(1 - p)$. The answer is easily found, by reasoning just like the above, to be

$$P(m) = \binom{n}{m}p^m(1 - p)^{n-m} \tag{14-20}$$

This probability distribution is known as the *binomial distribution*.

Two different limits of the binomial distribution for large n are of practical importance. In the first place, suppose that both n and pn are large. Clearly, $P(m)$ will be peaked in some way near $m = pn$. Using Stirling's approximation (3-79) for the factorials, the probability of m successes is

$$P(m) = \frac{1}{\sqrt{2\pi n}}\left(\frac{m}{n}\right)^{-m-1/2}\left(\frac{n-m}{n}\right)^{-n+m-1/2}p^m(1-p)^{n-m}$$

$$= \frac{1}{\sqrt{2\pi n}}\exp\left[-\left(m+\frac{1}{2}\right)\ln\frac{m}{n} - \left(n-m+\frac{1}{2}\right)\ln\frac{n-m}{n}\right.$$

$$\left. + m\ln p + (n-m)\ln(1-p)\right] \tag{14-21}$$

We now let $m = np + \xi$, where $\xi \ll np$. Keeping only the dominant terms we obtain

$$P(m) = \frac{1}{\sqrt{2\pi n}}\frac{1}{\sqrt{p(1-p)}}\exp\left[-\frac{1}{2}\frac{\xi^2}{np(1-p)}\right] \tag{14-22}$$

This is customarily written in the form

$$P(m) = \frac{1}{\sqrt{2\pi}\,\sigma}\exp\left(-\frac{1}{2}\frac{\xi^2}{\sigma^2}\right) \tag{14-23}$$

where $\sigma = \sqrt{np(1 - p)}$ is a measure of the width of the distribution. (14-23) is the so-called *Gaussian*,[1] or *normal*, distribution. It is a very good approximation even for quite small values of n. For example, in Table 14–2, we take $n = 10$, $p = 0.4$, and compare the binomial and Gaussian results for $P(m)$.

[1] The result (14-23), although usually associated with the name of Gauss, was probably discovered first by de Moivre. See Cramer (C12) Section 17.8 for an interesting historical summary.

The second limit of the binomial distribution which is of interest to us results when $n \to \infty$ and $p \to 0$ in such a way that the product $np = a$ remains finite. The probability of m successes is, by (14-20),

$$P(m) = \frac{n!}{m!(n-m)!} \, p^m (1-p)^{n-m}$$

Now under the present conditions, with $m \ll n$,

$$\frac{n!}{(n-m)!} \to n^m \qquad \text{and} \qquad (1-p)^{n-m} \to (1-p)^{a/p} \to e^{-a}$$

Therefore,

$$P(m) = \frac{a^m e^{-a}}{m!} \tag{14-24}$$

This is known as the *Poisson* distribution. Note that $\sum_{m=0}^{\infty} P(m) = 1$, as it should.

The Gaussian distribution (14-23) results when an experiment with a finite probability of success is repeated a very large number of times. It is shown in Figure 14–1. The distribution has a width of about $\sqrt{np(1-p)}$ (depending on how high up the peak you measure the width); the important thing is that the width is proportional to $\sqrt{n}$. The distribution gets broader and broader, but the *relative width*, or (width)/n, goes to zero as $n \to \infty$.

The Poisson distribution (14-24), on the other hand, applies when a very large number of experiments is carried out, but the probability of success in each is very small, so that a, the expected number of successes, is a finite number. A classic example of a Poisson distribution is the probability that m calls are made to a switchboard in some given length of time. We may

Table 14–2   Comparison of the Binomial and Gaussian Distributions for $n = 10$ and $p = 0.4$

m	$P(m)_{binomial}$	$P(m)_{Gaussian}$
0	0.0060	0.0092
1	0.0403	0.0395
2	0.1209	0.1119
3	0.2150	0.2091
4	0.2508	0.2575
5	0.2006	0.2091
6	0.1115	0.1119
7	0.0425	0.0395
8	0.0106	0.0092
9	0.0016	0.0014
10	0.0001	0.0001

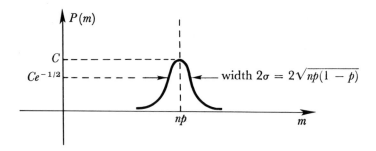

Figure 41–1   A Gaussian distribution. The maximum value of $P(m)$
    is $C = 1/(\sqrt{2\pi}\sigma)$

think of each second as denoting a separate experiment, success being a phone call during that second.   A more interesting example (at least for physicists) of the same kind is the distribution in time of cosmic-ray counts (or counts produced by any other random phenomenon).   For fairly small a, the Poisson distribution is quite asymmetric, as shown in Figure 14–2 for $a = 2$.   Of course, if $a \gg 1$, the distribution becomes Gaussian.   The half-width is then $\sigma = \sqrt{np(1 - p)} \approx \sqrt{a}$.

14–5   GENERAL PROPERTIES OF DISTRIBUTIONS

So far we have only discussed probabilities of discrete results, the probability of one success or two successes, and so forth.   One can also define continuous probability distributions.   For example, if x is a variable which can take on any real value, we may define $p(x)$, the probability distribution of x, by

$p(x)\, dx =$ probability that x lies in a small interval dx at x

When we drew the Gaussian distribution in Figure 14–1 as a smooth curve, we were really considering such a continuous probability distribution.

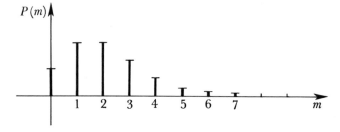

Figure 14–2   The Poisson distribution for $a = 2$

More precisely, if we replace the discrete variable m by the continuous variable x, whose average is $\bar{x} = np$, then, since m changes in unit increments, we must have in the limit of large n

$$P(m) = p(x)\Delta x \qquad \text{with} \qquad \Delta x = 1$$

Thus the Gaussian distribution of m becomes the continuous Gaussian distribution of x:

$$p(x) = \frac{1}{\sqrt{2\pi}\,\sigma} \exp\left[-\frac{1}{2}\frac{(x - \bar{x})^2}{\sigma^2}\right] \qquad (14\text{-}25)$$

where this is just Eq. (14-23) with $\xi = m - np \to x - \bar{x}$. As before, $\sigma = \sqrt{np(1 - p)}$.

The *mean* value (or *expectation* value) of a quantity x is

$$\langle x \rangle = \sum_x P(x)x \qquad (14\text{-}26)$$

if x takes on discrete values only, or more generally,

$$\langle x \rangle = \int p(x)x \, dx \qquad (14\text{-}27)$$

The mean of any function of x is

$$\langle f(x) \rangle = \int p(x)f(x) \, dx \qquad (14\text{-}28)$$

EXAMPLES

For the Gaussian distribution (14-25), the mean value of x is

$$\langle x \rangle = \int_{-\infty}^{\infty} \frac{1}{\sqrt{2\pi}\,\sigma} \exp\left[-\frac{1}{2}\frac{(x - \bar{x})^2}{\sigma^2}\right] x \, dx = \bar{x} \qquad (14\text{-}29)$$

as stated above, and the mean-square deviation of x from its average is

$$\langle (x - \bar{x})^2 \rangle = \int_{-\infty}^{\infty} \frac{1}{\sqrt{2\pi}\,\sigma} \exp\left[-\frac{1}{2}\frac{(x - \bar{x})^2}{\sigma^2}\right](x - \bar{x})^2 \, dx = \sigma^2 \qquad (14\text{-}30)$$

In general, $\langle (x - \bar{x})^2 \rangle$ is called the *variance* of x, and its square root is called the *standard deviation*, usually denoted by σ. We see that for a Gaussian distribution the standard deviation is the half-width σ.

The nth *moment* of a distribution $p(x)$ is defined as the mean value of x^n:

$$\langle x^n \rangle = \int p(x)x^n \, dx \qquad (14\text{-}31)$$

We note in passing a useful relation between the second moment, the mean value, and the variance.   It is

$$\langle(x - \langle x\rangle)^2\rangle = \langle x^2 - 2x\langle x\rangle + \langle x\rangle^2\rangle = \langle x^2\rangle - \langle x\rangle^2$$

that is,

$$\sigma^2 = \langle x^2\rangle - \langle x\rangle^2 \qquad (14\text{-}32)$$

One often characterizes a distribution by its moments.   This leads to the definition of a moment-generating function, or *characteristic function*, of a distribution.   This is just the Fourier transform of the probability density

$$\phi(k) = \int dx\, e^{ikx} p(x) \qquad (14\text{-}33)$$

Note what happens when we expand $\phi(k)$ as a power series in k:

$$\phi(k) = \int dx p(x)\left(1 + ikx - \frac{1}{2!} k^2 x^2 - \frac{i}{3!} k^3 x^3 + \cdots\right)$$

$$= 1 + ik\langle x\rangle - \frac{k^2}{2!}\langle x^2\rangle + \cdots + \frac{(ik)^n}{n!}\langle x^n\rangle + \cdots \qquad (14\text{-}34)$$

Thus the characteristic function is a generating function for the moments.

EXAMPLE

The characteristic function for a Gaussian distribution is

$$\phi(k) = \int_{-\infty}^{\infty} dx\, e^{ikx}\, \frac{1}{\sqrt{2\pi}\,\sigma} \exp\left[-\frac{(x - \bar{x})^2}{2\sigma^2}\right]$$

$$= \frac{1}{\sqrt{2\pi}\,\sigma} \int_{-\infty}^{\infty} \exp\left\{-\frac{1}{2}\left[\frac{x}{\sigma} - \left(\frac{\bar{x}}{\sigma} + ik\sigma\right)\right]^2\right\} \exp\left(ik\bar{x} - \frac{k^2\sigma^2}{2}\right) dx$$

$$\phi(k) = e^{ik\bar{x}} e^{-k^2\sigma^2/2} \qquad (14\text{-}35)$$

The Gaussian factor in $\phi(k)$ means that only small k are important; $\phi(k)$ is very small for $|k| \gg 1/\sigma$.   (Compare the uncertainty principles of Section 4–2.)

We shall now illustrate how useful characteristic functions can be.   Suppose x is a random variable with a probability distribution $p(x)$ and y is an independent random variable with a probability distribution $q(y)$.   Now consider some quantity z which is a function of x and y:

$$z = f(x, y)$$

What can we say about the probability distribution of z; for example, what is the probability that z lies in some interval dz at z_0?

Consider the various moments of the z-distribution. We have

$$\langle z \rangle = \iint f(x, y)p(x)\, dx\, q(y)\, dy$$

$$\langle z^2 \rangle = \iint [f(x, y)]^2 p(x)\, dx\, q(y)\, dy \tag{14-36}$$

etc.

Thus the characteristic function of the z-distribution is

$$\phi(k) = \iint e^{ikf(x,\,y)}p(x)\, dx\, q(y)\, dy \tag{14-37}$$

If $f(x, y)$ is the simple function $x + y$, we have

$$\phi(k) = \iint e^{ik(x+y)}p(x)\, dx\, q(y)\, dy$$

$$= \int e^{ikx}p(x)\, dx \int e^{iky}q(y)\, dy \tag{14-38}$$

This is a very important result. The characteristic function for the *sum* of two independent random variables is the *product* of the separate characteristic functions. Clearly, we can generalize from two to an arbitrary number.

The most important result of this theorem is the so-called *central limit theorem*. Suppose a random variable x has the probability distribution $p(x)$, with first and second moments $\bar{x}$ and $\langle x^2 \rangle$. Then suppose we make n "measurements" of x and form the average

$$a = \frac{1}{n}(x_1 + x_2 + \cdots + x_n) \tag{14-39}$$

What is the distribution $P(a)$ of the random variable a? We shall investigate instead of $P(a)$ the distribution $Q(a - \bar{x})$ of the variable $(a - \bar{x})$, whose average is zero. This is equivalent since $P(a) = Q(a - \bar{x})$.

The characteristic function of $Q(a - \bar{x})$ is

$$\Phi(k) = \int e^{ik(a-\bar{x})}Q(a - \bar{x})\, da$$

$$= \int \exp\left\{\frac{ik}{n}[(x_1 - \bar{x}) + (x_2 - \bar{x}) + \cdots + (x_n - \bar{x})]\right\}$$

$$\times\, p(x_1)\, dx_1 \cdots p(x_n)\, dx_n$$

$$= \int \exp\left[\frac{ik}{n}(x_1 - \bar{x})\right]p(x_1)\, dx_1 \int \exp\left[\frac{ik}{n}(x_2 - \bar{x})\right]p(x_2)\, dx_2 \cdots$$

$$= \left[\phi\left(\frac{k}{n}\right)\right]^n \tag{14-40}$$

where $\phi(k)$ is the characteristic function of the variable $(x - \bar{x})$, which has first and second moments 0 and σ^2 by definition of $\bar{x}$ and σ.   Thus

$$\phi(k) = \int e^{ik(x-\bar{x})} p(x)\, dx$$

$$= 1 - \tfrac{1}{2} k^2 \sigma^2 + O(k^3) \tag{14-41}$$

and from (14-40)

$$\Phi(k) = \left[1 - \frac{1}{2} \frac{k^2 \sigma^2}{n^2} + O\!\left(\frac{k^3}{n^3}\right) \right]^n$$

which, in the limit of large n, becomes

$$\Phi(k) \xrightarrow[n \to \infty]{} e^{-k^2 \sigma^2 / 2n} \tag{14-42}$$

By inverting this Fourier transform, or referring to (14-35), we find for large n

$$Q(a - \bar{x}) = P(a) = \frac{\sqrt{n}}{\sqrt{2\pi}\,\sigma} \exp\left[-\frac{n}{2} \frac{(a - \bar{x})^2}{\sigma^2} \right] \tag{14-43}$$

Thus, no matter what form the distribution of x might have, the distribution of the average of a large number of measurements of x is Gaussian, centered at $\bar{x}$ with a standard deviation $n^{-1/2}$ times that of the distribution of x.

14-6   MULTIVARIATE GAUSSIAN DISTRIBUTIONS

It is straightforward to generalize the concept of a probability distribution $p(x)$ for the random variable x to a *multivariate* probability distribution $p(x_1, x_2, \ldots, x_n)$ for the n random variables $x_1, \ldots, x_n$.   The probability that x_1 is in dx_1 about the value a_1, and simultaneously x_2 is in dx_2 about $a_2, \ldots$, is simply $p(a_1, a_2, \ldots, a_n)\, dx_1\, dx_2 \cdots dx_n$.

Of particular interest are multivariate Gaussian distributions, that is, distributions of the form[2]

$$p(x_1, x_2, \ldots, x_n) = N \exp\left[-\sum_{i \le j} a_{ij} x_i x_j \right] \tag{14-44}$$

The quadratic form $\sum a_{ij} x_i x_j$ must of course be positive definite (Why?); the

---

[2] We have, without loss of generality, assumed that the expectation values of all x_i are zero.   This accounts for the absence of linear terms in the exponent of (14-44).

evaluation of the normalization constant N is left as an exercise for the student (Problem 14-12).

The characteristic function of a multivariate Gaussian distribution is again Gaussian;

$$\phi(k_1, k_2, \ldots, k_n) = \int dx_1 \cdots dx_n N \exp\left[-\sum_{i \leq j} a_{ij} x_i x_j\right] e^{i(k_1 x_1 + \cdots + k_n x_n)}$$

$$= \exp\left[-\sum_{i \leq j} b_{ij} k_i k_j\right] \tag{14-45}$$

This important fact is an immediate generalization of the result (14-35) for a single-variable Gaussian distribution.   The absence of a normalizing factor in front of the exponential in (14-45) follows from the normalization of $p(x_1, \ldots, x_n)$ which requires $\phi(0, \ldots, 0) = 1$.

If the characteristic function of a multivariate Gaussian distribution is known, the momenta of the distribution are easily calculated [compare (14-34)].   For example, it follows from (14-45) that

$$\langle x_1^2 \rangle = -\left.\frac{\partial^2 \phi(k_1, \ldots, k_n)}{\partial k_1^2}\right|_{k_i = 0}$$

$$\langle x_1^3 x_4 \rangle = \left.\frac{\partial^4 \phi(k_1, \ldots, k_n)}{\partial k_1^3 \partial k_4}\right|_{k_i = 0}$$

and in general

$$\langle x_1^{\alpha_1} x_2^{\alpha_2} \cdots x_n^{\alpha_n} \rangle = (-i)^{\alpha_1 + \alpha_2 + \cdots + \alpha_n} \left.\frac{\partial^{\alpha_1 + \alpha_2 + \cdots + \alpha_n} \phi(k_1, \ldots, k_n)}{\partial k_1^{\alpha_1} \partial k_2^{\alpha_2} \cdots \partial k_n^{\alpha_n}}\right|_{k_i = 0} \tag{14-46}$$

Actually this device works for any multivariate distribution, but we shall only make use of it in connection with Gaussian distributions.

The "physical significance" of the coefficients b_{ij} in (14-45) is now apparent. For

$$\langle x_i^2 \rangle = -\frac{\partial^2 \phi}{\partial k_i^2}(0, \ldots, 0) = 2b_{ii}$$

and

$$\langle x_i x_j \rangle = -\frac{\partial^2 \phi}{\partial k_i \partial k_j}(0, \ldots, 0) = b_{ij}$$

In other words, the coefficient of k_i^2 in the exponent of the characteristic function (14-45) is half of $\langle x_i^2 \rangle$; the coefficient of the cross-term $k_i k_j$ is simply $\langle x_i x_j \rangle$.

As a simple application of this result, consider two Gaussian variables x_1 and x_2, with the characteristic function

$$\phi(k_1, k_2) = e^{-ak_1{}^2 - bk_1k_2 - ck_2{}^2}$$

We have just seen that $a = \frac{1}{2}\langle x_1^2 \rangle$, $b = \langle x_1 x_2 \rangle$, $c = \frac{1}{2}\langle x_2^2 \rangle$. Now suppose that $\langle x_1 x_2 \rangle = 0$, that is, the variables are *uncorrelated*.[3] Then $b = 0$, and the characteristic function is simply a product of separate characteristic functions

$$\phi(k_1, k_2) = \phi_1(k_1)\phi_2(k_2)$$

Taking the Fourier transform of this equation gives for the probability distribution

$$p(x_1, x_2) = p_1(x_1)p_2(x_2)$$

that is, the variables x_1 and x_2 are statistically independent. Hence the theorem: *Uncorrelated Gaussian random variables are statistically independent.*

As a less trivial illustration of the power of these techniques, a noise signal $I(t)$ is usually assumed to be a *Gaussian random process*: that is, the distribution $f[I(t)]$ is Gaussian, the multivariate, or joint, distribution $g[I(t), I(t')]$ is Gaussian, and so on. We shall in addition assume that our process is *stationary*, which means that the distribution of $I(t)$ is independent of t, the joint distribution of $I(t)$ and $I(t')$ depends only on the difference $t' - t$, and so on. Finally we assume $\langle I(t) \rangle = 0$; there is no D.C. background.

What is needed to characterize such noise? Two quantities that immediately come to mind are the *mean square signal* $P = \langle I^2(t) \rangle$, which by assumption is constant, and the *autocorrelation function* $\rho(\tau) = \langle I(t)I(t + \tau) \rangle$, again independent of t by our assumption that the noise is stationary. It now follows that the entire process is completely fixed by the constant P and the function $\rho(\tau)$—in fact, by $\rho(\tau)$ alone, since $P = \rho(0)$.

For example, let us evaluate the joint distribution of $I(t)$ at three times $I(t)$, $I(t + \delta)$, and $I(t + \delta')$. For brevity, define

$$x = I(t)$$
$$y = I(t + \delta)$$
$$z = I(t + \delta')$$

The joint distribution of these three variables is (by assumption!) Gaussian; hence the characteristic function $\phi(k, l, m)$ is Gaussian:

$$\phi(k, l, m) = \exp\left[-ak^2 - bl^2 - cm^2 - dkl - ekm - flm\right]$$

---

[3] Two random variables are said to be uncorrelated, or to have zero correlation, when the expectation value of their product equals the product of their expectation values.

All of the constants $a, \ldots, f$ are now calculable.   For

$$a = \tfrac{1}{2}\langle x^2 \rangle = \tfrac{1}{2}P$$
$$b = \tfrac{1}{2}\langle y^2 \rangle = \tfrac{1}{2}P$$
$$c = \tfrac{1}{2}\langle z^2 \rangle = \tfrac{1}{2}P$$
$$d = \langle xy \rangle = \rho(\delta)$$
$$e = \langle xz \rangle = \rho(\delta')$$
$$f = \langle yz \rangle = \rho(\delta' - \delta)$$

As another problem, what is the autocorrelation function $R(\tau)$ for the noise *power*, that is, for $I^2(t)$?   We must evaluate

$$R(\tau) = \langle I^2(t)I^2(t + \tau) \rangle \tag{14-47}$$

Let $x = I(t)$ and $y = I(t + \tau)$; the joint distribution of x and y has the characteristic function

$$\phi(k, l) = \exp\left[-\tfrac{1}{2}Pk^2 - \tfrac{1}{2}Pl^2 - \rho(\tau)kl\right]$$

The autocorrelation function $R(\tau)$ of (14-47) may now be evaluated by means of the general result (14-46):

$$R(\tau) = \frac{\partial^4 \phi}{\partial k^2\, \partial l^2}\bigg|_{k=l=0} = P^2 + 2[\rho(\tau)]^2$$

Relations of this type have been used by Purcell (P3) in his discussion of the photon counting experiments of Hanbury Brown and Twiss (H5).

14–7   FITTING OF EXPERIMENTAL DATA

In interpreting the significance of experimental results, we make use of the idea of *relative likelihood*, which may be "derived" from Bayes' theorem as follows:

Let A be a possible theory

Let B be an alternative theory

Let E be an experimental result (or results)

We wish to compare $P(A \mid E)$ and $P(B \mid E)$, the probabilities that A and B are true, given the experimental result E.   Bayes' theorem tells us

$$\frac{P(A \mid E)}{P(B \mid E)} = \frac{P(A)}{P(B)}\frac{P(E \mid A)}{P(E \mid B)} \tag{14-48}$$

Now let us suppose $P(A) = P(B)$; the two theories are a priori "equally likely," whatever that means. This is sometimes not true, in which case Eq. (14-48) must be used as it stands. However, with this assumption,

$$\frac{P(A|E)}{P(B|E)} = \frac{P(E|A)}{P(E|B)} \qquad (14\text{-}49)$$

The quantities on the right are simple; they are just the probabilities of getting the result E if theory A or theory B is right, respectively. Thus, the relative "likelihood" of two theories is taken to be the ratio of the probabilities, predicted by those two theories, of getting the result we actually do get. The most likely theory will, of course, be the one with the maximum likelihood.

Suppose we are trying to determine a parameter α in a theory. We do an experiment and get an answer. If the probability $L(\alpha)$ of getting that answer, as a function of α, looks like the curve in Figure 14-3a, we have a "good" experiment and α is well determined. If, on the other hand, the likelihood curve looks like Figure 14-3b, we are not likely to place much reliance in the result $\alpha = \alpha^*$.

If the curve is relatively narrow, it will often have a Gaussian shape. We can then describe the experimental results by giving the maximum likelihood value α^* and the *standard error*:

$$\Delta\alpha = \left[\frac{\int (\alpha - \alpha^*)^2 L(\alpha)\,d\alpha}{\int L(\alpha)\,d\alpha}\right]^{1/2}$$

Note that $\Delta\alpha$ is the same as the standard deviation σ if α were a random variable with mean α^* and distribution $L(\alpha)$. Of course, $L(\alpha)$ is not really the probability distribution of α, the true distribution being a δ-function if α is a parameter having a definite (though unknown) value.

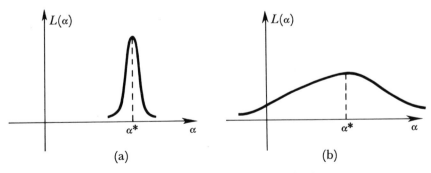

Figure 14–3   Relative likelihood functions $L(\alpha)$ for a parameter α obtained from (a) a "good" experiment and (b) a "poor" experiment

If $L(\alpha)$ is Gaussian,

$$L(\alpha) = \frac{1}{\sqrt{2\pi}\,\sigma} \exp\left[-\frac{(\alpha - \alpha^*)^2}{2\sigma^2}\right]$$

$$\frac{d}{d\alpha}\ln L(\alpha) = -\frac{(\alpha - \alpha^*)}{\sigma^2}$$

and the standard error is

$$\Delta\alpha = \sigma = \left[-\frac{d^2}{d\alpha^2}\ln L(\alpha)\right]^{-1/2} \tag{14-50}$$

If the experiment is statistically poor, or the likelihood function is non-Gaussian for some other reason, the values α^* and $\Delta\alpha$ do not describe the results completely, and it is best to give a plot of the likelihood function.

We shall illustrate the maximum likelihood principle by several examples. First, suppose we wish to determine the mean life τ of an unstable particle. Several such particles are observed experimentally. The first lasts for time t_1, the second for time t_2, and so on. What is the most likely value of τ we can deduce from these results?

The probability of a particle of mean life τ decaying at time t within dt is

$$p(t)\,dt = \frac{1}{\tau}e^{-t/\tau}\,dt$$

Thus the probability of our experiment turning out the way it did is proportional to

$$L(\tau) = \frac{1}{\tau}e^{-t_1/\tau}\frac{1}{\tau}e^{-t_2/\tau}\cdots\frac{1}{\tau}e^{-t_n/\tau}$$

$$L(\tau) = \exp\left(-\frac{1}{\tau}\sum_{i=1}^{n}t_i - n\ln\tau\right)$$

We want to maximize this probability; differentiating with respect to τ gives

$$\frac{d}{d\tau}\ln L(\tau) = \frac{1}{\tau^2}\sum_{i=1}^{n}t_i - \frac{n}{\tau} = 0$$

The maximum likelihood value of τ is thus

$$\tau^* = \frac{1}{n}\sum_{i=1}^{n}t_i \tag{14-51}$$

which is simply the average of the observed times t_i.

We may also find the standard error, if the number of observed particles is

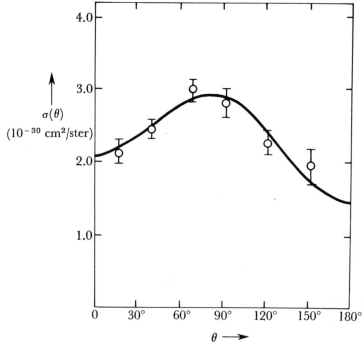

Figure 14–4 Typical angular distribution in a nuclear reaction showing experimental data and a least-squares fit

large enough so that $L(\tau)$ is approximately Gaussian. It is, by (14-50),

$$\Delta\tau = \left[-\frac{d^2}{d\tau^2} \ln L(\tau) \right]^{-1/2}$$

$$= \left(\frac{2}{\tau^3} \sum_{i=1}^{n} t_i - \frac{n}{\tau^2} \right)^{-1/2}$$

This expression is not independent of τ as it would be for a true Gaussian. Evaluating it at τ^*, we obtain[4]

$$\Delta\tau = \frac{\tau^*}{\sqrt{n}} \qquad (14\text{-}52)$$

Another application of the principle of maximum likelihood is the determination of a number of parameters in a theory when we have more experimental points than parameters, and, because of experimental errors, a perfect fit is not possible. As an illustration of this kind of problem, consider the data shown in Figure 14-4. The angular distribution in a nuclear

---

[4] This procedure will usually underestimate the error, and a better one is to average $[-(d^2/d\tau^2) \ln L(\tau)]^{-1/2}$ over the likelihood function. However, if the difference is large, it is probably better just to plot the likelihood function in the first place.

reaction has been investigated by measuring the differential cross section $\sigma(\theta)$ at a number of different angles, with errors as indicated.   We wish to fit the data with a "theoretical" curve of the form

$$\sigma(\theta) = a_0 + a_1 \cos \theta + a_2 \cos^2 \theta + a_3 \cos^3 \theta \qquad (14\text{-}53)$$

and find the maximum likelihood values of the parameters a_m.

More generally, suppose we measure N quantities for which we obtain the experimental values x_i with standard errors σ_i.   Suppose we also have "theoretical" expressions for these quantities and that these expressions contain a set of n parameters a_m whose values are to be adjusted to give a best fit to the data.   Call $\xi_i = \xi_i(a_1, a_2, \ldots, a_n)$ the "theoretical" expression for x_i.

We consider the case where the probability distributions for the experimental results x_i are Gaussian.   Then the likelihood function is proportional to

$$L(a_1, a_2, \ldots, a_n) = \frac{1}{(2\pi)^{N/2}\sigma_1\sigma_2 \cdots} \exp\left[-\sum_{i=1}^{N} \frac{(x_i - \xi_i)^2}{2\sigma_i^2}\right] \qquad (14\text{-}54)$$

and we wish to adjust the parameters a_m to give the maximum likelihood. This is equivalent to minimizing the exponent

$$\sum_{i=1}^{N} \frac{(x_i - \xi_i)^2}{2\sigma_i^2} = \frac{1}{2}\chi^2(a_m) \qquad (14\text{-}55)$$

and the result is therefore called a *least-squares* fit.   The quantity χ^2 defined by (14-55) will be used later [compare (14-71)].

To find the maximum likelihood values a_m^* of the a_m, we take the derivatives of (14-55) with respect to these parameters:

$$\sum_{i=1}^{N} \frac{(x_i - \xi_i)}{\sigma_i^2} \frac{\partial \xi_i}{\partial a_m} = 0 \qquad m = 1, 2, \ldots, n \qquad (14\text{-}56)$$

The solution of these equations can be carried out in an elegant form if the ξ_i are linear homogeneous functions of the parameters a_j; that is, if

$$\xi_i = \sum_{m=1}^{n} C_{im} a_m \qquad (14\text{-}57)$$

with known coefficients C_{im}.   The maximum likelihood conditions (14-56) then become

$$\sum_{i} \frac{C_{im}}{\sigma_i^2} x_i = \sum_{i,l} \frac{C_{im} C_{il}}{\sigma_i^2} a_l \qquad (14\text{-}58)$$

We now define a "data vector" X and a "measurement matrix" M with components as follows:

$$X_m = \sum_{i=1}^{N} \frac{C_{im}}{\sigma_i^2} x_i \tag{14-59}$$

$$M_{ml} = \sum_{i=1}^{N} \frac{C_{im} C_{il}}{\sigma_i^2} = M_{lm} \tag{14-60}$$

Note that X depends on the experimental results x_i and the errors σ_i, whereas M depends only on the errors. M is symmetric.

The conditions (14-58) are then the matrix equation[5]

$$X = Ma \tag{14-61}$$

with the solution

$$a = M^{-1}X \tag{14-62}$$

Next, we investigate the errors associated with the parameters a_m determined as above. There are two ways of thinking about this problem, both leading to the same algebra and the same results. One is to consider the likelihood function (14-54) to be the probability distribution of the parameters a_m and to compute from it the mean-square deviations $\langle (a_m - a_m^*)^2 \rangle$, etc. The other point of view is to imagine repeating the entire experiment a large number of times in the same way (that is, with the same errors σ_i) and then calculating the mean-square deviations of the a_m obtained in individual experiments from the grand average $\bar{a}_m$ over all the experiments. (We denote by an upper bar an average over all the imagined experiments. The value of such an average is not known, of course, but will not be needed.)

To find the errors in the parameters, we need to calculate the expected mean-square deviations $\langle (a_m - \bar{a}_m)^2 \rangle$, or more generally,

$$\langle (a_m - \bar{a}_m)(a_l - \bar{a}_l) \rangle$$

We have from (14-62) and (14-59)

$$(a_m - \bar{a}_m) = \sum_k (M^{-1})_{mk}(X_k - \bar{X}_k) = \sum_{kj} (M^{-1})_{mk} \frac{C_{jk}}{\sigma_j^2} (x_j - \bar{x}_j) \tag{14-63}$$

---

[5] Equations (14-57), (14-59), and (14-60) may also be written in matrix form, if we define a matrix D by $D_{ij} = C_{ji}/\sigma_j^2$. Then these equations are

$$\xi = Ca$$

$$X = Dx$$

and

$$M = DC$$

Note that C is not a square matrix and has no inverse—except in the case $n = N$. If the number of parameters equals the number of data points, a perfect fit $\xi_i = x_i$ is achieved by the parameters

$$a = C^{-1}x \quad \text{(case } N = n \text{ only)}$$

while

$$\langle (x_j - \bar{x}_j)(x_i - \bar{x}_i) \rangle = \sigma_i^2 \, \delta_{ij} \tag{14-64}$$

because the individual measurements x_i are assumed to be statistically independent.   Therefore,

$$\langle (a_m - \bar{a}_m)(a_l - \bar{a}_l) \rangle = \sum_{kjpi} (M^{-1})_{mk} \frac{C_{jk}}{\sigma_j^2} (M^{-1})_{lp} \frac{C_{ip}}{\sigma_i^2} \sigma_i^2 \, \delta_{ij}$$

$$= \sum_{kp} (M^{-1})_{mk}(M^{-1})_{lp} M_{pk} = (M^{-1})_{ml} \tag{14-65}$$

and the standard error in a_m is

$$\Delta a_m = \langle (a_m - \bar{a}_m)^2 \rangle^{1/2} = \sqrt{(M^{-1})_{mm}} \tag{14-66}$$

Note that the cross terms $\langle (a_m - \bar{a}_m)(a_l - \bar{a}_l) \rangle$ with $l \neq m$ are not zero in general.   This means that the parameters a_m are not statistically independent and their errors are correlated!   (One could, of course, diagonalize the matrix M^{-1} and thereby find a set of statistically independent combinations of the a_m.)

The error to be assigned to any quantity which is a function of the parameters a_m may be found from the *error matrix* M^{-1}.

EXAMPLE

Find the error in $z = a_1 + 2a_2$.   The variance, or mean-square deviation is

$$\langle (z - \bar{z})^2 \rangle = \langle [(a_1 - \bar{a}_1) + 2(a_2 - \bar{a}_2)]^2 \rangle$$

$$= \langle (a_1 - \bar{a}_1)^2 \rangle + 4\langle (a_1 - \bar{a}_1)(a_2 - \bar{a}_2) \rangle + 4\langle (a_2 - \bar{a}_2)^2 \rangle$$

$$= (M^{-1})_{11} + 4(M^{-1})_{12} + 4(M^{-1})_{22} \tag{14-67}$$

and

$$\Delta z = \sqrt{\langle (z - \bar{z})^2 \rangle}$$

So far, we have assumed that our theoretical expressions ξ_i are correct for some values of the parameters [which would be the grand average values $\bar{a}_m$ if there are no "systematic" experimental errors.   We would then have $\bar{x}_i = \xi_i(\bar{a}_m)$].   The errors Δa_m found above depend sensitively on this assumption and give no indication of its correctness!   We therefore consider next the problem of deciding whether the *form* of the theoretical expressions used is any good.   The problem is a one-sided one.   We cannot conclude that a given theory is right just because it gives a good fit to the data.   On the other hand, if the fit is bad, then either the theoretical expressions used are wrong or the experimenter has underestimated his errors (or made a mistake).

We shall discuss one quantitative criterion for a good fit, known as the *chi-square* (χ^2) *test.* It consists simply of deciding whether the exponent (14-55) in the likelihood function is larger than one should expect statistically.[6]

If $\bar{x}_i$ is the true mean of x_i, the (normalized) probability of observing x_1 in dx_1, x_2 in dx_2, etc., is

$$P(x_1, x_2, \ldots, x_N)\, dx_1\, dx_2 \cdots dx_N = \prod_{i=1}^{N} \frac{1}{\sqrt{2\pi}\,\sigma_i} \exp\left[-\frac{1}{2\sigma_i^2}(x_i - \bar{x}_i)^2 \right] dx_i$$

(14-68)

Define

$$q_i = \frac{1}{\sigma_i}(x_i - \bar{x}_i) \tag{14-69}$$

and

$$Q(q_i)\, dq_1 \cdots dq_N = P(x_i)\, dx_1 \cdots dx_N$$

Then

$$Q(q_i) = \left(\frac{1}{2\pi}\right)^{N/2} \exp\left(-\frac{1}{2} \sum_{i=1}^{N} q_i^2 \right) \tag{14-70}$$

and Q depends on the q_i only through

$$\sum_{i=1}^{N} q_i^2 = \chi^2 \tag{14-71}$$

where χ^2 is the square of the "radius," or distance in N-dimensional Euclidean space from the origin to the point with Cartesian coordinates $q_1 \cdots q_N$.

Since $Q(q_i)$ depends only on χ^2, it has the same value everywhere on a "spherical" surface in N dimensions and the probability distribution of χ^2 must have the form

$$F(\chi^2)\, d\chi^2 = F(\chi^2)2\chi\, d\chi$$

$$= \int_{\substack{\text{spherical} \\ \text{shell}\, \chi, d\chi}} Q(q_i)\, dq_1 \cdots dq_N = (2C)e^{-\chi^2/2}\chi^{N-1}\, d\chi$$

$$= Ce^{-\chi^2/2}\chi^{N-2}2\chi\, d\chi \tag{14-72}$$

We determine the constant C by the normalization

$$\int_0^\infty F(\chi^2)\, d\chi^2 = 1 = C \int_0^\infty e^{-z/2}z^{(N/2)-1}\, dz$$

so

$$C = [2^{N/2}\Gamma(N/2)]^{-1}$$

---

[6] For a more precise but somewhat cumbersome approach to this question, see Problem 14-7.

Thus, from (14-72),

$$F(\chi^2)\,d\chi^2 = \frac{1}{2^{N/2}\Gamma(N/2)}\,e^{-\chi^2/2}(\chi^2)^{(N/2)-1}\,d\chi^2 \qquad (14\text{-}73)$$

This distribution of χ^2 is known as the *chi-square distribution* with N *degrees of freedom*. The probability that χ^2 exceeds the observed value χ_0^2 is

$$P_N(\chi^2 > \chi_0^2) = \int_{\chi_0^2}^{\infty} F(\chi^2)\,d\chi^2 \qquad (14\text{-}74)$$

Tables of this function are given in the *Handbook of Chemistry and Physics* and in other places. If this chi-square probability is high, the fit is good, of course.

We have assumed in discussing the chi-square distribution that the true parameters $\bar{a}_m$, and thus the true mean values $\bar{x}_i$, are known. However, in practice, this is not the case; we adjust the parameters a_m to the maximum likelihood values a_m^*, which give the best fit to the data. Using these best values a_m^* to compute the observed χ^2 according to (14-55), we will, of course, obtain a smaller value than would be found with fixed $\bar{a}_m$. For example, if the number of adjustable parameters n is equal to the number of data points N a perfect fit is possible and $\chi^2 = 0$. If $N > n$, it turns out that we should use the number $N' = N - n$ instead of N as the number of degrees of freedom in the chi-square test (14-74). (See Problem 14-17.)

We conclude this section with a comment about the role of a priori probabilities. It should be kept in mind that the use of likelihood functions as defined above assumes equal a priori probabilities for the different "theories," that is $P(A)/P(B) \approx 1$. In general, this is a reasonable assumption, and, for a good experiment, it does not make much difference. However, in some cases, a given "theory" is known to be impossible so that the corresponding $P(A) = 0$. Also, previous experiments may make the a priori likelihood of some theories greater than others. The effects of unequal a priori probabilities may be included by returning to the fundamental equation (14-48) based on Bayes' theorem.

REFERENCES

There are many references to probability and statistics, both theoretical and applied. Two of the more useful are the books by Feller (F2) and by Hoel (H11). Both of these have many exercises and are easy to learn from.

A standard reference, at a somewhat higher mathematical level than Feller or Hoel, is the book by Cramer (C12).

Probability and statistics have many applications in physics. The whole theory of statistical mechanics is an example. This subject is treated in numerous books, a few of which are Landau and Lifshitz (L3); Huang (H12); Tolman (T7); and Davidson (D1). More elementary treatments are given by Margenau and Murphy (M2) Chapter 12 and Kittel (K2).

The application of probability theory to the analysis of measurements which are subject to random errors is treated in the classical reference by Whittaker and Robinson (W4). More recent work, as applied to the determination of the fundamental atomic constants, is to be found in Cohen (C5, C6). The notes of Orear (O2) are also very useful for physicists.

PROBLEMS

14-1 Consider the three drawers A, B, C of page 375 of the text. A drawer is chosen at random, and a coin is chosen from it at random. The coin is golden. The coin is returned to the drawer, the drawer is shaken thoroughly, and another coin is drawn at random from the same drawer. Again the coin is golden. The coin is returned to the drawer, the drawer is once more shaken, and a final random draw is made from it. Find the probability that the last coin drawn will be golden, if

(a) the gold coins are identical

(b) the gold coins are distinguishable, and the first two were different

(c) the gold coins are distinguishable, and the first two were the same coin.

14-2 A counting circuit, or scaler, consists of a series of binary stages, each of which gives a pulse out for every second input pulse. Suppose random cosmic-ray counts are fed into a scale-of-four (two binary stages) whose output pulses are counted by a mechanical register. If the mechanical register remains insensitive or "dead" for 0.1 sec after each pulse it receives (whether recorded or not), calculate the fraction of counts missed by it if the average cosmic-ray rate is 10 per sec. (Assume counting losses in the scaling circuit are negligible.)

14-3 In a simple one-dimensional "random walk," steps of unit length are made in the positive or negative x-direction with equal probability.

(a) Find the probability distribution for the position x_n after n steps. Find $\langle x_n \rangle$ and $\langle x_n^2 \rangle$.

(b) Find the expected number of returns to the origin if a total of N steps are taken.

(c) Find the probability that the walk returns to the origin at the nth step, but not before.

14-4 Consider an asymmetric random walk along the x-axis, beginning at the origin; when each step is taken, there is a probability p that the step is $+1$, and a probability $q(= 1 - p)$ that the step is -1.

(a) Let $f(m)$ be the probability of *ever* reaching the point $x = m$ ($m > 0$). Show that $f(m + 1) = f(m)f(1)$, and $f(1) = p + qf(2)$, and hence calculate $f(m)$.

(b) Let $g(m, n)$ be the probability of reaching the point $x = +m$ *before* reaching the point $x = -n$. Show that

$$f(m) = g(m, n) + [1 - g(m, n)]f(m + n)$$

and hence evaluate $g(m, n)$.

(c) A physicist with some telekinetic ability is able to call correctly the flip of a coin 60% of the time. With two marbles in his possession, he challenges a man with an infinite supply of marbles to a coin-tossing competition; each time a coin is tossed, the loser pays the winner one marble. What is the probability of the physicist ultimately losing all his marbles?

14-5 The random variable x has the probability distribution

$$f(x) = e^{-x} \qquad (0 < x < \infty)$$

(a) Find $\langle x \rangle$.

(b) Two values x_1 and x_2 are chosen independently. Find $\langle x_1 + x_2 \rangle$ and $\langle x_1 x_2 \rangle$.

(c) What is the probability distribution $P(a)$ of the random variable $a = \frac{1}{2}(x_1 + x_2)$?

14-6 The random variable x has the probability distribution

$$f(x) = \frac{1}{\sqrt{2\pi}} e^{-x^2/2} \qquad (-\infty < x < \infty)$$

(a) n independent measurements x_i are made. Let

$$a = \frac{1}{n}(x_1 + x_2 + \cdots + x_n)$$

Find $P(a)$, the probability distribution of a. Find $\langle a \rangle$ and $\langle a^2 \rangle$.

(b) Let $s = (1/n)(x_1^2 + x_2^2 + \cdots + x_n^2)$. Find $Q(s)$, the probability distribution of s. Find $\langle s \rangle$.

14-7 The n quantities $x_1, x_2, \ldots, x_n$ are statistically independent; each has a Gaussian distribution with zero mean and a common variance:

$$\langle x_i \rangle = 0 \qquad \langle x_i^2 \rangle = \sigma^2 \qquad \langle x_i x_j \rangle = 0$$

(a) n new quantities y_i are defined by $y_i = \sum_j M_{ij} x_j$, where M is an orthogonal matrix. Show that the y_i have all the statistical properties of the x_i; that is, they are independent Gaussian variables with

$$\langle y_i \rangle = 0 \qquad \langle y_i^2 \rangle = \sigma^2 \qquad \langle y_i y_j \rangle = 0$$

(b) Choose $y_1 = \sqrt{\dfrac{1}{n}}(x_1 + x_2 + \cdots + x_n)$, and the remaining y_i arbitrarily, subject only to the restriction that the transformation be orthogonal. Show thereby that the mean $\bar{x} = \dfrac{1}{n}(x_1 + \cdots + x_n)$ and the quantity $s = \sum_{i=1}^{n}(x_i - \bar{x})^2$ are statistically *independent* random variables. What are the probability distributions of $\bar{x}$ and s?

(c) One often wishes to test the hypothesis that the (unknown) mean of the Gaussian distribution is really zero, without assuming anything about the magnitude of σ. Intuition suggests $\tau = \bar{x}/\sqrt{s}$ as a useful quantity. Show that the probability distribution of the random variable τ is

$$p(\tau) = \sqrt{\frac{n}{\pi}} \, \frac{\Gamma\left(\dfrac{n}{2}\right)}{\Gamma\left(\dfrac{n-1}{2}\right)} \, \frac{1}{(1 + n\tau^2)^{n/2}}$$

The crucial feature of this distribution (essentially the so-called *Student t-distribution*[7]) is that it does not involve the unknown parameter σ.

14-8 In Section 14–7 the χ^2 test for plausibility, unlike the test described in Problem 14–7, depended on knowing σ^2 for each experimental point. The following question is thereby raised: How do we estimate σ^2 from experimental data? For some measurements, σ^2 may be estimated from our knowledge of the accuracy of the experimental procedures; we consider here the alternative situation, where σ^2 must be estimated from the experimental results themselves.

Imagine that n measurements of the random variable x yield the results $x_1, x_2, \ldots, x_n$, and that x is assumed to have a Gaussian

---

[7] Student's t is actually equal to $\sqrt{n(n-1)}$ times our τ, and therefore the distribution of t is

$$\frac{1}{\sqrt{\pi(n-1)}} \, \frac{\Gamma\left(\dfrac{n}{2}\right)}{\Gamma\left(\dfrac{n-1}{2}\right)} \left(1 + \frac{t^2}{n-1}\right)^{-(n/2)}$$

distribution

$$p(x) = \frac{1}{\sqrt{2\pi\sigma^2}} \, e^{-(x-a)^2/2\sigma^2}$$

with the two unknown parameters a and σ^2.

(a) Give maximum likelihood estimates for a and σ^2 in terms of the data $x_1, \ldots, x_n$.

(b) An estimate of a parameter is said to be *unbiased* if the expectation value of the estimate equals the parameter. Show that the estimate of a in Part a is unbiased, but that the estimate of σ^2 is biased.

(c) Show that

$$\sigma^2 = \frac{1}{n-1} \sum_{i=1}^{n} (x_i - \bar{x})^2$$

where

$$\bar{x} = \frac{1}{n} \sum_{i=1}^{n} x_i$$

is an unbiased estimate of σ^2. The appearance of $(n-1)$, instead of n, in this estimate is closely related to a similar appearance of $(n-1)$ in the solution of Part b of Problem 14-7.

Moral: Do not use maximum likelihood estimates blindly.

14-9 The quantity y is believed theoretically to depend linearly on the quantity x; that is $y = Ax + B$. Experimental results are

x	1	2	3
y	5 ± 2	9 ± 1	15 ± 2

(a) Evaluate A and B, with probable errors for each.

(b) Evaluate $y(4)$ and its probable error.

14-10 Measurements of the differential cross section for a nuclear reaction at several angles yield the following data:

θ	30°	45°	90°	120°	150°
$\sigma(\theta)$	11	13	17	17	14
error	± 1.5	± 1.0	± 2.0	± 2.0	± 1.5

The units of cross section are 10^{-30} cm^2/steradian.

(a) Make a least-squares fit to $\sigma(\theta)$ of the form

$$\sigma(\theta) = A + B \cos \theta + C \cos^2 \theta$$

Give values and errors for A, B, C.

(b) Find the total cross section $\sigma = \int \sigma(\theta) \, d\Omega$ and its error.

(c) Find the differential cross section at 0° and its error.

14-11 In the decay of polarized Λ-particles, $\Lambda \to \pi^- + p$; the angular distribution in the Λ rest system is

$$w(\theta)\, d\Omega = (1/4\pi)(1 + a \cos \theta)\, d\Omega$$

$w(\theta)\, d\Omega$ is the probability that the π^- will be emitted within solid angle $d\Omega$ at an angle θ from the direction of polarization. (The parameter $a = \alpha P$, where P is the polarization of the Λ particles, and α is an intrinsic asymmetry parameter for Λ decay.)

If an experiment is proposed to measure the parameter a with a 5 percent standard error, how many decay events will have to be observed, assuming $a \approx 0.4$, and assuming all decays are detected, regardless of the angle?.

14-12 Evaluate the normalization constant N in (14-44).

14-13 There is a fairly good one-to-one correspondence between the energy of a fast proton and its *range* in a given material; that is, the distance traveled in that material before the proton is slowed down and stops.   There is, of course, some *straggling*, or variation in path lengths, because the slowing-down process is a statistical one.

Suppose a beam of monoenergetic protons is stopped in a bubble chamber and N individual tracks are found to have lengths x_1, x_2, $\ldots$, x_N.   Assuming the probability distribution for track length is Gaussian, use the data to find maximum likelihood values of the range R (mean length) and the straggling parameter (standard deviation) δ.   Find the errors in both R and δ.

14-14 Calculate the mean and the standard deviation for the variable χ^2 for N degrees of freedom and thereby obtain a rough limit for "reasonable" values of χ^2.

14-15 In the statistical theory of nuclear reactions, a typical cross section is assumed to be given by

$$\sigma(E) = \left| \sum_i \frac{\gamma_i}{E - E_i - i(\Gamma_i/2)} \right|^2$$

where the sum runs over many resonances, each having a resonance energy E_i, partial width γ_i, and total width Γ_i.   With the assumptions
(1) all Γ_i are equal to a (real) constant Γ
(2) the γ_i are independent randomly distributed real numbers with

$$\langle \gamma_i \rangle = 0 \qquad \langle \gamma_i^2 \rangle = \alpha^2 = \text{constant}$$

(3) the E_i are equally spaced, with spacing D

(4) $D \ll \Gamma$

evaluate $\langle \sigma(E) \rangle$, $\langle \sigma^2(E) \rangle$, and $\langle \sigma(E) \sigma(E + k) \rangle$ and show that

$$\frac{\langle \sigma(E) \rangle^2}{\langle \sigma^2(E) \rangle} = \frac{1}{2} \qquad \frac{\langle [\sigma(E) - \sigma(E + k)]^2 \rangle}{\langle \sigma^2(E) \rangle} = \frac{k^2}{k^2 + \Gamma^2}$$

14-16 Two integers are selected at random, say from the integers 1 through 10^9. Calculate the probability of their being relatively prime. *Hint*: The probability of their not both being even is $1 - 1/2^2$.

14-17 (a) Consider the function $F(\mathbf{q}) = (\mathbf{q} - \lambda \mathbf{a} - \mu \mathbf{b})^2$, where $\mathbf{a}$ and $\mathbf{b}$ are constant vectors, while for each choice of the vector $\mathbf{q}$ the scalars λ and μ are adjusted so as to minimize $F(\mathbf{q})$. Show that this minimum $F(\mathbf{q}) = \mathbf{q}_\perp^2$, where we define $\mathbf{q}_\parallel$ and $\mathbf{q}_\perp$ by $\mathbf{q} = \mathbf{q}_\parallel + \mathbf{q}_\perp$, with $\mathbf{q}_\parallel$ and $\mathbf{q}_\perp$ parallel and perpendicular, respectively, to the plane containing $\mathbf{a}$ and $\mathbf{b}$.

(b) Suppose a variable y is known to be a linear function of x, $y = \alpha x + \beta$, with α and β unknown constants. In order to determine these constants experimentally, y is measured at N different values of the variable x, and a least squares fit of these data is made. If the experimental values of y have equal standard errors σ, the fit is made by choosing a and b to minimize the quantity

$$\chi^2 = \sum_{i=1}^{N} \frac{1}{\sigma^2} (y_i - ax_i - b)^2$$

so that a and b are least squares estimates of α and β. Show that the random variable χ^2 has the chi-square distribution (14-73) but with $N - 2$ degrees of freedom.

FIFTEEN

TENSOR ANALYSIS AND DIFFERENTIAL GEOMETRY

In Chapter 6 we discussed general linear vector spaces and linear operators which operate in these spaces. In this chapter we investigate in greater detail the properties of vectors, operators, and related objects which exist in specific linear vector spaces. In physics, the spaces most commonly considered are the ordinary three-dimensional Euclidean space and the four-dimensional space of relativity. In fact, it is in the theory of relativity that tensor analysis finds one of its most extensive applications.

We begin this chapter with a study of Cartesian tensors in ordinary three-dimensional space. This is followed by a short digression on the analytical description of curves in three-space, including the Frenet formulas. Section 15–3 then contains the general treatment of tensor analysis using arbitrary coordinate systems. It ends with a discussion of covariant differentiation of tensors and a brief description of the problem of geodesics.

15–1   CARTESIAN TENSORS IN THREE-SPACE

We shall begin by reviewing the effect of a rotation of Cartesian coordinates on vectors in ordinary three-dimensional space.   Consider what happens if our coordinate system is rotated through an angle of θ in a positive sense about the z-axis, as indicated in Figure 15–1.   The new components of a vector are given in terms of the old ones by

$$\overline{A}_x = A_x \cos \theta + A_y \sin \theta$$
$$\overline{A}_y = - A_x \sin \theta + A_y \cos \theta$$
$$\overline{A}_z = A_z$$

which may be written $\overline{A} = MA$, with the matrix M defined by

$$M = \begin{pmatrix} \cos \theta & \sin \theta & 0 \\ -\sin \theta & \cos \theta & 0 \\ 0 & 0 & 1 \end{pmatrix} \tag{15-1}$$

Matrices describing rotations about the x- or y-axis are easily written down by analogy.

If more than one rotation is being considered, a more specific notation is useful; for example, $M_z(\theta)$ may be used to denote a rotation through the angle θ about the z-axis.

An arbitrary orientation of our coordinate system can be achieved by three successive rotations about coordinate axes.   Conventionally, these are made by

1. Rotating (xyz) through angle α about the z-axis to give $(x'y'z')$.   The matrix describing this rotation is $M_z(\alpha)$.

2. Rotating $(x'y'z')$ through angle β about the y'-axis to give $(x''y''z'')$.   The matrix corresponding to this rotation is $M_{y'}(\beta)$.

3. Rotating $(x''y''z'')$ through angle γ about the z''-axis to give $(\overline{x}\,\overline{y}\,\overline{z}')$.   The associated matrix is $M_{z''}(\gamma)$.

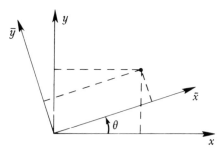

Figure 15–1   Two coordinate systems, one rotated by an angle θ about the z-axis relative to the other

These successive rotations are illustrated in Figure 15–2. Note that β and α are the conventional polar coordinates giving the direction of the final $\bar{z}$-axis relative to the original coordinates (x, y, z). The angles (α, β, γ) are called *Euler angles*.[1] The transformation matrix describing the overall rotation is[2]

$$M(\alpha, \beta, \gamma) = M_{z''}(\gamma)M_{y'}(\beta)M_z(\alpha)$$

$$= \begin{pmatrix} \cos\gamma & \sin\gamma & 0 \\ -\sin\gamma & \cos\gamma & 0 \\ 0 & 0 & 1 \end{pmatrix} \begin{pmatrix} \cos\beta & 0 & -\sin\beta \\ 0 & 1 & 0 \\ \sin\beta & 0 & \cos\beta \end{pmatrix} \begin{pmatrix} \cos\alpha & \sin\alpha & 0 \\ -\sin\alpha & \cos\alpha & 0 \\ 0 & 0 & 1 \end{pmatrix}$$

$$= \begin{pmatrix} \cos\beta\cos\alpha\cos\gamma & \cos\beta\sin\alpha\cos\gamma & -\sin\beta\cos\gamma \\ \quad -\sin\alpha\sin\gamma & \quad +\cos\alpha\sin\gamma & \\ -\cos\beta\cos\alpha\sin\gamma & -\cos\beta\sin\alpha\sin\gamma & \sin\beta\sin\gamma \\ \quad -\sin\alpha\cos\gamma & \quad +\cos\alpha\cos\gamma & \\ \sin\beta\cos\alpha & \sin\beta\sin\alpha & \cos\beta \end{pmatrix} \quad (15\text{-}2)$$

The components of a vector relative to the new system $(\bar{A})$ are related to the components in the old system (A) by

$$\bar{A} = MA \quad (15\text{-}3)$$

Note that M is orthogonal; $M^{-1} = \tilde{M}$. Also, det $M = 1$.

We may now define as a *vector* in our three-dimensional space *any* object A which has three components in a Cartesian coordinate system, which transform according to

$$\bar{A} = MA \quad (15\text{-}4)$$

when we rotate our coordinates, M being the matrix (15-2).

The transformation matrix M defined here is the inverse (or transpose) of the matrix γ used in Chapter 6, as may be easily seen by comparing Eq. (15-4) with Eq. (6-30) of Chapter 6. Whether we call M or its inverse the "transformation matrix" is, of course, only a question of definition. The definition which is most "natural" depends on the way in which the transformation is defined. In Chapter 6 this was done by giving the new base vectors $\mathbf{e}'_i$ as

---

[1] Different conventions are sometimes used for the rotations [compare Goldstein (G4)]. We have used the convention which is most common in present-day quantum mechanical discussions of rotations, as, for example, in Rose (R3).

[2] For some purposes, it is convenient to describe the reorientation of the coordinate system in terms of rotations about the original (unprimed) axes. It may be shown that the same reorientation of coordinates as that shown in Figure 15–2 may be accomplished by the three rotations (read from right to left)

$$M(\alpha, \beta, \gamma) = M_z(\alpha)M_y(\beta)M_z(\gamma)$$

(The student will find it worthwhile to verify this.)

linear combinations of the old ones $\mathbf{e}_j$.   Here, on the other hand, the transformation is defined by giving the new components $\bar{A}_i$ of a vector $\mathbf{A}$ as linear combinations of the old components A_j.

Notice the difference between our definition (15-4) of a vector and the postulates given in Section 6–1 for an abstract vector space.   For example, consider the equation

$$\mathbf{E} = \frac{e}{r^3}\,\mathbf{r} \qquad \text{(Coulomb's law)} \qquad (15\text{-}5)$$

giving the electric field $\mathbf{E}$ at the point $\mathbf{r}$, produced by a point charge e at the origin.   The quantities $\mathbf{E}$ and $\mathbf{r}$ are vectors for our present purposes, since they obey (15-4), but it is inconvenient to think of them as part of an abstract vector space; their (dimensionally inconsistent) sum, for example, has no physical meaning.

Next we consider objects with transformation properties different from vectors.   For example, if A_i and B_i are the components of two vectors, we may form a thing (the "outer" product of two vectors) with nine components,

$$T_{ij} = A_i B_j \qquad (15\text{-}6)$$

The way these components transform when the coordinate system is rotated follows directly from (15-4)

$$\bar{T}_{ij} = \sum_{kl} M_{ik} M_{jl} T_{kl} \qquad (15\text{-}7)$$

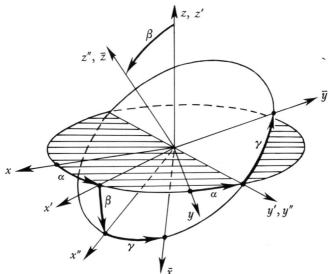

Figure 15–2   Diagram illustrating the three successive rotations through the Euler angles α, β, and γ

Any object with nine components which transform in this way is defined to be a *second-rank tensor.*

EXAMPLE

The tensor of inertia of a rigid body has components[3]

$$I_{jk} = \delta_{jk} \int_{\text{body}} r^2 \rho \, d^3x - \int_{\text{body}} r_j r_k \rho \, d^3x \tag{15-8}$$

where ρ is the density and r is the length of the position vector **r** whose components are r_i. Notice that the second integral contains the outer product of **r** with itself and is thus a second-rank tensor.

What about the Kronecker delta which multiplies the first (scalar) integral in I_{jk}?

$$\delta_{ij} = \begin{cases} 0 & \text{if} \quad i \neq j \\ 1 & \text{if} \quad i = j \end{cases} \quad \text{(in } \textit{all} \text{ coordinate systems)} \tag{15-9}$$

Is δ_{ij}, defined in this way, a tensor? The transformation (15-7) requires

$$\bar{\delta}_{ij} = \delta_{ij} \overset{?}{=} \sum_{kl} M_{ik} M_{jl} \delta_{kl} = \sum_k M_{ik} M_{jk} = (M\tilde{M})_{ij} = \delta_{ij}$$

Thus δ_{ij} *is a second-rank tensor.*

Tensors of third and higher rank are defined by an obvious extension of the definition for rank 2. A tensor of rank n has 3^n components which transform under coordinate rotations like the products $A_i B_j \cdots G_l$ of the components of n vectors; that is, the transformation contains n factors $M_{ik} \cdots$. The "outer product" of two tensors of ranks n and m will clearly give a tensor of rank $n + m$. A vector is a tensor of rank 1 and a *scalar* is a tensor of rank 0.

Now consider the so-called *Levi-Civita antisymmetric symbol*

$$\varepsilon_{ijk} = \begin{cases} 1 & \text{if} \quad ijk = 123, 231, 312 \\ -1 & \text{if} \quad ijk = 321, 213, 132 \\ 0 & \text{otherwise} \end{cases} \quad \text{(in all coordinate systems)} \tag{15-10}$$

Is ε_{ijk} a tensor? That is,

$$\bar{\varepsilon}_{ijk} = \varepsilon_{ijk} \overset{?}{=} \sum_{lmn} M_{il} M_{jm} M_{kn} \varepsilon_{lmn}$$

We observe that for an arbitrary 3×3 matrix A_{ij}

$$\sum_{lmn} A_{il} A_{jm} A_{kn} \varepsilon_{lmn} = \varepsilon_{ijk} \det A \tag{15-11}$$

[3] See, for example, Goldstein (G4) Chapter 5.

Thus, since det $M = 1$ for all rotations, ε_{ijk} *is* in fact a (third-rank) tensor.

One often defines as an *improper rotation* a transformation whose matrix M is orthogonal but has determinant -1.   An improper rotation equals a *proper rotation* (det $M = +1$) followed by a reflection or inversion.   A right-handed coordinate system becomes left-handed if we apply an improper rotation to it.

We now see that ε_{ijk} is not really a tensor if we admit the possibility of det M being -1.   A tensor whose transformation law is

$$\bar{A}_{ij...} = \sum_{kl...} M_{ik} M_{jl} \cdots A_{kl...} \times \det M \qquad (15\text{-}12)$$

is called a *pseudotensor*. ε_{ijk} is a pseudotensor.   The outer product of a tensor and a pseudotensor is a pseudotensor, and the outer product of two pseudotensors is a tensor since $(\det M)^2 = 1$.

The transformation law (15-7) for a second-rank tensor can be written

$$\bar{T} = MTM^{-1} \qquad (15\text{-}13)$$

in matrix notation.   Thus, if A is a vector

$$\bar{T}\bar{A} = (MTM^{-1})(MA) = MTA \qquad (15\text{-}14)$$

so that TA is a vector.   This is an example of *contraction*; if two indices in a tensor are set equal to each other and summed from 1 to 3, a new tensor with two less indices results.   For

$$\left(\sum_i \bar{T}_{iijk...} \right) = \sum_i \sum_{lmnp...} M_{il} M_{im} M_{jn} M_{kp} \cdots T_{lmnp...}$$

$$= \sum_{lmn...} \delta_{lm} M_{jn} M_{kp} \cdots T_{lmnp...}$$

$$= \sum_{n...} M_{jn} M_{kp} \cdots \left(\sum_l T_{llnp...} \right) \qquad (15\text{-}15)$$

and we may write $\left(\sum_l T_{llnp...} \right) = S_{np...}$ where S is a tensor with two less indices than T.   For example, the vector TA is formed from the tensor T and the vector A by

1. Multiplying together to get a third-rank tensor $T_{ij} A_k$, and
2. Contracting the second and third indices

$$(TA)_i = \sum_j T_{ij} A_j \qquad (15\text{-}16)$$

The " dot product " AB of two vectors A and B is invariant under rotations

$$\bar{A}\bar{B} = (MA)(MB)$$

$$= A\tilde{M}MB$$

$$= AB \qquad (15\text{-}17)$$

Such an object is of course a scalar, or tensor of zero rank. The object (cross product)

$$\sum_{ij} \varepsilon_{ijk} A_i B_j = (\mathbf{A} \times \mathbf{B})_k \tag{15-18}$$

is a *pseudovector*, and the object

$$\sum_{ijk} \varepsilon_{ijk} A_i B_j C_k = \mathbf{A} \cdot \mathbf{B} \times \mathbf{C} \tag{15-19}$$

is a *pseudoscalar*.

A second-rank tensor $T_{ij} = A_i B_j$ formed by the outer product of two vectors is sometimes represented by a symbol called a *dyad*, $\mathbf{AB}$, and a linear combination of dyads $\sum \lambda_k \mathbf{A}_k \mathbf{B}_k$ is called a *dyadic*. Note that the order of vectors in a dyad is important; $\mathbf{AB} \neq \mathbf{BA}$. If $T = \mathbf{AB}$, the contracted product

$$V_i = \sum_{j=1}^{3} T_{ij} C_j \tag{15-20}$$

may then be written

$$\mathbf{V} = \mathbf{AB} \cdot \mathbf{C} = (\mathbf{B} \cdot \mathbf{C})\mathbf{A} \tag{15-21}$$

Similarly, a *triad* $\mathbf{ABC}$ is a third-rank tensor with components $T_{ijk} = A_i B_j C_k$, and so on for higher-order *polyads*. Developments of tensor analysis using polyadics may be found in Block (B4) and Drew (D7).

15-2 CURVES IN THREE-SPACE; FRENET FORMULAS

In the preceding section, we were discussing a particular type of coordinate transformation, namely, rotations of Cartesian coordinates in three-dimensional space. In the next section, we shall go on to discuss general coordinate transformations. Before we leave the familiar world of three dimensions, however, we shall digress briefly on the analytical description of curves in three dimensions.

A curve may be described by an equation

$$\mathbf{x} = \boldsymbol{\phi}(t) \tag{15-22}$$

where $\mathbf{x}$ is the vector from the origin of coordinates to a point on the curve and t is a parameter that varies along the curve.

The squared element of arc length is

$$ds^2 = d\mathbf{x}^2 = \left(\frac{d\boldsymbol{\phi}}{dt}\right)^2 (dt)^2 \tag{15-23}$$

We can solve this equation for $t(s)$ or $s(t)$ and thus use the arc length s along our curve as a parameter.   We will usually do this as it makes most formulas simpler; the generalization to an arbitrary parameter is straightforward, if occasionally a little complicated.

Consider then our curve in the form

$$\mathbf{x} = \boldsymbol{\phi}(s) \tag{15-24}$$

and form the vector

$$\boldsymbol{\alpha}(s) = \frac{d\boldsymbol{\phi}}{ds} \tag{15-25}$$

It is easily seen that $\boldsymbol{\alpha}$ is just a unit vector along the tangent to the curve.

From $\boldsymbol{\alpha}^2 = 1$, we deduce

$$\boldsymbol{\alpha} \cdot \frac{d\boldsymbol{\alpha}}{ds} = 0 \tag{15-26}$$

Thus, $d\boldsymbol{\alpha}/ds$ is perpendicular to $\boldsymbol{\alpha}$; let us define

$$\frac{d\boldsymbol{\alpha}}{ds} = \kappa\boldsymbol{\beta} \tag{15-27}$$

where $\boldsymbol{\beta}$ is a unit vector and κ is positive. $\boldsymbol{\beta}$ is called the *principal normal* of our curve, and κ is the *curvature*.

Now what about $d\boldsymbol{\beta}/ds$?   From $\boldsymbol{\beta} \cdot \boldsymbol{\beta} = 1$, we see by analogy with (15-26) that $d\boldsymbol{\beta}/ds$ is perpendicular to $\boldsymbol{\beta}$.   We therefore write

$$\frac{d\boldsymbol{\beta}}{ds} = C_1\boldsymbol{\alpha} + C_2\boldsymbol{\gamma} \qquad (\boldsymbol{\gamma} = \boldsymbol{\alpha} \times \boldsymbol{\beta}) \tag{15-28}$$

The vector $\boldsymbol{\gamma}$ is called the *binormal* to our curve.   Note that $\boldsymbol{\alpha}$, $\boldsymbol{\beta}$, $\boldsymbol{\gamma}$ form a right-handed coordinate system at each point of our curve.

From $\boldsymbol{\alpha} \cdot \boldsymbol{\beta} = 0$, we deduce

$$\boldsymbol{\alpha} \cdot \frac{d\boldsymbol{\beta}}{ds} + \boldsymbol{\beta} \cdot \frac{d\boldsymbol{\alpha}}{ds} = 0 \tag{15-29}$$

or

$$C_1 + \kappa = 0$$
$$\therefore \ C_1 = -\kappa \tag{15-30}$$

The quantity $(-C_2)$ is conventionally written τ and called the *torsion*.

Finally, we should consider $d\gamma/ds$.

$$\boldsymbol{\alpha} \cdot \boldsymbol{\gamma} = 0 \Rightarrow \boldsymbol{\alpha} \cdot \frac{d\boldsymbol{\gamma}}{ds} + \boldsymbol{\gamma} \cdot \frac{d\boldsymbol{\alpha}}{ds} = 0 \Rightarrow \boldsymbol{\alpha} \cdot \frac{d\boldsymbol{\gamma}}{ds} = 0$$

$$\boldsymbol{\beta} \cdot \boldsymbol{\gamma} = 0 \Rightarrow \boldsymbol{\beta} \cdot \frac{d\boldsymbol{\gamma}}{ds} + \boldsymbol{\gamma} \cdot \frac{d\boldsymbol{\beta}}{ds} = 0 \Rightarrow \boldsymbol{\beta} \cdot \frac{d\boldsymbol{\gamma}}{ds} = \tau$$

$$\boldsymbol{\gamma} \cdot \boldsymbol{\gamma} = 1 \Rightarrow \boldsymbol{\gamma} \cdot \frac{d\boldsymbol{\gamma}}{ds} = 0$$

Therefore,

$$\frac{d\boldsymbol{\gamma}}{ds} = \tau\boldsymbol{\beta} \tag{15-31}$$

To summarize, we have deduced the three equations

$$\frac{d\boldsymbol{\alpha}}{ds} = \kappa\boldsymbol{\beta}$$

$$\frac{d\boldsymbol{\beta}}{ds} = -\kappa\boldsymbol{\alpha} - \tau\boldsymbol{\gamma} \tag{15-32}$$

$$\frac{d\boldsymbol{\gamma}}{ds} = \tau\boldsymbol{\beta}$$

These are called the *Frenet formulas.*

15–3   GENERAL TENSOR ANALYSIS

We shall now generalize the concepts introduced in the first section of this chapter, by considering *arbitrary* coordinate transformations, rather than just linear orthogonal transformations (rotations) in three dimensions. That is, we now let the new coordinates $\bar{x}$ be arbitrary functions of the old coordinates x:

$$\bar{x}_i = \bar{x}_i(x_1, x_2, \ldots) \tag{15-33}$$

or, more briefly,

$$\bar{x} = \bar{x}(x)$$

Consider a point P with coordinates x and a neighboring point Q with coordinates $x + dx$. We may suppose the quantities dx are the components of a little vector from P to Q; this picture is not necessary but it motivates the following discussion.

Suppose we change coordinates. P now has coordinates $\bar{x}$, Q has coordinates $\bar{x} + d\bar{x}$, and

$$d\bar{x}_i = \sum_j \frac{\partial \bar{x}_i}{\partial x_j} dx_j$$

We shall hereafter use the summation convention (due to Einstein); any index which occurs twice in the same term is to be summed.   Thus

$$d\bar{x}_i = \frac{\partial \bar{x}_i}{\partial x_j} dx_j \tag{15-34}$$

Any object A^i which has the transformation law

$$\bar{A}^i = \frac{\partial \bar{x}_i}{\partial x_j} A^j \tag{15-35}$$

is said to be a *contravariant vector*.   The coordinate differentials dx_i form a contravariant vector, according to (15-34).

The transformation law (15-35) is the same as the law $\bar{A} = MA$ (15-3) for Cartesian systems except that now the transformation matrix $M_{ij} = \partial\bar{x}_i/\partial x_j$ may vary with the position in space.

A second type of vector is provided by the quantities

$$\frac{\partial\phi}{\partial x_i} \tag{15-36}$$

where ϕ is a *scalar*, that is, a quantity which does not change when we change coordinates.   More precisely, the value of ϕ *at every point* does not change; the functional dependence of ϕ on the coordinates of course varies from coordinate system to coordinate system.   Now, by the chain rule of calculus,

$$\frac{\partial\phi}{\partial\bar{x}_i} = \frac{\partial\phi}{\partial x_j}\frac{\partial x_j}{\partial\bar{x}_i} \tag{15-37}$$

Any object B_i which has this transformation law

$$\bar{B}_i = \frac{\partial x_j}{\partial\bar{x}_i} B_j \tag{15-38}$$

is said to be a *covariant vector*.   Note that contravariant indices go upstairs; covariant ones downstairs.   In order to conform more completely to this convention, we shall henceforth use *upper* indices on all coordinates, since the differentials dx^i form a contravariant vector.

Let us compare Eq. (15-38) with Eq. (6-36),

$$\alpha'_i = \alpha_j \gamma_{ji}$$

Remembering that now $\gamma = M^{-1}$ [see the paragraph following (15-4)] the transformation laws are the same provided the inverse matrix M^{-1} is

$$(M^{-1})_{ji} = \frac{\partial x^j}{\partial\bar{x}^i} \tag{15-39}$$

But this is true since

$$M_{kj}(M^{-1})_{ji} = \frac{\partial \bar{x}^k}{\partial x^j}\frac{\partial x^j}{\partial \bar{x}^i} = \frac{\partial \bar{x}^k}{\partial \bar{x}^i} = \delta_{ki}$$

and

$$(M^{-1})_{kj}M_{ji} = \frac{\partial x^k}{\partial \bar{x}^j}\frac{\partial \bar{x}^j}{\partial x^i} = \frac{\partial x^k}{\partial x^i} = \delta_{ki} \tag{15-40}$$

We have thus shown that the components α_i of a linear scalar function, which were seen in Section 6–4 to transform contragrediently to the components of a vector, can be said with our present nomenclature to comprise a covariant vector.

Tensors of higher rank are defined as in Section 15–1, as having the same transformation laws as products of vectors.   For example, the three different transformation laws that are possible for second-rank tensors are

$$\text{contravariant tensor:} \quad \bar{T}^{ij} = \frac{\partial \bar{x}^i}{\partial x^k}\frac{\partial \bar{x}^j}{\partial x^l} T^{kl}$$

$$\text{mixed tensor:} \quad \bar{T}^i_j = \frac{\partial \bar{x}^i}{\partial x^k}\frac{\partial x^l}{\partial \bar{x}^j} T^k_l \tag{15-41}$$

$$\text{covariant tensor:} \quad \bar{T}_{ij} = \frac{\partial x^k}{\partial \bar{x}^i}\frac{\partial x^l}{\partial \bar{x}^j} T_{kl}$$

We could clearly go on and define (by their transformation laws) tensors with arbitrary numbers of covariant and contravariant indices.

We recall that Cartesian tensors could be *contracted*, by setting two indices equal to each other and summing.   The same can be done with our more general tensors, *provided one index is upstairs and the other is downstairs*. For example, let A^{ij} and B_k be tensors, and

$$C^i = A^{ij}B_j$$

Then

$$\bar{C}^i = \bar{A}^{ij}\bar{B}_j$$

$$= \frac{\partial \bar{x}^i}{\partial x^k}\frac{\partial \bar{x}^j}{\partial x^l} A^{kl} \frac{\partial x^m}{\partial \bar{x}^j} B_m$$

$$= \frac{\partial \bar{x}^i}{\partial x^k} A^{kl}B_l = \frac{\partial \bar{x}^i}{\partial x^k} C^k \tag{15-42}$$

so that C^i is indeed a contravariant vector.

As with Cartesian coordinates, the Kronecker delta turns out to be a tensor provided we write it δ^i_j with one index upstairs and the other down-stairs (verification is left to the student).   Also, as before, the antisymmetric symbol $\varepsilon_{ij...}$ turns out to be "not quite" a tensor.

In this section we are considering an arbitrary number of dimensions, but ε_{ijk} was only defined by (15-10) for three dimensions.   The general alternating symbol is defined by

$$\underbrace{\varepsilon_{ij...}}_{n \text{ indices}} = \begin{cases} +1 & \text{if indices are an } even \text{ permutation of } 12 \cdots n \\ -1 & \text{if indices are an } odd \text{ permutation of } 12 \cdots n \\ 0 & \text{otherwise} \end{cases} \qquad (15\text{-}43)$$

An ordering of the numbers $12 \cdots n$ is said to be an even (odd) permutation of $12 \cdots n$ if it can be made from the standard order by an even (odd) number of interchanges of pairs of numbers.

An object with the transformation law

$$\bar{A}^{ij...}_{kl...} = \frac{\partial \bar{x}^i}{\partial x^\alpha} \frac{\partial \bar{x}^j}{\partial x^\beta} \cdots \frac{\partial x^\gamma}{\partial \bar{x}^k} \frac{\partial x^\delta}{\partial \bar{x}^l} \cdots A^{\alpha\beta...}_{\gamma\delta...} \left| \frac{\partial x}{\partial \bar{x}} \right|^w \qquad (15\text{-}44)$$

(where $|\partial x/\partial \bar{x}|$ is the *Jacobian*

$$\left| \frac{\partial x}{\partial \bar{x}} \right| = \begin{vmatrix} \dfrac{\partial x_1}{\partial \bar{x}_1} & \dfrac{\partial x_2}{\partial \bar{x}_1} & \cdots & \dfrac{\partial x_n}{\partial \bar{x}_1} \\ \dfrac{\partial x_1}{\partial \bar{x}_2} & \cdots & & \\ \cdots & & & \dfrac{\partial x_n}{\partial \bar{x}_n} \end{vmatrix} \qquad (15\text{-}45)$$

of the transformation) is called a *relative tensor* of weight w.   It will be left to the reader to verify that $\varepsilon_{ijk...}$ is a relative covariant tensor of weight -1, and that $\varepsilon^{ijk...}$ (which is numerically equal to $\varepsilon_{ijk...}$) is a relative contravariant tensor of weight $+1$.

EXAMPLE

The "volume element"

$$d^n x = dx^1 \, dx^2 \cdots dx^n$$

transforms[4] according to the law

$$d^n\bar{x} = d^n x \left| \frac{\partial \bar{x}}{\partial x} \right| = d^n x \left| \frac{\partial x}{\partial \bar{x}} \right|^{-1} \qquad (15\text{-}46)$$

so that $d^n x$ is a relative scalar of weight -1.

---

[4] See any book on advanced calculus, for example, Apostol (A5) Section 10-9.

A relative tensor of weight $+1$ is called a *tensor density*. The reason is that

$$\int d^n x \, A(x)$$

is an *absolute scalar* (scalar of weight zero) if A is a scalar density, in view of (15-46).

Other tensors and relative tensors besides $\varepsilon_{ijk\ldots}$ may have definite symmetry properties. A tensor may be symmetric or antisymmetric with respect to the interchange of two or more indices, and, if these indices are all covariant or all contravariant, the symmetry property is invariant under coordinate transformations.

The alternating symbol $\varepsilon_{ijk\ldots}$ may be used to construct antisymmetric tensors. For example, if A_{ij} is a tensor

$$A_{ij}\varepsilon^{jkl\cdots} = R_i^{kl\cdots} \tag{15-47}$$

is a relative tensor of weight $+1$ which is antisymmetric in its contravariant indices. Or, if R^{ij} is a relative tensor of weight $+1$,

$$R^{ij}\varepsilon_{jkl\ldots} = T^i_{kl\ldots} \tag{15-48}$$

is an absolute tensor which is antisymmetric in its covariant indices.

EXAMPLE

In three dimensions, if V_i is a relative vector of weight -1 (a pseudo-vector),

$$V_i\varepsilon^{ijk} = T^{jk} \tag{15-49}$$

is an antisymmetric tensor. Thus we establish a one-to-one correspondence between covariant pseudovectors of weight -1 and second-rank antisymmetric contravariant tenors. The cross product $\mathbf{A} \times \mathbf{B}$ (15–18) may be considered either kind of object.

We next consider an important theorem, the so-called *quotient law*. We shall illustrate it by first considering a simple example. Let A_{ij} be an object such that $A_{ij}u^j$ is a covariant vector for *every* contravariant vector u^j. It follows that A_{ij} is a covariant tensor.

Proof:

$$\bar{A}_{ij}\bar{u}^j = (\partial x^k/\partial \bar{x}^i)A_{kl}u^l \qquad \text{by hypothesis}$$

Now transform the vector $\bar{u}^j$ in the left side:

$$\bar{A}_{ij}\frac{\partial \bar{x}^j}{\partial x^m}u^m = \frac{\partial x^k}{\partial \bar{x}^i}A_{kl}u^l$$

Since this is to be true for arbitrary u^m, it follows that

$$\bar{A}_{ij} \frac{\partial \bar{x}^j}{\partial x^m} = \frac{\partial x^k}{\partial \bar{x}^i} A_{km}$$

Multiply both sides by $\partial x^m / \partial \bar{x}^l$ to complete the proof:

$$\bar{A}_{il} = \frac{\partial x^k}{\partial \bar{x}^i} \frac{\partial x^m}{\partial \bar{x}^l} A_{km}$$

This proof is easily generalized to show the following. If the product $AB = C$ is a tensor, where B is an *arbitrary* tensor, then A is a tensor. The product may involve any number of contractions of pairs of upper and lower indices between A and B (but not *within* A or B; $A_i^i = $ scalar does not imply $A_j^i = $ tensor).

EXAMPLE

A generalized force is an object with n components such that for an *arbitrary* displacement dx^i of the point of application of the force, the product

$$f_i\, dx^i = W \tag{15-50}$$

is a scalar, the work done by the force.   Therefore, f_i is a covariant vector.

EXAMPLE

A second-rank covariant tensor A_{ij} may be thought of as a matrix, with the indices i and j labeling the rows and columns, respectively.   Let us denote the *inverse* of this matrix by B^{ij}; that is,

$$A_{ij} B^{jk} = \delta_i^k$$

We shall now show that B^{ij} is in fact a second-rank contravariant tensor— which is, of course, why we put its indices upstairs.

Note first that we cannot apply the quotient law immediately to the above equation; A_{ij} and δ_i^k are indeed tensors, but A_{ij} is a *particular* tensor, not an *arbitrary* one.   We consider instead an arbitrary contravariant vector u^i.   Then $A_{ij}u^j$ is an arbitrary covariant vector v_i.   But

$$B^{ij}v_j = B^{ij}A_{jk}u^k = \delta_k^i u^k = u^i$$

Thus B times the arbitrary vector v gives a vector u.   Therefore, by the quotient law, B is a tensor.

Let us choose some particular symmetric nonsingular covariant tensor g_{ij} and call it the *fundamental* covariant tensor.   It doesn't really have to be symmetric, but if it isn't, things get a little messy.   The inverse of g_{ij} will be written g^{ij}:

$$g_{ij}g^{jk} = \delta_i^k \tag{15-51}$$

By using the fundamental tensor g_{ij}, we can set up a one-to-one correspondence between covariant and contravariant vectors.   For every covariant vector u_i there is a contravariant vector u^i defined by

$$u^i = g^{ij}u_j \tag{15-52}$$

Similarly,

$$u_i = g_{ij}u^j \tag{15-53}$$

Because of this correspondence, one generally calls u_i and u^j the covariant and contravariant *components* of the same vector u.

This *raising* and *lowering* of indices can also be applied to tensors:

$$A_{ij} = g_{ik}A_j^k = g_{ik}g_{jl}A^{kl} \tag{15-54}$$

We can use the fundamental tensor to obtain a relative tensor from a tensor and vice versa or, in general, to change the weight of a relative tensor. For this, we make use of the transformation character of det g, which we now find.   Calling $\partial x^i/\partial \bar{x}^j = \gamma_{ij}$, we have

$$\bar{g}_{lm} = \frac{\partial x^i}{\partial \bar{x}^l}\frac{\partial x^j}{\partial \bar{x}^m}g_{ij} = \gamma_{il}\gamma_{jm}g_{ij}$$

or, in matrix notation,

$$\bar{g} = \tilde{\gamma}g\gamma$$

Therefore,

$$\det \bar{g} = (\det \gamma)^2 \det g$$

$$(\det \bar{g})^{1/2} = \det \gamma(\det g)^{1/2} = \left|\frac{\partial x}{\partial \bar{x}}\right|(\det g)^{1/2} \tag{15-55}$$

Thus, $(\det g)^{1/2}$ is a relative scalar of weight $+1$, and it may be applied as a factor to increase the weight of a relative tensor by 1.

In *Riemannian geometry*, the fundamental tensor is connected with the so-called *interval ds* by

$$ds^2 = g_{ij}\,dx^i\,dx^j \tag{15-56}$$

a scalar.   That is, g_{ij} is chosen to be a particular tensor, the *metric tensor*.

We see by (15-56) that, for an ordinary Cartesian coordinate system, the metric tensor g_{ij} equals δ_{ij}, so that no distinction between contravariant and covariant tensors need be made.

We remarked earlier [see (15-37)] that the derivative of a scalar is a (covariant) vector.   But it is easy to see that differentiating a tensor other than a scalar does not generally produce a tensor.   For example, consider a covariant vector v_i:

$$\bar{v}_i = \frac{\partial x^j}{\partial \bar{x}^i} v_j$$

$$\frac{\partial \bar{v}_i}{\partial \bar{x}^k} = \frac{\partial^2 x^j}{\partial \bar{x}^i \, \partial \bar{x}^k} v_j + \frac{\partial x^j}{\partial \bar{x}^i} \frac{\partial x^l}{\partial \bar{x}^k} \frac{\partial v_j}{\partial x^l} \tag{15-57}$$

If only the second term on the right were present,

$$\frac{\partial v_i}{\partial x^k}$$

would be a tensor, but the first term complicates things.   This term arises because the transformation matrix $\partial x^i / \partial \bar{x}^j$ varies with the position in space. It, and thus our difficulty, are absent in a Cartesian coordinate system.

We now show how, by use of the fundamental tensor g_{ij}, the difficulty may be overcome.   Consider the derivative of the fundamental tensor

$$\bar{g}_{ij} = \frac{\partial x^k}{\partial \bar{x}^i} \frac{\partial x^l}{\partial \bar{x}^j} g_{kl}$$

$$\frac{\partial \bar{g}_{ij}}{\partial \bar{x}^m} = \frac{\partial^2 x^k}{\partial \bar{x}^i \, \partial \bar{x}^m} \frac{\partial x^l}{\partial \bar{x}^j} g_{kl} + \frac{\partial x^k}{\partial \bar{x}^i} \frac{\partial^2 x^l}{\partial \bar{x}^j \, \partial \bar{x}^m} g_{kl} + \frac{\partial x^k}{\partial \bar{x}^i} \frac{\partial x^l}{\partial \bar{x}^j} \frac{\partial x^n}{\partial \bar{x}^m} \frac{\partial g_{kl}}{\partial x^n} \tag{15-58}$$

Now we write down two more equations by permuting the indices in the last equation (we shall just write the indices)

$$\begin{array}{c} mi \\ j \end{array} = \begin{array}{cc} k & l \\ mj & i \end{array} kl + \begin{array}{cc} k & l \\ m & ij \end{array} kl + \begin{array}{cc} k & l & n \\ m & i & j \end{array} kl$$

$$\begin{array}{c} jm \\ i \end{array} = \begin{array}{cc} k & l \\ ji & m \end{array} kl + \begin{array}{cc} k & l \\ j & mi \end{array} kl + \begin{array}{cc} k & l & n \\ j & m & i \end{array} kl$$

If we subtract the last equation from the sum of the first two and define the *Christoffel symbol of the first kind*

$$[ij, k] = \frac{1}{2}\left(\frac{\partial g_{ik}}{\partial x^j} + \frac{\partial g_{jk}}{\partial x^i} - \frac{\partial g_{ij}}{\partial x^k} \right) \tag{15-59}$$

we obtain, after some manipulation of indices,

$$\overline{[ij, k]} = \frac{\partial x^l}{\partial \bar{x}^i} \frac{\partial x^m}{\partial \bar{x}^j} \frac{\partial x^n}{\partial \bar{x}^k} [lm, n] + g_{lm} \frac{\partial x^l}{\partial \bar{x}^k} \frac{\partial^2 x^m}{\partial \bar{x}^i \, \partial \bar{x}^j} \tag{15-60}$$

We may solve this equation for the second derivative; the result is

$$\frac{\partial^2 x^s}{\partial \bar{x}^i \, \partial \bar{x}^j} = \frac{\partial x^s}{\partial \bar{x}^p} \begin{Bmatrix} p \\ ij \end{Bmatrix} - \frac{\partial x^l}{\partial \bar{x}^i} \frac{\partial x^m}{\partial \bar{x}^j} \begin{Bmatrix} s \\ lm \end{Bmatrix} \tag{15-61}$$

where we have introduced the *Christoffel symbol of the second kind*

$$\left\{ {i \atop jk} \right\} = g^{il}[jk, l] \tag{15-62}$$

Now we are in a position to define the so-called *covariant derivative* of a covariant vector. The result of substituting the expression (15-61) for the second derivative of x in terms of the $\bar{x}$ into our previous equation (15-57) for $\partial \bar{v}_i / \partial \bar{x}^k$ gives

$$\frac{\partial \bar{v}_i}{\partial \bar{x}^k} = \frac{\partial x^j}{\partial \bar{x}^i} \frac{\partial x^l}{\partial \bar{x}^k} \frac{\partial v_j}{\partial x^l} + v_j \left[\frac{\partial x^j}{\partial \bar{x}^l} \overline{\left\{ {l \atop ik} \right\}} - \frac{\partial x^l}{\partial \bar{x}^i} \frac{\partial x^m}{\partial \bar{x}^k} \left\{ {j \atop lm} \right\} \right]$$

which may be written

$$\frac{\partial \bar{v}_i}{\partial \bar{x}^k} - \overline{\left\{ {l \atop ik} \right\}} \bar{v}_l = \frac{\partial x^j}{\partial \bar{x}^i} \frac{\partial x^l}{\partial \bar{x}^k} \left[\frac{\partial v_j}{\partial x^l} - \left\{ {m \atop jl} \right\} v_m \right] \tag{15-63}$$

Thus

$$v_{i;\,j} = \frac{\partial v_i}{\partial x^j} - \left\{ {k \atop ij} \right\} v_k \tag{15-64}$$

is a covariant tensor of the second rank, the so-called *covariant derivative* of v_i.

The covariant derivative of a contravariant vector turns out to be

$$v^i_{;\,j} = \frac{\partial v^i}{\partial x^j} + \left\{ {i \atop jk} \right\} v^k \tag{15-65}$$

Verification of this result is left as an exercise (Problem 15-7).

The covariant derivative of a product is formed by covariantly differentiating the factors one at a time, just as in ordinary calculus. The covariant derivative of a tensor is formed by analogy with the expressions given above for covariant derivatives of vectors, treating one index at a time:

$$A^{ij\cdots}_{kl\cdots;\,\alpha} = \frac{\partial A^{ij\cdots}_{kl\cdots}}{\partial x^\alpha} + \left\{ {i \atop \alpha\beta} \right\} A^{\beta j\cdots}_{kl\cdots} + \left\{ {j \atop \alpha\beta} \right\} A^{i\beta\cdots}_{kl\cdots} + \cdots$$

$$- \left\{ {\beta \atop \alpha k} \right\} A^{ij\cdots}_{\beta l\cdots} - \left\{ {\beta \atop \alpha l} \right\} A^{ij\cdots}_{k\beta\cdots} - \cdots \tag{15-66}$$

One defines the covariant derivative of a *relative* tensor in the following way. Let $A^{ij\cdots}_{kl\cdots}$ be a relative tensor of weight w. Define the *absolute* tensor

$$B^{ij\cdots}_{kl\cdots} = A^{ij\cdots}_{kl\cdots} (\det g)^{-w/2}$$

The covariant derivative $B^{ij\cdots}_{kl\cdots;\,\alpha}$ of $B^{ij\cdots}_{kl\cdots}$ is now formed according to (15-66), and the covariant derivative of $A^{ij\cdots}_{kl\cdots}$ is then defined by

$$A^{ij\cdots}_{kl\cdots;\,\alpha} = B^{ij\cdots}_{kl\cdots;\,\alpha}\,(\det g)^{w/2}$$

We see that, in general, the covariant derivative of a tensor is complicated by the presence of Christoffel symbols. However, there are some covariant derivatives which are not complicated in this way. Some of these simple derivative forms are listed below.

1. The covariant derivative of a scalar is the ordinary partial derivative introduced before, in (15-36),

$$\phi_{;k} = \frac{\partial \phi}{\partial x^k} \tag{15-67}$$

2. If A_i is a covariant vector, the operation "curl A," defined by

$$(\text{curl } A)_{ij} = A_{i;\,j} - A_{j;\,i} = \frac{\partial A_i}{\partial x^j} - \frac{\partial A_j}{\partial x^i} \tag{15-68}$$

gives an antisymmetric covariant tensor. The Christoffel symbol terms from $A_{i;\,j}$ and $A_{j;\,i}$ cancel because of the symmetry of these symbols. The fact that $\partial A_i/\partial x^j - \partial A_j/\partial x^i$ is a tensor may also be seen directly from Eq. (15-57), since the second derivative terms cancel.

3. If R^i is a relative vector of weight $+1$, the contracted covariant derivative, which is a relative scalar of weight $+1$, has the simple form

$$R^i_{;\,i} = \frac{\partial R^i}{\partial x^i}\, (= \text{``div } R\text{''}) \tag{15-69}$$

4. The covariant derivative of the fundamental tensor is especially simple, namely, zero:

$$g_{ij;\,k} = 0 \tag{15-70}$$

The student should verify properties 3 and 4.

Suppose we differentiate a covariant vector twice covariantly:

$$u_{i;\,jk} = (u_{i;\,j})_{;\,k}$$

This is not equal to $u_{i;\,kj}$; in fact,

$$u_{i;\,jk} - u_{i;\,kj} = R^l_{ijk}\, u_l \tag{15-71}$$

where

$$R^l_{ijk} = \begin{Bmatrix} l \\ mj \end{Bmatrix}\begin{Bmatrix} m \\ ik \end{Bmatrix} - \begin{Bmatrix} l \\ mk \end{Bmatrix}\begin{Bmatrix} m \\ ij \end{Bmatrix} + \frac{\partial}{\partial x^j}\begin{Bmatrix} l \\ ik \end{Bmatrix} - \frac{\partial}{\partial x^k}\begin{Bmatrix} l \\ ij \end{Bmatrix} \tag{15-72}$$

It follows from the quotient law that R^l_{ijk} must be a tensor. It is called the *Riemann curvature tensor*.

Let us now consider the problem of finding a curve $x_i = x_i(t)$ which minimizes the integral

$$I = \int \sqrt{g_{ij}\, dx^i\, dx^j}$$

$$= \int \sqrt{g_{ij}\, \frac{dx^i}{dt}\, \frac{dx^j}{dt}}\, dt \tag{15-73}$$

If, as in (15-56), $\sqrt{g_{ij}\, dx^i\, dx^j}$ is considered an element of interval, our integral measures the "total interval," or extension or distance or whatever you want to call it, along the path. In this case, the name *geodesic* is appropriate for such a curve. However, we shall just consider the analytic problem of minimizing I, whatever its geometrical or physical significance.

We must write down the Euler–Lagrange equations for the integrand

$$F\left(x^i, \frac{dx^i}{dt}\right) = \sqrt{g_{ij}(x)\, \frac{dx^i}{dt}\, \frac{dx^j}{dt}}$$

[Compare Eqs. (12-17).] It is straightforward to verify that the result may be written

$$\frac{d^2 x^i}{dt^2} + \left\{ \begin{matrix} i \\ jk \end{matrix} \right\} \frac{dx^j}{dt}\, \frac{dx^k}{dt} = \frac{s''}{s'}\, \frac{dx^i}{dt} \tag{15-74}$$

where s is "arc length," defined by

$$s' = \frac{ds}{dt} = \sqrt{g_{ij}\, \frac{dx^i}{dt}\, \frac{dx^j}{dt}} \tag{15-75}$$

If we take s to be our parameter, $s' = 1$, $s'' = 0$, and the differential equation (15-74) for our geodesics becomes

$$\frac{d^2 x^i}{ds^2} + \left\{ \begin{matrix} i \\ jk \end{matrix} \right\} \frac{dx^j}{ds}\, \frac{dx^k}{ds} = 0 \tag{15-76}$$

Christoffel symbols have now appeared both in covariant differentiation and in the differential equation for a geodesic. This suggests that there may be a connection between the two concepts, covariant differentiation and geodesics. In fact, there are many connections. For example, consider the covariant derivative of the "tangent" vector

$$u^i = \frac{dx^i}{ds}$$

along a geodesic.   The covariant derivative along the geodesic is

$$u^i_{;j} \frac{dx^j}{ds} = \frac{d^2 x^i}{ds^2} + \begin{Bmatrix} i \\ jk \end{Bmatrix} \frac{dx^k}{ds} \frac{dx^j}{ds}$$

$$= 0 \quad \text{[by (15-76)]} \tag{15-77}$$

Thus, a geodesic is a curve with a "constant" tangent, that is, a tangent vector with zero covariant derivative along the curve. Geometrically, this just means that the geodesics are the "straight lines" in the space.

REFERENCES

The mathematical aspects of differential geometry and tensor analysis are discussed in numerous books, for example Lass (L5); Spain (S10); Weatherburn (W2); Levi-Civita (L7); and Eisenhart (E2).   The preceding books are listed (roughly) in the order of increasing mathematical complexity and rigor.

Treatments using polyadics are given by Block (B4) and Drew (D7).

The most important applications to physics of the mathematical apparatus of differential geometry have been in the general theory of relativity; see Anderson (A4); Einstein (E1); or Møller (M8).   The second of these is especially recommended, both for a concise description of tensor analysis and for discussions of its basic applications to the special and general theories of relativity.

The Euler angle description of rotations and orthogonal transformations in three dimensions are discussed in a classical framework by Goldstein (G4) and in connection with the quantum theory of angular momentum by Rose (R3).

PROBLEMS

15-1   A line is drawn in the direction defined by the conventional polar angles θ, ϕ, and our coordinate system is rotated through an angle ω with this line as axis of rotation.   Find the transformation matrix associated with this rotation.

15-2   A coordinate system is attached to a regular tetrahedron so that
(1) the origin is at the center of the tetrahedron
(2) vertex 1 lies on the positive z-axis
(3) vertex 2 lies in the xz-plane
(4) vertex 3 has $y > 0$.
The tetrahedron is now rotated (and the coordinate system with

it) so that

> vertex 1 goes where vertex 3 was
> vertex 3 goes where vertex 4 was
> vertex 4 goes where vertex 1 was
> vertex 2 stays put

Give the transformation matrix associated with this rotation.

15-3 (a) Show that $\sum_i \varepsilon_{ijk} \varepsilon_{ilm} = \delta_{jl} \delta_{km} - \delta_{jm} \delta_{kl}$
(b) Evaluate $\sum_{ij} \varepsilon_{ijk} \varepsilon_{ijl}$
(c) Evaluate $\sum_{ijk} \sum_{lmn} \varepsilon_{ijk} \varepsilon_{lmn} M_{il} M_{jm} M_{kn}$

where M is an arbitrary matrix.

15-4 Find the curvature and torsion, as functions of the parameter t, for the so-called twisted cubic

$$x = at \qquad y = bt^2 \qquad z = ct^3 \qquad (a, b, c \text{ constant})$$

15-5 A charged particle moves through space under the influence of a magnetic field which varies with position but not with time. Show that at each point of the particle's trajectory the direction of the magnetic field lies in the plane determined by the tangent and the binormal to the trajectory at that point.

15-6 When a particular curve $\phi(s)$ ($s = $ arc length) is given, it is sometimes convenient to label an arbitrary point $\mathbf{x}$ in space, not necessarily on the curve $\phi(s)$, by three parameters s, ξ, and η. A perpendicular is dropped from the point $\mathbf{x}$ to the curve, and s is the arc length along the curve to the point of intersection. The principal normal $\boldsymbol{\beta}$ and binormal $\boldsymbol{\gamma}$ are constructed at this point, and ξ and η are defined by $\mathbf{x} - \phi(s) = \xi\boldsymbol{\beta} + \eta\boldsymbol{\gamma}$.

In this way s, ξ, and η are defined for every point in space. Calculate ∇s, $\nabla \xi$, and $\nabla \eta$.

15-7 Justify the definition (15-65).

15-8 Consider an arbitrary symmetric covariant tensor A_{ij}. By means of A_{ij}, we may define an operation of covariant differentiation. Show that the covariant derivative, defined in this way, of A_{ij} vanishes. [This constitutes a verification of (15-70).]

15-9 Consider the Christoffel symbol $\begin{Bmatrix} i \\ jk \end{Bmatrix}$ formed with the aid of an arbitrary symmetric covariant tensor A_{lm}. Show that

$$\begin{Bmatrix} i \\ ik \end{Bmatrix} = \frac{\partial}{\partial x^k} \ln \sqrt{\det A}$$

15-10 Consider three vectors A^i, A_i, and $A(i)$, the last consisting of the "ordinary" or "physical" components of $\mathbf{A}$, familiar from ordinary vector analysis. The three vectors are numerically equal in a Cartesian coordinate system (xyz). A change of coordinates $(xyz) \rightarrow (r\theta\phi)$ is made in the usual way.

(a) Show that

$$A(r) = \sqrt{A_r A^r}$$
$$A(\theta) = \sqrt{A_\theta A^\theta}$$
$$A(\phi) = \sqrt{A_\phi A^\phi}$$

(b) A vector $\mathbf{v}$ and the tensor T in Cartesian coordinates are connected by the relation

$$\mathbf{v} = \nabla \cdot T \qquad \text{that is} \qquad v_i = \frac{\partial}{\partial x^j} T_{ji}$$

Express $v(r)$, $v(\theta)$, $v(\phi)$, the "ordinary" components of $\mathbf{v}$ in polar coordinates, in terms of the "ordinary" polar components of T and their derivatives with respect to r, θ, ϕ.

15-11 Coordinates α, β, γ are defined by

$$x = \beta\gamma \qquad y = \alpha\gamma \qquad z = \alpha\beta$$

where x, y, z are the usual Cartesian coordinates. Find $\nabla^2 u$ in terms of derivatives of u with respect to α, β, γ.

15-12 A line is drawn from a point to the origin of Cartesian coordinates. A plane passing through the point perpendicular to the line will intersect the three coordinate axes, determining three numbers (a, b, c) by the points of intersection. These three numbers may be associated with the point and will change if the coordinate axes are rotated while the point remains fixed.

(a) Are (a, b, c) components of a Cartesian tensor?

(b) If *yes*, give details.

If *no*, form three quantities from (a, b, c) that are a tensor.

15-13 The position of a point in the upper-half plane is determined by giving the distances x_1, x_2 of the point from two fixed reference points on the x-axis a unit of distance apart. Find the metric tensors g_{ij}, g^{ij} for these coordinates.

15-14 Give the formula for the covariant derivative of a scalar density.

SIXTEEN

INTRODUCTION TO GROUPS AND GROUP REPRESENTATIONS

The theory of groups is a mathematical discipline which has come to have quite varied and powerful applications to physics. In this chapter we do full justice neither to the mathematical theory, which is beautiful in its own right, nor to the usefulness of group theory in physics. Nevertheless, we present a brief summary of the definitions and properties of groups and their matrix representations, together with several illustrative applications. In the last two sections we discuss some aspects of several continuous groups of physical importance, without developing the physical motivation for studying these groups; this would take us too far afield.

16-1 INTRODUCTION; DEFINITIONS

A group G is a set of elements $g_1, g_2, \ldots$ and a rule for combining ("multiplying") any two of them to form a "product," subject to the following four conditions.

1. The product $g_i g_j$ of any two group elements must be a group element

2. Group multiplication is *associative*: $(g_i g_j)g_k = g_i(g_j g_k)$.

3. There is a unique group element $g_1 = I$, called the *identity*, such that

$$Ig_i = g_i I = g_i$$

for all g_i in G.

4. Each element has a unique *inverse*; that is, for each g_i there is a unique group element g_i^{-1} such that

$$g_i g_i^{-1} = g_i^{-1} g_i = I$$

A group with a finite number of elements is called a *finite* group; the number of elements in a finite group is the *order* of the group. We shall begin by confining our attention to finite groups; in the last two sections of this chapter a few infinite groups will be briefly discussed.

Group multiplication need not be *commutative*; that is, $g_i g_j$ need not equal $g_j g_i$. If $g_i g_j = g_j g_i$ for every pair of elements in our group, it is said to be a *commutative* or *Abelian* group.

If S is any subset of our group G, Sg_i will denote the set of elements which results if each element in S is multiplied by g_i on the right. A similar notation will be used for multiplication on the left. Note that the number of elements in Sg_i is the same as the number in S. For if

$$ag_i = bg_i \qquad (a \text{ and } b \text{ in } S)$$

then, multiplying on the right by g_i^{-1}, $a = b$. In particular, $Gg = G$; if the entire group is multiplied on the right (or left) by any element, the result is some rearrangement of the group.

One often extends a group to form an *algebra*. An algebra is a set of elements which form a linear vector space (Chapter 6) in which, besides addition, an operation of multiplication is defined in such a way as to obey the group postulates, except that the zero of the algebra has no inverse. For example, given a group G with elements $g_i (i = 1, \ldots, h)$, the linear combinations $\sum_{i=1}^{h} c_i g_i$ of group elements with real coefficients c_i form an algebra. The product is defined in an obvious way:

$$\left(\sum_{i=1}^{h} c_i g_i \right) \left(\sum_{j=1}^{h} d_j g_j \right) = \sum_{i=1}^{h} \sum_{j=1}^{h} c_i d_j g_i g_j$$

Since G is a group, each product $g_i g_j$ is an element g_k of G, and the product is therefore an element of the algebra.

We shall conclude this introduction by giving some examples of finite groups.

1. The group of rearrangements of n objects. This group, known as the *symmetric group on n objects*, will be denoted by S_n. A typical element of S_5 might be written [2 4 1 5 3], which means: put the second object first, the fourth object second, and so on. Two elements are multiplied by

performing first the rearrangement on the right, then the rearrangement on the left. For example,

$$[2\,4\,1\,5\,3][5\,1\,2\,3\,4]a\,b\,c\,d\,e = [2\,4\,1\,5\,3]e\,a\,b\,c\,d$$
$$= a\,c\,e\,d\,b$$
$$= [1\,3\,5\,4\,2]a\,b\,c\,d\,e$$

Therefore,

$$[2\,4\,1\,5\,3][5\,1\,2\,3\,4] = [1\,3\,5\,4\,2] \qquad (16\text{-}1)$$

The order of S_n is obviously $n!$ A concept which is very useful for the symmetric group S_n is the decomposition of a permutation into *cycles*. For example, the permutation $[3\,1\,2\,5\,4]$ puts the first object second, the second object third, and the third object first; this constitutes one cycle which might be written $(1\,2\,3)$.   In addition, the exchange of the fourth and fifth objects constitutes a cycle of two $(4\,5)$, so that in cycle notation our permutation would be written $(1\,2\,3)(4\,5)$.   As another example, the permutation $[\,1\,3\,5\,4\,2]$ appears in cycle notation as $(1)(2\,5\,3)(4)$.

It should be clear that the order in which the cycles are written is immaterial, and that the numbers within any cycle may be cyclically rotated.

2. The group of integers 0, 1, 2, 3.   Group "multiplication" is addition (mod 4).   For example,

$$1 \times 2 = 3 \qquad 2 \times 3 = 1$$

3. The group of transformations of an equilateral triangle onto itself (the *symmetry group* of the equilateral triangle).   This group contains six elements (see Figure 16–1):

I the identity; triangle is left alone
A triangle is rotated through 120° clockwise
B triangle is rotated through 240° clockwise
C triangle is reflected about line P
D triangle is reflected about line Q
E triangle is reflected about line R

This group is the same as S_3, with different *names* for the group elements. We say that this group is *isomorphic* to S_3.   They have the same multiplication table, or *group table*.

16–2 SUBGROUPS AND CLASSES

A subset of a group G, which is itself a group, is called a *subgroup* of G. Let S be a subgroup of G and consider the sets $Sg_1, Sg_2, \ldots, Sg_n$ obtained by multiplying S successively on the right by each element of our group. These sets are called *right cosets* of S.[1]

---

[1] Some authors, for example, Wigner (W6) exclude S itself from the set of cosets.

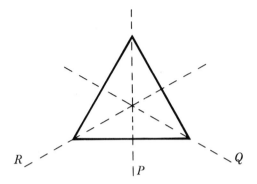

Figure 16–1   An equilateral triangle, which is transformed into itself by the group of transformations listed in example 3, p. 426

We now observe that two cosets Sg_1 and Sg_2 are either identical or have no element in common.   Suppose they have an element in common, that is,

$$s_1 g_1 = s_2 g_2 = g \qquad (s_1 \text{ and } s_2 \text{ in } S)$$

Then $g_2 = s_2^{-1} s_1 g_1$, and $Sg_2 = Ss_2^{-1} s_1 g_1 = Sg_1$, since $s_2^{-1} s_1$ is in S and $Ss = S$ if s is in S (because S is a group).   Also, if $Sg_1 = Sg_2$, we see that $g_2 g_1^{-1} = s_2^{-1} s_1$, which is in S.   In fact, we may summarize this paragraph as follows: The two cosets Sg_1 and Sg_2 are either identical or have no member in common, according to whether $g_2 g_1^{-1}$ is or is not a member of the subgroup S.

If we observe two further facts:

1. Each group element g appears in at least one coset of S, and
2. Each coset has h' elements, where h' is the order of S,

we see that the entire group G splits up into cosets of S, each having h' elements.   Therefore, if h is the order of G, h must be a multiple of h'.   *The order of a subgroup is a factor of the order of the full group.*

If A is any element of a finite group, the sequence of elements (called the *period* of A)

$$A, A^2, \ldots, A^p = I \qquad (16\text{-}2)$$

forms an Abelian subgroup of order p. p is the smallest integer for which $A^p = I$.   It is called the *order* of A.   Note that all elements in this sequence are distinct, for if $A^l = A^k$ for $l < k < p$, then multiplying by $(A^l)^{-1}$ gives $A^{k-l} = I$ contrary to the hypothesis that p was the smallest integer for which

$$A^p = I.$$

EXAMPLE

Consider the element A of the symmetry group of the triangle on page 426; the period of A contains the elements

$$A \quad A^2 = B \quad A^3 = I \tag{16-3}$$

so that A is of order 3. The elements C, D, and E are each of order 2.

Another way of dividing a group into subsets is of interest. Let us call two elements a, b of G *equivalent* if there exists a group element g such that

$$g^{-1}ag = b \tag{16-4}$$

We can now divide our group up into subsets, such that all elements in any subset are equivalent to each other. These subsets are called *classes*.

If we think of our group elements as transformations, Eq. (16-4) has a simple "physical" significance. The transformation b is obtained by first carrying out the transformation g, then doing a, and finally "undoing" g. In other words, b is the same as a but in a different coordinate system, or looked at differently, or something. Thus the word "equivalent" is not inappropriate.

EXAMPLE

It is instructive to consider what equivalence means in the symmetric group S_n. Let us take, for example, $n = 6$, define $a = [4\,6\,1\,3\,5\,2]$ and $g = [3\,5\,1\,4\,6\,2]$, and calculate $b = g^{-1}ag$, an element which is equivalent to a.

$$b = [3\,6\,1\,4\,2\,5][4\,6\,1\,3\,5\,2][3\,5\,1\,4\,6\,2]$$
$$= [3\,6\,1\,4\,2\,5][4\,2\,3\,1\,6\,5]$$
$$= [3\,5\,4\,1\,2\,6]$$

The relationship, or "similarity," between a and b can be made clear by describing the respective permutations in detail. After permutation a,
the first object has become the third object
the second object has become the sixth object
the third object has become the fourth object
the fourth object has become the first object
the fifth object has become the fifth object
the sixth object has become the second object
or, in brief, a might be written as

$$a = \begin{pmatrix} 1 & 2 & 3 & 4 & 5 & 6 \\ \downarrow & \downarrow & \downarrow & \downarrow & \downarrow & \downarrow \\ 3 & 6 & 4 & 1 & 5 & 2 \end{pmatrix} \tag{16-5}$$

Similarly, after permutation b,

> the first object has become the fourth object
> the second object has become the fifth object
> the third object has become the first object
> the fourth object has become the third object
> the fifth object has become the second object
> the sixth object has become the sixth object

or, in brief,

$$b = \begin{pmatrix} 1 & 2 & 3 & 4 & 5 & 6 \\ \downarrow & \downarrow & \downarrow & \downarrow & \downarrow & \downarrow \\ 4 & 5 & 1 & 3 & 2 & 6 \end{pmatrix} \tag{16-6}$$

It can now be seen that if we replace, in the form (16-5) for a, each number by another according to the rule *which we read off from g or g^{-1}*, namely

$$1 \rightarrow 3,\, 2 \rightarrow 5,\, 3 \rightarrow 1,\, 4 \rightarrow 4,\, 5 \rightarrow 6,\, 6 \rightarrow 2 \tag{16-7}$$

then the result is

$$\begin{pmatrix} 3 & 5 & 1 & 4 & 6 & 2 \\ \downarrow & \downarrow & \downarrow & \downarrow & \downarrow & \downarrow \\ 1 & 2 & 4 & 3 & 6 & 5 \end{pmatrix}$$

which, by comparison with (16-6), is seen to be b.   In other words, the permutations a and b may be thought of as the *same* permutation, with different labels for the objects being permuted.

This equivalence can be rendered much more visible by using the cycle notation described just after Eq. (16-1).   In this notation the permutation a may be written (134)(26)(5), and b may be written (314)(52)(6); again we see that the substitutions (16-7) induced by g convert the one to the other. Furthermore, since relabeling the objects in this way changes the numbers written inside the parentheses denoting the cycles but cannot change either the number of cycles or the length of each cycle, we see that *two elements of S_6 (or, indeed, of any S_n) are equivalent if and only if they have exactly the same cycle structure.*   For example, consider the following three elements of S_5:

$$\begin{aligned} g_1 &= [5\,4\,1\,2\,3] \\ g_2 &= [2\,1\,5\,3\,4] \\ g_3 &= [3\,5\,1\,4\,2] \end{aligned} \tag{16-8}$$

If these are written in cycle notation, namely

$$\begin{aligned} g_1 &= (135)(24) \\ g_2 &= (12)(345) \\ g_3 &= (13)(25)(4) \end{aligned}$$

then it is clear that g_1 and g_2 are equivalent since each contains a cycle of length three and a cycle of length two while g_3, which contains two cycles of length two and one of length one, is not equivalent to g_1 or g_2. Incidentally, it should be clear from the above discussion how to write down by inspection an element g of S_5 such that the equivalence is made explicit:

$$g^{-1}g_1g = g_2$$

This will be left as an exercise (Problem 16-2).

Note that the identity I is in a class by itself. Also, two elements of different orders cannot be in the same class, for $a^h = I$ implies $b^h = (g^{-1}ag)^h = I$. If a subgroup S has the property

$$g^{-1}Sg = S$$

for every element g in our group G, S is said to be an *invariant*, or *normal*, subgroup. In other words, an invariant subgroup is made up of classes; if it contains part of any class it contains the whole class.

16–3   GROUP REPRESENTATIONS

A *representation* of a group G is a set of square nonsingular matrices $M_1, M_2, \ldots$, one matrix M for each group element g, such that

$$g_ig_j = g_k \quad \text{implies} \quad M_iM_j = M_k$$

In other words, it is a set of matrices which multiply just like the group elements and which, therefore, also constitute a group. If the matrices are $n \times n$, we have an *n-dimensional representation*. Every group has a trivial one-dimensional representation in which each group element is represented by the number 1.

If the matrices corresponding to different elements of G are themselves different, the two groups are *isomorphic*, and the representation is said to be *faithful*. If one matrix M represents more than one group element of G, the group G is said to be *homomorphic* to the matrix group and the representation is unfaithful.

Consider a particular representation D. The matrix associated with the group element g will be written $D(g)$. We can form another representation D' in a quite trivial way by defining

$$D'(g) = S^{-1}D(g)S$$

where S is any nonsingular matrix. $D(g)$ and $D'(g)$ are connected by an equivalence transformation. Such representations are said to be *equivalent*, and will, in practice, be considered to be the *same* representation.

Consider two representations $D^{(1)}(g)$ and $D^{(2)}(g)$. We can form a new representation

$$D(g) = D^{(1)}(g) \oplus D^{(2)}(g) = \begin{pmatrix} D^{(1)}(g) & 0 \\ 0 & D^{(2)}(g) \end{pmatrix} \qquad (16\text{-}9)$$

If $D^{(1)}$ and $D^{(2)}$ have dimensionality n_1 and n_2, the dimensionality of $D^{(1)} \oplus D^{(2)}$ is clearly $n_1 + n_2$. The representation D is said to be *reducible*[2] and splits into the two smaller representations $D^{(1)}$ and $D^{(2)}$.

A representation $D(g)$, which is *not* of the form (16-9) and cannot be brought into this form by an equivalence transformation, is called an *irreducible* representation. The irreducible representations of a group are the "building blocks" for the study of that group's representations since an arbitrary representation can be decomposed into a linear combination of irreducible representations. For example, if $D(g)$ can be brought by a similarity transformation into the form

$$\begin{pmatrix} D^{(1)} & 0 & 0 & 0 \\ 0 & D^{(1)} & 0 & 0 \\ 0 & 0 & D^{(2)} & 0 \\ 0 & 0 & 0 & D^{(3)} \end{pmatrix}$$

we often write

$$D(g) = 2D^{(1)}(g) \oplus D^{(2)}(g) \oplus D^{(3)}(g)$$

We shall now state without proof several very important theorems about group representations. These are true for finite groups and for those infinite groups we will consider in this chapter. Proofs may be found in mathematics books[3] dealing with the theory of group representations.

I. Every representation is equivalent to a *unitary* representation, that is, a representation in which each group element is represented by a unitary matrix. Thus, without loss of generality, we shall consider only unitary representations in what follows. See Problem 16-22.

---

[2] Strictly speaking, a representation is said to be *reducible* if it is equivalent to a representation of the form

$$\begin{pmatrix} D^{(1)}(g) & Y(g) \\ 0 & D^{(2)}(g) \end{pmatrix}$$

and *fully reducible*, or *decomposable*, if it is equivalent to the form (16-9). For a unitary representation there is no distinction between these two characteristics.

[3] See, for example, Hamermesh (H3) Chapter 3 or Wigner (W6) Chapter 9.

II. A matrix which commutes with every matrix of an irreducible representation is a multiple of the unit matrix. That is, if $D(g)$ is irreducible, and, if

$$D(g)A = AD(g)$$

for all g in our group, then A is a multiple of the unit matrix.

III (*Schur's lemma*). Let $D^{(1)}$ and $D^{(2)}$ be two irreducible representations of dimensionality n_1 and n_2, and suppose a matrix A exists[4] such that

$$AD^{(1)}(g) = D^{(2)}(g)A$$

for all g in the group G. Then *either* $A = 0$ *or* $n_1 = n_2$, det $A \neq 0$, and the two representations $D^{(1)}$ and $D^{(2)}$ are equivalent to each other.

IV (*Orthogonality theorem*). Let the group G contain h elements, and let its inequivalent irreducible (unitary) representations be $D^{(1)}$ (dimensionality n_1), $D^{(2)}$ (dimensionality n_2), etc. $D^{(i)}(g)$ is a matrix; we denote a typical matrix element by $D_{\alpha\beta}^{(i)}(g)$. Then

$$\sum_g [D_{\alpha\beta}^{(i)}(g)]^* D_{\gamma\delta}^{(j)}(g) = \frac{h}{n_i} \delta_{ij} \delta_{\alpha\gamma} \delta_{\beta\delta} \qquad (16\text{-}10)$$

The sum runs over all g in G.

The orthogonality theorem is a very powerful and useful result. Let us consider the $\alpha\beta$ matrix element in the ith irreducible representation. As g runs through the h elements of our group G, $D_{\alpha\beta}^{(i)}(g)$ takes on h values; we can think of it as a vector with h components. The orthogonality theorem (16-10) tells us that this vector "dotted into" itself equals h/n_i, while it is orthogonal to any similar vector from another representation, or even from a different location in the same representation.

How many such vectors are there? A representation $D^{(i)}$ of dimensionality n_i is made up of matrices with n_i rows and n_i columns and therefore contains n_i^2 such vectors. Thus the total number of vectors is

$$n_1^2 + n_2^2 + \cdots = \sum n_i^2 \qquad (16\text{-}11)$$

where the sum runs over all (inequivalent) irreducible representations. But it is obvious that there cannot be more than h independent (let alone orthogonal) vectors with h components. Therefore,

$$\sum n_i^2 \leqslant h \qquad (16\text{-}12)$$

It can be shown that this is, in fact, an equality.

A simple example which can be used to "check" the above relations is provided by the group S_3, the symmetric group on 3 objects. This group has three irreducible representations, which are shown in Table 16–1.

---

[4] Note that A must have n_1 rows and n_2 columns.

**Table 16–1   The Irreducible Represen-
tations of the Group S_3**

g	$D^{(1)}$ $n_i = 1$	$D^{(2)}$ 1	$D^{(3)}$ 2
[1 2 3]	1	1	$\begin{pmatrix} 1 & 0 \\ 0 & 1 \end{pmatrix}$
[2 3 1]	1	1	$\frac{1}{2}\begin{pmatrix} -1 & \sqrt{3} \\ -\sqrt{3} & -1 \end{pmatrix}$
[3 1 2]	1	1	$\frac{1}{2}\begin{pmatrix} -1 & -\sqrt{3} \\ \sqrt{3} & -1 \end{pmatrix}$
[3 2 1]	1	−1	$\frac{1}{2}\begin{pmatrix} 1 & \sqrt{3} \\ \sqrt{3} & -1 \end{pmatrix}$
[2 1 3]	1	−1	$\frac{1}{2}\begin{pmatrix} 1 & -\sqrt{3} \\ -\sqrt{3} & -1 \end{pmatrix}$
[1 3 2]	1	−1	$\begin{pmatrix} -1 & 0 \\ 0 & 1 \end{pmatrix}$

16–4   CHARACTERS

The characterization of representations by explicitly exhibiting (as in Table 16-1) the matrices which represent the various group elements is not only cumbersome but also somewhat inappropriate. We could make a similarity transformation

$$D(g) \to D'(g) = SD(g)S^{-1} \qquad (16\text{-}13)$$

which would lead to a completely different set of matrices, but the two representations are equivalent, that is, the same for most physical purposes. It would seem preferable to identify our representation by something that is invariant under similarity transformations.

It turns out that the *traces* of our representation matrices are a very convenient set of numbers for this purpose. They are certainly unchanged by similarity transformations, and they still serve to distinguish the different representations.

We shall, therefore, define the *character* $\chi^{(i)}(g)$ as the trace of the matrix $D^{(i)}(g)$:

$$\chi^{(i)}(g) = \sum_\alpha D^{(i)}_{\alpha\alpha}(g) \qquad (16\text{-}14)$$

Table 16–2   Character Table for S_3

g	$\chi^{(1)}$	$\chi^{(2)}$	$\chi^{(3)}$
[1 2 3]	1	1	2
[2 3 1]	1	1	−1
[3 1 2]	1	1	−1
[3 2 1]	1	−1	0
[2 1 3]	1	−1	0
[1 3 2]	1	−1	0

From the matrices given in Table 16–1 for the three irreducible representations of the group S_3, we can form the character table shown in Table 16–2.

We may note several features of this table. The character of the identity element—which in this particular group is [1 2 3]—gives the dimensionality of the particular representation since the trace of the $n \times n$ unit matrix is n. Thus, from the first row of our character table, we can read off the dimensionalities of the various irreducible representations.

Another important feature of Table 16–2 is that certain sets of group elements have the same characters. That is, [2 3 1] and [3 1 2] have the same character in each representation, and so do [3 2 1], [2 1 3], and [1 3 2]. The general result is that two group elements which belong to the same class have the same character in any representation.

Proof: Let g_1 and g_2 belong to the same class. Then

$$g_1 = hg_2 h^{-1}$$

where h is some element of our group G. Therefore, in any representation

$$D(g_1) = D(h)D(g_2)D(h^{-1})$$

$$\begin{aligned}
\chi(g_1) = \text{Tr } D(g_1) &= \text{Tr } [D(h)D(g_2)D(h^{-1})] \\
&= \text{Tr } [D(h^{-1})D(h)D(g_2)] \\
&= \text{Tr } [D(I)D(g_2)] \\
&= \text{Tr } D(g_2) \\
&= \chi(g_2)
\end{aligned} \tag{16-15}$$

where I denotes the unit element of the group G.

Thus, characters are *class functions*; for a given representation there is one character for each class, and the character table for S_3 can be written as shown in Table 16–3. We have divided S_3 into its three classes:

$$\underset{C_1}{[1\ 2\ 3]} \quad \underset{C_2}{\underbrace{[2\ 3\ 1][3\ 1\ 2]}} \quad \underset{C_3}{\underbrace{[3\ 2\ 1][2\ 1\ 3][1\ 3\ 2]}} \tag{16-16}$$

An important orthogonality relation for group characters follows immediately from the one stated previously (16-10) for matrix elements of the
irreducible representations.   If $\chi^{(1)}$, $\chi^{(2)}$, ... are the characters of the irreducible representations, C_1, C_2, ..., C_s are the various classes, and p_k denotes
the number of group elements in the kth class C_k, then

$$\sum_{k=1}^{s} p_k \chi^{(i)}(C_k)^* \chi^{(j)}(C_k) = h\,\delta_{ij} \tag{16-17}$$

h being, as usual, the number of elements in our group.

If we think of the characters in each irreducible representation as an
s-dimensional vector, one component for each class, the fact that they are
all orthogonal means there cannot be more than s different irreducible
representations.   In fact (it can be shown) there are always exactly s, so our
character table is square; there are just as many irreducible representations as
classes.

A compact way of expressing the orthogonality relation for characters is to
say that the matrix whose (i, j) element is

$$\sqrt{p_j/h}\,\chi^{(i)}(C_j) \tag{16-18}$$

is a unitary matrix.

Now consider a *reducible* representation D, that is, one which after a
suitable change of coordinates (similarity transformation) looks like this

$$D(g) = \begin{pmatrix} D^{(a)}(g) & & & 0 \\ & D^{(b)}(g) & & \\ & & D^{(c)}(g) & \\ 0 & & & \ddots \end{pmatrix} \tag{16-19}$$

That is,

$$D(g) = D^{(a)}(g) \oplus D^{(b)}(g) \oplus \cdots \tag{16-20}$$

where $D^{(a)}$, $D^{(b)}$, ... are various irreducible representations.   The trace
of D is obviously the sum of the traces of the various submatrices in D, that is,

$$\chi(C) = \chi^{(a)}(C) + \chi^{(b)}(C) + \cdots \tag{16-21}$$

Suppose D contains the first irreducible representation $D^{(1)}$ c_1 times, the

Table 16–3   Character Table for S_3

Class	$\chi^{(1)}$	$\chi^{(2)}$	$\chi^{(3)}$
C_1	1	1	2
C_2	1	1	−1
C_3	1	−1	0

second c_2 times, and so on.   We might write

$$D = c_1 D^{(1)} \oplus c_2 D^{(2)} \oplus \ldots \tag{16-22}$$

Then certainly we have for the character of D

$$\chi(C_k) = \sum_i c_i \chi^{(i)}(C_k) \tag{16-23}$$

Multiplying both sides by $p_k \chi^{(j)}(C_k)^*$ and summing over k gives, by virtue of the orthogonality relation for group characters,

$$c_i = \frac{1}{h} \sum_k p_k \chi^{(i)}(C_k)^* \chi(C_k)$$

$$= \overline{\chi^{(i)}(g)^* \chi(g)} \tag{16-24}$$

where the bar denotes an average over the whole group.

This last result illustrates the usefulness of characters.   Given any representation D, all we need to do in order to decompose D into irreducible representations is to calculate χ, the traces of $D(g)$.   Formula (16-24) then tells how many times each irreducible representation $D^{(i)}$ is contained in D.

As an example, we shall decompose a representation of S_3.   The representation D will be formed by representing the rearrangements (which make up S_3) by matrices.   For example, [1 2 3], the identity element, is represented by the unit 3×3 matrix

$$\begin{pmatrix} 1 & 0 & 0 \\ 0 & 1 & 0 \\ 0 & 0 & 1 \end{pmatrix}$$

The group element [2 1 3], which exchanges the first and second objects, is represented by

$$\begin{pmatrix} 0 & 1 & 0 \\ 1 & 0 & 0 \\ 0 & 0 & 1 \end{pmatrix}$$

since

$$\begin{pmatrix} 0 & 1 & 0 \\ 1 & 0 & 0 \\ 0 & 0 & 1 \end{pmatrix} \begin{pmatrix} a \\ b \\ c \end{pmatrix} = \begin{pmatrix} b \\ a \\ c \end{pmatrix}$$

and so on.   The student is encouraged to write out all six matrices; we shall here give only the character table of our new representation:

Class	C_1	C_2	C_3	
χ	3	0	1	(16-25)

Now we can find the coefficients c_i in our decomposition

$$D = c_1 D^{(1)} \oplus c_2 D^{(2)} \oplus c_3 D^{(3)} \tag{16-26}$$

by applying the general formula (16-24) and using the character table for S_3 (Table 16–3):

$$c_1 = \overline{\chi^{(1)*}(g)\chi(g)} = \tfrac{1}{6}[1 \times 1 \times 3 + \quad 2 \times 1 \times 0 \quad + \quad 3 \times 1 \times 1] = 1$$

$$c_2 = \overline{\chi^{(2)*}(g)\chi(g)} = \tfrac{1}{6}[1 \times 1 \times 3 + \quad 2 \times 1 \times 0 \quad +3 \times (-1) \times 1] = 0$$

$$c_3 = \overline{\chi^{(3)*}(g)\chi(g)} = \tfrac{1}{6}[1 \times 2 \times 3 + 2 \times (-1) \times 0 + \quad 3 \times 0 \times 1] = 1$$

Therefore,

$$D = D^{(1)} \oplus D^{(3)} \tag{16-27}$$

As a second example, every group of order h has an $h \times h$ representation, the so-called *regular representation*, formed by considering the rearrangements of the group elements caused by left multiplication by each individual group element.   Again using S_3 for our model, we label, for convenience, the group elements

$$g_1 = [1\,2\,3] \qquad g_2 = [2\,3\,1] \qquad g_3 = [3\,1\,2]$$
$$g_4 = [3\,2\,1] \qquad g_5 = [2\,1\,3] \qquad g_6 = [1\,3\,2]$$

g_1 will, of course, be represented by the 6×6 unit matrix.   Upon multiplication by g_2 on the left,

$$\begin{array}{ccc} g_1 \rightarrow g_2 & g_2 \rightarrow g_3 & g_3 \rightarrow g_1 \\ g_4 \rightarrow g_5 & g_5 \rightarrow g_6 & g_6 \rightarrow g_4 \end{array} \tag{16-28}$$

Now think of each group element as a column vector

$$g_1 = \begin{pmatrix} 1 \\ 0 \\ 0 \\ 0 \\ 0 \\ 0 \end{pmatrix} \qquad g_2 = \begin{pmatrix} 0 \\ 1 \\ 0 \\ 0 \\ 0 \\ 0 \end{pmatrix} \qquad \text{etc.} \tag{16-29}$$

Then the matrix which performs the rearrangement (16-28) is

$$\begin{pmatrix} 0 & 0 & 1 & 0 & 0 & 0 \\ 1 & 0 & 0 & 0 & 0 & 0 \\ 0 & 1 & 0 & 0 & 0 & 0 \\ 0 & 0 & 0 & 0 & 0 & 1 \\ 0 & 0 & 0 & 1 & 0 & 0 \\ 0 & 0 & 0 & 0 & 1 & 0 \end{pmatrix} \tag{16-30}$$

and this will be the matrix which represents g_2 in the regular representation of S_3.   A more formal definition of this representation would be that

$$D_{ij}(g_k) = \begin{cases} 1 & \text{if} \quad g_k g_j = g_i \\ 0 & \text{otherwise} \end{cases} \tag{16-31}$$

What is the character $\chi(g)$ for this representation? Clearly, $\chi(g_1) = 6$, while all other $\chi(g)$ are zero. The character table is

$$
\begin{array}{cccc}
\textit{Class} & C_1 & C_2 & C_3 \\
\chi & 6 & 0 & 0
\end{array}
\qquad (16\text{-}32)
$$

If $D = c_1 D^{(1)} \oplus c_2 D^{(2)} \oplus c_3 D^{(3)}$, we find

$$
c_1 = \tfrac{1}{6}[1 \times 6] = 1 \qquad c_2 = \tfrac{1}{6}[1 \times 6] = 1 \qquad c_3 = \tfrac{1}{6}[2 \times 6] = 2
$$

Then

$$
D = D^{(1)} \oplus D^{(2)} \oplus 2D^{(3)} \qquad (16\text{-}33)
$$

The regular representation of an *arbitrary* group is just as easy to study as this simple example. We use two elementary facts:

1. The character χ of the regular representation for an arbitrary group of order h is

$$
\chi(g) = \begin{cases} h & \text{if} \quad g = I \\ 0 & \text{otherwise} \end{cases}
$$

2. The character $\chi^{(i)}(I)$ for the irreducible representation $D^{(i)}$ equals n_i, the dimensionality of $D^{(i)}$.

We deduce that the regular representation contains each irreducible representation $D^{(i)}$ a number of times equal to n_i, the dimensionality of $D^{(i)}$. Since the dimensionality of the regular representation is h, this verifies the statement we made earlier that

$$
\sum_i n_i^2 = h \qquad (16\text{-}34)
$$

We shall conclude this section by giving an example of the construction of a character table for a group. Consider A_4, the group of *even* permutations of 4 objects, often called the *alternating group* on 4 objects. We begin by labelling the $\tfrac{1}{2} \times 4! = 12$ elements

$$
\begin{array}{llll}
[1\,2\,3\,4] = I & [1\,3\,4\,2] = A & [1\,4\,2\,3] = B & [3\,2\,4\,1] = C \\
[4\,2\,1\,3] = D & [2\,4\,3\,1] = E & [4\,1\,3\,2] = F & [2\,3\,1\,4] = G \quad (16\text{-}35) \\
[3\,1\,2\,4] = H & [2\,1\,4\,3] = J & [3\,4\,1\,2] = K & [4\,3\,2\,1] = L
\end{array}
$$

Next, we decompose our group into classes of equivalent elements. This is straightforward and gives 4 classes:

$$
\begin{array}{ll}
C_1 = I & C_2 = A,\, D,\, E,\, H \\
C_3 = B,\, C,\, F,\, G & C_4 = J,\, K,\, L
\end{array}
\qquad (16\text{-}36)
$$

Since the number of irreducible representations equals the number of classes, there are four irreducible representations

$$
D^{(1)} \qquad D^{(2)} \qquad D^{(3)} \qquad D^{(4)}
$$

and four characters

$$\chi^{(1)} \qquad \chi^{(2)} \qquad \chi^{(3)} \qquad \chi^{(4)}$$

The dimensionalities n_i of these representations can be found from the condition

$$\sum_{i=1}^{4} n_i^2 = h = 12$$

Only one set of four squares adds up to 12, namely,

$$1^2 + 1^2 + 1^2 + 3^2 = 12 \tag{16-37}$$

so that

$$n_1 = n_2 = n_3 = 1 \qquad n_4 = 3$$

and these are the four characters for class C_1 (the identity element.) We also know that one representation (call it $D^{(1)}$) is trivial; $\chi^{(1)} = 1$ for all C_i. Thus we have the beginning of a character table as shown in Table 16–4.

Another result which aids the finding of characters may be found as follows. Suppose g is a group element of order m [see (16-2)] and $D(g)$ is a matrix representing g in an n-dimensional representation. Then

$$g^m = I$$

implies

$$[D(g)]^m = 1 \tag{16-38}$$

If we make a similarity transformation to diagonalize $D(g)$, the last equation becomes

$$\begin{pmatrix} \lambda_1^m & 0 & 0 & \cdots & 0 \\ 0 & \lambda_2^m & 0 & & \vdots \\ \vdots & & & & \\ 0 & \cdot & \cdot & \cdot & \lambda_n^m \end{pmatrix} = \begin{pmatrix} 1 & 0 & 0 & \cdots & 0 \\ 0 & 1 & 0 & \cdots & 0 \\ \vdots & & & & \vdots \\ 0 & \cdot & \cdot & \cdot & 1 \end{pmatrix} \tag{16-39}$$

so that the λ_k, the eigenvalues of $D(g)$, are all mth roots of unity. Then the character is

$$\chi(g) = \mathrm{Tr}\, D(g) = \sum_{k=1}^{n} \lambda_k \tag{16-40}$$

or $\chi(g)$ is the sum of n mth roots of unity.

Table 16–4

Class	$\chi^{(1)}$	$\chi^{(2)}$	$\chi^{(3)}_*$	$\chi^{(4)}$
C_1	1	1	1	3
C_2	1			
C_3	1			
C_4	1			

Table 16–5

Class	$\chi^{(1)}$	$\chi^{(2)}$	$\chi^{(3)}$	$\chi^{(4)}$
C_1	1	1	1	3
C_2	1	ω	ω^2	
C_3	1	ω^2	ω	
C_4	1	1	1	

Let us now apply this result to the one-dimensional representations $D^{(2)}$ and $D^{(3)}$ of our group A_4. The orders of elements in the classes C_1, C_2, C_3, C_4 are readily found to be 1, 3, 3, 2, respectively. Thus $\chi(C_4)$ can only be ± 1, while $\chi(C_2)$ and $\chi(C_3)$ can have values 1, ω, or ω^2, where ω is the cube root of unity, $\omega = e^{2\pi i/3}$. These facts, together with the requirement (16-17) of orthogonality with $\chi^{(1)}$, determine $\chi^{(2)}$ and $\chi^{(3)}$. Specifically, if $\chi^{(2)}(C_2) = \omega$, $\chi^{(2)}(C_3)$ must be ω^2 to cancel the imaginary part in the orthogonality relation. The choice $\chi^{(2)}(C_2) = \chi^{(2)}(C_3) = 1$ cannot satisfy orthogonality, so we must have $\chi^{(2)}(C_2) = \omega$ and $\chi^{(2)}(C_3) = \omega^2$ (or vice versa), and, furthermore, $\chi^{(2)}(C_4) = 1$. Similar conclusions hold for $\chi^{(3)}$ so our character table now reads as shown in Table 16–5.

The last column is most easily filled in by using the orthogonality between it and the first three columns. If we let the last column be 3, x, y, z, then [note $\omega^* = \omega^2$ and $(\omega^2)^* = \omega$]

$$3 + 4x + 4y + 3z = 0$$
$$3 + 4\omega x + 4\omega^2 y + 3z = 0 \qquad (16\text{-}41)$$
$$3 + 4\omega^2 x + 4\omega y + 3z = 0$$

The solution of these equations is straightforward, and our character table finally reads as shown in Table 16–6. A completely different derivation of this same character table is in Margenau and Murphy (M2) Chapter XV.

16–5 PHYSICAL APPLICATIONS

We shall only illustrate the usefulness of group-theoretic methods in a few situations; many other applications could be mentioned if space and time allowed.[5]

Let us consider the eigenvalue problem associated with some partial differential operator in two or three dimensions. We have seen in Chapter 8 that *degeneracy* frequently occurs; several modes or eigenfunctions may belong to the same eigenvalue. While this may sometimes be accidental, it more often is associated with some symmetry of the system under consideration.

[5] See, for example, Tinkham (T1); Hamermesh (H3); or Wigner (W6).

For example, the second and third modes of a square drum head are degenerate; any linear combination of

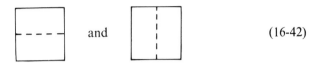 and   (16-42)

is a mode with the same frequency.   This is clearly not an accident, but rather follows from the fact that the drum head is completely unchanged if we rotate it 90°.

To make this idea of symmetry more precise, we consider the *symmetry group* of our physical system, namely, the set of all transformations of the system which leave it *dynamically* unchanged.   One side of our drum could, of course, be painted red and the top painted green, so that a rotation would not leave it unchanged; the *dynamics*, however, would not be affected by the rotation.

Now consider a particular eigenfunction.   If we apply some element of our symmetry group to this eigenfunction, we must get another eigenfunction with the same eigenvalue.   Although this fact can be proved mathematically, it should be obvious.   If a group element acting on an eigenfunction did not yield another eigenfunction with the same eigenvalue, our system would not be invariant under that transformation, and we included some transformation in the symmetry group that did not belong there.

We may think of all these degenerate eigenfunctions as a vector space, with the group elements making linear transformations in this space.   By choosing a coordinate system, we can represent these transformations by matrices, and in fact we have a *representation* of our symmetry group.   We can, therefore, use the results from the theory of group representations developed earlier in this chapter.   Two points should be mentioned.

1. We should use normalized, orthogonal wave functions for base vectors so that our representations come out unitary, otherwise most of the results of this chapter are inapplicable.

2. If several different choices for an orthonormal set of base vectors are available, it does not matter which we use since the different possible representations will only differ by an equivalence transformation.

Table 16–6

Class	$\chi^{(1)}$	$\chi^{(2)}$	$\chi^{(3)}$	$\chi^{(4)}$
C_1	1	1	1	3
C_2	1	ω	ω^2	0
C_3	1	ω^2	ω	0
C_4	1	1	1	-1

Let us apply these ideas to the lowest eigenfunctions of a square drum head. The lowest eigenfunction

which has no nodes, is left unchanged by all elements of the symmetry group. This mode, therefore, generates the representation which represents every group element by the number 1.

The next two modes

are mixed up by the group elements, and generate a two-dimensional representation. (This symmetry group has only one two-dimensional representation.)

The next mode

is nondegenerate, so that it generates a one-dimensional representation, but this representation has -1 in several places and is not the trivial representation generated by the lowest mode. The student is urged at this point to write this representation down as well as the entire character table; the result is given in Table 16–7.

As another example, consider the Schrödinger equation for n identical particles, such as the electrons in an n-electron atom. Let $\psi(x_1, x_2, \ldots, x_n)$ be an eigenfunction with a certain energy E. Then $\psi(x_{i_1}, x_{i_2}, \ldots, x_{i_n})$, where $i_1, \ldots, i_n$ is an arbitrary permutation of $1, \ldots, n$, must also be an eigenfunction with the same energy E, since we have merely exchanged particles which are dynamically equivalent. Thus the set of all eigenfunctions with energy E generates a representation of the symmetric group on n objects. If the particles in fact are electrons, the Pauli exclusion principle tells us that only one representation is allowed, the totally antisymmetric one in which every exchange of two electrons changes the sign of the wave function ψ.

How can we apply all this formalism to a specific problem? The simplest sort of problem we can discuss concerns the splitting of degeneracies by a perturbation which lowers the symmetry of our physical system. Consider a highly symmetric system which gives rise to the eigenvalue problem

$$L^0\psi = \lambda\psi \qquad (16\text{-}43)$$

The set of eigenfunctions ψ_1, ψ_2, ..., ψ_n which belong to some eigenvalue λ generates a representation of the symmetry group S of L^0.

We wish to investigate how this degenerate set splits up when we apply a perturbation Q having a symmetry group S' which is a subgroup of S. The procedure should be clear from the preceding discussion. The set $\psi_1, \ldots, \psi_n$ generates some representation of S'. If we decompose this representation into irreducible representations, the base vectors for each irreducible representation are still degenerate under the operator $L^0 + Q$, but eigenfunctions belonging to different irreducible representations need not have the same perturbed eigenvalue.

Rather than dwell any further on such generalities, we shall do a specific example. Consider a round drum head, perturbed by four masses at the vertices of a concentric square,

In the absence of the perturbation, the modes of the drum head are either nondegenerate or twofold degenerate (see p. 233). There is nothing group theory can tell us about the nondegenerate modes, but an interesting question is the following. Does the perturbation break the degeneracy of a pair of degenerate modes?

We begin by discussing the symmetry group and its character table. The

Table 16–7

Class	$\chi^{(1)}$	$\chi^{(2)}$	$\chi^{(3)}$	$\chi^{(4)}$	$\chi^{(5)}$
C_1	1	1	1	1	2
C_2	1	1	−1	−1	0
C_3	1	1	1	1	−2
C_4	1	−1	−1	1	0
C_5	1	−1	1	−1	0

symmetry group is the symmetry group of a square and contains eight elements:

I (identity) R (rotation through 90°) R^2 R^3
P (reflection about line through center) PR PR^2 PR^3

The classes are easily found to be

$$
\begin{array}{ll}
C_1: I & C_4: P, PR^2 \\
C_2: R, R^3 & C_5: PR, PR^3 \\
C_3: R^2 &
\end{array}
\tag{16-44}
$$

It is now straightforward, by methods just like those used earlier in this chapter, to write down the character table for our group (Table 16-7).

Now consider a degenerate pair of unperturbed eigenfunctions

$$
\psi_1 \sim \sin m\theta \qquad \psi_2 \sim \cos m\theta
\tag{16-45}
$$

We must decide whether the two-dimensional representation D of our symmetry group generated by ψ_1 and ψ_2 is reducible or irreducible.

To begin with, $I\psi_1 = \psi_1$, $I\psi_2 = \psi_2$, so that I is represented by the matrix

$$
\begin{pmatrix} 1 & 0 \\ 0 & 1 \end{pmatrix}
\tag{16-46}
$$

and $\chi(C_1) = 2$.

Next, R operating on ψ_1 or ψ_2 is equivalent to decreasing θ by $\pi/2$. Thus

$$
\begin{aligned}
R\psi_1 &\sim \sin m\left(\theta - \frac{\pi}{2}\right) = \left(\cos \frac{m\pi}{2}\right)\psi_1 - \left(\sin \frac{m\pi}{2}\right)\psi_2 \\
R\psi_2 &\sim \cos m\left(\theta - \frac{\pi}{2}\right) = \left(\sin \frac{m\pi}{2}\right)\psi_1 + \left(\cos \frac{m\pi}{2}\right)\psi_2
\end{aligned}
\tag{16-47}
$$

R is therefore represented by the matrix

$$
\begin{pmatrix}
\cos \dfrac{m\pi}{2} & \sin \dfrac{m\pi}{2} \\[2ex]
-\sin \dfrac{m\pi}{2} & \cos \dfrac{m\pi}{2}
\end{pmatrix}
\tag{16-48}
$$

and its trace is $\chi(C_2) = 2 \cos (m\pi/2)$. Continuing in this way, we find that our representation D has the following character:

$$
\begin{aligned}
&\chi(C_1) = 2 & &\chi(C_2) = 2 \cos\frac{m\pi}{2} \\
&\chi(C_3) = 2 \cos m\pi & &\chi(C_4) = \chi(C_5) = 0
\end{aligned}
\tag{16-49}
$$

We can now decompose D into irreducible representations; the result is

$$D = \tfrac{1}{4}\left[1 + 2\cos\frac{m\pi}{2} + \cos m\pi\right][D^{(1)} + D^{(2)}]$$

$$+ \tfrac{1}{4}\left[1 - 2\cos\frac{m\pi}{2} + \cos m\pi\right][D^{(3)} + D^{(4)}]$$

$$+ \tfrac{1}{2}[1 - \cos m\pi]D^{(5)}$$

$$= \begin{cases} D^{(1)} + D^{(2)} & m = 4, 8, 12, \ldots \\ D^{(3)} + D^{(4)} & m = 2, 6, 10, \ldots \\ D^{(5)} & m = 1, 3, 5, 7, \ldots \end{cases} \qquad (16\text{-}50)$$

Thus, if m is *even*, the representation is reducible and is the sum of two one-dimensional representations, so that the perturbation can split the degeneracy. If, on the other hand, m is *odd*, the representation is irreducible and no splitting is possible since the symmetry connects all states.

Notice that in this example we have drawn *qualitative* conclusions from group theory; the *amounts* of splitting due to the perturbation are not explored. This is a general feature of group-theoretic arguments. However, we shall now do a problem where we actually calculate numbers by using group theory.

Consider three point masses m, at the vertices of an equilateral triangle, connected by springs of spring constant k (see Figure 16–2). What are the normal modes of this mechanical system? We suppose the masses are only allowed to move in the plane of the page.

We shall number the masses as shown in the figure. Let the coordinates of m_1 relative to its equilibrium position be x_1, y_1, and similarly for the other two masses. We represent the configuration of the system by a six-dimensional "state vector" ξ. In an "elementary coordinate system," such as the one on page 155, ξ has components

$$\xi = (x_1, y_1, x_2, y_2, x_3, y_3) \qquad (16\text{-}51)$$

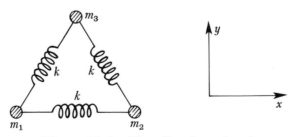

Figure 16–2 The vibrating triangle

The kinetic and potential energies of our system are [compare Eqs. (6-58) and (6-60)]

$$T = \tfrac{1}{2}m \sum_i \dot{\xi}_i^2$$

$$
\begin{aligned}
V &= \tfrac{1}{2}k\{(x_2 - x_1)^2 + [-\tfrac{1}{2}(x_3 - x_2) + \tfrac{1}{2}\sqrt{3}(y_3 - y_2)]^2 \\
&\quad + [\tfrac{1}{2}(x_1 - x_3) + \tfrac{1}{2}\sqrt{3}(y_1 - y_3)]^2\} \\
&= \tfrac{1}{2}k \sum_{ij} V_{ij}\xi_i\xi_j
\end{aligned}
\tag{16-52}
$$

Newton's second law gives

$$m\ddot{\xi}_i = -\frac{\partial V}{\partial \xi_i} = -k \sum_j V_{ij}\xi_j \tag{16-53}$$

For vibration in a normal mode, we will have $\xi \sim e^{-i\omega t}$. Then (16-53) becomes

$$\sum_j V_{ij}\xi_j = \lambda\xi_i \qquad \text{where} \qquad \lambda = \frac{m\omega^2}{k} \tag{16-54}$$

The normal modes are the eigenvectors of the matrix V, with the eigenvalues giving the frequencies.

Let us see what group theory can do to find these eigenvalues. In the first place, each eigenvector generates an irreducible representation when we act on it with all the elements of our symmetry group. Thus, in a coordinate system which diagonalizes V,

$$
V = \begin{pmatrix}
\lambda_a & & & & & \\
& \ddots & & & & \\
& & \lambda_a & & & \\
& & & \lambda_b & & \\
& & & & \ddots & \\
& & & & & \lambda_b \\
& & & & & & \text{etc.}
\end{pmatrix}
\begin{matrix} \Big\} D^{(a)} \\ \\ \Big\} D^{(b)} \end{matrix}
\tag{16-55}
$$

The first n_a of our new coordinate vectors are eigenvectors belonging to the eigenvalue λ_a, and transforming among each other according to the irreducible representation $D^{(a)}$, the next n_b eigenvectors belong to λ_b and transform according to $D^{(b)}$, and so on.

As usual, we must discuss the symmetry group briefly. It consists of six elements

$$I \quad R \quad R^2 \quad P \quad PR \quad PR^2$$

where I is the identity, R rotates the triangle 120° in a positive sense (counterclockwise), and P reflects about a vertical line through the center. The character table is shown in Table 16–8. This symmetry group is just S_3; compare Tables 16–1 and 16–3.

When various group elements act on the triangle, they induce linear transformations of the ξ_i. For example, if R operates on the triangle,

$$\xi' = D(R)\xi \tag{16-56}$$

where the matrix $D(R)$ is

$$D(R) = \begin{pmatrix} 0 & 0 & r \\ r & 0 & 0 \\ 0 & r & 0 \end{pmatrix} \qquad r = \begin{pmatrix} -\dfrac{1}{2} & -\dfrac{\sqrt{3}}{2} \\ \dfrac{\sqrt{3}}{2} & -\dfrac{1}{2} \end{pmatrix} \tag{16-57}$$

Similarly,

$$D(P) = \begin{pmatrix} 0 & p & 0 \\ p & 0 & 0 \\ 0 & 0 & p \end{pmatrix} \qquad p = \begin{pmatrix} -1 & 0 \\ 0 & 1 \end{pmatrix} \tag{16-58}$$

What representations does this representation D contain? Its character is

$$\chi(C_1) = 6 \qquad \chi(C_2) = 0 \qquad \chi(C_3) = 0$$

and this gives [compare (16-32) and (16-33)]

$$D = D^{(1)} \oplus D^{(2)} \oplus 2D^{(3)} \tag{16-59}$$

Note that D is just (equivalent to) the regular representation of our symmetry group. This particular feature is an accident, and not very general.

Now we can be more specific and write, in a coordinate system with the eigenvectors of V as basis,

$$V = \begin{pmatrix} \lambda_1 & & & & & \\ & \lambda_2 & & & & \\ & & \lambda_{31} & & & \\ & & & \lambda_{31} & & \\ & & & & \lambda_{32} & \\ & & & & & \lambda_{32} \end{pmatrix} \begin{matrix} D^{(1)} \\ D^{(2)} \\ \left.\vphantom{\begin{matrix}a\\b\end{matrix}}\right\} D^{(3)} \\ \\ \left.\vphantom{\begin{matrix}a\\b\end{matrix}}\right\} D^{(3)} \end{matrix} \tag{16-60}$$

where we have indicated the transformation properties of the various eigenvectors. Note that the two pairs of eigenvectors transforming like $D^{(3)}$ need not have the same eigenvalue.

Table 16–8

Class	$\chi^{(1)}$	$\chi^{(2)}$	$\chi^{(3)}$
$C_1(I)$	1	1	2
$C_2(R, R^2)$	1	1	-1
$C_3(P, PR, PR^2)$	1	-1	0

What does the matrix $D(G)V$, where G is an arbitrary element of our symmetry group, look like in this special coordinate system? The answer is clearly

$$D(G)V = \begin{pmatrix} \lambda_1 D^{(1)}(G) & & & \\ & \lambda_2 D^{(2)}(G) & & \\ & & \lambda_{31} D^{(3)}(G) & \\ & & & \lambda_{32} D^{(3)}(G) \end{pmatrix} \qquad (16\text{-}61)$$

This is not directly useful, of course, because we do not know what this coordinate system is; that is, we do not know the eigenvectors. However, the trace of this matrix is invariant to coordinate transformations, so that

$$\mathrm{Tr}\, D(G)V = \lambda_1 \chi^{(1)}(G) + \lambda_2 \chi^{(2)}(G) + (\lambda_{31} + \lambda_{32})\chi^{(3)}(G) \qquad (16\text{-}62)$$

in *any* coordinate system.

Now it is straightfoward to compute the following traces by making use of the specific forms (16-52), (16-57), and (16-58) of V, $D(R)$, and $D(P)$, respectively.

$$\mathrm{Tr}\, D(I)V = 6$$
$$\mathrm{Tr}\, D(R)V = \tfrac{3}{2} \qquad\qquad (16\text{-}63)$$
$$\mathrm{Tr}\, D(P)V = 3$$

Our eigenvalues therefore obey the equations

$$\lambda_1 + \lambda_2 + 2(\lambda_{31} + \lambda_{32}) = 6$$
$$\lambda_1 + \lambda_2 - (\lambda_{31} + \lambda_{32}) = \tfrac{3}{2}$$
$$\lambda_1 - \lambda_2 = 3$$

from which

$$\lambda_1 = 3 \qquad \lambda_2 = 0 \qquad \lambda_{31} + \lambda_{32} = \tfrac{3}{2} \qquad (16\text{-}64)$$

We could determine λ_{31} and λ_{32} separately by looking at things like

$$\mathrm{Tr}\, V^2 = \lambda_1^2 + \lambda_2^2 + 2(\lambda_{31}^2 + \lambda_{32}^2)$$

but a simpler way is to note that there must be three degrees of freedom having zero eigenvalue, two translational and one rotational. Thus

$$\lambda_{31} = 0 \qquad \lambda_{32} = 3/2 \qquad (16\text{-}65)$$

and we have determined all the eigenvalues without solving any secular equation.

16–6   INFINITE GROUPS

We shall conclude this chapter by discussing briefly groups with an infinite number of elements. We will give several examples of infinite groups encountered in physics and discuss their irreducible representations. Proofs are not given; the reader desiring a more complete treatment and proofs should consult the references listed at the end of this chapter.

We shall consider infinite groups whose elements may be labeled by real parameters, which vary continuously; a typical group element will be written $g(x_1, x_2, \ldots, x_n) = g(x)$. If n such parameters are required, we call our group an n-parameter group. Obviously, we cannot write down a multiplication table in the ordinary sense for an infinite group. If the product of $g(x)$ and $g(y)$ is $g(z)$, that is,

$$g(x_1, x_2, \ldots, x_n)g(y_1, y_2, \ldots, y_n) = g(z_1, z_2, \ldots, z_n)$$

then the n parameters $z_1, z_2, \ldots, z_n$ are functions of the parameters $x_1, \ldots, x_n, y_1, \ldots, y_n$. That is, the "multiplication table" for our infinite group consists of n real functions, each with $2n$ real arguments

$$z_1 = f_1(x_1, \ldots, y_n)$$
$$\cdot \quad \cdot \quad \cdot \quad \cdot \quad \cdot \quad \cdot \qquad \text{(16-66)}$$
$$z_n = f_n(x_1, \ldots, y_n)$$

Of course, an infinite group must satisfy the general group postulates given in Section 16-1. The most severe requirement is that imposed by associativity:

$$[g(x)g(y)]g(z) = g(x)[g(y)g(z)] \qquad \text{(16-67)}$$

In terms of the functions $f_i(x, y)$ of (16-66), (16-67) requires

$$f_i(f(x, y), z) = f_i(x, f(y, z)) \qquad \text{(16-68)}$$

to hold for all x, y, z.

If the functions $f_i(x, y)$ of (16-66), besides satisfying the group postulates, are continuous and possess derivatives of all orders, we speak of a *Lie group*. Some examples are given below.

1. The group of two-dimensional linear coordinate transformations of the form

$$\begin{aligned} x' &= ax + by + c \\ y' &= dx + ey + f \end{aligned} \qquad \text{(16-69)}$$

The requirement that an inverse exist means that the parameters $a, \ldots, f$ appearing in (16-69) must obey the condition

$$ae - bd \neq 0$$

Our group is clearly a six-parameter group.

2. The group of all unimodular unitary $n \times n$ matrices, that is, the group of all $n \times n$ unitary matrices with determinant 1. This group is usually called $SU(n)$. It can be shown that $SU(n)$ has $n^2 - 1$ parameters. For example, the general element of the group $SU(2)$ may be written

$$
\begin{pmatrix}
e^{i\xi} \cos \eta & e^{i\zeta} \sin \eta \\
-e^{-i\zeta} \sin \eta & e^{-i\xi} \cos \eta
\end{pmatrix}
\tag{16-70}
$$

with the three real parameters ξ, η, ζ.

3. The group of all real orthogonal $n \times n$ matrices, generally referred to as $O(n)$. If we wish to restrict ourselves to those matrices of $O(n)$ with determinant $+1$ (rather than -1), we speak of the group $O^+(n)$. For example, $O(3)$ is the group of rotations in three-dimensional space; $O^+(3)$ is the group of *proper* rotations. Since three parameters are required to specify a rotation (for example, the three Euler angles), $O(3)$ is a three-parameter group. In general, the number of parameters in $O(n)$ is $\frac{1}{2}n(n-1)$.

It will be noted that the examples above are all *transformation groups* that is, groups consisting of transformations of variables. When such transformations are linear and homogeneous, the group becomes a group of matrices. We shall often hereafter think of our infinite groups as transformation groups.[6]

It is of interest to consider the neighborhood of the origin in our infinite group. That is, let us label the identity element of our group by $g(0)$, with all parameters zero, so that

$$
g(0)g(x) = g(x)
$$

for all x. Now consider those group elements with all parameters infinitesimal. This leads us to the idea of the *generators* of our group. There is one generator for each parameter; they are defined by

$$
X_1 = \lim_{\varepsilon \to 0} \frac{g(\varepsilon, 0, 0, \ldots, 0) - g(0, 0, \ldots, 0)}{\varepsilon}
$$

$$
X_2 = \lim_{\varepsilon \to 0} \frac{g(0, \varepsilon, 0, \ldots, 0) - g(0, 0, \ldots, 0)}{\varepsilon}
$$

$$
\cdot \quad \cdot \quad \cdot \quad \cdot \quad \cdot \quad \cdot \quad \cdot \quad \cdot \quad \cdot \quad \cdot \quad \cdot \quad \cdot \quad \cdot \quad \cdot \tag{16-71}
$$

$$
X_n = \lim_{\varepsilon \to 0} \frac{g(0, 0, \ldots, \varepsilon) - g(0, 0, \ldots, 0)}{\varepsilon}
$$

[6] This is really not a restrictive assumption. If our group is not defined as a transformation group, we can always think of it as effecting transformations in its n-dimensional parameter space, that is, the element $g(x)$ produces the transformation $y \to z$ given by (16-66).

The existence of the "derivatives" in (16-71) is assured by the assumption that our group is a Lie group.

In order to give meaning to the *differences* of group elements which appear in the definitions (16-71), we must discuss the idea of coordinate transformations in some detail.   Consider a typical transformation R, which changes the coordinates x into coordinates $x' = Rx$.   For example, if the transformation is a rotation of the coordinate axes through $90°$ in a positive sense about the x-axis, the coordinates x, y, z become

$$x' = x$$
$$y' = z \qquad\qquad (16\text{-}72)$$
$$z' = -y$$

However, we do not need to restrict ourselves to *linear* coordinate transformations such as rotations.

In any case, with every transformation R we can associate an operator O_R which operates on *functions* of the variables x on which R operates. The definition of O_R is that O_R operating on an arbitrary function $f(x)$ gives a new function $g(x)$ which is numerically equal to $f(Rx)$, that is,

$$O_R f(x) = f(Rx) \qquad\qquad (16\text{-}73)$$

Let us consider the physical meaning of (16-73) in the specific case where R is an ordinary spatial rotation.   Equation (16-73) tells us that the new function $O_R f$, evaluated at any point, equals the original function f evaluated at the transformed coordinates.   For example, if R is the rotation of $90°$ about the x-axis defined by (16-72), (16-73) gives

$$O_R f(x, y, z) = f(x, z, -y)$$

Suppose that $f(x, y, z)$ has some particular feature along the positive z-axis ($x = y = 0$, $z > 0$).   Then $O_R f$ exhibits this same feature when $x = z = 0$, $y < 0$, that is, along the negative y-axis (see Figure 16–3).   We may thus think of our operator O_R as having rotated the *function* f through $90°$ about the x-axis, leaving the coordinate axes unaffected.

Our notation will not always distinguish between coordinate transformations R and operators O_R in what follows, but the reader must keep the distinction clear, and we shall try to indicate at all times which type of object we are considering.   To see that some care is necessary, consider the effect of successive operations O_R and O_S.   We obtain by repeated use of (16-73)

$$O_R O_S f(x) = O_S f(Rx) = f(SRx) \qquad\qquad (16\text{-}74)$$

Equation (16-74) tells us that the operator $O_R O_S$ is associated with the coordinate transformation SR; the order of operation is reversed when one goes from one to the other.

We do not mean to belabor this point, but we should point out a fallacy which often arises when one is performing manipulations such as those in (16-74). We might reason as follows. It follows from (16-73) that

$$O_S f(x) = f(Sx) \qquad (16\text{-}75)$$

Let us operate on both sides of (16-75) with the operator O_R. On the left, we get $O_R O_S f(x)$; on the right, since (16-73) tells us that operating on a function with O_R simply operates on the argument with R, we should get $f(RSx)$; thus

$$O_R O_S f(x) = f(RSx) \qquad (16\text{-}76)$$

which disagrees with (16-74) *and is wrong.*

The error lies in the last step, in which we operated on both sides of (16-75) with the operator O_R. O_R operates on *functions*, not on numbers; the two sides of (16-75) may be equal numerically, but they are different functions. A simple example is the following. Define the operator O by

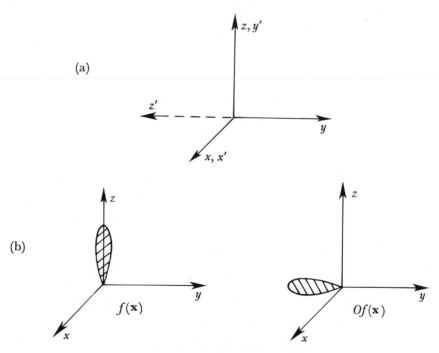

Figure 16–3 (a) A rotation of coordinate axes through 90° about the *x*-axis. (b) The effect of the associated rotation operator O on a function $f(x)$, which has a "lobe" along the positive *z*-axis

$Of(x) = f(x^2)$ and consider the two functions $g(x) = x$, $h(x) = 2x$.   Now it is certainly true that

$$h(x) = g(2x) \tag{16-77}$$

but operating on both sides of (16-77) with the operator O yields the incorrect answer

$$Oh(x) = Og(2x) = g(4x^2) = 4x^2$$

The correct answer, of course, is $Oh(x) = h(x^2) = 2x^2$.

The transposition of operators of (16-74) has been encountered previously. For example, if R is a matrix describing a rotation of coordinates, we can consider the associated matrix T_R which rotates a vector in the same way, leaving the coordinate axes fixed.   It is well known that $T_R = R^{-1}$, so that if two such rotations are performed, we have

$$T_R T_S = R^{-1} S^{-1} = (SR)^{-1} = T_{SR} \tag{16-78}$$

The transposition of order between the matrices of (15-2) and the matrices in the second footnote on p. 404 is a closely related phenomenon.

We now return to our discussion of generators, as defined by (16-71).   In view of the digression just concluded, we must decide whether our transformations are to be taken as operating on coordinates, or on functions of those coordinates.   Let us adopt the latter convention.   We can then exhibit the generators explicitly for some typical Lie groups.

1. Consider the six-parameter group defined by (16-69).   Note that the identity element of this group has the parameters

$$a = 1 \qquad b = 0 \qquad c = 0 \qquad d = 0 \qquad e = 1 \qquad f = 0$$

If we label a typical element of the group as $g(a, b, c, d, e, f)$, then the first generator is, according to (16-71),

$$X_a = \lim_{\varepsilon \to 0} \frac{g(1 + \varepsilon, 0, 0, 0, 1, 0) - g(1, 0, 0, 0, 1, 0)}{\varepsilon}$$

Thus, allowing X_a to operate on an arbitrary function $f(x, y)$ we have

$$X_a f(x, y) = \lim_{\varepsilon \to 0} \frac{1}{\varepsilon} \{ f[(1 + \varepsilon)x, y] - f(x, y) \}$$

$$= x \frac{\partial}{\partial x} f(x, y)$$

The generator X_a is thus the *differential* operator $x(\partial/\partial x)$:

$$X_a = x \frac{\partial}{x\partial} \tag{16-79}$$

Similarly,

$$X_b f(x, y) = \left[\lim_{\varepsilon \to 0} \frac{g(1, \varepsilon, 0, 0, 1, 0) - g(1, 0, 0, 0, 1, 0)}{\varepsilon}\right] f(x, y)$$

$$= \lim_{\varepsilon \to 0} \frac{1}{\varepsilon} [f(x + \varepsilon y, y) - f(x, y)]$$

$$= y \frac{\partial}{\partial x} f(x, y)$$

Therefore,

$$X_b = y \frac{\partial}{\partial x} \tag{16-80}$$

Proceeding in this way, we find the remaining generators:

$$X_c = \frac{\partial}{\partial x} \quad X_d = x\frac{\partial}{\partial y} \quad X_e = y\frac{\partial}{\partial y} \quad X_f = \frac{\partial}{\partial y} \tag{16-81}$$

2. Consider the group $SU(2)$, whose typical member is given by (16-70). Let us label the two variables which this 2×2 matrix transforms as u_1, u_2. Then the two generators X_ξ, X_η are immediately constructed by applying (16-71):

$$X_\xi f(u_1, u_2) = \left[\lim_{\varepsilon \to 0} \frac{g(\varepsilon, 0, 0) - g(0, 0, 0)}{\varepsilon}\right] f(u_1, u_2)$$

$$= \lim_{\varepsilon \to 0} \frac{1}{\varepsilon} \{f[(1 + i\varepsilon)u_1, (1 - i\varepsilon)u_2] - f(u_1, u_2)\}$$

$$= i\left(u_1 \frac{\partial}{\partial u_1} - u_2 \frac{\partial}{\partial u_2}\right) f(u_1, u_2) \tag{16-82}$$

and, similarly,

$$X_\eta = u_2 \frac{\partial}{\partial u_1} - u_1 \frac{\partial}{\partial u_2} \tag{16-83}$$

However, we run into trouble if we try to calculate X_ζ directly by the use of (16-71). The trouble is that ζ is not a good parameter to use near the origin, since when $\eta = 0$, $g(\xi, \eta, \zeta)$ becomes independent of ζ. This is a familiar disease; for example, consider the specification of directions in space by the conventional polar angles θ and ϕ. Along the polar axis ($\theta = 0$), ϕ loses its significance. More precisely, if we take gradients on the unit sphere, $\partial/\partial\theta$ is a perfectly reasonable operator near $\theta = 0$, but $\partial/\partial\phi$ will always vanish there. To get a finite gradient operator, we must divide $\partial/\partial\phi$ by $\sin \theta$ (or θ, which is the same thing near $\theta = 0$).

Thus, we have two alternatives. We could start over with a different choice of parameters in (16-70), so that nothing goes singular near the group identity. Instead, however, we shall just follow the heuristic suggestions of the last paragraph. That is, we compute X_ζ with η not quite zero. We obtain

$$X_\zeta = i \sin \eta \left(u_2 \frac{\partial}{\partial u_1} + u_1 \frac{\partial}{\partial u_2} \right)$$

We shall take for our third generator

$$X_\zeta' = \frac{X_\zeta}{\sin \eta} = i \left(u_2 \frac{\partial}{\partial u_1} + u_1 \frac{\partial}{\partial u_2} \right) \tag{16-84}$$

3. Finally, we consider $O(3)$, the group of rotations in three dimensions. In order to avoid the difficulty we encountered with $SU(2)$, we shall not use the Euler angles to characterize our infinitesimal rotations. Instead, we shall observe that any infinitesimal rotation may be composed of a rotation θ_x about the x-axis, a rotation θ_y about the y-axis, and a rotation θ_z about the z-axis. Let us calculate the respective generators:

$$X_x f(x, y, z) = \lim_{\varepsilon \to 0} \frac{1}{\varepsilon} \left[f(x, y \cos \varepsilon + z \sin \varepsilon, z \cos \varepsilon - y \sin \varepsilon) - f(x, y, z) \right]$$

$$= \left(z \frac{\partial}{\partial y} - y \frac{\partial}{\partial z} \right) f(x, y, z)$$

Thus

$$X_x = \left(z \frac{\partial}{\partial y} - y \frac{\partial}{\partial z} \right) \tag{16-85}$$

Cyclic permutation of x, y, z gives the other two generators:

$$X_y = x \frac{\partial}{\partial z} - z \frac{\partial}{\partial x} \qquad X_z = y \frac{\partial}{\partial x} - x \frac{\partial}{\partial y} \tag{16-86}$$

We conclude this section with some remarks on the importance of the concept of the generators of a group. Lie proved some remarkable theorems concerning the relations between the generators and the group. For example, the *commutator*

$$[X_i, X_j] = X_i X_j - X_j X_i$$

of two generators is always a linear combination of generators:

$$[X_i, X_j] = \sum_k c_{ij}^k X_k \tag{16-87}$$

The real constants c_{ij}^k in (16-87) are called the *structure constants* of the Lie group.

The relation (16-87) leads to the concept of the *Lie algebra* associated with a given Lie group. This algebra is composed of all linear combinations of generators

$$c_1 X_1 + c_2 X_2 + \ldots + c_n X_n \tag{16-88}$$

with real coefficients c_i. This is an algebra as defined on p. 425 if we take for the "product" of two elements their commutator. For example, the "product" of the two elements $X_1 + X_2$ and $X_3 - 2X_4$ is defined to be

$$[X_1, X_3] - 2[X_1, X_4] + [X_2, X_3] - 2[X_2, X_4] \tag{16-89}$$

It follows from (16-87) that (16-89) is of the form (16-88), that is, is itself an element of the algebra.

It is clear that the Lie group completely determines the structure of the associated Lie algebra. Sophus Lie proved the converse; the local structure (that is, the structure in some neighborhood of the identity) of a Lie group is completely determined by its Lie algebra, that is, by the structure constants c_{ij}^k in (16-87).

Finally, we observe that for a matrix group such as $SU(2)$ or $O(3)$, it is easy to construct the generators under the alternative picture noted on p. 451 in which the group elements are taken to be the matrices themselves rather than the associated transformation operators defined by (16-73). In this representation, we simply insert the appropriate matrices directly into the defining equations (16-71); the difference, and in fact the derivative, are well-known operations on matrices. For example, from the form (16-70) of the typical member of $SU(2)$, we deduce immediately by differentiation

$$X_\xi = \begin{pmatrix} i & 0 \\ 0 & -i \end{pmatrix} \qquad X_\eta = \begin{pmatrix} 0 & 1 \\ -1 & 0 \end{pmatrix} \tag{16-90}$$

Again for X_ζ, we must remember to divide by $\sin \eta$; the result is

$$X_\zeta' = \begin{pmatrix} 0 & i \\ i & 0 \end{pmatrix} \tag{16-91}$$

We shall leave the following points to be verified by the reader.

1. The Lie algebras defined by (16-82), (16-83), and (16-84) for $SU(2)$, and by (16-85) and (16-86) for $O(3)$ are *isomorphic*. That is, if we make the one-to-one correspondence

$$X_\zeta' \leftrightarrow 2X_x \qquad X_\eta \leftrightarrow 2X_y \qquad X_\xi \leftrightarrow 2X_z \tag{16-92}$$

the two algebras are identical. This results from the commutation rules in the two algebras:

$$[X_\xi, X'_\zeta] = 2X_\eta \qquad [X_x, X_y] = X_z$$
$$[X'_\zeta, X_\eta] = 2X_\xi \qquad [X_y, X_z] = X_x \qquad (16\text{-}93)$$
$$[X_\eta, X_\xi] = 2X'_\zeta \qquad [X_z, X_x] = X_y$$

Various alternatives were available in the correspondence (16-92); our choices were made to agree with current physical convention.

2. The *matrix* generators (16-90) and (16-91) of $SU(2)$ do *not* obey the commutation relations (16-93) which were deduced for the differential operator representations (16-82), (16-83), and (16-84). In fact, we can predict from the reversal of order dictated by (16-74) for the two types of objects that the signs of all commutators must be reversed in going from one picture to the other.   The commutators of the matrices (16-90), (16-91) are the negatives of the commutators in (16-93).

3. Matrix representations, analogous to (16-90), (16-91) of the generators of $O(3)$, interpreted as coordinate transformations, are

$$X_x = \begin{pmatrix} 0 & 0 & 0 \\ 0 & 0 & 1 \\ 0 & -1 & 0 \end{pmatrix} \qquad X_y = \begin{pmatrix} 0 & 0 & -1 \\ 0 & 0 & 0 \\ 1 & 0 & 0 \end{pmatrix}$$

$$(16\text{-}94)$$

$$X_z = \begin{pmatrix} 0 & 1 & 0 \\ -1 & 0 & 0 \\ 0 & 0 & 0 \end{pmatrix}$$

Again, the commutators of the matrices (16-94) are the negatives of (16-93).

16-7   IRREDUCIBLE REPRESENTATIONS OF $SU(2)$, $SU(3)$, AND $O^+(3)$

$SU(2)$

We shall begin by discussing the irreducible matrix representations of $SU(2)$.   There is, to begin with, the trivial one-dimensional representation D^1, in which each group element is represented by the number 1.

Next, there is the two-dimensional representation D^2 provided by the matrices (16-70) themselves.   As in the preceding section, we may consider two-dimensional vectors on which these matrices act; such a vector will be written u_α ($\alpha = 1, 2$).

We may also consider a representation consisting of the matrices which are the complex conjugates of the original matrices (16-70).   That is, each matrix in $SU(2)$ is represented by its complex conjugate.   A vector which transforms with these conjugate matrices will be written with an *upper* index, say v^α.

Now we may go on and define *tensors*; if, for example, u_α and v_α are vectors transforming with (16-70), and w^α is a *conjugate* vector transforming with the complex conjugate of (16-70), then an object $u^\gamma_{\alpha\beta}$ which transforms like the products $u_\alpha v_\beta w^\gamma$ is such a tensor. Clearly, tensors can be defined with an arbitrary number of upper and lower indices.

The transformations of the various components of a tensor into each other for each element of the group $SU(2)$ constitute a representation of the group. We will say that the representation is *generated* by the tensor. For example, a vector u_α generates the representation D^2, and a scalar generates the representation D^1. All tensors generate representations, but some of these are reducible, and some are equivalent to other representations.

Consider the two-subscript object $\varepsilon_{\alpha\beta}$, defined by

$$\varepsilon_{12} = 1 \qquad \varepsilon_{21} = -1 \qquad \varepsilon_{11} = \varepsilon_{22} = 0 \qquad (16\text{-}95)$$

This is just the two-dimensional Levi–Civita alternating symbol; compare (15-10). Is $\varepsilon_{\alpha\beta}$ a tensor? Let A be an arbitrary matrix of $SU(2)$. Then $\varepsilon_{\alpha\beta}$ transforms into

$$A_{\alpha\gamma} A_{\beta\delta}\, \varepsilon_{\gamma\delta} = A_{\alpha 1} A_{\beta 2} - A_{\alpha 2} A_{\beta 1} \qquad \text{(sum over repeated indices)}$$

$$= \begin{cases} 0 & \text{if } \alpha = \beta, \\ A_{11}A_{22} - A_{12}A_{21} = \det A = 1 & \text{if } \alpha = 1,\ \beta = 2 \\ A_{21}A_{12} - A_{22}A_{11} = -\det A = -1 & \text{if } \alpha = 2,\ \beta = 1 \end{cases}$$

$$= \varepsilon_{\alpha\beta}$$

Thus, $\varepsilon_{\alpha\beta}$ is a tensor under $SU(2)$. Similarly, $\varepsilon^{\alpha\beta}$, which is numerically equal to $\varepsilon_{\alpha\beta}$, is a tensor.

Next, we observe that an upper index and a lower index may be summed (*contracted*) to give a tensor with two fewer indices, just as with the contravariant and covariant indices of Section 15–3. For example, if u_α and v^β are vectors, the quantity $u_\alpha v^\alpha$ becomes, after any transformation A of $SU(2)$,

$$A_{\alpha\beta} u_\beta A^*_{\alpha\gamma} v^\gamma = A^\dagger_{\gamma\alpha} A_{\alpha\beta} u_\beta v^\gamma = \delta_{\beta\gamma} u_\beta v^\gamma = u_\alpha v^\alpha$$

so that $u_\alpha v^\alpha$ is invariant, as alleged. Similarly, from a tensor $u^\gamma_{\alpha\beta}$ we may form a vector $v_\alpha = u^\beta_{\alpha\beta}$, from the tensors $u_{\alpha\beta}$ and $v^{\gamma\delta}$ we may form the tensor $w^\beta_\alpha = u_{\alpha\gamma} v^{\gamma\beta}$, and so on.

We are now in a position to construct the irreducible representations of $SU(2)$. In the first place, we may restrict ourselves to tensors with all indices downstairs, without omitting any representations. For example, the tensor $u^{\gamma\delta}_{\alpha\beta}$ is composed of linear combinations of the components of the tensor $v_{\alpha\beta\gamma\delta} = \varepsilon_{\gamma\lambda}\varepsilon_{\delta\mu} u^{\lambda\mu}_{\alpha\beta}$, and vice versa, so that the representations generated by $u^{\gamma\delta}_{\alpha\beta}$ and $v_{\alpha\beta\gamma\delta}$ are equivalent. Note that the tensor $\varepsilon_{\alpha\beta}$ serves as a "lowering operator," just as the fundamental tensor g_{ij} did in Section 15–3.

Table 16-9   Irreducible Representations of $SU(2)$

Representation	Tensor (totally symmetric)
D^1	1
D^2	u_α
D^3	$u_{\alpha\beta}$
D^4	$u_{\alpha\beta\gamma}$
$\vdots$	$\vdots$

Next we observe that a tensor $u_{\alpha\beta\gamma\ldots}$ must be completely symmetric in all its indices if it is to generate an irreducible representation. For suppose the tensor $u_{\alpha\beta\gamma\ldots}$ were not symmetric in its first two indices,

$$u_{\alpha\beta\gamma\ldots} \neq u_{\beta\alpha\gamma\ldots}$$

Then we could form the tensor

$$v_{\gamma\ldots} = \varepsilon^{\alpha\beta} u_{\alpha\beta\gamma\ldots}$$
$$= u_{12\gamma\ldots} - u_{21\gamma\ldots}$$

with two fewer indices. The representation generated by $u_{\alpha\beta\gamma\ldots}$ would then be reducible, since it would contain the (smaller) representation generated by $v_{\gamma\ldots}$.

Thus, the general irreducible representation of $SU(2)$ is characterized by the totally symmetric tensor $u_{\alpha\beta\gamma\ldots}$ which generates it. If the tensor u has n subscripts, there are $n + 1$ independent components (why?), and therefore the associated representation has dimensionality $n + 1$. The results of this analysis are summarized in Table 16-9.

$SU(3)$

Much of the analysis which we carried through for the group $SU(2)$ remains valid for $SU(3)$. Again we introduce vectors u_α, where α now runs from 1 to 3. A vector u_α is operated on by the unitary unimodular 3×3 matrices that make up $SU(3)$. Thus we may say that a vector u_α generates the three-dimensional representation D^3 in which each group element is represented by itself.

Once more, we can also define vectors u^α with upper indices, which transform like the complex conjugates of vectors with lower indices. These generate a representation $\overline{D^3}$ in which each matrix of $SU(3)$ is represented by its complex conjugate. In contrast to the analogous situation in $SU(2)$, this representation $\overline{D^3}$ is *not* equivalent to the representation D^3.

Again we define tensors with arbitrary numbers of upper and lower indices. For example, when a vector u_α undergoes the transformation

$$u'_\alpha = A_{\alpha\beta} u_\beta \qquad (16\text{-}96)$$

a tensor $u^\gamma_{\alpha\beta}$ undergoes the transformation

$$u^{\gamma'}_{\alpha\beta} = A_{\alpha\rho} A_{\beta\sigma} A^*_{\gamma\tau} u^\tau_{\rho\sigma}$$

A being an arbitrary matrix in $SU(3)$.

The alternating symbol is once more a tensor, but of course it now has three indices. That is, the two numerically equal objects

$$\varepsilon^{\alpha\beta\gamma} = \varepsilon_{\alpha\beta\gamma} = \begin{cases} 1 & \text{if } \alpha\beta\gamma = 123,\ 231,\ \text{or } 312 \\ -1 & \text{if } \alpha\beta\gamma = 321,\ 213,\ \text{or } 132 \\ 0 & \text{otherwise} \end{cases} \qquad (16\text{-}97)$$

are tensors under $SU(3)$, because of the unimodularity condition; the proof is left to the reader.

Finally, contraction of indices is again a legitimate operation, as with $SU(2)$, provided one index being summed is upper and the other is lower. The proof, which we omit, follows immediately from the unitarity of the matrices of $SU(3)$, and is completely analogous to the corresponding calculation in the case of $SU(2)$.

We can now enumerate the irreducible representations of $SU(3)$. Upstairs and downstairs indices are no longer equivalent, because the alternating symbol now has three indices instead of two, and can no longer be used to raise or lower a single index. Therefore, we consider tensors with indices both above and below. If such a tensor $u^{\gamma\delta\cdots}_{\alpha\beta\cdots}$ is to generate an irreducible representation of $SU(3)$, it must fulfill two conditions.

1. It must be separately *symmetric* in all its upper indices and all its lower indices. If, for example,

$$u^{\gamma\delta\cdots}_{\alpha\beta\cdots} \neq u^{\delta\gamma\cdots}_{\alpha\beta\cdots}$$

we could form the tensor

$$v^{\cdots}_{\lambda\alpha\beta\cdots} = \varepsilon_{\lambda\gamma\delta} u^{\gamma\delta\cdots}_{\alpha\beta\cdots}$$

with one fewer index, and our representation would be reducible. Of course, there is no symmetry requirement connecting upper and lower indices.

2. It must be *traceless*; every contraction of an upper index with a lower index must give zero. Otherwise we could form a tensor with two fewer indices, and our representation would be reducible. Traces formed by summing two upper indices or two lower indices, of course, need not vanish, since this is not an invariant contraction.

Table 16–10   Some Irreducible Representations of $SU(3)$. $m =$ Number of Upper Indices; $n =$ Number of Lower Indices

m	n	Representation	Tensor
0	0	D^1	1
0	1	D^3	u_α
1	0	$\overline{D^3}$	u^α
0	2	D^6	$u_{\alpha\beta}$
1	1	D^8	u^β_α
2	0	$\overline{D^6}$	$u^{\alpha\beta}$
0	3	D^{10}	$u_{\alpha\beta\gamma}$
1	2	D^{15}	$u^\alpha_{\beta\gamma}$
2	1	$\overline{D^{15}}$	$u^{\alpha\gamma}_\beta$
3	0	$\overline{D^{10}}$	$u^{\alpha\beta\gamma}$

Thus, the general irreducible representation of $SU(3)$ is generated by a tensor with m upper indices and n lower indices; the tensor must be totally symmetric in all its upper indices, totally symmetric in all its lower indices, and all contractions of one upper index with one lower index must vanish. In Table 16–10 we enumerate some irreducible representations of $SU(3)$. It can be shown that the dimensionality of the general irreducible representation of $SU(3)$ is $\frac{1}{2}(m+1)(n+1)(m+n+2)$. Representations D^3 and $\overline{D^3}$, D^6 and $\overline{D^6}$, and so forth are said to be *conjugate representations*.

$O^+(3)$

Finally, we take up the important question of finding the irreducible representations of $O^+(3)$, the group of real unimodular orthogonal 3×3 matrices, which describe rotations in Euclidean three-space. The problem is again solved by constructing tensors, in this case just the Cartesian tensors of Section 15–1.

In fact, it should be clear, after our considerations for $SU(2)$ and $SU(3)$, that the general irreducible representation of $O^+(3)$ is generated by a symmetric, traceless tensor of rank n. For example, a scalar induces the trivial representation D^1. A vector ($n = 1$) generates the three-dimensional representation D^3 in which each rotation is represented by the usual 3×3 orthogonal matrix. A symmetric tensor T_{ij} has six independent components; removing the trace leaves five. These five quantities generate the five-dimensional irreducible representation D^5, and so on. It can be shown that a symmetric tensor of rank n has $2n + 1$ independent components, so that the associated representation has dimensionality $2n + 1$.

These representations are most elegantly thought of as operating on the spherical harmonics $Y_{lm}(\theta, \phi)$ of Section 7–1.   That is, the trivial representation D^1 consists of the transformations of $Y_{00}(\Omega)$ into itself, the "vector representation" D^3 describes the transformations of the three spherical harmonics $Y_{1m}(\Omega)$ $(m = -1, 0, 1)$ into linear combinations of each other under rotations, and, in general, the transformations of the $2n + 1$ functions $Y_{nm}(\Omega)$ with given n into each other generate the irreducible representation D^{2n+1} of $O^+(3)$.

Since we are now discussing the effects of rotations on *functions*, namely, the spherical harmonics, rather than on coordinates, we define our rotation operators by (16-73).   Thus, the effect of the rotation R on a particular spherical harmonic $Y_{lm}(\Omega)$ is defined to be

$$O_R Y_{lm}(\Omega) = Y_{lm}(R\Omega)$$

where $R\Omega$ denotes the new values of θ, ϕ associated with a direction Ω after the coordinate axes have undergone the rotation R, holding the direction Ω fixed in space.   In other words, the new function $O_R Y_{lm}(\Omega)$ is the function obtained by rotating $Y_{lm}(\Omega)$ instead of the coordinate axes (compare Figure 16–3).

One conventionally defines the *rotation matrix* $D^l_{m'm}(R)$ by

$$Y_{lm}(R\Omega) = \sum_{m'} D^l_{m'm}(R) Y_{lm'}(\Omega) \tag{16-98}$$

Notice the order of the subscripts on $D^l_{m'm}$ in (16-98).   This is chosen deliberately; to see why,[7] let us consider the effect of two successive rotation operators O_R, O_S applied to a spherical harmonic:

$$O_R O_S Y_{lm}(\Omega) = Y_{lm}(SR\Omega) \qquad \text{[from (16-74)]}$$

$$= \sum_{m'} D^l_{m'm}(S) Y_{lm'}(R\Omega)$$

$$= \sum_{m'} D^l_{m'm}(S) \sum_{m''} D^l_{m''m'}(R) Y_{lm''}(\Omega)$$

$$= \sum_{m''} \left[\sum_{m'} D^l_{m''m'}(R) D^l_{m'm}(S) \right] Y_{lm''}(\Omega) \tag{16-99}$$

Equation (16-99) shows that the rotation matrices $D^l_{m'm}(R)$ constitute a matrix representation of the rotation operators O_R.

In Table 16–11, we give the elements of the rotation matrix $D^1_{m'm}(\alpha\beta\gamma)$ as functions of the Euler angles α, β, γ.   Other properties of the $D^l_{m'm}(\alpha\beta\gamma)$ as well as a general formula are given in Rose (R3); we shall only make some observations.

---

[7] As an alternative to the algebra that follows, one might prefer to notice that (16-98) is just a special case of the relation (6-8); here the spherical harmonics are the basis vectors of our representation.

1. Inspection of Table 16–11 suggests the general rule, which is in fact true, that the dependence of $D^l_{m'm}(\alpha\beta\gamma)$ on α and γ is essentially trivial. To be specific, one can show that

$$D^l_{m'm}(\alpha\beta\gamma) = e^{-im'\alpha}\, d^l_{m'm}(\beta)\, e^{-im\gamma} \tag{16-100}$$

where the "reduced rotation matrix" $d^l_{m'm}(\beta)$ is given, for example, by Rose (R3). Property (16-100) actually follows directly from the definition (16-98) and the fact that the spherical harmonic $Y_{lm}(\theta, \phi)$ depends on ϕ only through a factor $e^{im\phi}$ [see (7-32) or (7-33).]

2. If we set $m = 0$ in (16-98), we obtain

$$\sqrt{\frac{2l+1}{4\pi}}\, P_l(R\Omega) = \sum_{m'} D^l_{m'0}(\alpha\beta\gamma) Y_{lm'}(\Omega) \tag{16-101}$$

Now consider the argument $R\Omega$ of the Legendre polynomial P_l on the left of (16-101). Let the direction Ω have polar coordinates θ, ϕ. After the coordinate axes are subjected to the rotation R, the z-axis points in the direction with polar coordinates β, α (referred to the *original* axes). Thus, the polar angle $R\Omega$ is the angle χ between these two directions:

$$\cos\chi = \cos\theta \cos\beta + \sin\theta \sin\beta \cos(\phi - \alpha) \tag{16-102}$$

Therefore, (16-101) becomes

$$\sqrt{\frac{2l+1}{4\pi}}\, P_l(\cos\chi) = \sum_{m'} D^l_{m'0}(\alpha\beta\gamma) Y_{lm'}(\theta, \phi) \tag{16-103}$$

A comparison of (16-103) and the addition theorem (7-46) for spherical harmonics gives immediately

$$D^l_{m0}(\alpha\beta\gamma) = \sqrt{\frac{4\pi}{2l+1}}\, Y^*_{lm}(\beta\alpha) \tag{16-104}$$

3. The fact that the spherical harmonics are orthonormal [see (7-35)] on the

Table 16–11   The Rotation Matrix $D^1_{m'm}(\alpha\beta\gamma)$

m'	$m = 1$	0	-1
1	$e^{-i\alpha}\dfrac{1+\cos\beta}{2}e^{-i\gamma}$	$-e^{-i\alpha}\dfrac{\sin\beta}{\sqrt{2}}$	$e^{-i\alpha}\dfrac{1-\cos\beta}{2}e^{i\gamma}$
0	$\dfrac{\sin\beta}{\sqrt{2}}e^{-i\gamma}$	$\cos\beta$	$-\dfrac{\sin\beta}{\sqrt{2}}e^{i\gamma}$
-1	$e^{i\alpha}\dfrac{1-\cos\beta}{2}e^{-i\gamma}$	$e^{i\alpha}\dfrac{\sin\beta}{\sqrt{2}}$	$e^{i\alpha}\dfrac{1+\cos\beta}{2}e^{i\gamma}$

unit sphere tells us that our representations are *unitary*:

$$D^l_{m'm}(R^{-1}) = D^l_{mm'}(R)^* \tag{16-105}$$

We shall conclude this chapter by exploring briefly the isomorphism we observed in the last section, between the Lie algebras associated with $O^+(3)$ and $SU(2)$. We begin with $O^+(3)$.

An arbitrary element near the origin (identity element) of $O^+(3)$ may be written $1 + \theta_x X_x + \theta_y X_y + \theta_z X_z$, where X_x, X_y, X_z are the generators, obeying the commutation relations (16-93) and the θ_α are infinitesimal. We shall define $X_\alpha = -iJ_\alpha$ ($\alpha = 1, 2, 3$) so that the group element becomes

$$1 - i\theta_x J_x - i\theta_y J_y - i\theta_z J_z = 1 - i\boldsymbol{\theta} \cdot \mathbf{J} \tag{16-106}$$

What is the significance of the (infinitesimal) vector $\boldsymbol{\theta}$? A little thought will convince the reader that the element (16-106) describes an infinitesimal rotation whose magnitude is the magnitude of the vector $\boldsymbol{\theta}$, while the axis lies along the direction of $\boldsymbol{\theta}$.

How do we obtain a finite rotation operator? That's easy; a finite rotation may be thought of as a sequence of infinitesimal rotations. The rotation operator for the finite rotation $\boldsymbol{\theta}$ is therefore

$$\lim_{n \to \infty} \left(1 - \frac{i}{n} \boldsymbol{\theta} \cdot \mathbf{J}\right)\left(1 - \frac{i}{n} \boldsymbol{\theta} \cdot \mathbf{J}\right) \cdots \left(1 - \frac{i}{n} \boldsymbol{\theta} \cdot \mathbf{J}\right) \quad (n \text{ factors}) \tag{16-107}$$

$$= e^{-i\boldsymbol{\theta} \cdot \mathbf{J}}$$

Of course, the exponential in (16-107) is a formal device only, since the argument is an operator. One may think of the exponential as an abbreviation for its Taylor series or product representation.

We now turn to $SU(2)$ and make the identification (16-92). That is, for J_x in (16-107) we take, instead of iX_x, the operator $(i/2) X'_\zeta$, and similarly for J_y and J_z. Thus, the element (16-107) of $O^+(3)$ is identified with the element

$$e^{-(i/2)\boldsymbol{\theta} \cdot \mathbf{S}} \quad (\mathbf{S} = iX'_\zeta, iX_\eta, iX_\xi) \tag{16-108}$$

of $SU(2)$. This enables us to make a one-to-one association of the elements of $O^+(3)$ with the elements of $SU(2)$, at least in some neighborhood of the identity. As we will see in a moment, the correspondence is two-to-one rather than one-to-one when we consider the entire groups.

This correspondence immediately suggests a matrix representation for $O^+(3)$ which differs from the tensor representations found earlier in this section. We have already remarked that the matrices (16-90), (16-91) almost provide a representation for the generators of $SU(2)$, except that the signs of the commutators are reversed. Thus, the *negatives* of the matrices (16-90), (16-91) do provide us with a representation. Remembering the i in the definition of **S** [see (16-108)], we obtain

$$S_x = \begin{pmatrix} 0 & 1 \\ 1 & 0 \end{pmatrix} \qquad S_y = \begin{pmatrix} 0 & -i \\ i & 0 \end{pmatrix} \qquad S_z = \begin{pmatrix} 1 & 0 \\ 0 & -1 \end{pmatrix} \qquad (16\text{-}109)$$

These matrices are known as the *Pauli spin matrices*.

Consider a rotation about one of the coordinates axes, for example, a rotation θ about the x-axis. According to (16-108), this is to be identified with the matrix $e^{-(i/2)\theta S_x}$ in $SU(2)$. This exponential is easy to evaluate. The Pauli matrices (16-109) give unity when squared, $S_x^2 = S_y^2 = S_z^2 = 1$, so that

$$e^{-(i/2)\theta S_x} = 1 - i\frac{\theta}{2}S_x - \frac{1}{2!}\left(\frac{\theta}{2}\right)^2 + \frac{i}{3!}\left(\frac{\theta}{2}\right)^3 S_x + \cdots$$

$$= \left[1 - \frac{1}{2!}\left(\frac{\theta}{2}\right)^2 + \frac{1}{4!}\left(\frac{\theta}{2}\right)^4 - + \cdots\right] - iS_x\left[\frac{\theta}{2} - \frac{1}{3!}\left(\frac{\theta}{2}\right)^3 + - \cdots\right]$$

$$= \cos\frac{\theta}{2} - iS_x \sin\frac{\theta}{2}$$

$$= \begin{pmatrix} \cos\dfrac{\theta}{2} & -i\sin\dfrac{\theta}{2} \\ -i\sin\dfrac{\theta}{2} & \cos\dfrac{\theta}{2} \end{pmatrix} \qquad (16\text{-}110)$$

By combining three such rotations, we can obtain the element of $SU(2)$ associated with an arbitrary rotation, in terms of the Euler angles α, β, γ. However, we must be careful in which order we perform the rotations. If we think of our rotations as *coordinate transformations*, we should use (15-2). However, we have been specifically working with *rotation operators*, because of our use of the commutation relations (16-93). Therefore, the order of rotations must be that given in footnote 2 on p. 404, so that with a rotation $R(\alpha\beta\gamma)$ of $O^+(3)$, we associate the matrix

$$e^{-(i/2)\alpha S_z}e^{-(i/2)\beta S_y}e^{-(i/2)\gamma S_z} \qquad (16\text{-}111)$$

of $SU(2)$. The evaluation of this product is left as an exercise; the result is

$$\begin{pmatrix} e^{-i(\alpha/2)}\cos\dfrac{\beta}{2}e^{-i(\gamma/2)} & -e^{-i(\alpha/2)}\sin\dfrac{\beta}{2}e^{i(\gamma/2)} \\ e^{i(\alpha/2)}\sin\dfrac{\beta}{2}e^{-i(\gamma/2)} & e^{i(\alpha/2)}\cos\dfrac{\beta}{2}e^{i(\gamma/2)} \end{pmatrix} \qquad (16\text{-}112)$$

Note that (16-112) is indeed a general element of $SU(2)$; compare (16-70).

We have apparently achieved something quite remarkable, a two-dimensional representation of $O^+(3)$ which was completely omitted from our previous enumeration! We just use the matrices (16-112), that is, the vector representation D^2 of $SU(2)$. This is correct, with one difficulty which is closely connected with the half-angles which occur in (16-112). There are *two* matrices, each the negative of the other, which must be associated with each rotation of $O^+(3)$. For example, consider a rotation α about the z-axis. According to (16-112), that is represented by the matrix

$$\begin{pmatrix} e^{-i(\alpha/2)} & 0 \\ 0 & e^{i(\alpha/2)} \end{pmatrix} \qquad (16\text{-}113)$$

If α increases from 0 to 2π, the rotation operator clearly returns to the identity, while (16-113) becomes the matrix

$$\begin{pmatrix} -1 & 0 \\ 0 & -1 \end{pmatrix}$$

that is, the negative of the identity element in $SU(2)$.

Within the limitations of this two-for-one-ness, we have actually obtained much more than a single new representation of $O^+(3)$. All the representations of $SU(2)$ given in Table 16-9 can now serve as representations of $O^+(3)$, since we have established a correspondence[8] between the group elements of $SU(2)$ and $O^+(3)$. The representations D^1, D^3, D^5, ..., of $SU(2)$ turn out to be just the scalar, vector, symmetric traceless tensor, and so forth, representations of $O^+(3)$ we identified earlier. The others are called *spinor representations*, the two-component "vector" u_α being often called a *spinor*. These two-valued representations are of equal physical significance to the single-valued representations in quantum mechanics, where the rotation operators are identified with angular momentum.

REFERENCES

Several references to the theory of finite groups are the books by Carmichael (C1); Hall (H1); Ledermann (L6); and Birkhoff and Mac Lane (B2).

The classical reference to applications of group theory in atomic physics is the book by Wigner (W6); this book contains clear and extensive discussions both of the mathematical apparatus and its application. Tinkham (T1) is a more recent reference on the same subject. A short treatment of applications of group theory in physics is given by Margenau and Murphy (M2) Chapter 15.

A recent reference which contains particularly thorough coverage of the

---

[8] In mathematical language, this correspondence is a homomorphism. If it were one-to-one, it would be an isomorphism.

symmetric group, continuous groups, and applications to atomic and nuclear spectroscopy is the book by Hamermesh (H3).

The theory of the group $O^+(3)$ and its representations is thoroughly explored, in the context of quantum mechanics, by Rose (R3).

PROBLEMS

16-1   Which of the following are groups?
  (a) All real numbers (group multiplication = ordinary multiplication)
  (b) All real numbers (group multiplication = addition)
  (c) All complex numbers except zero (group multiplication = ordinary multiplication)
  (d) All positive rational numbers ("product" of a and b is a/b)

16-2   Consider the following two elements of the symmetric group S_5:

$$g_1 = [54123] = (135)(24)$$

$$g_2 = [21534] = (12)(345)$$

Find a third element g of this group such that

$$g^{-1}g_1g = g_2$$

16-3   Denote the elements in one class of a group by $A_1, A_2, \ldots, A_{n_A}$, those in another class by $B_1, B_2, \ldots, B_{n_B}$, and so on.   Let A denote the element $\sum_{i=1}^{n_A} A_i$ of the group algebra, and similarly for $B, \ldots$.
  (a) Consider a particular n-dimensional irreducible representation D of the group.   Show that $\sum_{i=1}^{n_A} D(A_i)$, which may be abbreviated $D(A)$, equals a constant multiple of the n-dimensional unit matrix.   Evaluate the constant, in terms of n, n_A, and $\chi(A)$, the character of the class A in the representation D.
  (b) If the two elements A and B of the group algebra are multiplied together, show that the product consists of complete classes; that is, we may write
$$AB = \sum_c s_c C$$

where the s_c are nonnegative integers.
  Hint: First show that $g^{-1} ABg = AB$ for all g in the group.
  (c) Show that $n_A \chi(A) n_B \chi(B) = n \sum_c s_c n_c \chi(C)$.   Such algebraic relations among characters are often useful in evaluating characters.

16-4   Consider the symmetry group of a regular tetrahedron.
  (a) What is the order of this group?
  (b) Decompose it into classes.
  (c) Construct its character table.

16-5   Repeat Problem 16-4 for the symmetry group of a regular hexagon.

16-6   The elements of the group in Problem 16-4 can be represented by 3×3 orthogonal matrices which describe the various rotations and reflections. Decompose this representation into irreducible representations.

16-7   Show that a representation $D(g)$ is irreducible, if, and only if,

$$\overline{\chi(g)^*\chi(g)} = 1$$

where $\chi(g)$ is the character of $D(g)$. Suppose $\overline{\chi(g)^*\chi(g)} = 2$; what does this tell us about $D(g)$?

16-8   An eigenvalue problem has spherical symmetry, and a particular eigenvalue is 5-fold degenerate, the eigenfunctions being

$$P_2^2(\cos\theta)e^{\pm 2i\phi} \qquad P_2^1(\cos\theta)e^{\pm i\phi} \qquad P_2(\cos\theta)$$

where we have omitted a common radial function.

(a) The coordinate axes (or eigenfunctions) are rotated through an angle ω about some axis. Find the trace of the 5×5 transformation matrix giving the new eigenfunctions in terms of the old ones.

(b) The original problem is now perturbed so that the symmetry drops to the symmetry of a triangle. Give the characters of the 5×5 representation of the triangle symmetry group which our five eigenfunctions generate. (To avoid problems with the reflections of p. 426, you may replace them by the equivalent rotations.)

(c) Discuss the splitting of our 5-fold degenerate system by the perturbation; that is, does it remain degenerate, does it split into a 4-fold degenerate system and a separate nondegenerate eigenfunction, or what?

16-9   Four masses m are connected by six springs of spring constant k in such a way that the equilibrium positions of the masses are at the corners of a regular tetrahedron. Find the frequencies of all the normal modes, without solving a secular equation.

16-10   Do the following matrices form a group? (Group multiplication = matrix multiplication.)

$$\begin{pmatrix} 1 & 0 \\ 0 & 1 \end{pmatrix} \qquad \begin{pmatrix} \omega & 0 \\ 0 & \omega^2 \end{pmatrix} \qquad \begin{pmatrix} 0 & 1 \\ 1 & 0 \end{pmatrix}$$

where $\omega = e^{2\pi i/3}$ ($\omega^3 = 1$).

If not, add to them other 2×2 matrices needed to complete a group (of smallest order possible).

Divide the elements of the group into classes.

16-11 Consider the symmetry group of the equilateral triangle. When the coordinates undergo some transformation of the group, the quadratic form

$$a\,\frac{x^2}{\sqrt{2}} + bxy + c\,\frac{y^2}{\sqrt{2}}$$

is transformed into another quadratic form with different a, b, c which are related to the original ones by a 3×3 matrix. The 3×3 matrices so obtained form a representation of the group. Find the characters of this representation.

16-12 Consider the group of all *displacements* in three-dimensional space:

$$x' = x + a \qquad y' = y + b \qquad z' = z + c$$

(*a*) How many parameters does this group have?
(*b*) Construct the infinitesimal operators (in differential form).
(*c*) Show that all the infinitesimal operators commute with each other.

16-13 (*a*) Show that $[X_i, X_j] = -[X_j, X_i]$ implies $c_{ij}^k = -c_{ji}^k$.
(*b*) Show that the Jacobi identity

$$[[X_i, X_j], X_k] + [[X_j, X_k], X_i] + [[X_k, X_i], X_j] = 0$$

implies

$$c_{ij}^l c_{lk}^m + c_{jk}^l c_{li}^m + c_{ki}^l c_{lj}^m = 0 \qquad \text{(summation on repeated indices is implied)}$$

[Conditions (*a*) and (*b*) turn out to be the *only* conditions on the structure constants; *any* set of real numbers c_{ij}^k obeying these two conditions defines a Lie algebra.]

16-14 Consider the association of the matrix $-c_{ij}^k$ ($j = $ row label, $k = $ column label) with each generator X_i. Show that this is a *representation* of the Lie algebra; that is, the matrices have the same commutation relations as the respective generators. This representation is called the *regular representation*, or *adjoint representation*, of the Lie algebra.

16-15 Give the commutation relations for the infinitesimal operators (16-79), (16-80), and (16-81).

16-16 Verify the result (16-112).

16-17 If three coordinate rotations (matrices) R, R', R'' are related by $R'R = R''$, show that the associated rotation matrices obey the transposed equation $D^l(R)D^l(R') = D^l(R'')$. Show that this implies that the $D^l(R)$ give a representation of the *inverse* matrices R^{-1}.

16-18 Use (16-112) to show that the representation D^3 of $SU(2)$ is exactly the representation $D^1_{m'm}(R)$ of Table 16–11, provided the basis for the representation D^3 is taken to be $u_{11}, \sqrt{2}\, u_{12}, u_{22}$.

16-19 If σ_x, σ_y, and σ_z denote the Pauli spin matrices (16-109), show that

$$\{\sigma_i, \sigma_j\} = 2\delta_{ij}$$

where $\{\sigma_i, \sigma_j\} = \sigma_i \sigma_j + \sigma_j \sigma_i$ is the *anticommutator* of σ_i and σ_j.

16-20 Show that every irreducible representation of an Abelian group is one-dimensional.

16-21 Consider a group G and a nonfaithful representation D. Let G' be the set of all group elements which are respresented by the *unit* matrix.
(a) Show that G' is a *subgroup* of G.
(b) Show that G' is in fact a *normal* subgroup of G.
(c) If h' is the order of G', show that *every* matrix in D is the representative of h' distinct group elements.

16-22 (a) Let M_1, M_2, ..., M_n be an arbitrary set of matrices, and let a Hermitian matrix K have the property

$$M_i^+ K M_i = K$$

for all i.   Then, if all the eigenvalues of K are positive, show that a Hermitian matrix H, with $H^2 = K$, exists such that HM_iH^{-1} is unitary for all i.
(b) If $D(g)$ is a representation of a finite group of order n, show that $K = \sum_{i=1}^{n} D^+(g_i)D(g_i)$ has the properties

(1) $K = K^+$
(2) All eigenvalues of K are positive
(3) $D^+(g_k)KD(g_k) = K$ for all k

and hence the representation $D(g)$ can be made unitary by a similarity transformation.

APPENDIX

SOME PROPERTIES OF FUNCTIONS OF A COMPLEX VARIABLE

Throughout this book we make frequent use of certain properties of functions of a complex variable. A summary of these properties is presented for reference and review in this appendix. We recommend that the student unfamiliar with this material do some outside reading in a mathematics book on the subject. For example, the two little books by Knopp (K4) are quite short and very readable.

A–1 FUNCTIONS OF A COMPLEX VARIABLE. MAPPING

A complex number has the form

$$z = x + iy = re^{i\theta} \tag{A-1}$$

where x, y, r, and θ are real, $i^2 = -1$, and $e^{i\theta} = \cos\theta + i\sin\theta$. x and

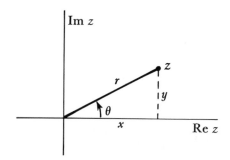

Figure A–1   A point in the complex plane

y are the *real* (Re z) and *imaginary* (Im z) parts of z, respectively, $r = |z|$ is the *magnitude*, and θ is the *phase* or *argument* arg z. Such a number may be represented geometrically by a point on the *complex z-plane*, or *xy*-plane, as shown in Figure A–1. The *complex conjugate* of z will be denoted by z^*; $z^* = x - iy$.

A function $W(z)$ of the complex variable z is itself a complex number whose real and imaginary parts U and V depend on the position of z in the *xy*-plane.

$$W(z) = U(x, y) + iV(x, y) \qquad \text{(A-2)}$$

Two different graphical representations of the function $W(z)$ are useful. One is simply to plot the real and (or) imaginary parts $U(x, y)$ and $V(x, y)$ as surfaces above the *xy*-plane (see, for example, Section 3–6, Figure 3–13). The other is to represent the complex number $W(z)$ by a point in the complex " *W*-plane," or *UV*-plane, so that to each point in the *z*-plane corresponds one (or more) points in the *W*-plane. In this way, the function $W(z)$ produces a *mapping* of the *xy*-plane onto the *UV*-plane.

EXAMPLE

$$W(z) = z^2 = (x + iy)^2 = x^2 - y^2 + 2ixy \qquad \text{(A-3)}$$
$$U = x^2 - y^2 \qquad V = 2xy$$

Alternatively,

$$W = z^2 = r^2 e^{2i\theta}$$

The mapping of a number of points and two curves from the *z*-plane onto the *W*-plane is shown in Figure A–2. For example, the line $x = 1$ becomes the parabola $4U = 4 - V^2$.

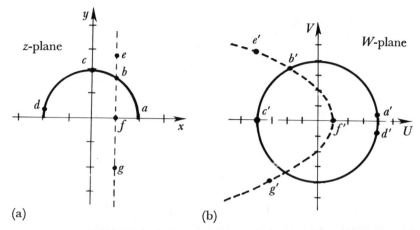

(a)   (b)

**Figure A–2   Illustration by some points and curves of the mapping
produced by the function $W(z) = z^2$. The points $a, b, \ldots$
in the z-plane are mapped into $a', b', \ldots$ in the W-plane**

In the above example, two points z and $-z$ go into the same point $W(z)$.
The upper half of the z-plane maps onto the entire W-plane, and so does the
lower half z-plane.   Clearly, this situation presents difficulties for the inverse
mapping, which is produced by the square root

$$W(z) = z^{1/2} = \sqrt{r}\, e^{i\theta/2} \tag{A-4}$$

This is a *multivalued* function, one point p in the xy-plane going into two
points p' and p'' in the UV-plane.   [These are the two square roots corre-
sponding to the phases $\theta_{p'} = \frac{1}{2}\theta_p$ and $\theta_{p''} = \frac{1}{2}(\theta_p + 2\pi)$.]   This situation is
illustrated in Figure A–3.

Suppose we try to make the mapping single valued by agreeing that a

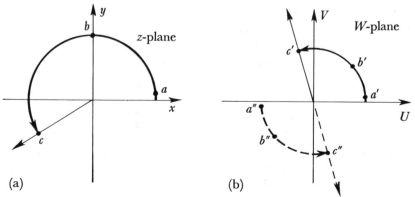

(a)   (b)

Figure A–3   Illustration of the mapping produced by $W(z) = z^{1/2}$

point p corresponds to p' and not p''.   We must make sure that if we start at p and trace a closed curve in the z-plane, the mapping will produce a closed curve in the W-plane starting at p' and returning to p', not p''.   This is true provided the closed curve in the xy-plane does not encircle the origin. However, if the curve encircles the origin once, θ changes by 2π and the mapped curve in the W-plane will not return to its starting point.

Thus the multivalued feature can be avoided only if we agree never to encircle the origin $z = 0$.   To ensure this, we draw a so-called *branch line* or *branch cut* from $z = 0$ to infinity and agree not to cross it.   The singular point $z = 0$ is called a *branch point*.   The branch line may be drawn from $z = 0$ to infinity in any way but it is usually convenient to take it along the positive or negative real axis.

The z-plane, when cut in this way, is called a *sheet*, or *Riemann sheet*, of the function $W(z)$.   This sheet maps in a single-valued manner onto a portion (in our example, half) of the W-plane, this portion being called a *branch* of the function.   A second sheet, similarly cut, is needed to map onto the other half of the W-plane.   We may now cross the branch line without getting into multivalue troubles if we transfer from one sheet to the other, when crossing the cut.   To picture this, imagine that the edges of the sheets along the cut are joined to each other in the manner indicated in Figure A–4.   The sheets so joined form a *Riemann surface* which maps in a single-valued manner onto the entire W-plane.   If we now go around the branch point $z = 0$ twice, once

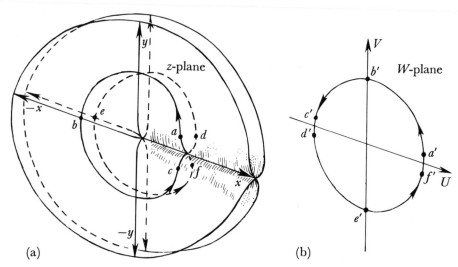

(a)   (b)

Figure A–4   Riemann surface and mapping for the function $W(z) = z^{1/2}$. The dashed part of the curve in the z-plane lies on the lower sheet

on each sheet, we come back to the starting point in the W-plane, as indicated in Figure A–4.

Other roots may be described in the same way.

EXAMPLE

$$W = z^{1/3} \tag{A-5}$$

The mapping produced by this function is indicated in Figure A–5.   The origin is again a branch point, said to be of order 2 because the Riemann surface contains $3(= 2 + 1)$ sheets.

Another example is $W(z) = \ln z$:

$$z = re^{i\theta} \tag{A-6}$$

$$\ln z = \ln r + i\theta$$

The origin is again a branch point, this time of infinite order because the Riemann surface has an infinite number of sheets.   Each sheet maps onto a horizontal strip in the W-plane of width $\Delta V = 2\pi$ in the "imaginary" direction.   By continued circling of the origin $z = 0$ in the same sense, we never return to the starting point on the map.

Another important type of function is one containing two branch points arising from square roots.   Consider

$$W(z) = \sqrt{(z - a)(z - b)}$$

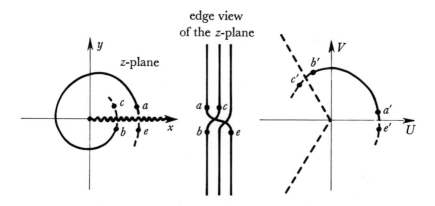

Figure A–5   Riemann surface and mapping for $W = z^{1/3}$.   This sketch is a less pictorial way of conveying information similar to that in Figure A–4

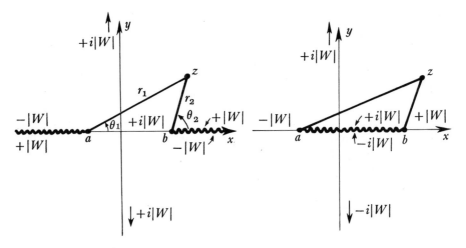

Figure A–6   Two of the ways of drawing branch lines for the function $W(z) = \sqrt{(z-a)(z-b)}$, with the behavior of W indicated in various regions of one Riemann sheet covering the z-plane. In both drawings, the sheet is that one for which $W(z)$ is positive along the "upper side" of the real axis to the right of b. The symbol $-i|W|$, for example, means that $W(z)$ is pure negative imaginary at the place indicated

with branch points at $z = a$ and $z = b$. The Riemann surface of this function may be formed by drawing branch cuts from each of the two branch points to infinity in arbitrary directions, or by making a single cut connecting the two points. The resulting Riemann sheets and the branch of the function corresponding to a given sheet depend on the choice of cuts, as shown for an example in Figure A–6.

In order to sketch the mapping, it is convenient to introduce polar coordinates of z centered at each branch point, that is,

$$z - a = r_1 e^{i\theta_1} \qquad z - b = r_2 e^{i\theta_2}$$

$$W(z) = \sqrt{(z-a)(z-b)} = (r_1 r_2)^{1/2} e^{i\frac{1}{2}(\theta_1 + \theta_2)}$$

The behavior of this function is indicated in Figure A–6.[1]

The mappings produced by more complicated functions may be investigated by extensions of the above procedures.

---

[1] In the right-hand drawing of Figure A–6, as the point z moves around the cut, the radius vectors from a and b to z will of course sweep across the cut. This is perfectly all right, but the point z itself must not cross the cut—if it does, it will find itself on the other Riemann sheet.

A–2   ANALYTIC FUNCTIONS

In this section we review those properties of analytic functions of a complex variable which are needed in this book. For the general mathematical theory, see any of the numerous books on the subject; for example, Apostol (A5), especially Chapter 16; Knopp (K4); Copson (C8); Whittaker and Watson (W5); or Titchmarsh (T4).

1. A function is *analytic* at a point z if it has a derivative there; that is, if

$$f'(z) = \lim_{h \to 0} \frac{f(z + h) - f(z)}{h} \tag{A-7}$$

exists and is independent of the path by which the complex number h approaches zero.   If a function is both analytic and single valued throughout a region R, we shall call it *regular* in R.   A region of regularity of a multivalued function should be specified on a cut Riemann sheet.

2. If $W(z) = U(x, y) + iV(x, y)$ is an analytic function and we write

$$h = h_x + ih_y$$

then two paths for $h \to 0$ are along the horizontal and vertical directions, for which $h_y = 0$ and $h_x = 0$, respectively.   The limits (A-7) obtained for these paths must be equal:

$$\frac{\partial U}{\partial x} + i \frac{\partial V}{\partial x} = \frac{1}{i} \frac{\partial U}{\partial y} + \frac{\partial V}{\partial y} \quad (= W'(z))$$

Equating real and imaginary parts of this equation gives the *Cauchy–Riemann differential equations*:

$$\frac{\partial U}{\partial x} = \frac{\partial V}{\partial y} \qquad \frac{\partial V}{\partial x} = -\frac{\partial U}{\partial y} \tag{A-8}$$

These equations may be shown to be sufficient as well as necessary for the function $W = U + iV$ to be regular in a region, provided the four partial derivatives exist and are continuous there.

EXAMPLE

$$W = z^2$$

$$U = x^2 - y^2 \qquad V = 2xy$$

$$\frac{\partial U}{\partial x} = 2x = \frac{\partial V}{\partial y} \qquad \frac{\partial V}{\partial x} = 2y = -\frac{\partial U}{\partial y}$$

The following example shows that some functions are not analytic anywhere.

EXAMPLE

$$W = z^* \quad \text{(complex conjugate)}$$

$$U = x \qquad V = -y$$

$$\frac{\partial U}{\partial x} = +1 \qquad \frac{\partial V}{\partial y} = -1$$

3. *Integration.* The integral

$$\int_{z_1}^{z_2} f(z)\, dz$$

is a line integral which depends in general on the path followed from z_1 to z_2 (Figure A–7). However, the integral will be the same for two paths if $f(z)$ is regular in the region bounded by the paths. An equivalent statement is *Cauchy's theorem*:

$$\oint_C f(z)\, dz = 0 \qquad\qquad\qquad \text{(A-9)}$$

if C is any closed path lying within a region in which $f(z)$ is regular. A kind of converse is also true; if $\oint_C f(z)\, dz = 0$ for every closed path C within a region R, where $f(z)$ is continuous and single valued, then $f(z)$ is regular in R.

4. If $f(z)$ is regular in a region, its derivatives of all orders exist and are regular there.

5. If $f(z)$ is regular in a region R, the value of $f(z)$ at any point within R may be expressed by *Cauchy's integral formula*

$$f(z) = \frac{1}{2\pi i} \int_C \frac{f(\zeta)\, d\zeta}{\zeta - z} \qquad\qquad \text{(A-10)}$$

where C is any closed path within R encircling z once in the counterclockwise direction. This formula follows directly from the theorem of residues,

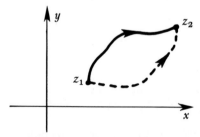

Figure A–7   Paths of integration in the complex plane

item 8 below.   The remarkable property of analytic functions implied by Eq. (A-10) should be noted.   The values of an analytic function throughout a region are completely determined by the values of the function on the boundary of that region.   See Section 5–2 for an application of this property.

Cauchy's formula may be differentiated any number of times to obtain

$$f'(z) = \frac{1}{2\pi i} \int_C \frac{f(\zeta)\, d\zeta}{(\zeta - z)^2}$$

$$f^{(n)}(z) = \frac{n!}{2\pi i} \int_C \frac{f(\zeta)\, d\zeta}{(\zeta - z)^{n+1}} \tag{A-11}$$

6.   A power series expansion (Taylor's series) is possible about any point z_0 within a region where $f(z)$ is regular:

$$f(z) = a_0 + a_1(z - z_0) + a_2(z - z_0)^2 + \cdots$$

$$a_0 = f(z_0) \qquad a_n = \frac{1}{n!}\, f^{(n)}(z_0) \tag{A-12}$$

The region of the z-plane in which the series converges is a circle.   This *circle of convergence* extends to the nearest singularity of $f(z)$, that is, to the nearest point where $f(z)$ is not analytic.

The converse is also true.   Any power series convergent within a circle R represents a regular function there.

7. *The Laurent expansion.*   If $f(z)$ is regular in an annular region between two concentric circles with center z_0, then $f(z)$ may be represented within this region by a *Laurent expansion*

$$f(z) = \sum_{n=-\infty}^{\infty} a_n(z - z_0)^n$$

where the coefficients a_n are

$$a_n = \frac{1}{2\pi i} \oint_C \frac{f(z)\, dz}{(z - z_0)^{n+1}} \tag{A-13}$$

C is any closed path encircling z_0 counterclockwise within the annular region. Note that the coefficient a_{-1} is

$$a_{-1} = \frac{1}{2\pi i} \oint_C f(z)\, dz \tag{A-14}$$

If $f(z)$ is regular in the annulus, no matter how small we make the inner circle, and yet $f(z)$ is not regular throughout the larger circle, we say that z_0 is an *isolated singularity* of $f(z)$.   For such an isolated singularity, there are three possibilities:

(a) The Laurent series for $f(z)$ may contain *no* terms with negative powers of $(z - z_0)$. This is a trivial case, and is called a *removable singularity*. By redefining $f(z)$ at the point $z = z_0$, the singularity may be removed. For example, the function

$$f(z) = \begin{cases} z & |z| > 0 \\ 1 & z = 0 \end{cases}$$

has a removable singularity at $z = 0$.

(b) The Laurent series for $f(z)$ may contain *a finite number* of terms with negative powers of $(z - z_0)$. In this case z_0 is called a *pole of order m*, where $-m$ is the lowest power of $(z - z_0)$ appearing in the Laurent series. For example, the function $f(z) = (1/\sin z)^2$ has poles of order two at $z = 0, \pm\pi, \pm 2\pi, \ldots$. If $f(z)$ has a pole of order m at z_0, the function $(z - z_0)^m f(z)$ is regular in the neighborhood of z_0.

(c) The Laurent series for $f(z)$ may contain *infinitely many* terms with negative powers of $(z - z_0)$. In this case, $f(z)$ is said to have an *essential singularity* at $z = z_0$. For example, $e^{1/z}$ has an essential singularity at $z = 0$ (and therefore e^z has an essential singularity at $z = \infty$).

If z_0 is an isolated singularity, the coefficient a_{-1} in the Laurent expansion is called the *residue* of $f(z)$ at z_0. It has special importance, because of the relation (A-14), as will now be discussed.

8. The *theorem of residues* allows us to evaluate easily the integral of a function $f(z)$ along a closed path C such that $f(z)$ is regular in the region bounded by C except for a finite number of poles and (isolated) essential singularities in the interior of C. By Cauchy's theorem, the path, or *contour*, C may be deformed without crossing any singularities until it is reduced to little circles surrounding each singular point. The integral around each little circle is then given by (A-14), so that we have the theorem of residues

$$\int_C f(z)\, dz = 2\pi i \sum \text{residues} \tag{A-15}$$

where the sum is over all the poles and essential singularities inside C. This theorem is of enormous practical importance in the evaluation of integrals, and a number of examples of its application are given in Section 3–3.

What if a pole lies on the contour? The first thing to do is to look into the physics of the problem to see if this awkward location of the pole results from some approximation. If so, one can decide on which side of the path the pole really lies and thus see whether its residue should be included or not.

A mathematical integral with a pole on the contour strictly does not exist, but, for a simple pole on the real axis, one defines the *Cauchy principal value* as

$$P \int_a^b \frac{f(x)}{x - x_0}\, dx = \lim_{\delta \to 0} \left[\int_a^{x_0 - \delta} \frac{f(x)}{x - x_0}\, dx + \int_{x_0 + \delta}^b \frac{f(x)}{x - x_0}\, dx \right] \tag{A-16}$$

where δ is positive.

The path for the Cauchy principal value integral can form part of a closed contour in which the ends $x_0 \pm \delta$ are joined by a small semicircle centered at the pole (see Figure A-8). Along this semicircle the integral is easy to evaluate; if we let the radius approach zero, $f(z) \to a_{-1}(z - x_0)^{-1}$. Let

$$z - x_0 = re^{i\theta} \qquad dz = ire^{i\theta}\, d\theta$$

Then

$$\int_{\text{semicircle}} f(z)\, dz \to -\int_0^\pi a_{-1} i\, d\theta = -\pi i a_{-1}$$

and if, as is usually the case, the large semicircle gives no contribution,

$$\oint_C f(z)\, dz = P \int f(z)\, dz - \pi i(\text{residue at } z_0)$$

$$= 2\pi i(\textstyle\sum \text{residues inside } C)$$

This gives the result

$$P \int f(z)\, dz = 2\pi i(\tfrac{1}{2}\, \text{residue at } x_0 + \textstyle\sum \text{residues inside } C) \qquad \text{(A-17)}$$

Thus the Cauchy principal value is the average of the two results obtained with the pole inside and outside of the contour.

We often have an integral along the real axis with a simple pole just above (or just below) the axis at x_0. We may consider the pole to be on the axis if we make the path of integration miss the pole by going around x_0 on a little semicircle below (or above). Then it follows by reasoning similar to that leading to (A-17) that the integral may be expressed in terms of the Cauchy principal value as follows:

$$\int \frac{f(x)}{x - x_0 \mp i\varepsilon}\, dx = P \int \frac{f(x)}{x - x_0}\, dx \pm i\pi f(x_0)$$

We may express this result in the somewhat symbolic form

$$\frac{1}{x - x_0 \mp i\varepsilon} = P\, \frac{1}{x - x_0} \pm i\pi\, \delta(x - x_0) \qquad \text{(A-18)}$$

where $\delta(x - x_0)$ is the Dirac delta-function defined in (4-19).

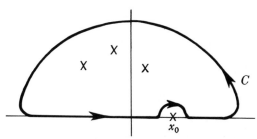

Figure A-8   Illustration of a pole on the real axis

9. The *identity theorem* states that if two functions are each regular in a region R, and have the same values for all points within some subregion or for all points along an arc of some curve within R, or even for a denumerably infinite number of points having a limit point within R, then the two functions are identical everywhere in the region.   For example, if $f(z) = 0$ all along some arc in R, then $f(z)$ is the regular function 0 everywhere in R.

This theorem is useful in extending into the complex plane functions defined on the real axis.   For example,

$$e^z = 1 + z + \frac{1}{2!} z^2 + \cdots$$

is the unique function $f(z)$ which is equal to e^x on the real axis.

10. Consider a function $f(z)$ which is analytic in a region R of the complex plane, and assume that a finite part of the real axis is included in R.   If the function $f(z)$ assumes only *real* values on that part of the real axis in R, then it can be shown that $f(z^*) = [f(z)]^*$ throughout R.   That is, going from a point z to its "image" in the real axis, namely, z^*, just carries the *value f* of the function over into *its* image f^*.   This is known as the *Schwartz reflection principle.*

The identity theorem forms the basis for the procedure of *analytic continuation.*   A power series about z_1 represents a regular function $f_1(z)$ within its circle of convergence, which extends to the nearest singularity.   If an expansion of this function is made about a new point z_2, the resulting series will converge in a circle which may extend beyond the circle of convergence of $f_1(z)$.   The values of $f_2(z)$ in the extended region are uniquely determined by $f_1(z)$—in fact, by the values of $f_1(z)$ in the common region of convergence of $f_1(z)$ and $f_2(z)$. $f_2(z)$ is said to be the analytic continuation of $f_1(z)$ into the new region. This process may be repeated (with limitations mentioned below) until the entire plane is covered except for singular points by these *elements* of a single function $F(z)$.

EXAMPLE

$$f_1(z) = 1 + z + z^2 + z^3 + \cdots$$

converges in a circle of radius 1 to

$$F(z) = \frac{1}{1-z}$$

But $F(z)$ is analytic everywhere except at the simple pole $z = 1$, and no other function analytic outside $|z| = 1$ can coincide with $f_1(z)$ within $|z| < 1$. $F(z)$ is the unique analytic continuation of $f_1(z)$ into the entire plane.

Not all functions can be continued indefinitely.   The extension may be blocked by a barrier of singularities.

It may also happen that the function $F(z)$ obtained by continuation is multivalued.   For example, suppose that after repeating the process described above a number of times, the nth circle of convergence partially overlaps the first one.   Then the values of the element $f_n(z)$ in the common region may or may not agree with $f_1(z)$.   If they do not agree, then the function $F(z)$ is multivalued, and the "path" along which the continuation was made has encircled one or more branch points.

A power series which converges everywhere defines a single-valued analytic function with no singularities in the entire plane (excluding ∞).   Such a function is called an *entire function*.   Examples are polynomials, e^z, and $\sin z$.   A single-valued function which has no singularities other than poles in the entire plane (excluding ∞) is called a *meromorphic* function.   Examples are rational functions, that is, ratios of polynomials.

We conclude by mentioning *Liouville's theorem*; if the function $f(z)$ is regular *everywhere* in the z-plane, including the point at infinity, then $f(z)$ is a constant.

REFERENCES

A very nice treatment of the theory of functions of a complex variable may be found in the two small volumes by Knopp (K4).   This subject is treated in many other books, for example, Copson (C8); Whittaker and Watson (W5); Apostol (A5); Nehari (N2); and Titchmarsh (T4).

PROBLEMS

A-1   Describe the mapping produced by the function

$$W(z) = \frac{1}{\sqrt{(z^2 + 1)(z - 2)}}$$

A-2   Describe the mapping produced by the function

$$W(z) = \frac{1}{\sqrt{z - 1 - i\sqrt{2}}}$$

A-3   Which of the following are analytic functions of the complex variable z?
   (*a*) $|z|$
   (*b*) Re z
   (*c*) $e^{\sin z}$

A-4 The function $f(z)$ has a pole of order n at $z = z_0$. Show that the function $f'(z)/f(z)$ has a simple pole at z_0. What is the residue?

A-5 In the examples below, $U(x, y)$ and $V(x, y)$ denote the real and imaginary parts, respectively, of the analytic function $W(z)$. Find the missing quantities.

(a) $U(x, y) = e^x \cos y$, $V(x, y) = ?$, $W(z) = ?$

(b) $U(x, y) = ?$, $V(x, y) = y(3x^2 - y^2 - 1)$, $W(z) = ?$

(c) $U(x, y) = ?$, $V(x, y) = ?$, $W(z) = \tan^{-1} z$.

A-6 Find the residue of the function $z^2 e^{1/\sin z}$ at the isolated (essential) singularity $z = \pi$.

BIBLIOGRAPHY

A1   Abramowitz, M., and Stegun, I., *Handbook of Mathematical Functions*, Dover, New York (1965).

A2   Aitken, A. C., *Determinants and Matrices*, Wiley, New York (1956).

A3   Akhiezer, N. I., *The Calculus of Variations*, Blaisdell, Waltham, Massachusetts (1962).

A4   Anderson, J. L., *Principles of Relativity Physics*, Academic Press, New York (1967).

A5   Apostol, T. M., *Mathematical Analysis*, Addison-Wesley, Reading, Massachusetts (1964).

A6   Artin, E., *The Gamma Function*, Holt, Rinehart and Winston, New York (1964).

B1   Bédard, G., *Phys. Rev.* **151**, 1038 (1966).

B2   Birkhoff, G., and MacLane, S. *A Survey of Modern Algebra*, Macmillan, New York (1966).

B3   Birkhoff, G., and Rota, G., *Ordinary Differential Equations*, Blaisdell, Waltham, Massachusetts (1962).

B4   Block, H. D., *Introduction to Tensor Analysis*, Merrill, Columbus, Ohio (1962).

B5   Boole, G., *A Treatise on the Calculus of Finite Differences*, Dover, New York (1872).

B6   Booth, A. D., *Numerical Methods*, Plenum, New York (1957).

B7   Bowman, F., *Introduction to Elliptic Functions with Applications*, Dover, New York (1961)

B8   Bromwich, T. J. I., *An Introduction to the Theory of Infinite Series*, St. Martin's, New York (1959).

B9   Burkill, J. C., *The Theory of Ordinary Differential Equations*, Wiley (Interscience), New York (1956).

C1   Carmichael, R. D., *Introduction to the Theory of Groups of Finite Order*, Dover, New York (1956).

C2   Carslaw, H. S., and Jaeger, J. C., *Operational Methods in Applied Mathematics*, Dover, New York (1963).

C3   Castillejo, L., Dalitz, R. H., and Dyson, F. J., *Phys. Rev.* **101**, 453 (1955).

C4   Cohen, A., *An Elementary Treatise on Differential Equations*, D. C. Heath, Boston, Massachusetts (1933).

C5   Cohen, E. R., *Phys. Rev.* **101**, 1641 (1956).

C6   Cohen, E. R., *Rev. Mod. Phys.* **25**, 709 (1953).

C7   Conkwright, N. B., *Introduction to the Theory of Equations*, Blaisdell, Waltham, Massachusetts (1957).

C8   Copson, E. T., *An Introduction to the Theory of Functions of a Complex Variable*, Oxford University Press, New York (1962).

C9   Corinaldesi, E., *Nuovo Cimento Suppl.* **14**, 369 (1959).

C10   Courant, R., and Hilbert, D., *Methods of Mathematical Physics* (2 Vols.), Wiley (Interscience), New York, (1966).

C11   Courant, E., Livingston, M. S., and Snyder, H., *Phys. Rev.* **88**, 1190 (1952).

C12   Cramér, H., *Mathematical Methods of Statistics*, Princeton University Press, Princeton, New Jersey (1966).

D1   Davidson, N., *Statistical Mechanics*, McGraw-Hill, New York (1962).

D2   Davis, H. T., *The Summation of Series*, Trinity University Press, San Antonio, Texas (1962).

D3   De Alfaro, V., Fubini, S., Rossetti, G., and Furlan, G., *Phys. Letters* **21**, 576 (1966).

D4   De Bruijn, N. G., *Asymptotic Methods in Analysis*, Wiley (Interscience), New York (1961).

D5   Dennery, P., and Krzywicki, A., *Mathematics for Physicists*, Harper and Row, New York (1967).

D6   Dirac, P. A. M., *The Principles of Quantum Mechanics*, Oxford University Press, New York (1967).

D7   Drew, T. B., *Handbook of Vector and Polyadic Analysis*, Reinhold, New York (1961).

D8   Dwight, H. B., *Tables of Integrals and Other Mathematical Data*, Macmillan, New York (1968).

E1   Einstein, A., *The Meaning of Relativity*, Princeton University Press, Princeton, New Jersey (1956).

E2   Eisenhart, L. P., *An Introduction to Differential Geometry*, Princeton University Press, Princeton, New Jersey (1947).

E3   Erdélyi, A., *Asymptotic Expansions*, Dover, New York (1956).

E4   Erdélyi, A., *Operational Calculus and Generalized Functions*, Holt,

Rinehart and Winston, New York (1962).

E5   Erdélyi, A., Magnus, W., Oberhettinger, F., and Tricomi, F. G., *Higher Transcendental Functions* (3 Vols.), McGraw-Hill, New York (1953).

E6   Erdélyi, A., Magnus, W., Oberhettinger, F., and Tricomi, F. G., *Tables of Integral Transforms* (2 Vols.), McGraw-Hill, New York (1954).

F1   Faddeyeva, V. N., and Terent'ev, N. M., *Tables of Values of the Function*

$$w(z) = e^{-z^2}\left[1 + \frac{2i}{\sqrt{\pi}}\int_0^z e^{t^2}\,dt\right] \text{ for Complex Argument.}$$ Pergamon , New York (1961).

F2   Feller, W., *An Introduction to Probability Theory and Its Application* (2 Vols.), Wiley, New York (1961, 1966).

F3   Feynman, R. P., *Phys. Rev.* **76**, 769 (1949).

F4   Forsyth, A. R., *Calculus of Variations*, Dover, New York (1927).

F5   Forsyth, A. R., *A Treatise on Differential Equations*, St. Martin's, New York (1961).

F6   Frazer, R. A., Duncan, W. J., and Collar, A. R., *Elementary Matrices*, Cambridge University Press, New York (1963).

F7   Fried, B. D. and Conte, S. D., *The Plasma Dispersion Function*, Academic Press, New York (1961).

F8   Friedman, B., *Principles and Techniques of Applied Mathematics*, Wiley, New York (1962).

G1   Gel'fand, I. M., *Lectures* on *Linear Algebra*, Wiley (Interscience), New York (1963).

G2   Gell-Mann, M., Goldberger, M. L., and Thirring, W., *Phys. Rev.* **95**, 1612 (1954).

G3   Goertzel, G. H., and Tralli, N., *Some Mathematical Methods of Physics*, McGraw-Hill, New York (1960).

G4   Goldstein, H., *Classical Mechanics*, Addison-Wesley, Reading, Massachusetts (1965).

G5   Golomb, M. and Shanks, M., *Elements of Ordinary Differential Equations*, McGraw-Hill, New York (1965).

G6   Gradshteyn, I. S., and Ryzhik, I. M., *Tables of Integrals, Series and Products*, Academic Press, New York (1967).

G7   Green, J. A., *Sequences and Series*, Dover, New York (1964).

G8   Grobner, W., and Hofreiter, N., *Integraltafel*, Springer, New York (1957).

H1   Hall, M., *The Theory of Groups*, Macmillan, New York (1967).

H2   Halmos, P. R., *Finite-Dimensional Vector Spaces*, Van Nostrand, Princeton, New Jersey (1958).

H3   Hamermesh, M., *Group Theory*, Addison-Wesley, Reading, Massachusetts (1964).

H4   Hamming, R. W., *Numerical Methods for Scientists and Engineers*, McGraw-Hill, New York (1962).

H5   Hanbury Brown, R., and Twiss, R. Q., *Nature* **177**, 27 (1956).

H6   Hancock, H., *Elliptic Integrals*, Dover, New York (1958).

H7   Henrici, P., *Elements of Numerical Analysis*, Wiley, New York (1964).

H8   Hildebrand, F. B., *Introduction to Numerical Analysis*, McGraw-Hill, New York (1956).

H9   Hildebrand, F. B., *Methods of Applied Mathematics*, Prentice-Hall, Englewood Cliffs, New Jersey (1965).

H10 Hirschman, I. I., *Infinite Series*, Holt, Rinehart and Winston, New York (1962).

H11 Hoel, P. G., *Introduction to Mathematical Statistics*, Wiley, New York (1966).

H12 Huang, K., *Statistical Mechanics*, Wiley, New York (1967).

H13 Hyslop, J. M., *Infinite Series*, Wiley (Interscience), New York (1959).

I1   Ince, E. L., *Integration of Ordinary Differential Equations*, Wiley (Interscience), New York (1963).

I2   Ince, E. L., *Ordinary Differential Equations*, Dover, New York (1956).

I3   Irving, J., and Mullineux, N., *Mathematics in Physics and Engineering*, Academic Press, New York (1966).

J1   Jackson, J. D., *Classical Electrodynamics*, Wiley, New York (1967).

J2   Jacobson, N., *Lectures in Abstract Algebra, Volume II—Linear Algebra*, Van Nostrand, Princeton, New Jersey (1953).

J3   Jahnke, E., Emde, F., and Losch, F., *Tables of Higher Functions*, McGraw-Hill, New York (1960).

J4   Jeffreys, H., and Jeffreys, B. S., *Methods of Mathematical Physics* Cambridge University Press, New York (1966).

K1   Kemble, E. C., *The Fundamental Principles of Quantum Mechanics*, Dover, New York (1958).

K2   Kittel, C., *Elementary Statistical Physics*, Wiley, New York (1958).

K3   Knopp, K., *Theory and Application of Infinite Series*, Hafner, New York (1947).

K4   Knopp, K., *Theory of Functions* (2 Vols.), Dover, New York (1945).

K5   Kramers, H. A., *Z. Physik* **39**, 828 (1926).

L1   Landau, L. D., and Lifshitz, E. M., *The Classical Theory of Fields*, Addison-Wesley, Reading, Massachusetts (1965).

L2   Landau, L. D., and Lifshitz, E. M., *Quantum Mechanics; Non-Relativistic Theory*, Addison-Wesley, Reading, Massachusetts (1965).

L3   Landau, L. D., and Lifshitz, E. M., *Statistical Physics*, Addison-Wesley, Reading, Massachusetts (1958).

L4   Langer, R. E., *Bull. Amer. Math. Soc.* **39**, 696 (1933).

L5   Lass, H., *Vector and Tensor Analysis*, McGraw-Hill, New York (1950).

L6   Ledermann, W., *Introduction to the Theory of Finite Groups*, Wiley,

New York (1964).

L7   Levi-Civita, T., *The Absolute Differential Calculus*, Hafner, New York (1950).

L8   Lighthill, M. J., *Introduction to Fourier Analysis and Generalized Functions*, Cambridge University Press, New York (1964).

L9   Lovitt, W. V., *Linear Integral Equations*, Dover, New York (1950).

M1   Magnus, W., Oberhettinger, F., and Soni, R. P., *Formulas and Theorems for the Special Functions of Mathematical Physics*, Springer-Verlag, New York (1966).

M2   Margenau, H., and Murphy, G. M., *Mathematics of Physics and Chemistry* (2 Vols.), Van Nostrand, Princeton, New Jersey (1967, 1964).

M3   McLachlan, N. W., *Theory and Application of Mathieu Functions*, Dover, New York (1964).

M4   Merzbacher, E., *Quantum Mechanics*, Wiley, New York (1967).

M5   Mikusinski, J., *Operational Calculus*, Pergamon, New York (1959).

M6   Milne–Thomson, L. M., *The Calculus of Finite Differences*, St. Martin's New York (1960).

M7   Milne–Thomson, L. M., *Jacobian Elliptic Function Tables*, Dover, New York (1950)

M8   Moller, C., *The Theory of Relativity*, Oxford University Press, New York (1966).

M9   Morse, P. M., and Feshbach, H., *Methods of Theoretical Physics* (2 Vols.), McGraw-Hill, New York (1953).

N1   National Bureau of Standards, *Tables of Lagrangian Interpolation Coefficients*, Columbia University Press, New York (1944).

N2   Nehari, Z., *Introduction to Complex Analysis*, Allyn and Bacon, Boston, Massachusetts (1964).

N3   Noble, B., *Methods Based on the Wiener-Hopf Technique for the Solution of Partial Differential Equations*, Pergamon, New York (1959).

N4   Noble, B., *Numerical Methods* (2 Vols.), Wiley, (Interscience), New York (1964).

O1   Omnes, R., *Nuovo Cimento* **8**, 316 (1958).

O2   Orear, J., *Notes on Statistics for Physics*, UCRL-8417, U.C. Radiation Laboratory, Berkeley, California (1958).

O3   Ostrowski, A. M., *Solution of Equations and Systems of Equations*, Academic Press, New York (1960).

P1   Pierce, B. O. and Foster, R. M., *A Short Table of Integrals*, Blaisdell, Waltham, Massachusetts, (1957).

P2   Poincaré, H., *Les Méthodes nouvelles de la Mécanique Céleste*, Dover, New York (1957).

P3   Purcell, E. M., *Nature* **178**, 1449 (1956).

R1   Rabenstein, A. L., *Introduction to Ordinary Differential Equations*, Academic Press, New York (1966).

R2   Ralston, A., *A First Course in Numerical Analysis*, McGraw-Hill, New York (1965).

R3   Rose, M. E., *Elementary Theory of Angular Momentum*, Wiley, New York (1957).

S1   Sagan, H., *Boundary and Eigenvalue Problems in Mathematical Physics*, Wiley, New York (1961).

S2   Schiff, L. I., *Quantum Mechanics*, McGraw-Hill, New York (1968).

S3   Schwartz, J. T., *Introduction to Matrices and Vectors*, McGraw-Hill, New York (1961).

S4   Seshu, S., and Balabanian, N., *Linear Network Analysis*, Wiley, New York (1959).

S5   Smith, L. P., *Mathematical Methods for Scientists and Engineers*, Dover, New York (1953).

S6   Smythe, W. R., *Static and Dynamic Electricity*, McGraw-Hill, New York (1968).

S7   Sneddon, I. N., *Special Functions of Mathematical Physics and Chemistry*, Wiley (Interscience), New York (1956).

S8   Sokolnikoff, I. S., and Redheffer, R. M., *Mathematics of Physics and Modern Engineering*, McGraw-Hill, New York (1966).

S9   Sommerfeld, A., *Partial Differential Equations in Physics*, Academic Press, New York (1967).

S10 Spain, B., *Tensor Calculus*, Wiley (Interscience), New York (1960).

S11 Stanaitis, O. E., *An Introduction to Sequences, Series and Improper Integrals*, Holden-Day, San Francisco, California (1967).

S12 Synge, J. L., and Griffith, B. A., *Principles of Mechanics*, McGraw-Hill, New York (1959).

T1   Tinkham, M., *Group Theory and Quantum Mechanics*, McGraw-Hill, New York (1964).

T2   Titchmarsh, E. C., *Eigenfunction Expansions Associated with Second-Order Differential Equations*, Oxford University Press, New York (1962).

T3   Titchmarsh, E. C., *Introduction to the Theory of Fourier Integrals*, Oxford University Press, New York (1948).

T4   Titchmarsh, E. C., *The Theory of Functions*, Oxford University Press, New York (1964).

T5   Todd, J., *Survey of Numerical Analysis*, McGraw-Hill, New York (1962).

T6   Toll, J. S., *Phys. Rev.* **104**, 1760 (1956).

T7   Tolman, R. C., *The Principles of Statistical Mechanics*, Oxford University Press, New York (1967).

T8   Tranter, C. J., *Integral Transforms in Mathematical Physics*, Methuen, London (1966).

T9   Turnbull, H. W., *Theory of Determinants, Matrices and Invariants*, Dover, New York (1960).

W1   Watson, G. N., *Proc. Roy. Soc. (London)* **95**, 83 (1918).

W2   Weatherburn, C. E., *An Introduction to Riemannian Geometry and the Tensor Calculus*, Cambridge University Press, New York (1950).

W3   Weisner, L., *Introduction to the Theory of Equations*, Macmillan, New York (1956).

W4   Whittaker, E. T., and Robinson, G., *The Calculus of Observations*, Dover, New York (1967).

W5   Whittaker, E. T., and Watson, G. N. *A Course of Modern Analysis*, Cambridge University Press, New York (1965).

W6   Wigner, E., *Group Theory and its Application to the Quantum Mechanics of Atomic Spectra*, Academic Press, New York (1964).

W7   Wilf, H. S., *Mathematics for the Physical Sciences*, Wiley, New York (1962).

INDEX

a posteriori probability, 373
a priori probability, 373
Abelian group, 425
Absolute convergence, 44
Absolute tensor, 414, 418
Absorptive part of scattering amplitude, 316
Addition theorem for spherical harmonics, 176
Adjoint representation, 469
Airy integral, 117
Algebra of a group, 425
Alternating gradient synchrotron, 125
Alternating group A_n, 438
Analytic continuation, 482
Analytic function, 477
Anticommutator, 470
Antilinearity, 149
Antisymmetric matrix, 145
Antisymmetric symbol of Levi–Civita $e_{ijk\ldots}$,
  as Cartesian tensor, 406
  as a general tensor, 413
  as a tensor under $SU(2)$, 458
  as a tensor under $SU(3)$, 460
Argument of a complex number, 472
Associated Legendre equation, 21, 175, 228
Associated Legendre function $P_n^m(x)$, 22, 175, 228

Asymptotic series,
  definition of, 81
  for $Ei(-x)$, 82
  for erf x, 81
  for gamma function, 84, 89
Autocorrelation function, 386

$B(x, y)$, beta function, 76
Bashforth–Adams–Milne method, 354
Basis, of vector space, 141
  orthonormal, 149
Bayes' theorem, 375
$\text{ber}_n x$, $\text{bei}_n x$, 187
Bernoulli differential equation, 4
Bernoulli numbers, 48, 49
Bessel's equation, 12, 16, 178
  second solution, 17, 180
Bessel functions,
  $\text{ber}_n x$, $\text{bei}_n x$, 187
  generating function, 183
  Hankel functions $H_n^{(1)}(x)$, $H_n^{(2)}(x)$, 185
  integral representations, 184, 185
  $J_m(x)$, 179
    recursive calculations, 356
    zeros, 233
  modified Bessel functions $I_n(x)$, $K_n(x)$, 187
  orthogonality and normalization integrals, 181, 182

492

recursion relations, 179
spherical Bessel functions $j_l(x)$, $n_l(x)$, 187, 229
$Y_m(x)$, 180
Bessel's inequality, 166
Bessel's integral, 184
Beta function $B(x, y)$, 76
Binomial coefficient, 376
Binomial distribution, 378
Binormal, 409
Block matrices, 146
Block wave functions, 200
Bohr–Sommerfeld quantization condition, 37
Born approximation, 303
Boundary conditions for partial differential equations, 219, 226
Boundary-valve problems, 229, 266
Bounded variation, 97
Brachistrochrone, 325
Branch of a function, 474
Branch line, cut, and point, 474

Calculus of variations, 322
    applied to eigenvalue problems, 333
    applied to integral equations, 340
    isoperimetric problems of, 331
    variable endpoints, 329
Cartesian tensors, 406
Cauchy boundary condition, 219
Cauchy principal value integral, 480
Cauchy–Riemann differential equations, 124, 477
Cauchy–Schwarz inequality, 165
Cauchy's integral formula, 478
    applied to dispersion relations, 129
    used to derive integral representations, 169, 184
Cauchy's theorem, 478
Causality and dispersion relations, 130
$ce_n(x)$, Mathieu function, 200
Central limit theorem, 383
Characteristic function of a distribution, 382
Characteristics, 221
Characters of a group, 433
    orthogonality, 435
Charged conducting strip, 127
Chi-square distribution, 395
Chi-square test, 394

Christoffel symbol,
    of first kind, 417
    of second kind, 418
$Ci\ x$, cosine integral function, 61, 77
Circle of convergence, 479
Clairaut equation, 5
Classes in a group, 428
Closure, 266
cn x, elliptic function, 209
Cofactor, 144
Cogredient, 147
Combinations and permutations, 375
Commutative group, 425
Commuting operators, 142
Complementary function, 6
Completeness, 141, 266
    of eigenfunctions of Sturm–Liouville differential equations, 338
Complex conjugate, 472
Complex number, 471
Components,
    of a linear operator, 141
    of a linear scalar function, 148
    of a vector, 141
Conditional probability, 374
Confluent hypergeometric equation and function, 194
Conformal transformations, 123
    Schwartz transformation, 137
Congruent transformation, 156
Conjugate linearity, 149
Conjugate representations of $SU(3)$, 461
Conjugate vector or tensor, 458
Connection formulas for WKB method, 28, 34, 36, 116
Continued fraction, 203
Contour integration, 65
    applied to Green's function, 279
Contraction of tensor indices,
    in general coordinates, 412
    in $SU(2)$, 458
    in $SU(3)$, 460
    in three-space, 407
Contragredient, 147
Contravariant tensor, 412
Contravariant vector, 411
Convergence,
    absolute, 44
    criteria for infinite series, 44
    of hypergeometric series, 46
    "in the mean," 340

of series solution of Legendre equation, 47
Convolution theorems, 113
Coordinate system for a vector space, 141
Coordinate transformations, 146
  congruent transformation, 156
  similarity transformation, 148
  unitary transformation, 150
Coset of a group, 426
Cosine transform, 103
Cosine-integral fuction $Ci\ x$, 61, 77
Covariant differentiation, 418
Covariant tensor, 412
Covariant vector, 411
Curvature of a curve, 409
Curvature tensor of Riemann, 419
Cycle notation for symmetric group, 426
Cycloid, 327

$D^j_{m',m}$, rotation matrix, 462
Decomposable representation, 431
Degenerate eigenvalue, 152, 233, 265, 440
Degenerate kernel of integral equation, 300
Degenerate perturbation theory, 293
  as an application of group theory, 443
Degree of differential equation, 1
Delta function,
  Dirac, 102
  in three dimensions, 106
Delta, Kronecker, 100
Density function, 265
Diagonalization of matrices, 153, 158
Difference equation, linear, 355
Differential equations,
  numerical solution, 353
  ordinary, 1
    exact, 2
    graphical discussion of, 22
    homogeneous, 4, 6, 12
    isobaric, 5, 12
    linear, 3, 6
    separable, 2
    series solutions of, 13
    WKB method of, 27
  partial, 217
    integral transform methods, 239
    separation of variables, 226
    Wiener–Hopf method, 245

Differential geometry, 402
  Frenet formulas, 410
Differentiation, numerical, 349
Diffusion equation, 218, 234, 235, 240, 242, 276
Dimensionality of a vector space, 140
Dirac delta function, 102
  in three dimensions, 106
Dirichlet boundary condition, 219
Dispersion relations, 129
  as integral equations, 316
  subtracted, 135
Displacement kernel, 312
dn x, elliptic function, 209
Double factorial $n!!$, 172
Double periodicity of elliptic functions, 206
Dual space, 149
Dyad, dyadic, 408

$E(\phi, k)$, $E(k)$, elliptic integrals, 79
Eigenfunctions,
  of a Hermitian operator, 263
  of Schrödinger equation, 21
Eigenvalues,
  degenerate, 152, 233, 265, 440
  group-theoretic approach to, 443
  of a Hermitian matrix, 152
  of a Hermitian operator, 263
  of an infinite-dimensional matrix, 163
  of integral equation, 299, 301
  of a matrix, 151, 260
  of Schrödinger equation, 21, 25
  variational approach to, 333
Eigenvectors,
  of a linear operator, 150
  of a matrix, 261
  orthogonality properties of, 152, 263
Elastic waves, differential equation for, 255
Elements of a matrix, 143
Elliptic functions sn x, cn x, dn x, 204, 209
Elliptic integrals $F(\phi, k)$, $E(\phi, k)$, $\Pi(\phi, n, k)$, $K(k)$, $E(k)$, $\Pi(n, k)$, 78, 79
Elliptic partial differential equations, 221
Entire function, 483
Envelope of solutions of a differential equation, 6
Equilateral triangle, normal modes of, 237
Equivalence,
  of group elements, 428
  of group representations, 431

Error estimates in interpolation, 348
  for Bashforth–Adams–Milne method, 354
Error function erf x, 77
  asymptotic expansion of, 81
Error matrix, 393
Essential singularity, 480
Euler angles, 404
Euler numbers, 48, 57
Euler–Lagrange equation, 324
Euler–Maclaurin formula, 365
Euler's transformation, 54
Exact differential equation, 2
Expectation value, 381
Exponential integral $Ei(x)$, 77
  asymptotic expansion of, 82

$F(\phi, k)$, elliptic integral, 78
$_1F_1(a; c; x)$, confluent hypergeometric function, 195
$_2F_1(a, b; c; x)$, hypergeometric series, 187
Faithful representation, 430
Floquet's theorem, 198, 199
Fourier series, 96
  complex form, 100
  for cos kx, 52
  Gibbs' phenomenon, 98, 120
Fourier transforms, 101
  applied to integral equation, 312
  applied to a partial differential equation, 242
  in complex plane, 246
  convolution theorem for, 113
  properties and applications, 110
  sine and cosine transforms, 103
  in three dimensions, 105
Fourier–Bessel transform, 109
Fredholm integral equation, 300
Fredholm solution of integral equation, 303, 320
Frenet formulas, 410
Fresnel integrals $C(x)$, $S(x)$, 77
Fully reducible representation, 431
Functional, 323
Functional derivative, 324
Fundamental solution, 274
Fundamental tensor, 415

Gamma function $\Gamma(x)$, 59, 75
  asymptotic expansion of, 84, 89

Gaussian distribution, 378, 381, 384
Gaussian function, 106
Gaussian integration, 351
Gaussian random process, 386
Generating function,
  for Bessel functions, 183
  for Hermite polynomials, 211
  for Legendre polynomials, 170
  for moments of a probability distribution, 382
Generators of a group, 450
Geodesic, 420
Gibbs' phenomenon, 98, 120
Gram–Schmidt orthogonalization, 152, 153
Graphical approach to the solution of a differential equation, 22
Green's functions, 267
  for diffusion equation, 243
  for Helmholtz equation, 283, 284
  for Laplace's equation and wave equation, 277
  for ordinary differential equation, 39
  for vibrating drum, 271
  for vibrating string, 269
Gregory's series, 99, 367
Group, 424
  Abelian or commutative, 425
  alternating, 438
  characters of, 433
  class, 428
  coset, 426
  infinite, 449
  $O(n)$, $O^+(n)$, 450
  order, 425
  physical applications of, 440
  representation, 430
  $SU(n)$, 450
  subgroup, 426
  symmetric, 425

$h_l^{(1)}(x)$, $h_l^{(2)}(x)$, spherical Hankel functions, 229
$H_n(x)$, Hermite polynomials, 21, 196
$H_n^{(1)}(z)$, $H_n^{(2)}(z)$, Hankel functions, 185
Hamilton's variational principle, 324
Hankel functions $H_n^{(1)}(z)$, $H_n^{(2)}(z)$, 185
  spherical Hankel functions $h_l^{(1)}(x)$, $h_l^{(2)}(x)$, 229
Hankel transform, 109, 212

Heat conduction in a cube, 234, 245
  in an infinite medium, 242
  in a semi-infinite medium, 240, 245
  in a slab, 235
Heaviside step function, 107
Helmholtz equation, 227, 232
  in an equilateral triangle, 237
  Green's function, 283, 284
Hermite differential equation, 20
Hermite polynomial $H_n(x)$, 21, 196
  generating function of, 211
Hermitian conjugate of a matrix, 145
Hermitian kernel, 306
Hermitian linear operator, 263
Hermitian matrix, 145
Hilbert space, 162
Hilbert transform, 109, 316
Hill equation, 200
Homogeneous differential equation,
  ordinary, 4, 6, 12
  partial, 218
Homogeneous function, 4
Homographic transformation, 189
Homomorphic, 430
  relation of $SU(2)$ to $O^+(3)$, 466
Horner's method, for roots of polynomials,
  361
Hyperbolic partial differential equation,
  221
Hypergeometric differential equation, 187
  confluent hypergeometric differential
    equation, 194
  second solution to, 190
Hypergeometric series or function, 187
  confluent hypergeometric function, 195
  convergence of, 46
  integral representation of, 191

Idempotent matrix, 145
Identity theorem, 482
Images, method of, 245
Improper rotation, 407
Indicial equation, 16
Infinitesimal rotation, 455, 464
Inhomogeneous differential equation, 6
Inhomogeneous problems involving par-
  tial differential equations, 267
Inner product, 149
Integral equations, 299
  with degenerate kernel, 300
  in dispersion theory, 316

Fredhlom theorems for, 301
  Neumann and Fredholm series in, 302
  Schmidt–Hilbert theory for, 305
  variational approach to, 340
  in Volterra equations, 300, 311
  Wiener–Hopf method for, 312
Integral representations, 79
  of $\delta(x)$, 102
  of $\delta(\mathbf{x})$, 105
  of $_1F_1(a; c; z)$, 195
  of $_2F_1(a, b; c; z)$, 191
  of $J_n(z)$, 185
  of $P_n(z)$, 169, 170
Integral transforms, 96
  applied to integral equations, 312
  applied to partial differential equations,
    239
  Fourier transforms, 101
  Laplace transforms, 107
  other transforms, 109
  properties and applications, 110
Integrals, evaluation, 58
  by contour integration, 65
  by differentiation or integration with
    respect to a parameter, 60
  by expansion in series, 80
  by saddle-point methods, 82
  by symmetry arguments, 62
  use of complex variables in, 59
Integrating factor, 3
Integration, numerical, 349
Interpolation, 345
  error estimation, 348
  Lagrangian interpolation, 347
Interval in Riemann geometry, 416
Invariant subgroup, 430
Inverse,
  of a matrix, 144
  of an operator, 142
Inverse distance formula, 170
Irreducible representations, 431
  of $O^+(3)$, 461
  of $SU(2)$, 457
  of $SU(3)$, 459
Irregular singular point of a differential
  equation, 14
Isobaric differential equation, 5, 12
Isolated singularity, 479
Isomorphism, 430
  of Lie algebras of $SU(2)$ and $O(3)$, 456,
    464

Isoperimetric problems, 331
Iteration, applied to differential equations, 25

$J_m(x)$, Bessel function, 179
Jacobi polynomials, 194
Jacobian elliptic functions sn x, cn x, dn x, 204, 209
Joint probability, 374
Jordan's lemma, 67

$K(k)$, elliptic integral, 79
Kernel of an integral equation, 163, 299
   degenerate, 300
   displacement kernel, 312
   symmetric, Hermitian, 305, 306
Kronecker delta δ_{ij}, $\delta_j{}^i$, 100
   as a Cartesian tensor, 406
   as a general tensor, 412
Kummer function $_1F_1(a; c; z)$, 195
Kummer's transformation, 197

$L_n{}^\alpha(z)$, Laguerre polynomial, 196
Langrange multiplier, 331
Lagrangian interpolation, 347
Lagrange's equations of motion, 324
Laguerre polynomial, $L_n{}^\alpha(z)$, 196
Laplace transforms, 107
   applied to integral equation, 312
   applied to a partial differential equation, 240
   convolution theorems, 113
   properties and applications, 110
Laplace's equation, 218, 232
   Green's function, 277
   two-dimensional, 124
   variational approach to, 333
Laplace's integral representation of $P_n(z)$, 170
Laurent expansion, 479
Least-squares fit, 391
Legendre's differential equation, 14, 167
   associated equation, 21, 175, 228
   second solution $Q_n(z)$, 174
Legendre elliptic integrals $F(\phi, k)$, $E(\phi, k)$, $\Pi(\phi, n, k)$, $K(k)$, $E(k)$, $\Pi(n, k)$, 78, 79
Legendre functions $P_n(z)$, $Q_n(z)$, 167
   associated functions $P_n{}^m(z)$, $Q_n{}^m(z)$, 175, 228

Legendre polynomials $P_n(z)$, 16, 169
Length of a vector, 149
Levi–Civita antisymmetric symbol $e_{ijk}\ldots$,
   as a Cartesian tensor, 406
   as a general tensor, 413
   as a tensor under $SU(2)$, 458
   as a tensor under $SU(3)$, 460
Lie algebra, 456
Lie group, 449
Lienard–Wiechert potential, 282
Likelihood, 387
Linear difference equation, 355
Linear differential equation, with constant coefficients, 6
   first-order, 3
   series solution, 13
   with three regular singular points, 187, 190
Linear independence of vectors, 140
Linear integral equation, 299
Linear operators, 141, 163
   Hermitian, 263
Linear scalar function of a vector, 148
Linear vector function of a vector, 141
Linear vector space, 139
Liouville's theorem, 483
Loaded string, 160
Lowering of tensor indices, 416, 458

Magnitude of a complex number, 472
Mapping, 472
Mathieu functions $se_n(x)$, $ce_n(x)$, 200
   stability curves, 201
Matrices, 143
   diagonalization, 158
   normal, 160
   transformation matrix, 147
Mean life of unstable particle, 389
Mean value, 381
Mellin transform, 109
Meromorphic function, 483
Metric, 149
Metric tensor, 416
Mittag–Lefler theorem, 53
Mixed tensor, 412
Modified Bessel functions $I_n(z)$, $K_n(z)$, 187
Molecular vibrations, 154
Moments of a distribution, 381
Multivalued function, 473

Neumann boundary condition, 219
Neumann series for integral equation, 302
Newton's method for roots of equations, 360
Nonholonomic constraint, 332
Nonsingular operator, 142
Normal distribution, 378
Normal form of one-dimensional wave equation, 223
Normal frequencies, coordinates, modes, 157
   of circular drum, 233
   of equilateral triangle, 239
Normal matrix, 160
Normal subgroup, 430
Normalization,
   of associated Legendre functions, 175
   of Bessel functions, 182
   of Legendre polynomials, 173
Numerical methods,
   differentiation, 349
   integration, 349
   interpolation, 345
   roots of equations, 358
   solution of differential equations, 353
   summing series, 363

$O(n)$, $O^+(n)$ groups, 450
   irreducible representations of $O^+(3)$, 461
Operator, linear, 141, 163
   Hermitian, 263
Order,
   of differential equation, 1
   of a group, 425
   of a group element, 427
Ordinary point of a differential equation, 14
   at infinity, 19, 20
Orthogonality,
   of associated Legendre functions, 175
   of Bessel functions, 181
   of eigenfunctions of Hermitian operator, 263
   of eigenvectors of Hermitian matrix, 152
   of group characters, 435
   of group representations, 432
   of Legendre polynomials, 172
   of matrix, 145
   of vectors, 149
Orthonormal vectors, 149

Oscillations, small, 154
Osculating parameters, 9
Outer product of vectors and tensors, 405, 406
Outgoing wave, 230

P symbol, Riemann, 189
$P_n(z)$, Legendre polynomial, 16, 169
$P_n^m(z)$, associated Legendre function, 175
Parabolic partial differential equation, 221
Parameters of an infinite group, 449
Parseval's theorem, 103
Partial differential equations, 217
   boundary conditions of, 219, 226
   classification of, 221
   integral transform methods of, 239
   separation of variables in, 226
   Wiener–Hopf method of, 245
Partial fraction representation of cotangent, 52
Particular solution or integral, 7
Pauli exclusion principle, 442
Pauli spin matrices, 465
Period of a group element, 427
Permutations, 375
Perturbation theory, 286
   applied to circular drum, 290, 295
   applied to Schrödinger equation, 291
   degenerate theory, 293
   rearranged series, 292
Phase of a complex number, 472
Pochhammer contours, 192
Poisson distribution, 379
Pole, 480
Polyad, polyadic, 408
Principal normal, 409
Principal value integral, 480
Probability theory, 372
   distributions, 377, 380
Projection operator, 143
Proper rotation, 407
Pseudotensor, 407

$Q_n(z)$, Legendre function of the second kind, 174
Quadrupole magnet, 125
Quotient law, 414

Raising of tensor indices, 416
Random walk, 396, 397
Rank of a tensor, 406
Ratio test for convergence of an infinite
    series, 45
Rayleigh–Ritz method, 337
Rearranged perturbation series, 292
Reciprocity relation for Green's functions,
    270
Recursion relations,
    for Bessel functions, 179
    for coefficients in solution of differen-
        tial equations, 15, 16, 20
    for Legendre polynomials, 171
    three-term, avoidance of, 19, 21
    three-term, occurrence for Mathieu
        functions, 202
Reducible representation, 431, 435
Regula falsi, 360
Regular function, 477
Regular representation of a group, 437
    of a Lie algebra, 469
Regular singular point of a differential
    equation, 14
    at infinity, 19, 20
Relative likelihood, 387
Relative tensor, 413
Removable singularity, 480
Representation of a Lie algebra, 469
Representations of a group, 430
    characters in, 433
    orthogonality theorem of, 432
    reducible, irreducible, decomposable,
        431
    regular, 437
    Schur's lemma, 432
Residues, theorem, 65, 66, 480
Resolvent kernel, of integral equation, 303,
    309
Retarded potential, 280
Riemann curvature tensor, 419
Riemann geometry, 416
Riemann P symbol, 189
Riemann sheet, surface, 474
Riemann zeta function, 45, 53
Rodrigues' formula for Legendre poly-
    nomials, 169
Roots of equations, Horner's method, 361
    Newton's method, 360
    regula falsi, 360
Rotation matrix $D^l_{m'm}(R)$, 462

for $l = 1/2$, 465
for $l = 1$, 463
Runge–Kutta method, 353

Saddle-point methods, 82
    applied to Airy integral to derive WKB
        connection formula, 116
Scalar, as a tensor of rank zero, Cartesian,
    406
    general, 411
Scalar product, 149
    for infinite-dimensional vector space,
        162
Schläfli's integral representation of $P_n(z)$,
    169, 170
    of $J_n(z)$, 184
Schmidt–Hilbert theory of integral equa-
    tions, 305
Schrödinger equation, 218
    Born approximation in scattering the-
        ory, 303
    as example for perturbation theory, 291
    graphical discussion of, 24
    for harmonic oscillator, 19
    solution by WKB method, 36
Schur's lemma, 432
Schwartz reflection principle, 482
Schwartz transformation, 137
$se_n(x)$, Mathieu function, 200
Second solution, of Bessel's equation, 17,
    18, 180
    of confluent hypergeometric equation,
        196
    of hypergeometric equation, 191
    of Legendre's equation, 167
Secular equation, 146
Separable (ordinary) differential equa-
    tion, 2
Separation of variables, 226, 235
Series asymptotic, 81
    convergence criteria of, 44
    familiar series, 48
    infinite, 44
    miscellaneous devices for summing, 50
    for sec x, 57
    $\Sigma(1/n^2) = \mathscr{S}(2)$, 52, 363
    for tan x and ctn x, 50
Series solutions of differential equations,
    13
    associated Legendre equation, 21

Bessel's equation, 16
Hermite equation, 20
indicial equation, 16
irregular singular point, 14
Legendre's equation, 14
near an ordinary point, 14
near a regular singular point, 14, 16
recursion relations for the coefficients, 15, 16, 20
singularities at infinity, 19
Sheet, Riemann, 474
Similarity transformation, 148
Simpson's rule, 350
Sine transform, 103
Sine-integral function $Si\ x$, 61, 77
Singular points of a differential equation, at infinity, 19
regular and irregular, 14
Singular solution of a differential equation, 5
Small oscillations, 154
sn x, elliptic function, 204
Sommerfeld–Watson transformation, 75
Special functions, 167
Bessel functions, 17, 178
beta function, 67
confluent hypergeometric functions, 194
elliptic functions, 204
gamma function, 75
Hermite polynomial, 21, 188
hypergeometric functions, 187
Legendre functions, 16, 167
Mathieu functions, 198
Spherical Bessel functions $j_l(x)$, $n_l(x)$, 187, 229
asymptotic behavior, 230
Spherical harmonics $Y_{lm}(\Omega)$, 175, 178
addition theorem, 167
as basis for representations of $O^+(3)$, 462
Spinor, spinor representations of $O^+(3)$, 466
Spur of a matrix, 145
Stability, of numerical integration of differential equations, 354
of a parabolic equation, 226
Standard deviation, 381
Standard error, 388
Stationary phase, method of, 90
Stationary random process, 386
Statistical independence, 374
Statistics, 372

Steepest descent, method of, 82
Step function (Heaviside), 107
Stirling's formula, 84, 89
Stokes' phenomenon, 33
Stream function, 127
Structure constants of a Lie group, 456
Student t-distribution, 398
Sturm-Liouville differential equation, 264
completeness of eigenfunctions, 338
variational approach, 334
$SU(n)$ group, 450
irreducible representations of $SU(2)$, 457
irreducible representations of $SU(3)$, 459
Subgroup, 426
invariant or normal, 430
Subtracted dispersion relation, 135
Summing series numerically, 363
Superconvergence relation, 138
Symmetric group on n objects, S_n, 425
Symmetric kernel, 305
Symmetric matrix, 145
Symmetry arguments for evaluating integrals, 62
Symmetry of Green's functions, 270
Symmetry groups, 426, 441, 443
Synthetic division, 361

Tabulated integrals, 75
beta function $B(x, y)$, 76
cosine-integral $Ci\ x$, 61, 77
elliptic integrals $F(\phi, k)$, $E(\phi, k)$, $\Pi(\phi, n, k)$, $K(k)$, $E(k)$, $\Pi(n, k)$, 78, 79
error function erf x, 77
exponential integral $Ei(x)$, 77
gamma function $\Gamma(x)$, 59, 75
sine-integral $Si\ x$, 61, 77
Taylor's series, 14, 479
t-distribution of Student, 398
Tensor, Cartesian, 403, 406
general, 412
metric, 416
relative, 413
Riemann curvature, 419
symmetric, antisymmetric, 414
under $SU(2)$, 458
under $SU(3)$, 460

Tensor analysis, Cartesian tensors, 403
  general tensor analysis, 410
Tensor density, 414
Torsion, 409
Trace of a matrix, 145
Transfer matrix, 166
Transformation group, 450
Transformation matrix, 147
  congruent transformation, 156
  for rotations in three-space, 404
  similarity transformation, 148
  unitary transformation, 150
Transpose of a matrix, 145
Trapezoidal rule, 350
Triad, 408
Twisted cubic, 422

Unbiased estimate, 399
Uncertainty principle, 105, 107
Undetermined coefficients, method of, 7
Unimodular unitary group $SU(n)$, 450
Unitarity condition on scattering amplitude, 316 317
Unitary matrix, 145
Unitary transformation, 150

Variance, 381
Variation of parameters, 8
Variational derivative, 324, 327, 328
Variations, calculus of, 322
  applied to eigenvalue problems, 333
  applied to integral equations, 340
  isoperimetric problems in, 331
  variable end points, 329
Vector space, linear, 139
  basis, 141
  linear operators, 141

Vectors, 140
  Cartesian, 404
  components, 141
  contravariant and covariant, 411
  linearly independent, 140
  orthogonality, 149
  under $SU(2)$, 457
  under $SU(3)$, 459
Vibrating drum, 231, 266, 271
  with perturbation, 290, 295
Vibrating string, 262, 269
Vibrations, small, 154
Volterra integral equation, 300, 311

$W_{k,\,m}$, Whittaker functions, 197
Wave equation, 217, 218, 222, 227, 278
  Green's function, 278
Whittaker equation and functions, 197
Wiener–Hopf method,
  applied to integral equation, 312
  applied to partial differential equation, 245
WKB method, 27
  applied to Schrödinger equation, gives Bohr–Sommerfeld quantization condition, 37
  connection formulas, 28, 34, 36, 116
Wronskian, 31, 35

$Y_{lm}(\Omega)$, spherical harmonics, 175, 178
$Y_m(x)$, second solution of Bessel's equation, 180

Zeta function, Riemann, $\zeta(s)$, 45, 53